Theories and Application in Maize-Soybean Strip Intercropping System

Wenyu Yang • Feng Yang
Yushan Wu
Editors

Theories and Application in Maize-Soybean Strip Intercropping System

Springer

Science Press
Beijing

Editors
Wenyu Yang
College of Agronomy
Sichuan Agricultural University
Chengdu, China

Feng Yang
College of Agronomy
Sichuan Agricultural University
Chengdu, China

Yushan Wu
College of Agronomy
Sichuan Agricultural University
Chengdu, China

ISBN 978-981-97-8722-7 ISBN 978-981-97-8723-4 (eBook)
https://doi.org/10.1007/978-981-97-8723-4

Jointly published with Science Press
The print edition is not for sale in China (Mainland). Customers from China (Mainland) please order the print book from: Science Press.

This Springer imprint is published by the registered company Springer Nature Singapore Pte Ltd.
The registered company address is: 152 Beach Road, #21-01/04 Gateway East, Singapore 189721, Singapore

Introduction

"Among the various subjects related to the governance of a nation or state, food security stands as the foremost priority." China, as a densely populated nation, considers safeguarding food security a paramount priority in governance and national well-being. Maize and soybeans are pivotal staple crops, and the supply-demand dynamics significantly impact national food security. In comparison to maize, soybeans face a weaker competitive advantage due to a smaller planting area. The domestic production capacity is critically insufficient, resulting in a high dependence on imports. Moreover, soybean imports are susceptible to significant market fluctuations, rendering them a pivotal leverage point for hegemonic nations implementing anti-China policies. Both maize and soybeans are rain-fed crops cultivated in the same season, exacerbating land competition—a critical bottleneck in the realm of national food security. Coordinating the development of maize and soybeans on limited arable land presents a substantial challenge to national food security. Historically, the national policy to meet the demand for maize and soybeans involved developing maize and importing soybeans. Following China-U.S. trade friction, the country proposed reducing maize planting areas, expanding soybean planting areas, and mitigating the soybean supply-demand imbalance. The 2019 Central Document No.1 proposed to "stabilize maize production" and "implement the soybean revitalization plan, expanding planting areas through various channels." Ensuring absolute food security, stabilizing maize production, and significantly increasing soybean production capacity have become strategic imperatives for national food security. On the other hand, major grain crops, including maize and soybeans, have long adhered to a high-input, monoculture, and high-yield planting model, substantially contributing to the national food quantity. However, a host of ecological challenges, such as excessive resource consumption, precipitous declines in soil quality, water body eutrophication, and heightened greenhouse gas emissions, have gained prominence. The conflict between high-yield, high-efficiency planting, and agricultural sustainability underscores the acute contradictions between food security and resource constraints. China, with its large population and finite resources, can only surmount this impasse by pursuing a trajectory that integrates high yield and sustainability in modern agricultural development. A singular emphasis on either high yield or sustainable development constitutes a cul-de-sac for the Chinese nation.

Intercropping embodies a traditional agricultural practice in China and represents one of the effective measures to ensure food security. The global recognition of intercropping legumes with cereals underscores its status as a sustainable agricultural approach that maximizes resource utilization. Notably, the intercropping of maize and soybeans emerges as the preeminent model in this category. Widely applied in China before the 21st century, it made an enduring contribution to the reproduction and development of the Chinese nation. Traditional maize-soybean intercropping demonstrates advantages in fully utilizing arable land resources and maintaining ecological sustainability, serving as the maizeerstone for mitigating land competition and addressing the unsustainable high yields of maize and soybeans. However, traditional production techniques encounter challenges in meeting the demands of modern agriculture for high yield, efficiency, and large-scale production. Urgent innovation in traditional intercropping, employing new development concepts, modern technology, and equipment, is imperative for achieving the modernization of intercropping.

During the 1970s to the 1990s, the extensive adoption of the 'wheat-maize-sweet potato' double cropping system prevailed in the hilly regions of the Sichuan Basin, effectively augmenting grain yields and resolving farmers' food security concerns. Nevertheless, factors such as soil erosion, land depletion, labor intensity, and declining consumption prompted farmers to gradually abandon this practice. A novel planting model is required to supplant it. Toward the close of the 20th century, our research endeavors commenced on the 'wheat-maize-soybean' double cropping system, essentially substituting the 'wheat-maize-sweet potato' system with maize-soybean intercropping. Consequently, maize-soybean intercropping became the focal point of our research, involving production surveys, literature reviews, and field experiments on traditional intercropping practices, including maize with bean roots and single-row intercropping. Four key bottlenecks in application were identified: prolonged and irrational field configuration, severe soybean lodging, incongruent fertilization techniques, and a dearth of environmentally friendly pest control technology. These challenges resulted in unstable and meager yields, operational complications with machinery, limitations in in-field rotation, and impediments to integration with modern agricultural practices.

To address the challenges mentioned earlier, we established the goals of 'high yield, mechanization, and sustainability.' We adopted strip intercropping as our approach, grounded in innovative theories that focus on the efficient utilization of light and fertilizer resources, as well as the regulation of low-position crop types. A pivotal aspect of this endeavor was the breakthrough in core and complementary technologies, with a specific emphasis on the development of planting and harvesting machinery. Over a span of 23 years, our interdisciplinary collaborative research received support from initiatives such as the National '973' Program, National Key Research and Development Program, National Natural Science Foundation, Technological Innovation in Grain Production, and the National Modern Agricultural Industry Technology System. During this period, we systematically delved into theories related to the efficient utilization of light and fertilizer resources in the strip intercropping system of maize and soybeans. We also explored the theory concerning shade-tolerant and lodging-resistant crop type regulation. Additionally, we investigated corresponding cultivation techniques and pest control strategies. Our efforts resulted in the development of a series of planting and harvesting machinery tailored for the intercropping system, effectively achieving the integration of agricultural machinery and agronomy. Building upon this foundation, we successfully integrated the strip intercropping technology system of maize and soybeans, specifically adapted for modern agriculture, and implemented it on a large scale.

Over the past two decades, strip intercropping involving maize and soybeans has been widely implemented in both Sichuan and the southwestern regions. Trial demonstrations have also taken place in the northwest, Huang-Huai-Hai region, and the Yangtze River basin. This technique has led to heightened maize and soybean yields, enhanced overall productivity, minimized the application of chemical fertilizers and pesticides, and fostered harmonious development between the two crops. The method proves to be both high-yielding and sustainable, offering a novel technological avenue to ensure food security, foster environmentally friendly practices, and augment farmers' income. Acknowledged as a pivotal technology for ensuring national maize security and markedly enhancing soybean self-sufficiency, this technique has been consistently featured in the national key technology promotion for 13 successive years. It prominently appeared in the Central Document No.1 in 2020, 2022, and 2023, establishing itself as a fundamental core technology for the national initiative aimed at augmenting soybean and oilseed production capacity. In 2022, the Ministry of Agriculture and Rural Affairs orchestrated the nationwide promotion and implementation of strip intercropping involving maize and soybeans on 15.5 million mu (approximately 1.03 million hectares), surpassing the target with an actual implementation of 16.45 million mu. This initiative resulted in a 3-percentage-point rise in soybean self-sufficiency. In 2023, the promotion is scheduled to cover 20 million mu across 17 provinces, municipalities, and autonomous regions. As per the "14th Five-Year Plan" for National Crop Development issued by the Ministry of Agriculture and Rural Affairs, the aim is to promote strip intercropping involving maize and soybeans on 50 million mu by 2025.

This work serves as a comprehensive overview of the accomplishments attained in the development and application of the "Maize-Soybean Strip Intercropping Technology System." It stands as the inaugural systematic summary of the original monograph concerning the theory, technical exploration, and holistic applications of maize-soybean strip intercropping. The publication encompasses a collection of more than one hundred figures and tables, predominantly derived from academic papers either already published or currently under review within the author's research team. This compilation also encompasses the theses of graduate students nurtured by the team. A minor fraction is referenced from other research reports, and suitable citations have been diligently incorporated.

Given the author's constrained proficiency and temporal limitations, there may exist some imperfections. We welcome and highly value your constructive criticisms.

Contents

Challenges Faced by Grain Production in China and Advantages of Intercropping

1

Wenyu Yang, Feng Yang, and Xiaoling Wu

Abstract

Agriculture is a fundamental industry in a country, and food is an important material related to the national economy and people's livelihood. This chapter summarizes the basic situation, challenges, and solutions of China's grain production. Intercropping is an important production mode in agricultural production, and a detailed introduction to the concept, history, types, and applications of intercropping in China is provided in the text. In addition, the advantages and disadvantages of intercropping in increasing land resources, improving soil fertility, enhancing soil fertility, and increasing crop yield were presented. These results aim to provide reference and inspiration for the development of grain production.

Keywords

Intercropping · Crop production · Food security · Resource utilization

1.1 Current Situation of Agriculture Production in China

The current population of China exceeds 1.4 billion people, the largest in the world. National food security has always been a top priority for the prosperity and well-being of the people. In turn, agriculture is the foundation of the national economy and, particularly, cereal grains are important industrial commodities that affect the economy of the country and the livelihoods of people, which makes their production a fundamental strategic industry. Moreover, grain security concerns the world, and

W. Yang (✉) · F. Yang · X. Wu
College of Agronomy, Sichuan Agricultural University, Chengdu, China
e-mail: mssiyangwy@sicau.edu.cn; f.yang@sicau.edu.cn

W. Yang et al. (eds.), *Theories and Application in Maize-Soybean Strip Intercropping System*, https://doi.org/10.1007/978-981-97-8723-4_1

agricultural stability concerns society (Bao and Yao 2019). Under the complex and difficult current international circumstances, domestic agricultural development and grain production face unprecedented challenges due to strongly destabilizing factors, such as large price increases and market fluctuations of agricultural products.

1.1.1 The Situation of Agriculture Production in China

During the "13th Five-Year Plan" period, the area under grain-crop cultivation in China became stabilized at over 116 million ha (1.74 billion mu). In 2020, the overall grain yield reached a historic high of 669.5 million tons. As the largest staple crop in the country, rice is the main food for nearly 60% of the population. In 2020, the national rice-cultivated area was 30.07 million ha, with a total yield of 212 million tons and a comprehensive mechanization rate of over 80%. In turn, wheat is the second largest staple crop in China, and 40% of the population depends on it as main food. Meanwhile, in 2020, the wheat-cultivated area was 23.4 million ha, with a total yield of 134.25 million tons, and basic full mechanization was achieved. Additionally, maize is the largest grain crop and an important source of feed and industrial raw materials. Specifically, in 2020, the maize-cultivated area was 41.33 million ha, with a total yield of 260.65 million tons and a comprehensive mechanization rate of over 80%. Despite increased production, the gap between overall grain supply and demand has widened, with imports reaching 11.3 million tons. Soybean is an important source of plant protein, edible oil, and protein for feed. In 2020, the cultivation area was 9.867 million ha, with a yield of 19.6 million tons, and imports reaching 100 million tons. Thus, the increasing dependence on imports now poses a serious threat to food security in China.

Currently, the production and demand for rice and wheat are relatively balanced, whereas the gap between maize production and demand continues to expand, and the discrepancy between soybean supply and demand is also cause for alarm. Maize and soybean are both major agricultural products and same-season dryland crops in China, whereby, a trade-off between maize and soybean cultivation areas is inevitable, with the conflict of land competition becoming increasingly prominent. Thus, expanding the soybean-cultivated area inevitably reduces that of maize. Therefore, how to improve soybean production capacity on the premise of ensuring maize safety has become a major challenge facing the country. In the traditional crop production process, high-input cultivation technology has played an important role in increasing maize and soybean yields. However, excessive resource consumption, declining land quality, and increasing environmental pollution have made it increasingly difficult to achieve harmony between "high output" and "sustainability," which is a major challenge facing crop production not only nationally, but worldwide. Intercropping and relay intercropping have great internal advantages for ecological sustainability and intensive use of resources. Through theoretical research and technological innovation, realizing the integration and modernization of maize-soybean relay intercropping systems is deemed an effective way to solve the abovementioned threat to food security and the concomitant challenges (Yang and Yang 2019b).

1.1.2 The Challenges Facing Grain Production in China

1.1.2.1 The Need to Increase Land Productivity

Land productivity reflects the productive capacity of farmland in terms of the amount of grain produced per unit area during a production cycle, which is affected by natural environmental factors and the socio-economic level. According to the third national land survey, in 2019, the total area of cultivated land in China was 1.28 million km^2, indicating a decrease of 75,000 km^2 relative to the situation 10 years before, although with significant regional differences (Fig. 1.1). On the one hand, with the rapid increase of urbanization, the area of construction land has increased year by year, especially in urban areas, at the expense of high-quality arable land, some of which has been relocated to areas with relatively poor farmland quality. The area of arable land has been compressed, and the quality of arable land has degraded, consequently reducing land output. On the other hand, due to economic development and improvement, an increasing number of people have left the rural sector, and fewer people are now involved in agricultural production, resulting in low land productivity. Abandoned farmland is an increasing problem. Low land productivity and non-grain-production uses of farmland are on the rise in pursue of higher economic benefits, which adversely affects staple food production.

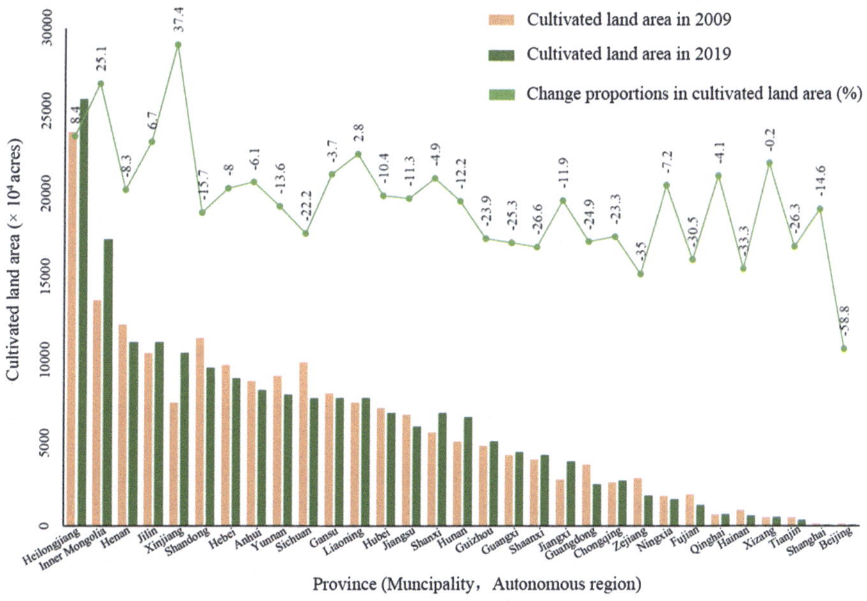

Fig. 1.1 Data from the second (2009) and third (2019) national land surveys

1.1.2.2 The Need to Increase Labor Productivity

Grain production costs comprise mainly three elements: material and service costs, labor costs, and land costs. Material and service costs have increased due to the growth of fertilizer and leasing operations. With the improvement of land property rights and the secondary land market, the value of land has increased. However, production costs remain high, while farmers' income associated with crop production continues to decline. The grain industry chain is relatively short, and the value added associated with processing remains low. Long-term low grain prices and low income from grain cultivation directly affect farmers' enthusiasm for cropping, leading to a large-scale reduction in cultivation and increasing difficulties in stabilizing grain production. Furthermore, although the proportion of arable land is large in the hilly and mountainous regions of southern China, the level of comprehensive agricultural mechanization is low, with prevailing high labor inputs and serious difficulties in large-scale management, leading to a major limitation for the development of modern agriculture in the region. Additionally, the difference in efficiency between grain crops and the overlapping between their seasonal cultivation cycles are prominent. For example, maize and soybeans are planted in the same season. Over the past decade, the soybean cultivation area in northeast China has sharply declined, whereas the maize area has increased rapidly. The comparative benefits of maize cropping are significantly higher than those of soybean cropping, such that the difference in labor efficiency and benefits between these two crops directly affects the respective cultivation area and its distribution.

1.1.2.3 The Need to Improve Land Quality

Improving the quality of cultivated land is an important measure to ensure food security in China. According to CCTV's "Focus Interview," more than two-thirds of the arable land in the country consist of low-yielding fields, and problems such as degradation of black soil in the northeast, acidification of cultivated land in the south, and salinization of land in the north are becoming increasingly prominent. Continuous long-term monoculture and excessive use of fertilizers and pesticides have led to an increase in harmful microorganisms in the soil and fewer beneficial microbial species and populations. Further, long-term heavy stubble continuous cropping has led to soil degradation, causing problems such as soil salinization and compaction. Adjusting cultivation structure, crop rotation, and intercropping can be utilized to effectively reduce the occurrence of soil-borne diseases and improve soil fertility, thereby contributing to crop yield increases achieved by breeding.

1.1.2.4 Increasing Frequency of Occurrence of Extreme Natural Disasters

Global climate change has had, and will continue to have, a significant impact on human activities, especially those involved in agricultural production. Agriculture in China is particularly sensitive to climate change due to the uniquely large population whose livelihoods are intricately associated with agriculture, high pressure on resources, complex geographic conditions, and distinct regional climate differences. Specifically, global climate change has increased the frequency of occurrence of extreme weather events such as high temperature, drought, and heavy rain, thus making disasters more extreme and sudden. Meanwhile, the main agricultural production areas have become more concentrated, which has a particularly strong impact on agriculture, especially on grain production. Concomitantly, climate change has disrupted any clear pattern of regional incidence of pests and diseases, making their prevention and control difficult, and further exacerbating the impact of climate change on grain production.

1.1.2.5 The Pressure for Self-Sufficiency in Grain and Oil Crop Production Has Intensified

China has a large population and an increasingly limited amount of arable land. The cost of grain production is generally above international average, and farmers' enthusiasm for grain cropping is not high, resulting in insufficient production to meet the national demand for agricultural products. Thus, some grain must be imported to ease the pressure on domestic production. As mentioned earlier, while supply and demand are balanced in the cases of rice and wheat, there is an enormous gap between supply and demand of maize and soybeans (Table 1.1), making imports absolutely necessary. Thus, for example, in 2021, soybean imports exceeded 96 million tons, accounting for 60% of the global soybean trade, with an external dependence rate of up to 85%. Similarly, maize imports exceeded 28 million tons, and the source countries for both products were highly concentrated. Further, under the influence of factors such as the COVID-19 pandemic and international geopolitics, scarce agricultural products such as soybeans, maize, and cotton are in high demand, while there are severe problems with their expedite entry in sufficient amounts. In addition, although the participation of foreign companies in grain production has allowed for some breakthroughs in the industry, the introduction of advanced production technology and production experience has gradually improved agricultural competitiveness and the level of production technology in China. However, there are still severe limitations, such as the fierce control of germplasm resources by foreign seed companies.

Table 1.1 Production and consumption of the four major grain crops

Crop	Year	Production (10^4x tons)	Consumption (10^4x tons)	Imports (10^4x tons)	Exports (10^4x tons)
Rice	2021	21,284	21,545	496	245
	2022	20,850	21,089	619	221
Wheat	2021	13,694	14,857	977	8
	2022	13,772	13,868	996	15
Maize	2021	27,255	28,205	2835	0.8
	2022	27,720	28,870	2062	0.1
Soybean	2021	1640	11,138	9647	8
	2022	2028	11,425	9108	8

1.1.3 Countermeasures to Ensure National Food Security

General Secretary Xi Jinping has repeatedly emphasized the importance of optimizing crop layout, stabilizing grain and maize production, expanding soybean and oilseed cultivation, and ensuring that annual grain output remains above 650 million tons. Additionally, he has underlined the need to ensure that the rice bowls of the Chinese people are filled with Chinese rice, thus asserting that food security is the foundation of national security. We must firmly hold this principle as the bottom line of national food security and take multiple measures to ensure it.

1.1.3.1 Ensuring the Fundamentality of Farmland to Guarantee Food Security

Farmland is a crucial foundation for food production and is essential for addressing the issue of feeding our population of 1.4 billion. To ensure food security, it is necessary to optimize and improve the necessary measures for supplying arable land, provide effective guarantees for the total amount of farmland, and continuously promote the improvement of agricultural land quality. Since 2019, the Chinese grain-sowing area stopped its previous declining trend and begun to show positive growth. In 2022, the national grain-sowing area was nearly 1.775 billion mu (117.7 million ha), and a certain amount of grain-sowing area must be ensured in different grain-producing areas. To this end, China has successively promulgated a number of laws and regulations, such as the "Land Management Law" and the "Regulations on the Protection of Basic Farmland," and introduced a series of strict policies and measures to protect arable land, safeguarding the production area for grain and firmly putting an end to non-agricultural activities on arable land.

1.1.3.2 Tapping the Potential of Grain Yield per Unit Area

Since the 18th National Congress of the Chinese Communist Party, the average national grain yield has significantly increased, markedly enhancing our international impact. The contribution of such improvement of grain yield per unit area to total production has exceeded 66%. Currently, and in the near future, enhancing large-scale yield per unit area will remain the focus of grain and oil crop production policy, especially in different regions, crop types, cultivation modes, and production methods. In short, a comprehensive plan is required to integrate improved farmland, quality seeds, sound regulations, effective machinery, and optimal management practices. High-quality and efficient varieties and technologies must be promoted to boost the yield per unit area in a timely manner.

1.1.3.3 Fostering Farmers' Enthusiasm for Growing Grain Crops

Promoting agricultural and rural reform, facilitating land transformation, effectively expanding the scale of operation, and improving the mechanism for protecting grain production are all absolutely necessary to safeguard vital farmer interests. Firstly, we must adhere to improving the market price system and implement relevant agricultural product-subsidy systems. Secondly, we must continuously enhance the status of farmers, improve their training, and upgrade their skills, such that they gradually become professional, literate grain producers. Thirdly, we must make the best possible use of grain-subsidy policies, fully implement production support policies for grain growers, such as large-scale crop cultivation households, family farms, professional farmers' cooperatives, and industrialization consortia. Those who do not produce or manage grain cropping properly will not be subsidized. The policy of "benefiting those who grow grain and prioritizing support to those who grow more grain" must be permanently reinforced.

1.1.3.4 Supporting Infrastructure Development and Equipment Construction

We need to enhance the development of high-standard farmland, prioritizing grain production functional areas and permanent basic farmland. Additionally, we should harmonize and consolidate relevant funds and projects, strongly support the establishment of high-standard farmland, and facilitate standardized and regulated farmland development. Consistently, in accordance with the "14th Five-Year Plan for National Cultivation Industry Development" issued by the Ministry of Agriculture and Rural Affairs, 716,667 ha of high-standard farmland are to be developed by 2025 and seven million ha of high-standard farmland are to be transformed and upgraded to secure and ensure a grain production capacity of over 550 million tons. Furthermore, innovation and development of agricultural machinery equipment must be accelerated. Thus, according to the "Fourteenth Five-Year Plan for the Advancement of Agricultural and Rural Modernization," the overall nationwide mechanization rate of cultivation and crop harvesting should reach 75% and 55%, respectively, for hilly and mountainous counties (cities, districts). The mechanization of the entire process of major grain-crop production must be swiftly promoted, and agricultural production efficiency markedly enhanced by furnishing robust conditions for the modernization of the development of the cultivation industry.

1.1.3.5 Accelerating the Application of Core Agricultural Technologies

While addressing the problems derived from the lack of original technological innovation, weak scientific research foundations, scattered research resources, and shortage of talent in the grain industry, we must concentrate on fostering new agricultural scientific and technological innovation forces and incessantly strengthen agricultural technological competitiveness to achieve independent innovation in various fields, such as agricultural production and operation. Firstly, we must expedite the development of high-quality products, elevate the level of modernization and mechanization in agriculture, and persistently upgrade technology innovation and construction promotion efforts to lay the foundation for the rapid development of agriculture. Secondly, we must increase our support to the construction of modern agricultural parks to provide strong support for innovative agricultural management entities. Finally, we must intensify the promotion of technological applications and fully exploit the leadership and driving role of provincial and ministerial-level technologies.

1.2 Development and Advantages of Relay/Intercropping

As the primary productive force, agriculture currently faces great challenges due to population growth, urban development, decreasing arable land, and ecological and environmental degradation. According to the "Chinese Grain Issue" white paper, by 2030, the Chinese total demand for grain will be approximately 640 million tons, and it will be characterized by a very prominent supply-demand imbalance. Currently, among the four major grain crops, the supply and demand of wheat and rice are balanced, whereas maize and soybeans show a rapidly expanding gap between supply and demand. This is particularly the case for soybeans. High crop yields and sustainability are difficult to reconcile. Regarding all inputs involved in agricultural productivity, the continuous reduction of arable land and the increasing scarcity of water resources are paramount. Hence, given its environmental friendliness, economic feasibility, and high land productivity characteristics, intercropping has become an attractive alternative model for pursuing sustainable agricultural development to ensure food grain and oil security in China (Li 2013; Yang et al. 2021).

1.2.1 The Concept of Intercropping

1.2.1.1 Intercropping

Intercropping refers to a cultivation method that involves cultivating two or more crops in alternating rows or bands on the same plot of land within a single growing season, such as maize-soybean intercropping in the Huang-Huai-Hai region, wheat-sesame intercropping in the northwest region, and cassava-soybean intercropping in the south China region.

1.2.1.2 Relay Cropping

Relay cropping refers to a cultivation method that involves sowing or transplanting a second crop in the rows or bands of a first crop during the late growth stage of the latter, such as wheat-maize relay cropping in the southwest region, maize-soybean relay cropping, and cassava-soybean relay cropping in the south China region.

1.2.2 The Historical Development of Intercropping

Intercropping stands as a treasure among traditional agricultural technologies conceived and created by the Chinese people, serving as an integral element of the meticulous agriculture in the country. According to historical records, intercropping in China started during the Han dynasty and, as early as the first century BC, it was concluded that intercropping or mixed cropping could boost income as long as the relationship between the different crops was properly managed. In his book, "*Fan Shengzhi shu,*" Fan Shengzhi, an outstanding agronomist of the Western Han dynasty, summarized the valuable experience of Guanzhong farmers in intercropping melons, garlic, and beans: "*Plant melons in compartments, and plant ten garlics around each compartment as well, surrounding the melon seeds. In May, when melons are ripe, garlic can be harvested and sold, avoiding the melons. You can also plant small beans among melons, four or five liters per Mu (a unit of area, 0.0667 hectares) and sell the vines. This method is suitable for flat land, and the income from melons is ten thousand coins per acre.*"

Intercropping began to take shape during the Wei, Jin, and Northern and Southern dynasties, and by the sixth century AD, Jia Sixie, a prominent agronomist of the Northern Wei period, documented various intercropping methods in his book "*Qimin Yaoshu*," including methods that marked the initiation of the intercropping of trees and grains, as illustrated by the following examples: "*Prune and dig under the mulberry trees frequently, and plant mung beans or adzuki beans,*" "*plant grains and beans close to the trees, without losing the land benefit and making the field fertile. Scatter turnips around the trees, no need to press them*"; *vegetable intercropping,* "*plant coriander between scallions for immediate consumption,*" *and mixed fodder production,* "*plant soybeans with millet in one Mu of land, leave them with grasses, no need to weed or till, harvest them as green fodder in August or September.*" In the chapter "*Planting Hemp*" he recorded the following recommendation; "*For relay cropping, scatter turnip seeds between hemp plants and weed them in June, intending to harvest their roots.*" These recommendations show that at that time, intercropping was applied to various aspects of agricultural production, while considering the combination of land use and fertilization, and the integration of agriculture and animal husbandry. In addition, the following recommendations: "*two beans are good and moist, benefiting mulberries,*" "*be careful not to plant hemp seeds mixed with soybeans,*" indicate that at the time, people had already recognized the dialectical relationship between crops, and between crops and the environment in intercropping, thus laying an enduring theoretical foundation for intercropping.

Intercropping and relay cropping continuously evolved and developed from the Northern Wei period to the Yuan dynasty, as recorded in ancient books such as "*Nong Sang Ji Yao*," "*Si Shi Zuan Yao*," "*Wang Zhen Nong Shu*," and "*Chen Fu Nong Shu*," all of which contain records of intercropping practices. Two types of intercropping were commonly applied: One was the co-cultivation of multiple crops in the same field, following a specific schedule, as exemplified by practices such as the following: "*the field should be plowed and dug down during idle periods, and spring barley is sown in January, yams and taros in February and March, millet, wheat, soybeans and mung beans in March, April and May, and winter wheat and peas in August. This should be done according to the season without being too greedy.*" The second method involved a more comprehensive and systematic approach toward the relationship between mulberry trees and other crops. For example, the method recommends mulberry-ramie intercropping near the farm, thus reflecting the principle of adaptability to local conditions; "*Grain crops can be grown between mulberries depending on their compatibility with them. For example, if grain is grown, it must make the soil dry and cracked, which will attract insects that bore roots and suck bark; if millet is grown, its branches and leaves are similar to mulberries, making them tangled up and hindering mulberry growth;*

However, crops such as mung beans, black beans, sesame, and melons and tubers can thrive under the shade of the mulberry tree and enhance its growth. Sorghum can also be planted with mulberry trees. As farmers say: 'mulberry trees thrive with sorghum, and sorghum thrives with mulberry trees, this is generally the case'." Furthermore, "*Mulberry roots grow deep while ramie roots are shallow, such that they do not interfere with each other, but can double their benefits.*" Hence, the method considers not only crop combinations above ground but root distribution underground as well.

As a result of the significant increase in population and the remarkable progress in agricultural technology, intercropping and relay cropping techniques developed rapidly during the Ming and Qing dynasties. Intercropping and relay cropping were combined with crop rotation and multiple cropping, thereby greatly improving crop yield index. There are various types of intercropping and relay cropping mentioned in "*Qun Fang Pu*," "*Qi Min Si Shu*," "*Nong Zheng Quan Shu*," "*Gu Cai Tong Zhen*", "*Nong Tian Yu Hua*," and "*San Nong Ji*," such as mulberry trees with peas, fava beans, mung beans, sesame, and other mulberry-grain intercropping, as well as intercropping modes wheat-bean, cotton-sweet potato, cotton-vegetable, grain-vegetable, cotton-bean, rice-coix, and relay cropping modes, including early rice-late rice, rice-bean, wheat-cotton, wheat-bean, and potato-taro, as well as cereal-sorghum with mung bean relay cropping, rice-taro, etc. By the mid-eighteenth century, "*Zhi Ben Ti Gang*" and "*Xiu Qi Zhi Zhi,*" by Yang Shen, recorded the methods of multiple harvests in one year and 13 harvests in two years, indicating an intensive cultivation system that made full utilization of land and seasons, and combined cultivation and fertilization in China (Yang and Yang 2019a). It was a specific manifestation of the comprehensive use of intercropping and multiple cropping technologies, which pushed intercropping and multiple cropping to a peak in China.

During the "Republic of China" era, agriculture experienced a strong decline but farmers still used intercropping and relay cropping techniques to maintain agricultural productivity for survival. Southern regions continued to use rice-bean, sesame-soybean intercropping, and relay cropping, while the northeast regions had "wheat-ditch-bean" and sorghum-soybean, maize-soybean intercropping, and mixed cropping, and the north China regions had maize and sorghum intercropped with black and yellow beans and other planting methods, while northwestern regions had less intercropping and relay cropping.

After the founding of the People's Republic of China, intercropping and relay cropping techniques were widely applied in an attempt to cope with the continuous population increase and food crisis. Due to the influence of climate, time, and space factors, different regions of the country had unique intercropping and relay cropping methods involving grain, vegetable, and fruit crops, in a total area of approximately 20 million ha. Searching Chinese databases, provinces with more than 20 types of intercropping and relay cropping modes were found, with Shandong, Henan, Hebei, Jiangsu, Hunan, Sichuan, Yunnan, and Guizhou among the provinces where these techniques were commonly used. Among various intercropping and relay cropping

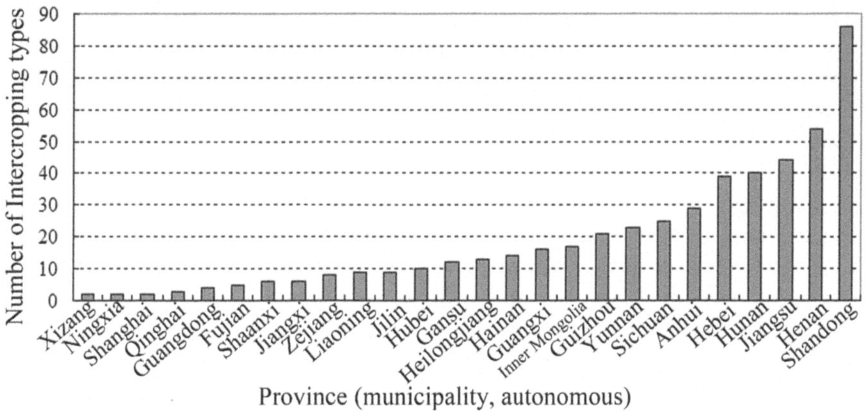

Fig. 1.2 Distribution of intercropping types in China based on published Chinese literature on intercropping

modes, *Gramineae-Leguminosae* and *Gramineae-Gramineae* intercropping are the main systems (Fig. 1.2).

At the dawn of the twenty-first century, traditional intercropping technology has been innovatively renovated by the development of high-yield, large-scale, mechanized, and sustainable agriculture. In the late twentieth century, the large-scale application of the three-crop relay cropping system "wheat-maize-sweet potato" in the hilly areas of Sichuan Province effectively increased grain yields. However, due to soil erosion, increased labor cost, and reduced consumption, the cultivation area has gradually decreased. To reverse this trend, since 2002, Sichuan Agricultural University has conducted research on the new three-crop rotation system of "wheat-maize-soybeans" to replace maize-sweet potato relay cropping with maize-soybean intercropping. From the core technology of "selecting and matching varieties, widening rows and increasing light incidence, shortening plants and keeping secrets" to the supporting technologies of "integrated fertilization with reduced fertilizer amounts, chemical lodging resistance, and green prevention and control," this research innovatively integrated maize-soybean strip intercropping technology, which improved soybean self-sufficiency rate on the basis of ensuring maize production capacity, and conducted demonstration and promotion events in the Southwest, Northwest, Huang-Huai-Hai, and the middle and lower reaches of the Yangtze River, with relevant content listed in the Central Government No. 1 Document for 2020, 2022, and 2023.

1.2.3 Types and Distribution of Intercropping

1.2.3.1 Types

According to the type of crop combination, the crops used for intercropping belong to: *Gramineae* (maize, sorghum, wheat, millet, barley, oat, sugarcane, etc.),

Leguminosae (soybean, faba bean, pea, peanut, alfalfa, kidney bean, chickpea, etc.), *Cruciferae* (rapeseed, mustard, etc.), *Compositae* (sunflower), *Malvaceae* (cotton, okra), *Convolvulaceae* (sweet potato), *Solanaceae* (potato), and *Euphorbiaceae* (cassava), among other plant families. The common combinations involve *Gramineae-Gramineae* and *Gramineae-Leguminosae* intercropping, such as wheat-maize relay cropping, maize-soybean intercropping, and sugarcane-soybean intercropping.

According to the carbon dioxide assimilation pathway, crops used for intercropping are C3 species (wheat, barley, millet, soybean, faba bean, pea, peanut, etc.) and C4 species (maize, sorghum, sugarcane), and the common combinations are C4-C3 and C3-C3, such as maize-soybean intercropping and wheat-soybean relay cropping.

Intercropping can be classified based on plant height. This includes combinations of tall and short crop plants, as well as combinations of crop plants with the same height. Examples of tall-short combinations include maize-soybean intercropping and maize-peanut intercropping, while examples of same-height combinations include wheat-broad bean intercropping and wheat-soybean intercropping.

Alternatively, intercropping can be classified based on whether the overall density of the combined crops is changed. Based on this criterion, intercropping can be divided into stacking-type and replacement-type intercropping. In stacking-type intercropping systems, the sum of the relative densities of the combined crops is greater than 1 but less than or equal to 2. On the other hand, in replacement-type intercropping systems, the sum of the relative densities of the combined crops is equal to 1.

1.2.3.2 Distribution

Intercropping is widely used in Asia, Europe, Africa, America, and Oceania. In Asia, it is mainly concentrated in countries such as China, India, and Iran, with *Poaceae-Legumes* and *Poaceae-Poaceae* relay- and intercropping as the main types, such as maize-soybean, wheat-soybean, maize-potato, wheat-pea, and other intercropping combinations. In Europe, intercropping in mainly distributed along the Mediterranean coast and Atlantic coast countries, mostly adopting *legume forage-Gramineae* combinations of mixed sowing, intercropping, and relay cropping. In Africa, intercropping is extensively used mainly in South Africa, Nigeria, Ethiopia, Kenya, and other countries, mostly adopting local crops for mixed sowing and intercropping, such as cassava-soybean, cassava-okra, okra-maize, potato-cowpea, and okra-cowpea combinations. In turn, in America, intercropping is mainly distributed in the United States of North America and Brazil in South America, with *Gramineae-Leguminosae* intercropping as the main type. Meanwhile, in Oceania, intercropping mainly concentrates in the Brisbane and Perth areas of Australia, with *Leguminosae-Gramineae* intercropping as the main type.

The main types of intercropping in China are *Gramineae-Gramineae* (maize-wheat, etc.) and *Gramineae-Leguminosae* (maize-soybean, wheat-faba bean, etc.) combinations. According to the climatic characteristics of the regions of crop distribution in China, the intercropping regions can be divided into four sectors:

northeast region, northwest region, Huang-Huai-Hai region, and southwest-south region. The northeast region has one crop cycle per year and has adopted intercropping systems such as maize-soybean intercropping; in turn, the northwest region has one crop cycle per year, or three crops in 2 years, and uses intercropping or relay cropping systems such as maize-soybean intercropping or wheat-soybean relay cropping. Meanwhile, the Huang-Huai-Hai region has two crop cycles per year and uses intercropping systems such as maize-soybean intercropping; finally, the southwest-south region has two- or three-crop cycles per year and uses relay cropping or intercropping such as wheat-maize-soybean relay cropping or maize-soybean intercropping after the wheat harvest (Yang and Yang 2019; Knrzer et al. 2009).

1.2.4 Advantages and Disadvantages of Intercropping

As the essence of traditional Chinese agriculture, intercropping is a recognized traditional cultivation pattern for intensive land utilization and sustainable agricultural development worldwide. Not only can it increase the multiple cropping index, and grain production, and improve farmers' income, but it effectively alleviates the problems of high input and output, low nutrient utilization rate, and environmental pollution as well. Combined with temporal crop rotation, intercropping achieves spatial crop rotation and is particularly suitable for densely populated countries with limited land resources, playing an important role in modern agricultural development.

1.2.4.1 Advantages in Intercropping

1.2.4.1.1 Intercropping Improves Land Use Efficiency

Extensive experiments and production practices have proved that reasonable intercropping can increase yields compared to monoculture, mainly by utilizing and exploiting the complementarity relationships between crops. Cultivating two or more crops on the same field can yield more products with fewer economic inputs. One Mu of land under intercropping can produce more agricultural products than one Mu under monoculture, thereby optimizing the use of available land, improving land use efficiency, and yield per unit area (Du et al. 2018; Li et al. 2023).

1.2.4.1.2 Intercropping Improves Resource Use Efficiency

Proper intercropping can intensively utilize light, temperature, water, fertilizer, and all other resources in time and space. The intercropping complex community can increase leaf area index (LAI) and extend the time of photosynthetic activity, thus improving light and heat energy use efficiencies (Willey 1990). Especially in the intercropping of legumes and non-legumes, the competition for nutrients between species is alleviated, with the former providing some nitrogen for the latter. The non-legume crops can absorb 25–155 kg N/ha from the nitrogen fixed by legumes. Concomitantly, due to the different root distribution and nutrient water demand of the intercropped species, a spatiotemporal intensive utilization of underground water and fertilizer resources results, which improves water and fertilizer use efficiency.

1.2.4.1.3 Intercropping Stabilizes Crop Yields

In natural ecosystems, the higher the biodiversity, the more stable the system. Intercropping uses the differences among crop species for ecological adaptability, resource utilization spatiotemporal-compensation mechanisms, and resistance to natural disasters to enhance the biodiversity and stability of the farmland ecosystem, thereby, improving the stability and market adaptability of farmland productivity. For example, the unique microclimate formed by the complex plant community in an intercropped field can prevent severe incidence and rapid spread of pests and diseases; similarly, the heat- or cold- and waterlogging tolerance of the crops in a complex community contribute to yield stability during drought or flood events.

1.2.4.1.4 Intercropping Coordinates Crop Competition for Land

Intercropping is a practice that consists in growing multiple crops in the same field, aiming to increase the annual yield and alleviate the competition for land among food crops, cash crops, fodder crops, green manure crops, and fruits and vegetables. It is conducive to the comprehensive development and optimization of crop diversity and structure, and promotes the overall sustainable development of agricultural production. While having little impact on major grain crops, intercropping achieves high-efficiency resource utilization and satisfies the diversified demand for agricultural products, playing an important role in optimizing cultivation structure.

1.2.4.2 Disadvantages in Intercropping

Currently, traditional intercropping does not suit the requirements of modern agricultural development, and urgently requires innovation. Major limitations include the following: (1) Improper Crop Selection: Traditional intercropping has not fully considered the complementary advantages between crops, such as the combinations of tall and short crops, C3 and C4 crops, monocotyledonous and dicotyledonous crops, nitrogen-fixing and nitrogen-consuming crops, in-season and off-season crops, plant type differences, shade-tolerant, and sun-loving crops, resulting in irrational fertilization leading to excessive stem elongation and plant lodging; (2) Irrational Field Configuration: Traditional intercropping shows insufficient exploitation of crop yield potential and edge effect, low crop yield with low ecological niche crop prone to lodging result in low yield, poor mechanical workability or inability to mechanize operations, using single-row alternate or substitutional intercropping rather than additive strip cultivation methods, resulting in low production efficiency; (3) Lack of Green Prevention and Control Technology: It is easy to implement integrated schemes for the prevention and control of diseases, insect pests, and weeds in single-crop cultivation, whereas in complex intercropping cultivation systems, different crops respond differently to pesticides and demand different application periods, which makes it very difficult to implement integrated prevention and control, especially for weeds; (4) lastly, agricultural machinery and equipment need to be significantly improved. Intercropping systems have long been considered troublesome to plant and difficult to mechanize. Therefore, it is necessary to further change public awareness and concepts such as to innovate and develop integrated intercropping seeding and fertilizing equipment, design separate control and adjustment plant-protection machines as well as multi-functional harvesting machines to improve the level of mechanization of intercropping cultivation systems.

The Connotation and Research Protocol of Maize-Soybean Strip Intercropping

2

Jiang Liu, Jianhua Chen, Taiwen Yong, Xiaochun Wang, Weiguo Liu, Feng Yang, and Wenyu Yang

Abstract

This chapter provides an introduction to the definition of maize-soybean strip intercropping and its various research areas. These areas encompass the efficient utilization of resources, nutrient complementarity, lodging resistance of low-height crop plants, biodiversity, and mechanization. The main research content focuses on designing the optimal field configuration, developing integrated fertilization techniques, exploring chemical regulation methods for shade tolerance and lodging resistance, constructing a green prevention and control technology system, and developing a planting, management, and harvesting machine that allows for full mechanization.

Keywords

Strip intercropping · Resource utilization · Nutrient complementarity · Lodging resistance · Mechanization

Intercropping and crop rotation are traditional Chinese agriculture technological treasures that have made indelible contributions to the growth and development of the Chinese nation. Intercropping of legumes with cereals is a cultivation pattern recognized worldwide as one involving sustainable intensive land use, with maize-soybean intercropping being the most widespread pattern.

J. Liu (✉) · J. Chen · T. Yong · X. Wang · W. Liu · F. Yang · W. Yang
College of Agronomy, Sichuan Agricultural University, Chengdu, China
e-mail: jiangliu@sicau.edu.cn; chenjianhua@stu.sicau.edu.cn; yongtaiwen@sicau.edu.cn; xchwang@sicau.edu.cn; f.yang@sicau.edu.cn; mssiyangwy@sicau.edu.cn

W. Yang et al. (eds.), *Theories and Application in Maize-Soybean Strip Intercropping System*, https://doi.org/10.1007/978-981-97-8723-4_2

2.1 The Meaning of Maize-Soybean Strip Intercropping

Intercropping encompasses rich biodiversity and allows for intensive utilization of resources. However, the traditional soybean-maize intercropping scheme uses an irrational field configuration, the technology of collaborative fertilization is deficient, soybean lodging is a severe problem, and disease, insect pest, and weed prevention and control are deficient. These are significant drawbacks that characterize the soybean-maize intercropping system, hampering its integration into modern agriculture. Hence, the theory and technology of this intercropping system is in urgent need of innovation and upgrading.

2.1.1 The Definition of Maize-Soybean Strip Intercropping

Developed based on traditional intercropping, maize-soybean strip intercropping adopts maize-strip and soybean-strip intercropping. Being taller than soybean, maize plants have the border advantages, while the former increase the space of exposure to light and reduce border disadvantages. The system effectively achieves the rotation of maize and soybean strips within the field, and it is well suited for the mechanization of operations. Consequently, a harmonious symbiotic relationship develops between the two crops, and we can harvest twice in one season in this planting system.

A typical maize-soybean strip intercropping system is illustrated in Fig. 2.2-1, which shows two rows form the taller maize-strip and 2–6 rows form the shorter soybean-strip. The two strips form a planting geometry. The width of the planting geometry determines population density, field configuration, and yield effectiveness of intercropping.

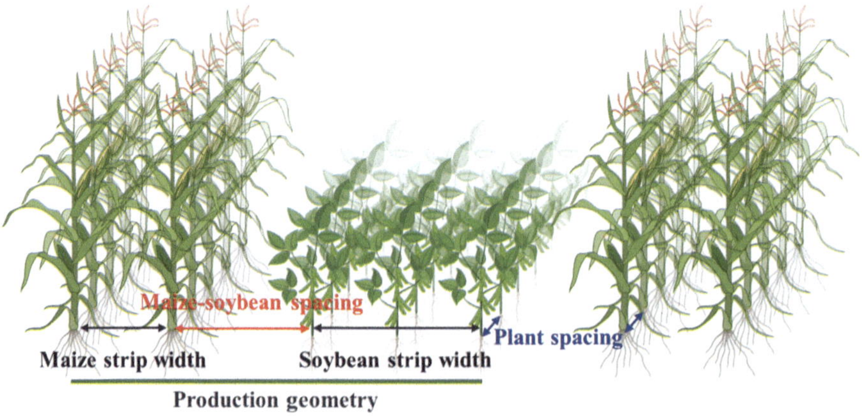

Fig. 2.2-1 Planting pattern in maize-soybean strip intercropping

2.1.2 Research Fields in Maize-Soybean Strip Intercropping

To truly become integrated into modern agriculture and spread widely for farmer adoption, maize-soybean strip intercropping systems must be provided with a sound theoretical basis and technological innovations leading to breakthroughs aimed at the efficient utilization of resources, nutrient complementarity, lodging resistance of low-height crops, biodiversity, and full mechanization, among other areas of interest.

2.1.2.1 Efficient Utilization of Resources

In intercropping systems, crop growth and yield are closely related to the spatiotemporal distribution of incident light, light interception, and light use efficiency. Plant species layout and row ratio configuration determine the amount of light interception. Thus, the analysis of the dynamic changes of light intensity and light use efficiency of a soybean canopy grown under different row ratio configurations in a maize-soybean intercropping system showed that the system effectively improved light use efficiency. Particularly, the low-height soybean canopy showed significantly higher light-use-efficiency rates than soybeans planted as monoculture. Conversely, other studies found no significant difference in light use efficiency at row ratios of 1:3 and 2:3 in maize-soybean intercropping. Such discrepancy is presumably because, in the latter case, no consideration was given to the effects of the spatial configuration of the system on the temporal or spatial changes of light intensity and light quality over the entire crop-growth period. Further, no in-depth discussion of the mechanisms underlying the use of light by crop plants growing in different ecological niches has been undertaken. Moreover, no study has evaluated a combination of different field configurations with the trafficability of sowing and harvesting machines, or the high output of symbiotic crops.

2.1.2.2 Nutrient Complementarity

The issue of soil nutrient competition among plants of different species planted together is a scientific challenge that must be faced while designing and implementing any type of intercropping system. Specifically, how to turn competition into complementarity is the focal point of current intercropping rhizosphere-nutrient ecology research. Thus, for example, legume-cereal intercropping cannot only activate nutrients and enhance nodule nitrogen-fixation ability but, additionally, it can achieve maximum utilization of limited nutrients through the separation of nutrient demand-peaks and an appropriate interspecific competition. Factors such as insufficient soil oxygen, high fertilization levels, and shading stress may inhibit root nodule formation and root respiration, accelerate root nodule aging, reduce beneficial root secretions, reduce nitrogenase activity, and shorten the physiological lifespan of root nodules. Concomitantly, rational intercropping can help improve root chemical characteristics and rhizosphere microecological environment, promote root nodule growth and development, and coordinate the relationship between root nutrient absorption and carbon and nitrogen accumulation in aboveground organs. Although previous studies have considered the complementary effect of underground rhizosphere nutrients in intercropping systems, studies on the coordinated relationship among environment, yield, and nutrient utilization are scarce. Especially, there are very few reports on the mechanism responsible for the changes in root nodule morphology and physiological functions regulating efficient nitrogen absorption under conditions of efficient utilization of light energy by aboveground organs. To date, research focused on fertilization theory has given no consideration to the need to optimize the layout of the intercropping species or the associated changes in light environment, which, in turn, regulate underground growth and nutrient complementarity; nor have theory-development efforts evaluated these factors in combination with fertilization technology to increase yield and save fertilizer in strip intercropping.

2.1.2.3 Lodging Resistance of Low-Height Crop Plants

Lodging of low-height crop plants is a common problem in intercropping systems. Under the traditional maize-soybean intercropping system, soybean lodging rates frequently exceed 80%, leading to yield reduction rates of over 22%, thus, seriously hindering the yield advantage offered by these traditional systems.

The degree of lodging is closely related to the crop species, specifically, to stem height and strength. The discovery and utilization of dwarfism-associated genes in plant breeding have helped reduce plant height in crops such as rice and wheat, thereby greatly improving their resistance to lodging. In order to achieve higher yields, crops must reach a threshold height to ensure the generation of sufficient photoassimilate. Nonetheless, while ensuring crops an adequate height is important for intercropping system-design purposes, the need to increase stem strength to improve lodging resistance is currently drawing increasing attention.

Much research on the genetics and the mechanism of stem growth and development in plants such as wheat, barley, rice, and peas has unequivocally established that stem strength is an important indicator to use in the evaluation of lodging resistance in wheat and rice. Specifically, lignin content, the main chemical component of plant cell walls, directly affects stem strength and lodging resistance in rice, wheat, rapeseed, and buckwheat. Further, the physicochemical characteristics of cellulose constitute the basis for the physiological functions of plant cell walls, as they largely determine the tensile strength of stems; thus, crop varieties with higher cellulose contents in the stem show lower rates of lodging. Particularly, in maize-soybean intercropping systems, whether the optimization of maize plant type and field configuration might affect soybean lodging resistance-associated traits remains to be elucidated. Furthermore, an important question to this respect is how to effectively evaluate such effects quantitatively. Specifically, a thorough understanding of how soybean varieties differing in shade tolerance and lodging resistance interact with the light environment, thereby affecting lodging resistance, would represent a significant breakthrough. Furthermore, what varieties and technologies can be selected to reduce soybean lodging rate? These issues warrant close experimental analysis, as they may severely restrict the large-scale adoption of maize-soybean strip intercropping.

2.1.2.4 Biodiversity

Intercropping can effectively prevent and control disease, insect pest, and weed incidence via a range of mechanisms, such as biodiversity, crop species layout, a heterogeneous light environment, spatial barrier effects, dilution effects, incidence of natural enemies, rhizosphere allelopathy, and interspecific competition. Thus, for example, wheat-broad bean intercropping reduced the occurrence of wheat powdery mildew and rust, and the improvement of broad bean rhizosphere-microflora inhibited soil-borne Fusarium wilt incidence. Similarly, maize-potato intercropping significantly reduced potato late blight and corn maize big (small) spot disease incidence. In turn, the change of field configuration in maize-soybean intercropping systems can effectively reduce the population density of pests such as bean weevils, turtle shells, common grasshoppers, ladybirds of genus Chrysanthemum, thrips, and corn stem borers. Further intercropping utilizes the spatial barrier effect of two or more high-/low-height crops from different families, genus, and species, to interfere the recognition of host plants by adult egg-laying corn borers, thereby reducing the number of specimens at any given time. Additionally, the intercropping of maize with leafy vegetables, peanuts, legumes, and wheat crops effectively reduces weed density, and inhibits weed growth. Indeed, over 20 types of intercropping systems have a similar effect.

Although extensive research has been conducted on the use of intercropping systems to control harmful organisms in major crops such as wheat and maize, previous studies have mainly focused on the incidence patterns of some insect pests and weeds, and the impact of a single abiotic environmental factor, while the main patterns of insect pest and weed incidence in maize-soybean strip intercropping systems have not yet been organically combined with prevention and control principles, strategies, and technologies; furthermore, the various prevention and control measures available have not been systematically optimized or applied in a coordinated manner. In short, to date, a lack of effective, cost-efficient, and ecologically friendly comprehensive prevention and control technologies prevails.

2.1.2.5 Mechanization

Advanced machinery is the basic equipment of modern agriculture, playing a crucial role in improving agricultural labor productivity and ensuring agricultural quality and efficiency. The United States is the country with the highest degree of agricultural mechanization and has achieved specialization, regionalization, mechanization, and network informatization. Similarly, in Western Europe, mechanization has been fully implemented for soil preparation, sowing, harvesting, and transportation of wheat, maize, and soybean. Further, numerous agricultural machines have even been equipped with GPS systems for precise operation. Similarly, in Japan also, mechanization has been fully implemented for field operations related to soil preparation, transplanting, and harvesting of rice. In this case, intensive, large-scale mechanized production has been fully implemented.

Meanwhile, although it lagged far behind that of developed western countries for a very long period, mechanization of agriculture in China is currently subject of an accelerated process of modernization, thereby developing much more rapidly than ever before. In terms of maize and soybean production, the highest level of mechanization has taken place in northeast China, basically realizing full mechanization of the production process. Indeed, mechanization technologies for plowing, planting, management practices, and harvesting have begun to catch up with current international standards. However, research on agricultural machinery, both at home and abroad, is mostly concerned with monoculture production, and most results obtained cannot be applied to intercropping due to the mismatched external size or machine performance. In China, our traditional intercropping systems cannot meet the requirements of modern agriculture due to the lack of supporting mechanized technology and the die-hard mentality whose view is that intercropping can only be conducted manually. Consequently, intercropping is gradually abandoned by farmers and threatened by obliteration by novel trends in modern agricultural production. The development of agricultural machinery and the use of mechanized technology suitable for maize-soybean strip intercropping will determine whether intercropping can be applied to solve the dual challenges of food security and sustainable agricultural development in the future.

2.2 Research Protocol for Maize-Soybean Strip Intercropping

2.2.1 Research Background

Maize and soybean are major bulk agricultural products in China, facing mutual conflict over land use. Soybean imports already rise to over 85% of total annual consumption. It is impossible to meet the ever-increasing demand by simple monoculture. How to improve soybean production capacity while ensuring the safety of maize yields has become a major challenge to national food security. High-input planting techniques and continuous cropping have brought a high output at the expense of excessive consumption of resources, a decline in the quality of cultivated land, and increasingly alarming levels of environmental pollution. How to achieve unity between "high output" and "sustainability" is a major challenge faced by crop production around the world. Intercropping and crop rotation show beneficial "genes," such as "ecological sustainability" and "intensive utilization of resources." Inheriting these beneficial "genes" and innovatively realizing the integration and modernization of maize intercropping with soybean is an effective way to tackle the abovementioned problems and challenges.

However, traditional maize-soybean intercropping systems lack a systematic theoretical support for high and stable yields, and the efficient use of resources. In addition, four major technological issues exist, including irrational field configuration, severe soybean lodging, lack of coordination in fertilization technology, and lack of disease, insect pest, or weed prevention and control. Altogether, these drawbacks result in low and unstable yield, poor trafficability of machines and tools, and difficulty in crop rotation, making the integration of intercropping systems into modern agriculture extremely difficult (Fig. 2.2-2).

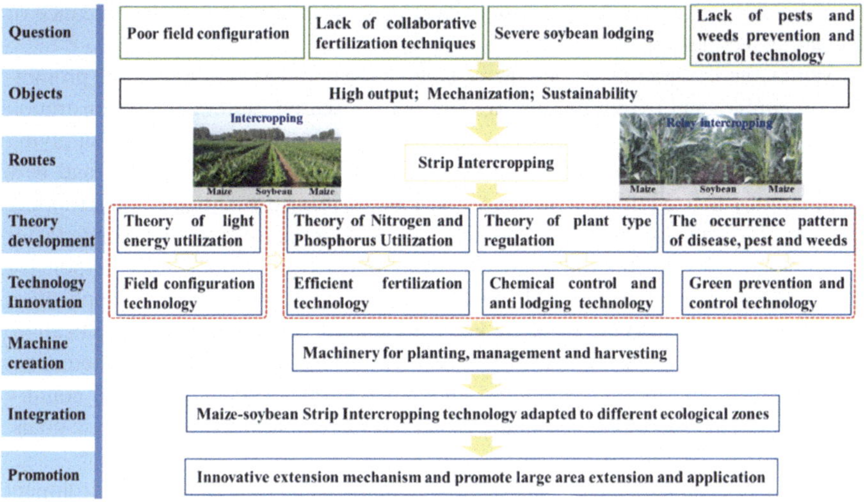

Fig. 2.2-2 Research needs and technological development route

2.2.2 Technology Route

Based on the preceding discussion, setting goals for high output, mechanization, and sustainability; taking a route of exploration of strip intercropping systems based on the theory of synergistic utilization of light and fertilizer resources by symbiotic crops, and plant-type regulation of low-height crops; focusing on achieving breakthroughs for the key technologies required by strip intercropping, such as collaborative fertilization, green prevention and control, chemical control, and counter-lodging technology; applying research ideas to develop a variety of machines for planting, performing management practices, and harvesting as to guarantee best results (Fig. 2.2-2), over the past 23 years, this project has comprehensively applied multidisciplinary theories and methods to innovate key theories, technologies, and machinery, aiming to model the maize-soybean strip intercropping systems best suited for modern agriculture, thereby providing novel insights for ensuring maize production capacity, attaining soybean self-sufficiency, and sustainable agricultural development in the country, with the support of projects such as the National 973 Program, the National Natural Science Foundation of China, the High-Yield and Science and Technology Innovations Program, and the National Modern Agricultural Industrial Technology System.

2.2.3 Technology Development Protocol

Focusing on high output, mechanization, and sustainability as technology goals, and aiming at the solution of the paramount scientific and technological problems faced by maize-soybean intercropping, the team in charge of the project was integrated by multidisciplinary researchers to jointly overcome difficulties under the guidance of the above technology route.

While conducting the project, the research team adopted the strategy of "promoting technological (machinery) and basic research through demonstration applications to achieve mutual promotion and coordinated development," applying multidisciplinary methods and theories, a new-development philosophy, and new technologies and equipment to innovate and modernize traditional intercropping systems. Thus, the project combined field and indoor experiments, from single factor to complex experimentation schemes, from phenotype to molecular responsive mechanisms, from experimental demonstrations to the promotion of application and to mutual verification, layer by layer in-depth and comprehensive application, achieving technology standardization, operation mechanization, theory systematization, high output, and sustainability, altogether in an attempt to lead the country toward modern agriculture and going global. In pursue of this set of strategic goals, we prioritized the following goals.

2.2.3.1 Focusing on the unclear regulation of the changes in the light environment of the maize-soybean intercropping systems, the rule of spatiotemporal variation in light intensity and quality at different plant-growth stages, and the spatial distribution of the population components were clarified through spacing and plant-type configuration testing. Based on the recorded findings, the project focused the response of leaf morphology to light environment and the mechanism underlying the efficient utilization of light energy in the system, which were both unclear due to the difference in ecological niche characteristics for maize and soybean plants in the maize-soybean intercropping strip system. The physiological mechanism responsive to light intensity and light quality, which regulates the photosynthetic rate of maize and soybean plants, was systematically analyzed in the field and indoors, from the phenotypic to the molecular level. Further, the difference in light energy interception and utilization between the two crops in the strip intercropping system was clarified and the mechanism of efficient utilization of light energy was explored. The key technologies for maize-soybean strip intercropping systems were developed using the following technical approaches: (1) selection of varieties with suitable plant types, (2) optimization of the configuration of strip width and row spacing, (3) reduction of the distance between plants to achieve optimum population density, and (4) adjustment and optimization of yield and quality as pivotal center of the research rationale.

2.2.3.2 Focusing on the serious difficulties derived from the non-synergy of light and fertilizer resource complementarity and use, and the low land productivity index and resource utilization rate due to interspecific competition, and considering the complementary utilization of the light resource and space aboveground, and rhizosphere-nutrient collaborative and efficient utilization underground as the principle in the maize-soybean strip intercropping systems, combined with field positioning experiments and the use of a ^{15}N isotopic tracer technique, we analyzed and compared the characteristics of crop nutrient absorption and utilization in monoculture, intercropping and reduced cultivating fertility, the evolution law of soil fertility, soil nitrogen, and phosphorus transformation capacity, and rhizosphere microbial community diversity, to determine soil nitrogen and phosphorus supply capacity and the ecological mechanism underlying strip intercropping systems. By using methods such as root box, pot, and pool cultivation medium, the interspecific competition compensation effect and the morphological and physiological response characteristics of root nodules in the maize-soybean strip intercropping were explored, revealing the physiological regulation mechanism responsible for root nutrient absorption. Based on the corresponding findings, fertilization technical parameters were duly proposed and the effect of technology on yield improvement and fertilizer savings, as well as the environmental effects associated, were verified through field experiments.

2.2.3.3 Focusing on the problem of severe soybean lodging caused by soybean shading by maize, which severely limits soybean yield improvement, researchers made inclusive collection of soybean germplasm resources in Southwest China, and conducted studies on the relationships among agronomic traits, lodging, and yield. Subsequently, they established an evaluation index system to identify the extent of shade tolerance and lodging resistance in the maize-soybean strip intercropping systems. Then, they selected specific soybean varieties suitable for the strip intercropping systems, and used physiological, biochemical, molecular, and other methods and theories to elucidate the mechanism responsible for soybean shade tolerance and lodging resistance, and developed lodging-regulatory technology.

2.2.3.4 Focusing on the severe problems of unclear patterns of disease, insect pest, and weed incidence, as well as uncertain prevention and control strategies, and lack of key prevention and control technologies, all of which together affect the yield and overall economic benefits offered by maize-soybean strip intercropping systems, researchers conducted studies of the species, incidence periods and intensities, and population characteristics of diseases, insect pests, and weeds, thus revealing the incidence patterns of major diseases, insect pests, and weeds, allowing for the elaboration of the key prevention and control strategies and technology required.

2.2.3.5 Focusing on the characteristics of the field configuration of the maize-soybean strip intercropping systems, researchers adopted the technology of "Reduce, Integrate, Increase, and Improve," to create a batch of machines suitable for planting, management, and harvesting in strip intercropping systems. In this formula, "Reduce" refers to using lightweight and simplification technology to compact the machines and reduce their sizes and weight. In turn, "Integrate" refers to using modularization technology to integrate multiple function machines. Meanwhile, "Increase" refers to increasing machine power and generated units. Lastly, "Improve" refers to improving and optimizing existing machines and tools. Based on these principles, integration with agricultural technology, adoption of key technologies, such as precision fertilization and sowing, zone-specific spraying of pesticides, and efficient product threshing and selection, will achieve mechanization and high-efficiency operation throughout the entire production process in maize-soybean strip intercropping systems.

2.2.3.6 With the support of key technology and matching technology, we considered the conditions of production and the characteristics of the climate in the Southwest, the Northwest, and the Huang-huai-hai Plain in China to integrate maize-soybean strip intercropping systems suitable for these different ecological regions. Aiming at the difficulties for large-scale application because of the fixed mindset that intercropping cannot be mechanized, a new application system suitable for new situations was constructed. Large grain producers, professional cooperatives, family farms, etc., were selected as demonstration targets. Through model demonstrations, on-site communication, and technical training, the demonstration and promotion application of the maize-soybean strip intercropping systems were accomplished.

2.2.4 Main Research Content

Taking the maize-soybean strip intercropping systems as the research object with the aim to address the most pressing issues of traditional intercropping of maize with soybean, the basic theory and key technologies were developed, and green prevention and control strategies and technology were devised, and suitable operation machines and tools were developed, thus modeling the most feasible maize-soybean strip intercropping systems best suited to reach modern agriculture standards. In this pursue, we itemized our goals as follows.

2.2.4.1 Researching the Mechanism of Efficient Utilization of Light Energy in the Maize-Soybean Strip Intercropping Systems to Devise the Core Technology for Best Field Configuration

Based on the characteristics of the climate and those of the crop varieties in different ecological regions, shade-tolerant and lodging-resistant soybean varieties were collected and screened for their suitability for planting in different regions, such as the Southwest, the Huang-huai-hai Plain, and the Northwest. Experiments were set up to test different strip widths, row ratios, and maize plant-type configurations to elucidate the control mechanism over the dynamic changes in light environment within the plant populations and the physiological and biochemical mechanisms responsible for the crop photosynthetic responses. Based on the standards of yield, quality, and machine trafficability, the basic requirements for selecting maize and soybean varieties in the maize-soybean strip intercropping system were clarified, and the dynamic light environment and efficient utilization mechanism of light energy under different field configurations were unraveled. Thus, the core technology for field configuration in the maize-soybean strip intercropping systems was innovatively formed.

2.2.4.2 Researching the Physiological and Ecological Mechanisms for Efficient Nutrient Utilization in the Maize-Soybean Strip Intercropping Systems, and Developing Integrated Fertilization Technology

Taking the maize-soybean strip intercropping systems that were optimized by field configuration of technical parameters as the research object, and using multi-year field experiments, pot, box, and other types of experiments, the evolution law of soil fertility, crop rhizosphere microenvironment, soil nutrient activation, plant nitrogen and phosphorus absorption characteristics, and the characteristics of the morphological and physiological responses of crop organs to the strip intercropping nutrient stress environment, were all systematically studied. Combined with the theories of interspecific competition and nitrogen and phosphorus transfer, the physiological and ecological mechanisms underlying efficient utilization of nitrogen and phosphorus in the maize-soybean strip intercropping systems were systematically explained. Furthermore, based on the theory developed, an integrated fertilization technology for the maize-soybean strip intercropping systems was developed to ensure that the systems achieve both high output and sustainability.

2.2.4.3 Revealing the Physiological Mechanism at Play in Soybean Plant-Type Regulation and Developing Chemical Regulation Techniques for Shade Tolerance and Lodging Resistance

Considering soybean, the low-height crop in the strip intercropping system under study, as the research object, field experiments combined with indoor pot experiments, and plant physiology and molecular biology research methods were used to conduct studies on more than 2000 soybean germplasm resources in the strip intercropping system. An evaluation index system for soybean shade tolerance and lodging resistance was established, and soybean varieties with different shade tolerance and lodging resistance were selected. Additionally, the relationship between lignin and cellulose accumulation, and stem strength, endogenous hormone metabolism and stem elongation were researched. Based on these studies, chemical regulation techniques for shade tolerance and lodging resistance of low-height soybean canopy were developed.

2.2.4.4 Clarifying the Patterns of Major Disease, Insect Pest, and Weed Incidence in Strip Intercropping, and Constructing a Green Prevention and Control Technology System

Focusing the maize-soybean strip intercropping systems as the research object, through years of field investigation, field positioning experiments, pot simulation experiments, and other means, the incidence and diversity of population structure of diseases, insect pests, and weeds in the maize-soybean strip intercropping systems were systematically investigated. These studies revealed the regulatory mechanism of field configuration on the disease, insect pest, and weed incidence. Combined with the principles of crop heterogeneity, spatial barriers, and rhizosphere microecological regulation, our research elucidated the incidence patterns of diseases, insect pests, and weeds in the maize-soybean strip intercropping systems. Subsequently, based on the findings, research and application of key technologies such as seed treatment technology, early monitoring and diagnosis technology, high-efficiency chemical pesticides and synergists technology, and physicochemical resistance-induction technology were performed to establish comprehensive prevention and control strategies and technologies for minimizing the damage caused by diseases, insect pests, and weeds in the maize-soybean strip intercropping systems, thereby ensuring green, high yield, and sustainable development of such systems.

2.2.4.5 Developing a Planting, Management, and Harvesting Machine Suitable for Realizing Full Mechanization

Based on the integration concept of agricultural machinery and agronomy, with the core technology devised for maize-soybean strip intercropping systems, and with the goal of realizing their full mechanization, in combination with the selected field configuration parameters, we developed a sowing machine and machine sowing technology, a plant protection machine and machine defense technology, a harvester and machine harvesting technology with adequate trafficability, high operation quality and efficiency, and excellent safety and stability, hence forming a mechanized technology for the entire cropping process from planting, through harvest in the maize-soybean strip intercropping systems, thereby realizing the integration of agricultural machinery and agronomy.

Theory of Efficient Use of Light and Water Resources in Maize-Soybean Strip Intercropping

3

Feng Yang, Wenyu Yang, and Xin Liu

Abstract

Intercropping has the advantage of fully utilizing light and water resources. This chapter mainly analyzes the effects of different field configurations and maize plant types on the light environment of intercropped system. The HHLA and ERCRT models are used to evaluate the light interception rate and light energy utilization rate in maize-soybean strip intercropping system. In addition, the physiological and biochemical mechanisms of light and water utilization were analyzed in intercropping system. These results will provide theoretical basis and technical support for the optimized configuration and resource efficient utilization of maize-soybean strip intercropping.

Keywords

Intercropping · Light use utilization · Water use utilization · Light interception model

Soybean-maize strip intercropping systems are strongly influenced by the spatiotemporal distribution of light and water. Effective growth and yield formation depend on the successful capture and utilization of these resources by the crop plant members in the system. The use of light energy and water is synergistically determined by a

F. Yang (✉) · W. Yang
College of Agronomy, Sichuan Agricultural University, Chengdu, China
e-mail: f.yang@sicau.edu.cn; mssiyangwy@sicau.edu.cn

X. Liu
College of Agronomy, Sichuan Agricultural University, Chengdu, China

College of Agronomy, Shandong Agricultural University, Tai'an, China
e-mail: liux@sdau.edu.cn

W. Yang et al. (eds.), *Theories and Application in Maize-Soybean Strip Intercropping System*, https://doi.org/10.1007/978-981-97-8723-4_3

combination of the variety and row ratio of the crop species, specifically, between relatively taller maize and relatively shorter soybean plants. In this chapter, we discuss the spatiotemporal distribution patterns of radiant light energy, the simulation of population light environment, the synergistic and efficient use of light energy, and the distribution and utilization of water resources in the system. The result of our study is a theory we call "higher crop canopy, high-low synergy," which explains the utilization of light energy and water resources in maize-soybean relay strip intercropping systems.

3.1 The Spatiotemporal Distribution of Light Radiation Energy in Soybean-Maize Relies Strip Intercropping Systems

3.1.1 Different Field Configurations Affect the Crop Canopy Light Environment

3.1.1.1 Methods for Describing the Light Environment

The growth, development, and yield of crops are directly influenced by the spatiotemporal dynamics of the light environment prevalent in maize-soybean strip intercropping systems (Chen et al. 2019). In the case of soybean-maize strip intercropping, such dynamics for soybean plants were measured after the emergence stage, when maize plants were at the tasseling stage. For the whole soybean-maize strip intercropping population, light radiation measurements were made at approximately 60 days after simultaneous sowing. At this stage, the higher maize canopy had already begun to affect the growth and development of the lower soybean canopy. During the measurement process, a canopy analyzer and spectrometer were used for making measurements every 10–15 days from 8:00 am to 6:00 pm, at two-hour intervals. Data were collected on selected sunny days. Readings were taken at each sampling point every 20 cm horizontally and vertically to calculate the different canopy extinction coefficients of maize and soybean plants.

3.1.1.2 Crop Canopy Light Environment

The intercropping of soybean and maize is directly influenced by the configuration of row and inter-row spacing, which impact the light environment of both the upper crop, maize, and the lower crop, soybean. In this study, the light environment was observed in a maize-soybean intercropping experimental production unit of 200 cm width in which six different spatial configurations were tested (Fig. 3.1), including 20:180, 40:160, 60:140, the traditional 1:1 intercropping, mono-cropping maize, and mono-cropping soybean, as well as different maize cultivars (Yang et al. 2015a; Gao et al. 2015).

The characteristics of the light environment in the wide-row maize area are shown in Figs. 3.2. The spatial distribution of photosynthetically active radiation (PAR) in the wide-row maize canopy showed an inverted pyramid shape, with PAR values decreasing from the top to the bottom of the plants as the distance from the maize row increased. The PAR values for the transverse and longitudinal amplitudes

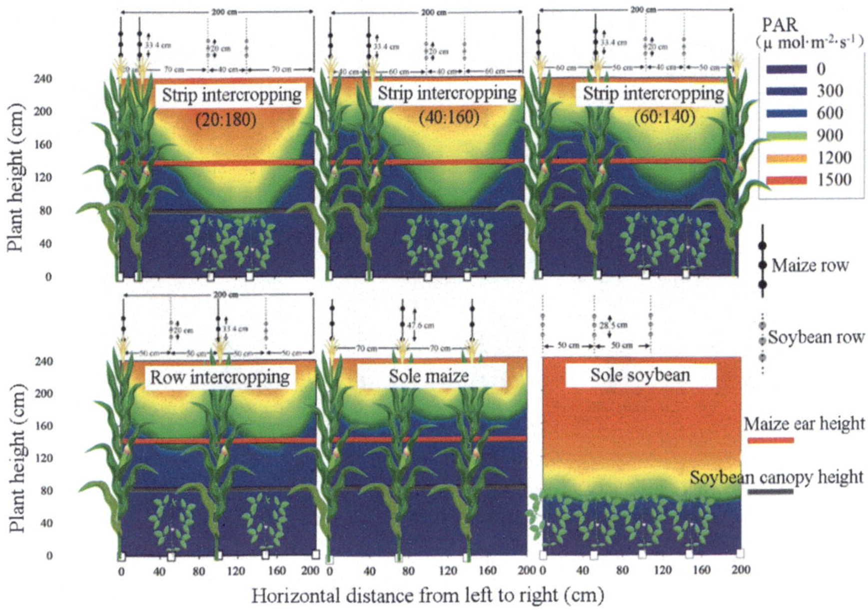

Fig. 3.1 Spatial variation of photosynthetically active radiation (PAR) in intercropped populations of maize at different row spacing during tassel and silk emergence stages

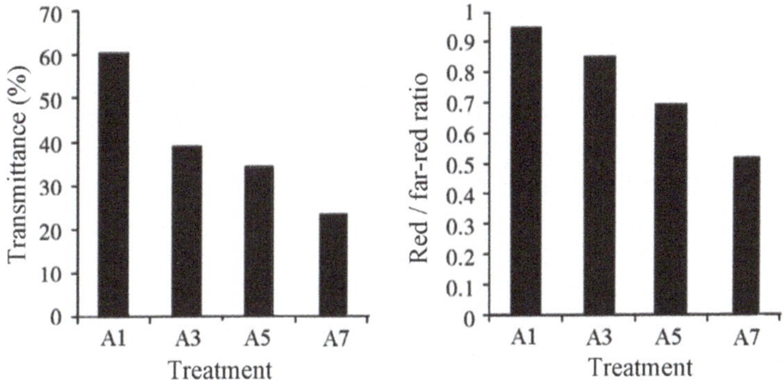

Fig. 3.2 Variation of canopy light transmittance and red/far-red ratio of soybean at different row-spacing configurations. Note: A1, A3, and A5 represent the irradiance environments of soybean canopies with maize intercropping at a row ratio of 2:2 and soybean row spacing of 70, 60, and 50 cm, and maize row spacings of 20, 40, and 60 cm, respectively. A7 represents the soybean canopy-irradiance environment in traditional maize-soybean intercropping at a row ratio of 1:1

greater than 1000 μmol·m^{-2}·s^{-1} gradually increased in the wide-row maize canopy of intercropping systems as the inter-row spacing between maize and soybean bands increased from 50 to 70 cm. These values were significantly higher than those of the traditional 1:1 row ratio or either mono-cropping system. Moreover, the average

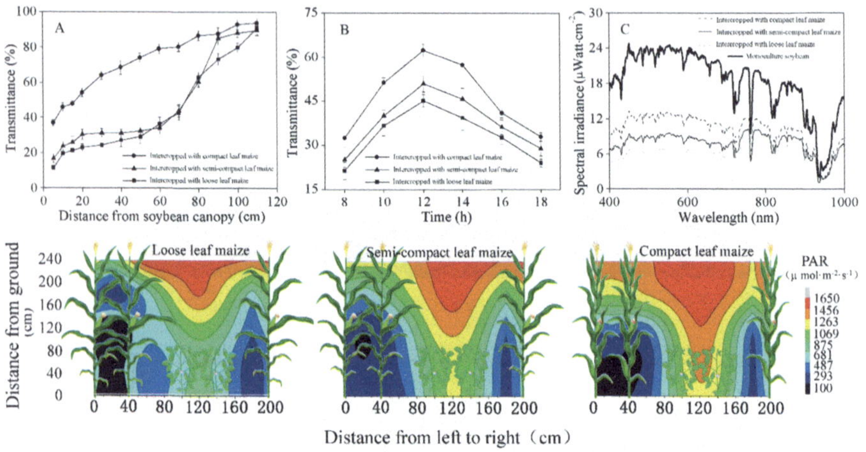

Fig. 3.3 Canopy light transmittance of soybean intercropped with different maize plant types. (**a**) at different heights; (**b**) at different time points; (**c**) spectral irradiance profiles

value of PAR incident on the maize ear leaf also increased from 454 to 518 μmol·m^{-2}·s^{-1}, and the red/far-red ratio (R/FR) increased from 0.58 to 0.99 as the inter-row spacing between maize and soybean bands increased. These values were significantly higher than those of the mono-cropping system, which showed a PAR value of 400 μmol·m^{-2}·s^{-1} and an R/FR value of 0.42.

The light environment of the lower soybean canopy was also influenced by the spatial configuration. Thus, as the inter-row spacing between maize and soybean bands increased, the average light intensity of the soybean canopy increased from 420 to 900 μmol·m^{-2}·s^{-1}, and the red/far-red ratio (R/FR) increased from 0.69 to 0.9. These values were significantly higher than those of the traditional 1:1 row intercropping system, for which soybean canopy light intensity was 300 μmol·m^{-2}·s^{-1} and the value for R/FR was 0.52 (Fig. 3.3). The average light transmission rate in the region above the soybean canopy, specifically 5–80 cm above the dense maize intercrop, was more than 20% higher than that of the half-dense and loose maize intercropping systems with soybean canopy when different maize cultivar configurations were tested. Additionally, the R/FR value was greater than 0.9, and the light transmission rate of the soybean canopy showed an alternating upward/downward trend.

3.2 Simulation of the Light Environment in a Maize-Soybean Strip Intercropping Population

3.2.1 Methods for Simulating the Light Environment

The distribution of light in strip intercropping systems not only affects light interception and utilization but it influences crop growth and yield as well. Further, because numerous measurement points are needed to obtain a reliable profile of light distribution in intercropping populations, a considerable amount of manpower is required, and significant errors in measurements may occur. Alternatively, by comparing the results of model simulation with actual measurements, not only can the number of measurements be reduced but, additionally, the reliability of the results is enhanced. Hence, using the instantaneous light transport model, this study aimed to investigate the distribution of light in soybean-maize strip intercropping and explore the distribution patterns of light in intercropping systems with different row-spacing configurations and row orientation configurations to provide a basis for studying crop responses to light.

3.2.1.1 Instantaneous Light Transmission Model

The distance of disturbance of light passing through a maize-soybean strip intercropping system involves maize and soybean leaves and air. According to Beer-Lambert's law, the method for calculating PAR at each point in the strip intercropping system was used after Tsubo and Walker (2002).

$$F_{\mathrm{PAR}} = 1 - \exp\left(-g_M LAD_M L_M - g_S LAD_S L_S - g_A LAD_A L_A\right) \tag{3.1}$$

where subscripts M, S, and A represent maize, soybean, and air, respectively. In turn, g, LAD, and L represent the extinction coefficient, leaf area density ($m^2\ m^{-3}$), and the distance of light passing through the disturbance medium (m), respectively.

Above the soybean canopy, the medium through which light passes consists only of maize leaves and air. As the impact of air as a disturbance medium can be ignored, PAR above the soybean canopy can be calculated as follows:

$$F_{\mathrm{PAR}} = 1 - \exp(-g_M LAD_M L_M) \tag{3.2}$$

where LAD can be calculated by the following equation:

$$LAD = \frac{WtLAI}{WsH} \tag{3.3}$$

where leaf area index (LAI) is the unitless ratio of one-sided leaf area per unit ground area, thus expressing the extent of ground coverage by the crop canopy. From 2013 to 2015, the value of LAI of maize grown in an east-west orientation ranged from 5.50 to 5.93, while that of maize grown in a north-south orientation ranged from 5.43 to 5.66; Wt represents the width of the production unit and has a value of 2 m; Ws refers to the width of the maize strip, measuring 0.8 m, at a maize row spacing of 0.4 m; each maize row extending an additional 0.2 m; lastly, H represents the difference in canopy height between maize and soybean plants, which remained constant at 1.7 m throughout the 3 years (Fig. 3.4).

Figure 3.5 illustrates θa, which is the angle between a horizontal projection of a light beam, and the direction of the strip; in turn, θb is the angle between the projection of the light beam on the cross section of the strip and the direction that is perpendicular to the ground. Lastly, θc is the angle between the light beam and the projection of the light beam on the cross section of the strip. The geometric model used calculates these angles as follows (Tsubo and Walker 2002; Munz et al. 2014):

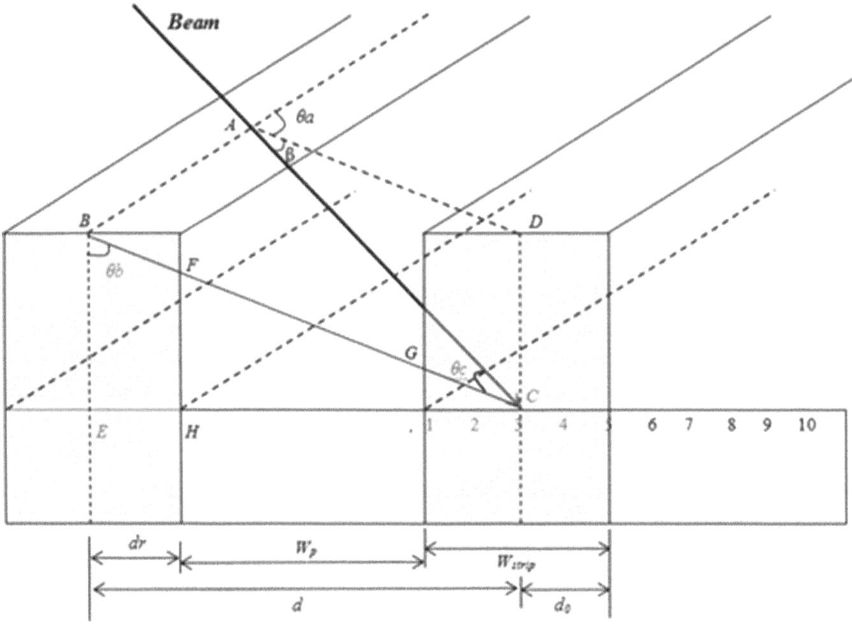

Fig. 3.4 Transmission of light beams above the soybean canopy. (Liu 2016)

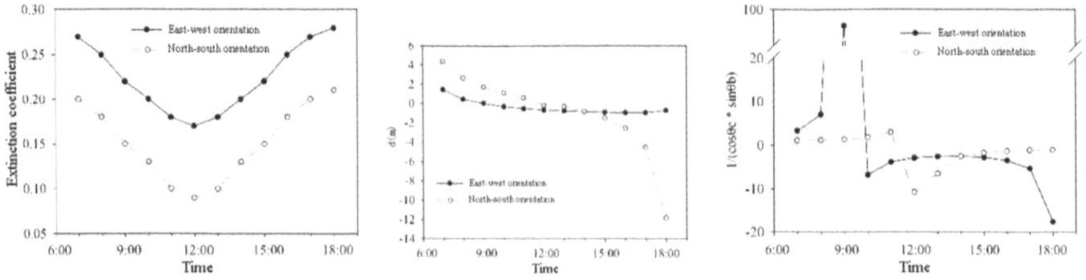

Fig. 3.5 Daily variation of the extinction coefficient, the transmission distance of light above the soybean canopy (d), and 1/(cos θc * sin θb), on August 15

$$\sin \beta = \sin \lambda \ \sin \delta + \cos \lambda \ \cos \delta \ \cos \tau \tag{3.4}$$

$$\cos \alpha = \frac{\sin \lambda \sin \beta - \sin \delta}{\cos \lambda \cos \beta} \tag{3.5}$$

$$\sin \theta c = \cos(\pi - \theta a) \ * \ \cos \beta \tag{3.6}$$

$$\cos \theta b = \frac{\sin \beta}{\cos \theta c} \tag{3.7}$$

In Fig. 3.6, α and β represent the solar azimuth and altitude angles, respectively. δ, λ, and τ are the solar declination (14.25° on August 15), the observation point latitude (35°15′09″N), and the hour angle, respectively.

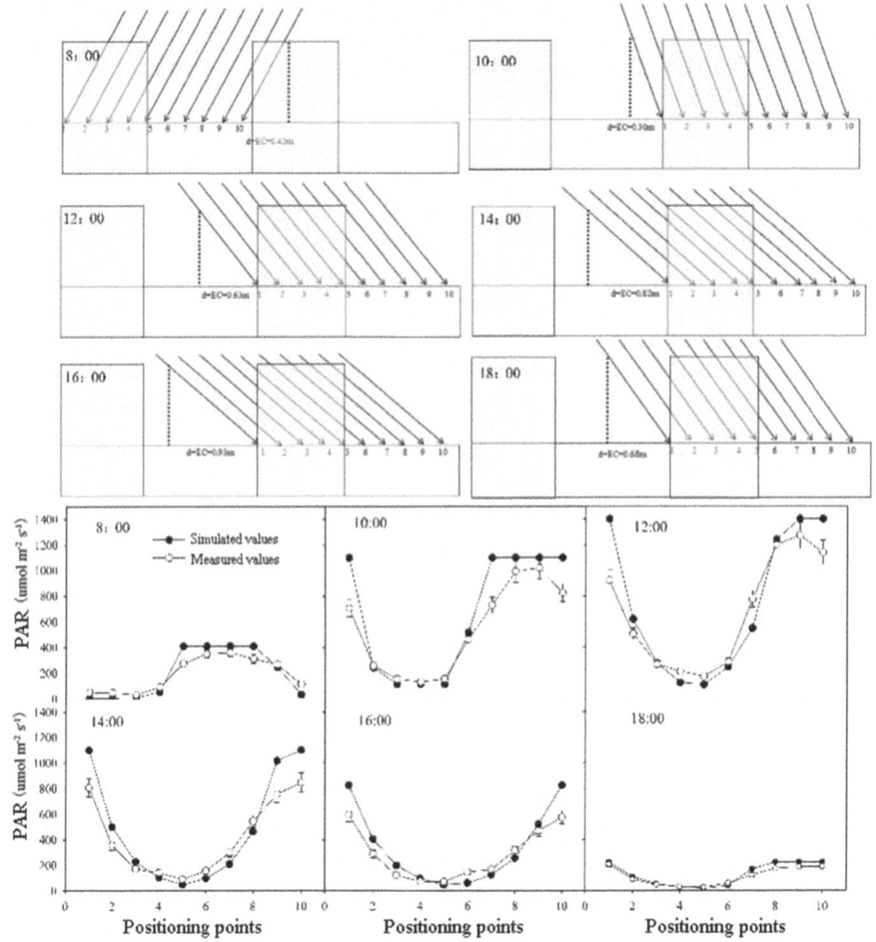

Fig. 3.6 Distribution of PAR at the top of the soybean canopy (east-west orientation)

The calculation method used for the extinction coefficient g is as follows:

$$g = \frac{\sqrt{\chi^2 \cos^2(\pi/2 - \beta) + \sin^2(\pi/2 - \beta)}}{\chi + 1.774(\chi + 1.182)^{-0.773}} \tag{3.8}$$

where χ represents the leaf distribution coefficient, with a value close to 0 indicating that plant leaves are vertically distributed, and a value close to $+\infty$ indicating a predominantly horizontal leaf distribution. According to the measurements obtained, the χ values for the east-west and north-south orientations in this experiment were 0.49 and 0.31, respectively. The calculation method for the transmission distance d of light in the system was as follows:

$$d = EC = H \tan \theta b \tag{3.9}$$

The calculation method for the distance d' of light passing through maize leaves was the following:

$$d' = \begin{cases} (N-1)\, W_{\text{strip}} + \left(W_{\text{strip}} - d_0\right) + d_r & d_0 \le W_{\text{strip}}, d_r \le W_{\text{strip}} \\ (N-1) W_{\text{strip}} + d_r & d_0 > W_{\text{strip}}, d_r \le W_{\text{strip}} \\ N\, W_{\text{strip}} + W_{\text{strip}} - d_0 & d_0 \le W_{\text{strip}}, d_r > W_{\text{strip}} \\ N\, W_{\text{strip}} & d_0 > W_{\text{strip}}, d_r > W_{\text{strip}} \end{cases} \tag{3.10}$$

In the formula, d_0 is the distance from the calculation point to the far right end of the last maize model that the light is transmitted through. In turn, d_r is the distance from the light point of incidence to the far right end of the first maize model and was calculated as follows:

$$d_\text{r} = d + d_0 - N\,\left(W_{\text{strip}} + W_p\right) \tag{3.11}$$

where N represents the number of maize bands that the light beam passes through, which is calculated by Eq. 3.2:

$$\frac{d + d_0}{W_{\text{strip}} + W_p} - 1 < N \le \frac{d + d_0}{W_{\text{strip}} + W_p} \tag{3.12}$$

The actual distance, L_M, that light is transmitted in a perturbed medium can be calculated by Eq. 3.13:

$$L_M = d' / (\cos \theta c * \sin \theta b) \tag{3.13}$$

At any point, PAR in the system can be calculated by Eq. 3.14:

$$I = (1 - F_{\text{PAR}}) \times I_0 \tag{3.14}$$

where I_0 refers to incident PAR on the control, unshaded plants.

3.2.1.2 Model Operation and Parameter Determination

The extinction coefficient, g, reflects the vertical or horizontal distribution of leaves and reflects the ability of leaves to intercept light. The daily change trend followed by g values showed a decreasing phase followed by an increasing one. Minimum g values for the east-west (0.17) and north-south (0.09) orientations were recorded at noon. In contrast, maximum g values were recorded at 18:00 pm, with values of 0.28 and 0.21, for east-west and north-south orientations, respectively. The daily variation of the straight-line distance perpendicular to the row direction, d, of the light transmission above the soybean canopy (Table 3.1) showed that d values for the east-west orientation were more stable than those for the north-south orientation (Fig. 3.6). The maximum d value for the east-west orientation was 1.45 m, at 7:

00 am, decreasing to 0 m at 9:00 am, and then switching to another orientation and slowly increasing to 0.92 at 17:00 pm. The d value for the north-south orientation was 4.36 m at 7:00 am, decreasing to 0 at noon, and then rapidly increasing in another orientation. 1/(cos θc * sin θb) represents the ability of light to pass diagonally through the intercropping system. The variation in the east-west direction was very large in the morning, with the value peaking at 95.41, at 9:00 am. Meanwhile, in the afternoon, it remained stable first and then rapidly decreased at 17:00 pm. In contrast, d values for the north-south orientation showed fluctuations of about 15 units at noon and were relatively stable at other times.

3.2.2 Results of Light Environment Simulation

3.2.2.1 Light Environment in the Soybean Canopy

Measurement and simulation of the light transmission above the soybean canopy was conducted from 8:00 am to 18:00 pm, using both east-west and north-south orientations. The distance that light traveled through disturbed media in different directions varied, resulting in different characteristics of light distribution above the soybean canopy (Figs. 3.7 and 3.8). Simulation results of the light intensity at ten points in the soybean canopy showed that PAR between maize rows (points 1 and 4) was lower than the overall PAR intensity, whereas, above the soybean canopy (points 5–10), PAR intensity was higher than the overall PAR intensity. The distribution of light intensity above the soybean canopy was also uneven, with higher light intensity at the middle points (points 7–9) compared to other points.

Comparison between two treatments showed that the average PAR value above the soybean canopy (points 6–10) was higher in the east-west than in the north-south orientation in the morning (8:00 am) and in the evening (4:00 pm and 6:00 pm), whereas PAR in the north-south orientation was higher than in the east-west orientation from 10:00–14:00. Meanwhile, at noon (12:00), the values for PAR in the north-south and east-west orientations were 1023 and 932 μmol m^{-2} s^{-1}, respectively. The daily average values for PAR at soybean canopy height (points 1–10) were 370.02 μmol m^{-2} s^{-1} in the east-west orientation and 360.68 μmol m^{-2} s^{-1} in the north-south orientation.

The RMSE values for the east-west and the north-south orientations were 0.1535 and 0.1584, respectively (Fig. 3.8). The regression lines of the measured values and the true values for the east-west and north-south orientations were as follows: $y = 0.7076x + 0.0846$, and $y = 0.6138x + 0.1286$, respectively. The deviation for estimating the transmittance from 0 to 0.1 was 0.1; the deviation for estimating the transmittance from 0.9 to 1 was 0.2; and the estimation error for the range of 0.1–0.9 was less than 0.1.

3.2.2.2 Simulating Crop Light Environment

Using the PAR estimation model, the graph for the simulated light transmission model in the strip intercropping system was compared with that of the actual data. The trends were similar, but the results of using the actual values measured were

Table 3.1 Calculation of transmission distance (considering the east-west orientation as an example)

	Hour angle	Height difference	Latitude	Declination	Altitude	Sun azimuth				Transmission distance
Time	t (°)	Δh (m)	λ (°)	δ (°)	β (°)	α (°)	θa (°)	θc (°)	θb (°)	d (m)
7:00	75	1.7	35.25	14.25	20.3	93.44	108.44	17.26	68.7	4.36
8:00	60	1.7	35.25	14.25	32.54	84.64	99.64	8.12	57.09	2.63
9:00	45	1.7	35.25	14.25	44.57	74.15	89.15	−0.6	45.43	1.73
10:00	30	1.7	35.25	14.25	55.85	59.68	74.68	−8.53	33.2	1.11
11:00	15	1.7	35.25	14.25	65.04	36.47	51.47	−15.24	20.01	0.62
12:00	0	1.7	35.25	14.25	69	0	15	−20.25	−5.68	−0.17
13:00	15	1.7	35.25	14.25	65.04	−36.47	−21.47	−23.12	−9.67	−0.29
14:00	30	1.7	35.25	14.25	55.85	−59.68	−44.68	−23.53	−25.5	−0.81
15:00	45	1.7	35.25	14.25	44.57	−74.15	−59.15	−21.43	−41.07	−1.48
16:00	60	1.7	35.25	14.25	32.54	−84.64	−69.64	−17.06	−55.77	−2.5
17:00	75	1.7	35.25	14.25	20.3	−93.44	−78.44	−10.84	−69.31	−4.5
18:00	90	1.7	35.25	14.25	8.17	−101.72	−86.72	−3.26	−81.82	−11.82

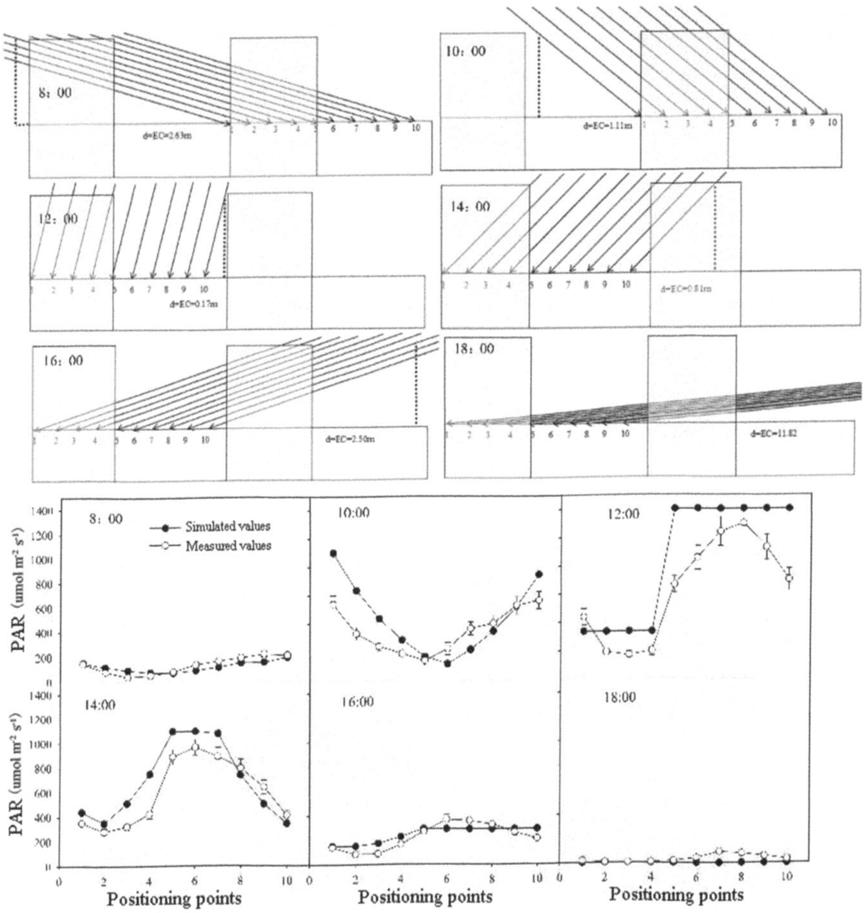

Fig. 3.7 Distribution of PAR at the top of the soybean canopy (north-south orientation)

smoother than those of the simulated data (Fig. 3.9). As light transmittance decreased at the interface of the two media from 0.8 to 0.2, the graph constructed with actual values showed a transition zone of approximately 40 cm, whereas the transition in the zone simulation graph was less than 20 cm, which may lead to error when simulating the light intensity between the two. The simulation graph considers plants as rigid geometric bodies, whereas the actual light distribution is influenced by the maize plant type, which explains the smoothness of the graph constructed with actual values. Additionally, the simulation results indicated lower PAR above the soybean canopy at noon on the east-west orientation line than on the north-south orientation line, consistent with results using actual measurements.

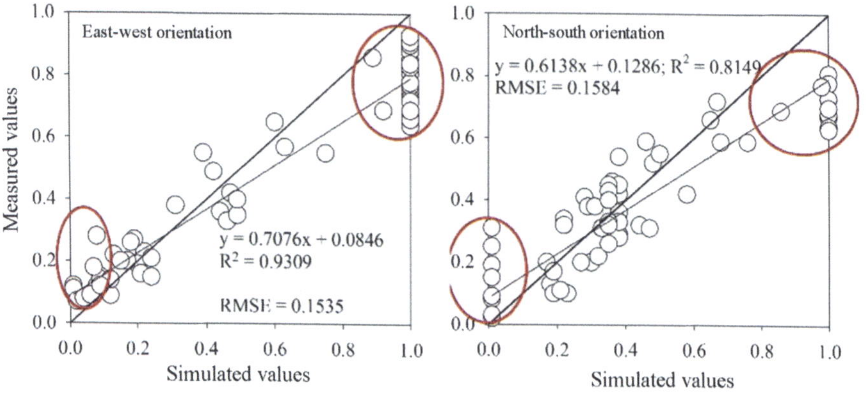

Fig. 3.8 Comparison between simulated and actual values of light transmission rate in a soybean canopy

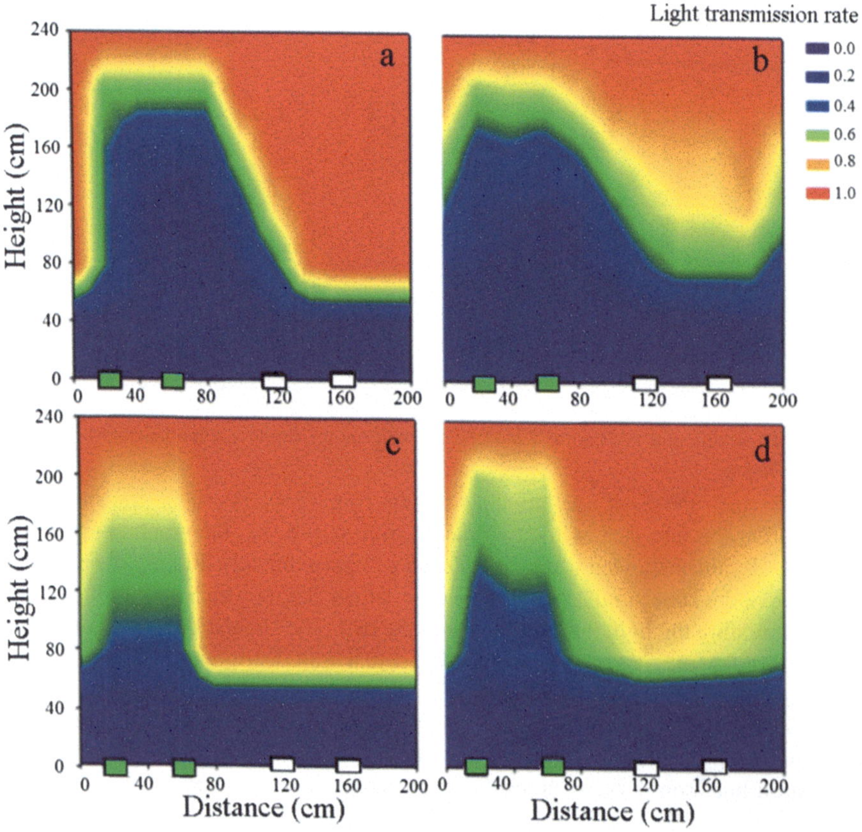

Fig. 3.9 Simulated and measured light distribution above the soybean canopy at noon. Note: **a** Simulation value of the east-west row; **b** simulation value of the north-south row; **c** measured value of the east-west row; **d** measured value of the north-south row. Empty rectangles on the horizontal axis indicate the position of maize rows

3.3 Light Interception and Synergistic Utilization in Maize-Soybean Strip Intercropping

Differences in light interception and utilization among different field configurations were the fundamental causes for dry matter and yield differences. For example, in strip intercropping, canopy structure differences prompted the selection of different light interception models. The goal was to optimize the field configuration for strip intercropping with respect to light interception and utilization and to achieve a coordinated utilization of light energy between the two member species in the system.

3.3.1 Models of Light Interception and Utilization

Current research mainly focuses on crop light interception under monoculture conditions. Assuming uniform distribution of crop leaves, the Beer–Lambert law can be applied to calculate the rate of light interception by crop canopies.

$$F = 1 - \exp(-k\mathrm{LAI}) \tag{3.15}$$

In Eq. 3.15, F and LAI represent the light interception rate and leaf area index of the crops, respectively. In turn, k represents the extinction coefficient of the crops, which is determined by the ratio of the projection of leaves on the ground to the leaf area, based on the leaf angle. When k equals 0 or 1, the leaves are completely upright or horizontal, respectively, with the impact of transmitted and reflected light having little effect on the k value. Crops such as maize, with relatively upright leaves, typically have k values below 0.5, while crops, such as soybeans and cotton, with relatively horizontal leaves, typically have k values above 0.5.

Canopy structure under strip intercropping is notably more intricate than that under conditions of monoculture, with field configuration having a substantial influence on crop light interception. The HHLA and ERCRT models were applied across different strip intercropping scenarios for a precise calculation of crop light interception under various field configurations (Munz et al. 2014; Tsubo and Walker 2002).

3.3.1.1 The HHLA Model

$$F_A = 1 - \exp(-k_A\mathrm{LAI}_A) \tag{3.16}$$

$$F_B = \exp(-k_A\mathrm{LAI}_A)[1 - \exp(-k_B\mathrm{LAI}_B)] \tag{3.17}$$

In strip intercropping scenarios where the height of B is significant, such as in maize (m) and soybean (s) strip intercropping, where soybean canopy height often exceeds 80 cm, the upper canopy is composed of upper maize leaves, while the lower mixed canopy is comprised of lower maize leaves and soybean leaves.

Fm-upper represents the proportion of PAR interception by upper maize leaves, and it was calculated as follows:

$$F_{m-\text{upper}} = 1 - \exp\left(-k_m L_{m-\text{upper}}\right) \tag{3.18}$$

In Eq. 3.18, $L_{\text{m-upper}}$ and $K_{\text{m-upper}}$ represent the leaf area and extinction coefficient of the upper maize leaves, respectively. The calculation method of the *k* value was as follows:

$$k = \frac{\ln(1-f)}{\text{LAI}} \tag{3.19}$$

According to the measurements, the extinction coefficients k_{m} and k_{s} of maize and soybean were 0.42 and 0.70, respectively. The calculation method of the intercepted proportion of PAR in the lower canopy of maize and soybean was as follows:

$$F_s = \exp\left(-k_m L_{m-\text{upper}}\right) \times \frac{k_s L_s}{k_m L_{m-\text{lower}} + k_s L_s} \times \left[1 - \exp(-k_m L_{m-\text{lower}} - k_s L_s)\right] \tag{3.20}$$

$$F_{m-\text{lower}} = \exp\left(-k_m L_{m-\text{upper}}\right) \times \frac{k_m L_{m-\text{lower}}}{k_m L_{m-\text{lower}} + k_s L_s} \times \left[1 - \exp(-k_m L_{m-\text{lower}} - k_s L_s)\right] \tag{3.21}$$

$$F_m = F_{m-\text{upper}} + F_{m-\text{lower}} \tag{3.22}$$

$L_{m\text{-lower}}$ and L_s are the LAI values of the lower parts of soybean and maize plants, respectively. LAI values $L_{m\text{-upper}}$ and $L_{m\text{-lower}}$ of maize can be calculated as:

$$L_{m-\text{upper}} = \frac{h_m - h_s}{h_m} \times L_m \tag{3.23}$$

$$L_{m\text{-lower}} = \frac{h_s}{h_m} \times L_m \tag{3.24}$$

where h_{m} and h_{s} are maize and soybean plant heights, respectively. L_m is the total leaf area of maize.

3.3.1.2 The ERCRT Model

In strip intercropping systems, the proportions of intercepted radiation by various crop components can be calculated using the ERCRT model. The mixed canopy is divided into four sections, namely im, iim, ib, and iib. Incident PAR is divided into nine sections, as depicted in Fig. 3.10. The proportions of intercepted radiation by the crop, F1-F9, are calculated using an equation from Wang et al. (2015).

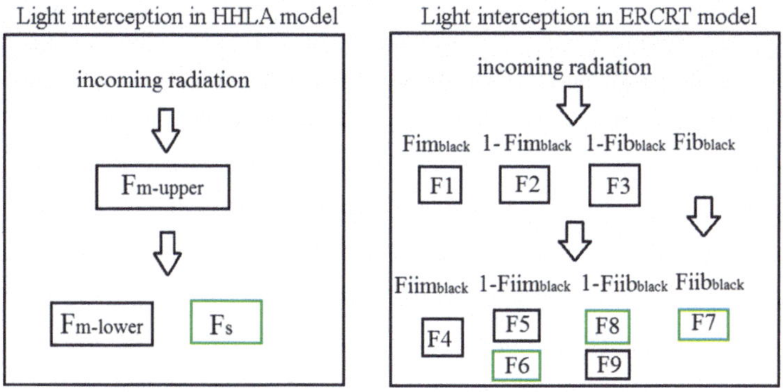

Fig. 3.10 Schematic diagram of the calculated photosynthetic effective-interception rate using the HHLA (left) and the ERCRT (right) models. Note: The black box represents the light interception by maize and the green box represents the light interception by soybean

$$F1 = f_m Fim_{\text{black}} \left[1 - \exp\left(-k_m L_{m-\text{upper}}/f_m\right)\right] \tag{3.25}$$

$$F2 = f_m\left\{(1 - Fim_{\text{black}}) \left[1 - \exp\left(-k_m L_{m-\text{upper}}\right)\right]\right\} \tag{3.26}$$

$$F3 = f_s\left\{(1 - Fib_{\text{black}}) \left[1 - \exp\left(-k_m L_{m\text{-upper}}\right)\right]\right\} \tag{3.27}$$

$$F4 = Fim\ Fiim_{\text{black}} \left[1 - \exp(-k_m L_{m-\text{lower}})/f_m\right] \tag{3.28}$$

$$F5 = Fim \left\{(1 - Fim_{\text{black}}) \left[1 - \exp(-k_m L_{m-\text{lower}} - k_s L_s)\right] \frac{k_m L_{m-\text{lower}}}{k_m L_{m-\text{lower}} + k_s L_s}\right\} \tag{3.29}$$

$$F6 = Fim \left\{(1 - Fim_{\text{black}}) \left[1 - \exp(-k_m L_{m-\text{lower}} - k_s L_s)\right] \frac{k_s L_s}{k_m L_{m-\text{lower}} + k_s L_s}\right\} \tag{3.30}$$

$$F7 = Fib\ Fiib_{\text{black}} \left[1 - \exp(-k_s L_s/f_s)\right] \tag{3.31}$$

$$F8 = Fib \left\{(1 - Fib_{\text{black}}) \left[1 - \exp(-k_m L_{m-\text{lower}} - k_s L_s)\right] \frac{k_s L_s}{k_m L_{m-\text{lower}} + k_s L_s}\right\} \tag{3.32}$$

$$F9 = Fib \left\{(1 - Fib_{\text{black}}) \left[1 - \exp(-k_m L_{m-\text{lower}} - k_s L_s)\right] \frac{k_m L_{m-\text{lower}}}{k_m L_{m-\text{lower}} + k_s L_s}\right\} \tag{3.33}$$

In the equations listed above, f_{m} and f_{s} represent the area ratio between the maize and the soybean strips, which is determined by field configuration. For example, the fm and fs values for the SI3 treatment were 0.5 and 0.3, respectively. Fimblack and

Fibblack represent the view angle below im and ib, respectively. Meanwhile, Fiimblack and Fiibblack represent the view angle below iim and iib, respectively. The view angle is the integral of the empty factor of each point on the plane. Thus, $\mathrm{Fim}_{\mathrm{balck}}$, $\mathrm{Fib}_{\mathrm{balck}}$, $\mathrm{Fiim}_{\mathrm{balck}}$, and $\mathrm{Fiib}_{\mathrm{balck}}$ were calculated as follows:

$$Fim_{\mathrm{balck}} = \frac{\sqrt{(h_m - h_s)^2 + {W_m}^2} - (h_m - h_s)}{W_m} \tag{3.34}$$

$$Fib_{\mathrm{balck}} = \frac{\sqrt{(h_m - h_s)^2 + {W_s}^2} - (h_m - h_s)}{W_s} \tag{3.35}$$

$$Fiim_{\mathrm{balck}} = \frac{\sqrt{{h_s}^2 + {W_m}^2} - h_s}{W_m} \tag{3.36}$$

$$Fiib_{\mathrm{balck}} = \frac{\sqrt{{h_s}^2 + {W_s}^2} - H_s}{W_s} \tag{3.37}$$

In turn, calculation of the proportion of light below the im and ib parts, Fim and Fib, was as follows (Wang et al. 2015):

$$\begin{aligned} Fim = & f_m\left[Fim_{\mathrm{blacck}} \times \exp\left(-k_m L_{m\,-\mathrm{upper}}/f_m\right)\right] \\ & + f_m \ (1 - Fim_{\mathrm{black}}) \\ & \times \exp\left(-k_m L_{m\,-\mathrm{upper}}\right)\right] + f_m f_s (1 - Fib_{\mathrm{black}}) \times \exp\left(-k_m L_{m\,-\mathrm{upper}}\right) \end{aligned} \tag{3.38}$$

$$\begin{aligned} Fib = & f_s\left[Fib_{\mathrm{black}} + f_s (1 - Fib_{\mathrm{black}}) \exp\left(-k_m L_{m\text{-upper}}\right)\right] \\ & + f_m f_s \ \left[(1 - Fim_{\mathrm{black}}) \times \exp\left(-k_m L_{m\text{-upper}}\right)\right] \end{aligned} \tag{3.39}$$

Additionally, the light interception ratio of maize and soybean was calculated as follows:

$$F_m = F1 + F2 + F3 + F4 + F5 + F9 \tag{3.40}$$

$$F_s = F6 + F7 + F8 \tag{3.41}$$

3.3.1.3 Photosynthetic Efficiency Model

Photosynthetic light use efficiency was calculated using Eq. 3.42:

$$\mathrm{LUE} = \frac{\mathrm{ADM}}{\Sigma I_{\mathrm{PAR}} F} \tag{3.42}$$

Accumulated dry matter (ADM) is measured in g m^{-2} and represents the amount of biomass produced by the crop. Incident photosynthetically active radiation

(IPAR) is the amount of radiation measured in megajoules per square meter (MJ m^{-2}) that accounts for 50% of the total radiation, with incident PAR obtained by multiplying the total radiation data by 0.5. The *F* value represents PAR interception rate of either maize or soybean on a given day. The amount of light intercepted by the crop on a certain day is represented by $I_{PAR}F$. The cumulative amount of light accumulated by the crop during a certain period of time is determined by the sum of the daily $I_{PAR}F$ values.

3.3.2 Pattern of Photosynthetic Effective Radiation Interception

3.3.2.1 LAI and Canopy Height Are the Main Parameters of the Light Interception Model

In a strip intercropping system, LAI and canopy height of maize and soybean are the crucial factors impacting the distribution of crop canopy light and the key parameters in the light interception model (Liu et al. 2017a). When two rows of maize and soybean were strip intercropped, as illustrated in Fig. 3.2, LAI of both maize and soybean increased rapidly within 40–60 days of sowing, reaching a maximum value at 70 days after sowing (Fig. 3.11). Specifically, LAI of maize under strip intercropping was equal to the net area index of maize when maize row spacing was above 20 cm (A3 and A5 treatments). In contrast, LAI of soybean under strip intercropping increased with a wider distance between the maize and soybean rows. Moreover, the spatial configuration in strip intercropping caused opposite changes in LAI values of maize and soybean, leading to mutual complementation.

Maize height in strip intercropping is insensitive to different spatial arrangements, with maximum plant height reaching 2.5 m at 60 days after sowing (Fig. 3.12). In contrast, soybean plant height was remarkably affected by the composite-planting spatial arrangement, increasing as the distance between maize and soybean rows

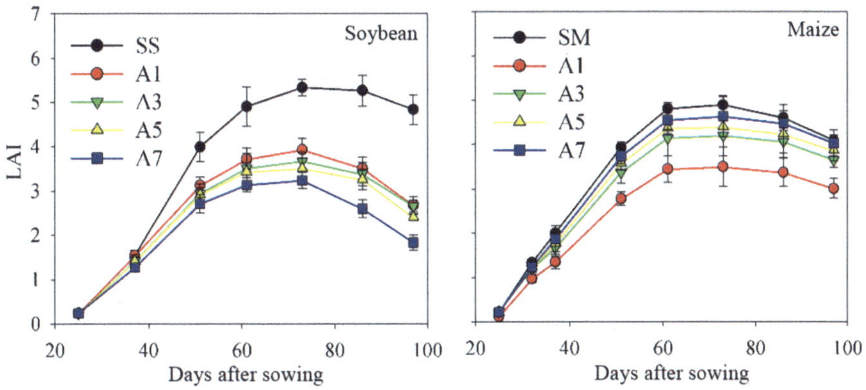

Fig. 3.11 Dynamic changes of leaf area in maize and soybeans

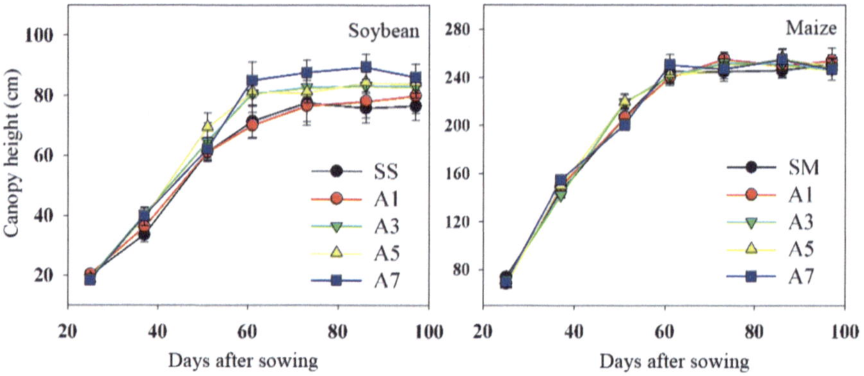

Fig. 3.12 Dynamic changes in canopy height of maize and soybeans. Note: A1, A3, and A5 represent the soybean canopy light environment under 70, 60, and 50 cm spacing (i.e., maize row spacing of 20, 40, and 60 cm), respectively, with a 2:2 maize: soybean row configuration in a 200 cm wide production unit; A7 represents the soybean canopy light environment in the traditional 1:1 of maize: soybeans configuration

decreased. In the traditional system, i.e., one row of maize and one row of soybean strip intercropping, maximum soybean plant height was 85 cm.

3.3.2.2 Light Interception Pattern

A strip intercropping system was implemented for maize and soybean, in which LAI and plant height served as input parameters for the light interception model. Light interception rates of maize and soybean peaked at 70 and 45 days after sowing, respectively. Different field configurations revealed a pattern of mutually complementary alternating rates for the two crops. Although the light interception rate of both crops in the strip intercropping system was lower than that of monoculture, the intercropping system captured significantly more light energy at an interception rate of 0.9 (Fig. 3.13), whereas the light interception rates of maize and soybean monocultures were 0.71 and 0.69, respectively. Moreover, daily light interception rates and radiation data computed from the system conformed to the pattern of light interception and accumulation for both crops under the strip intercropping system (Fig. 3.14).

3.3.3 Pattern of Collaborative Utilization of Light Energy

Crop biomass mostly derives from photosynthesis, accounting for 90–95% of the total. In the maize-soybean strip intercropping system, increasing the spacing between maize and soybean rows led to an increasing trend for soybean biomass accumulation, which peaked at a value of 405–434 g m^{-2} when maize rows were 20–40 cm apart from soybean rows (with a spacing of 60–80 cm). As for maize biomass, the opposite trend was observed, with a maximum recorded value of

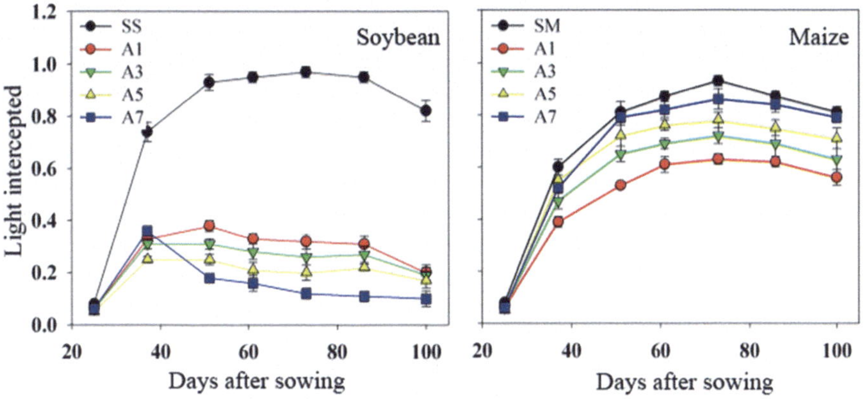

Fig. 3.13 Dynamic changes in the light interception rates of maize and soybean

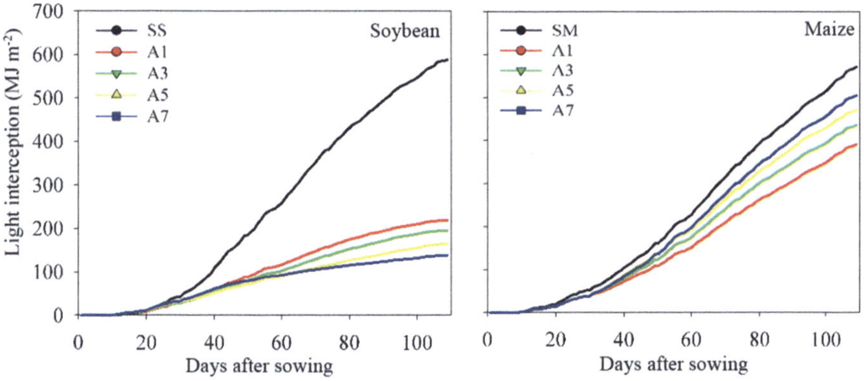

Fig. 3.14 Dynamic changes in the light interception of maize and soybean. Note: A1, A3, and A5 represent soybean canopy light environments in a 2:2 row configuration with a spacing of 70 cm, 60 cm, and 50 cm, respectively, when the production unit width was 200 cm (i.e., narrow row spacing for soybean was 20, 40, and 60 cm). A7 represents the soybean canopy light environment in the traditional 1:1 configuration with maize. SS and SM represent monoculture soybean and monoculture maize, respectively

2244 g m^{-2} (Table 3.2). Furthermore, strip intercropping facilitated collaborative light energy utilization in both crops. Soybean had the highest light energy conversion rate at a value of 2.36 g MJ^{-1} when maize spacing was 40 cm, whereas maize registered the highest light energy conversion rate at values between 4.7 and 4.79 g MJ^{-1}, at 20–40 cm spacing between maize plants. The intercropping system allowed notably higher biomass accumulation than monoculture systems of either soybean or maize. Additionally, the light energy utilization rate of the intercropping system (4.05 g MJ^{-1}) at a maize spacing of 40 cm was considerably higher than that

Table 3.2 Synergistic utilization of light energy in the maize-soybean strip intercropping system

	Soybean			Maize			Crop system (soybean + maize)		
	Dry matter	Light interception	LUE	Dry matter	Light interception	LUE	Dry matter	Light interception	LUE
Treatment	g m^{-2}	MJ m^{-2}	g MJ^{-1}	g m^{-2}	MJ m^{-2}	g MJ^{-1}	g m^{-2}	MJ m^{-2}	g MJ^{-1}
A1	434b	208b	2.10b	1926d	403d	4.79a	2360b	611c	3.88b
A3	405c	173c	2.36a	2115c	450c	4.70a	2520a	623b	4.05a
A5	310d	144d	2.16b	2191bc	475bc	4.46b	2501a	619bc	3.94ab
A7	292d	128d	2.29a	2244b	506b	4.31c	2536a	634a	3.91b
SS	749a	509a	1.48c				749c	509e	1.48c
SM				2364a	598a	3.95d	2364b	598d	3.95ab

Note: Different lowercase letters within columns indicate significant differences treatment means of other intercropping field configurations and those of either maize or soybean monoculture systems (Liu et al. 2017b, 2018).

3.4 Physiological and Biochemical Mechanisms of High-Efficiency Utilization of Light Energy in the Leaves of Maize Plants

3.4.1 Phenotypic and Photosynthetic Responses of Maize Leaves under Different Row Ratios

3.4.1.1 Maize Leaf Phenotypic Response

In a strip intercropping system consisting in two rows of maize plants alternating with two rows of soybean plants, the average number of green leaves per plant at the R4 and R5 maize growth stages was significantly higher ($P < 0.05$) than those in monoculture conditions. Specifically, the number of green leaves was 9.9 and 4.9, respectively, in comparison with monoculture of maize or maize-soybean intercropping in a 1:1 ratio. Additionally, the chlorophyll content of maize leaves was significantly higher ($P < 0.05$) in the strip intercropping system; specifically, there was an increase of 7% and 10%, and 9% and 15% (in the middle layer of leaves), respectively, compared to monoculture or traditional maize-soybean intercropping. Furthermore, these results imply that there was a significant impact from the different planting models ($P < 0.05$) on the green leaf area of maize plants at the R1, R3, R4, and R5 maize growth stages, with a decreasing trend at each stage. Additionally, when compared to monoculture or traditional intercropping, the green leaf area of maize plants at R3 and R4 stages under the strip intercropping system where two rows of maize alternate with two rows of soybean increased by 4% and 7%, and 9% and 16%, respectively. Therefore, it can be concluded that there is a direct correlation between the development of green leaf area in maize plants and the growth conditions of maize under different planting systems, as demonstrated by the results shown in Fig. 3.15.

3.4.1.2 Maize Leaf Photosynthesis

Three different treatment groups were compared to determine variations in photosynthesis, transpiration rate, and stomatal conductance in maize leaves among different row configurations under strip intercropping, relative to maize monoculture. Treatments included maize strip intercropping with two rows of soybeans (200 cm strip width, 40 cm maize row spacing, and 40 cm soybean row spacing), maize strip intercropping with one row of soybeans (100 cm maize row spacing, with soybeans planted between maize rows), and monoculture maize (60 cm row spacing) (Zeng 2018a). Photosynthetic (Pn) and transpiration rates (T_r), stomatal conductance (g_s), and mesophilic intercellular CO_2 (C_i) of the panicle leaf were recorded during the tassel emergence and silk emission stage. Results showed that rows of maize strip intercropped with two rows of soybeans had significantly higher values for Pn, T_r, and g_s in the panicle leaf, compared to the other two treatments, measuring at 25.4, 3.11, and 0.21 mmol m^{-2} s^{-1}, respectively. The results are summarized in Table 3.3.

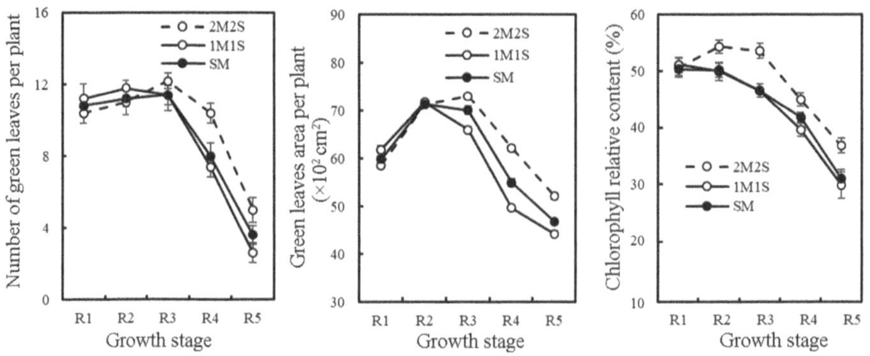

Fig. 3.15 Dynamic changes in the number of green leaves, leaf area, and chlorophyll content of maize plants under different configurations. Note: Maize silking stage, blister stage, milk stage, and wax stage of maize are represented by R1, R2, R3, and R4, respectively. Strip intercropping of two rows of maize with two rows of soybean, one row of maize with one row of soybean, and monoculture of maize are represented by 2M2S, 1M1S, and SM, respectively

Table 3.3 Photosynthetic parameters of maize under different row configurations (at the tassel emergence and silk emission stage)

Treatment	Pn	Tr	g_s	Ci
	mmol m^{-2} s^{-1}			mmol mol^{-1}
2M2S	25.4a	3.11a	0.21a	93.45b
1M1S	24.1ab	2.24b	0.14b	109.38a
SM	22.8b	2.13b	0.12b	116.29a

Note: 2M2S, 1M1S, and SM represent the intercropping of two rows of maize with two rows of soybeans, one row of maize with one row of soybeans, and monoculture maize planting, respectively. Different lowercase letters indicate significant differences among treatment means

3.4.2 Analysis of the Differences in Light Energy Utilization Among Maize Varieties

Maize-soybean intercropping varieties differed in light energy utilization responses. To investigate these differences and the underlying mechanisms, high-light efficient variety, Rongyu 1210, and low light-efficient variety, Zhongwangyu 18, were selected for analysis in strip intercropping at maize row spacings of 20 and 40 cm.

3.4.2.1 Photosynthetic Response of Maize Panicle Leaves at Varying Row Spacing

Maize varieties differed in their responses to changes in row spacing, as evidenced by the gas exchange parameters of panicle leaves during the grain filling period (Table 3.4). Thus, variety Rongyu 1210 showed higher apparent quantum efficiency, maximum Pn, and light saturation point compared to Zhongwangyu 18, with increases of 19.85–29.08%, 5.26–17.82%, and 3.91–28.40%, respectively, with a

Table 3.4 Photosynthetic parameters of panicle leaves at different row-spacing treatments in maize

Treatment		Apparent quantum efficiency (μmol·μmol·m $^{-1}$)	Maximum net photosynthetic rate (μmol ·m $^{-2}$·s $^{-2}$)	Light saturation point (μmol ·m $^{-2}$·s $^{-2}$)	Light compensation point (μmol ·m $^{-2}$·s $^{-2}$)	Dark respiratory rate (μmol ·m $^{-2}$·s $^{-2}$)
Rongyu 1210	Inside of 20 cm row spacing	0.040c	34.63c	658.96d	49.23a	1.14e
	Inside of 40 cm row spacing	0.052b	42.00b	979.69c	48.59a	1.45d
	Outside of 20 cm row spacing	0.075a	49.82a	1450.29a	22.26c	2.44a
	Outside of 40 cm row spacing	0.065a	49.50a	1180.60b	35.32b	2.23b
	Monoculture	0.071a	44.02ab	1030.13c	44.89a	1.89c
Average value		0.061a	43.99a	1059.93a	40.06b	1.83a
Zhongwangyu 18	Inside of 20 cm row spacing	0.032c	32.57b	594.32d	85.93a	1.36d
	Inside of 40 cm row spacing	0.039c	34.52b	701.47c	77.47b	1.43d
	Outside of 20 cm row spacing	0.077a	47.19a	1324.71a	40.29e	2.79a
	Outside of 40 cm row spacing	0.052b	42.63a	1134.44b	51.72d	2.37b
	Monoculture	0.050b	40.81a	1071.79b	63.22c	2.05c
Average value		0.050b	39.54a	965.35a	63.73a	1.99a
		A = 369.77**	6.30	9.28	447.33**	12.93
F value		B = 79.93**	11.27**	81.84**	96.93**	117.14**
		A*B = 101.87**	0.50	3.38*	8.63**	1.68

Note: The data in the table are means of three replicates. Different lowercase letters within columns indicate significant differences at the 5% level. * and ** denote significant differences at the 0.05 and 0.01 level, respectively. The average row is compared separately decrease of 28.99–44.74% in light compensation point. Reducing row spacing caused a variety-dependent reduction in apparent quantum efficiency, maximum Pn, light saturation point, and leaf dark-respiration rate. More precisely, panicle leaves at the wider-row spacing of intercropped maize were better at harnessing strong light intensities in alternating wider-row with short-row narrow-strip, while panicle leaves of monoculture maize were more effective in using weaker light intensities at narrow rows. The wider-row spacing in intercropping caused a 12.45% and 4.45% increase in maximum Pn of Rongyu 1210 and Zhongwangyu 18, respectively, compared to those of maize leaves under monoculture at 40 cm row spacing. Moreover, light saturation points increased by 14.6% and 5.85%, respectively. Conversely, the light compensation points of both varieties were 21.32% and 18.19% lower, respectively, than those of maize leaves under monoculture. Thus, strip intercropping requires maize varieties with high-light efficiencies, high light saturation points, and low light compensation points. These conditions characterized the effective use of the light energy range within panicle leaf blades, with strong light utilization at wider-row spacing intercropped with maize and weak light utilization at narrow row spacing.

3.4.3 Chlorophyll Content in Maize Panicle Leaves

The total chlorophyll content of different maize varieties during the silking and grain filling stages was significantly affected by changes in row spacing (Fig. 3.16). Particularly, variety Rongyu 1210 showed a higher total chlorophyll content than variety Zhongwangyu 18, with a 4.19–19.16% increase under different row-spacing treatments, regardless of cultivation system. Within each variety, a similar decreasing trend in total chlorophyll content was observed with increasing row spacing. Compared with monoculture, Rongyu 1210 and Zhongwangyu 18 showed a higher chlorophyll contents during silking and grain filling, with a 0.97–10.46% and 0.26–7.02% increase, respectively.

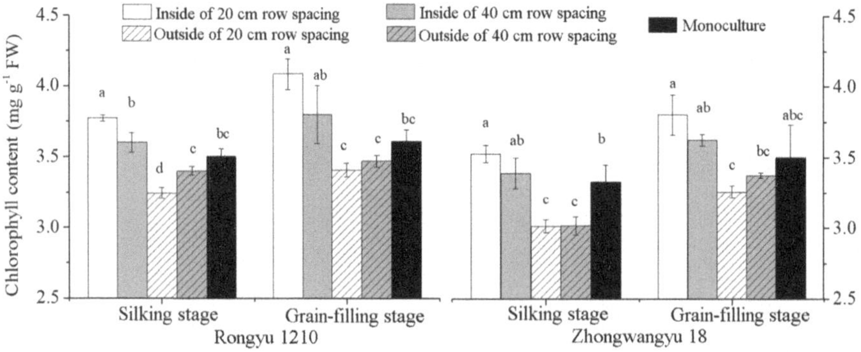

Fig. 3.16 Differences in chlorophyll content of maize panicle leaves under different row-spacing treatments

3.4.4 Photosynthetic Enzyme Activities in Maize Panicle Leaves

3.4.4.1 Ribulose-1, 5-Bisphosphate Carboxylase/Oxygenase Activity

Strip intercropping strongly influenced the activity of the RubisCO enzyme in maize panicle leaves during silking and grain filling, with row spacing as a significant factor (Fig. 3.17). Specifically, variety Rongyu 1210 showed significantly higher RubisCO activity than Zhongwangyu 18 during both silking and grain filling, increasing between 1.53% and 17.58% in various comparisons. However, neither variety showed a significant difference during the same growth stages when grown under monoculture. As plant growth proceeded, RubisCO activity decreased by 3.06–15.50% and 2.36–17.48% in Rongyu 1210 and Zhongwangyu 18, respectively. When row spacing was reduced from 40 to 20 cm, average RubisCO activity of narrow row panicle leaves decreased by 5.08% and 6.40% for Rongyu 1210 and Zhongwangyu 18, respectively, confirming that Rongyu 1210 has a higher photosynthetic carbon assimilation ability under strip intercropping conditions than Zhongwangyu 18.

3.4.4.2 Phosphoenolpyruvate Carboxylase (PEPC) Activity

Strip intercropping resulted in higher PEPC activity in panicle leaves of maize variety Rongyu 1210 than that recorded for those of variety Zhongwangyu 18 during silking and grain filling (Fig. 3.18). Further, in narrow rows, panicle leaf-PEPC activity decreased from silking to grain filling in both Rongyu 1210 and Zhongwangyu 18 under different black nylon shading (NP) and intercropping (IC) treatments, decreasing by 7.53–10.77% and 3.22–6.32%, respectively, whereas it showed an increasing trend in wide-row panicle leaves. The change in PEPC activity in panicle leaves ranked in order of narrow row IC > NP maize leaves > wide-row IC. Intercropping with maize at a row spacing of 40 cm led to an increase in PEPC activity by 2.56–23.52% and 4.77–19.37% for Rongyu 1210 and Zhongwangyu 18, respectively, relative to NP maize and wide-row IC panicle leaves. In narrow row panicle leaves, PEPC activity in Rongyu 1210 increased by 9.32% and 12.50% during the silking and grain filling periods, respectively, when

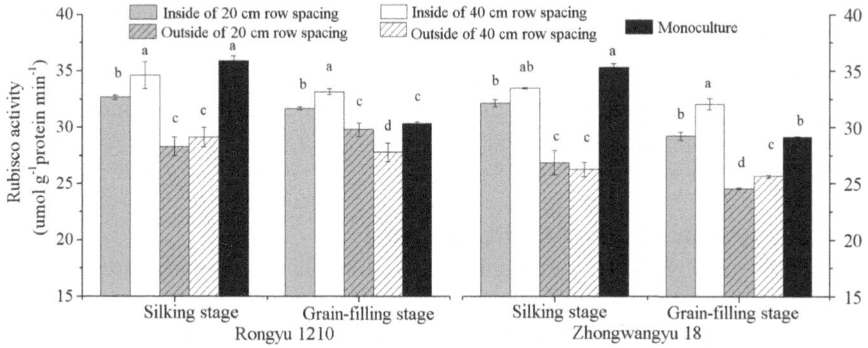

Fig. 3.17 RubisCO activity in maize panicle leaves under different row-spacing treatments

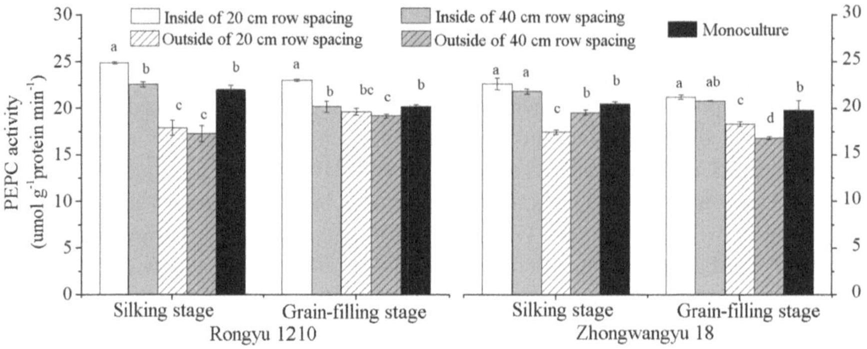

Fig. 3.18 Panicle leaf-PEPC activity in two maize varieties grown under different row-spacing treatments

row spacing was reduced to 20 cm, whereas that of Zhongwangyu 18 leaves increased only by 3.62% and 1.86%, respectively. These results confirm that intercropping maize with soybeans improved the adaptability of both maize varieties to low light via increased PEPC activity under strip intercropping conditions, particularly in variety Rongyu 1210, which demonstrated a greater adaptation ability.

3.5 Physiological and Biochemical Mechanisms for Efficient Utilization of Light Energy in Soybean Leaves

3.5.1 The Influence of Different Field Configurations on Photosynthetic Characteristics and Light Energy Utilization by Soybean Leaves

Alterations in the field configuration of soybean-maize intercropping influenced the exposure of the soybean canopy to light, causing strong variation in soybean photosynthetic performance. Two different row ratio configurations (1:1 and 2:2), in which the distance between maize and soybean is 50 and 40 cm, respectively, were established, and soybean was used as the control. The effects of the maize canopy, located at a higher position than that of the soybean canopy, on fluorescence parameters, anatomical structure, and ultrastructure of soybean leaves was analyzed for various maize-soybean row configurations.

3.5.1.1 Soybean Photosynthetic Pigments

Different maize-soybean row configurations resulted in substantially lower photosynthetic pigment content per unit leaf area in soybean plants grown under intercropping, compared to soybean monoculture (Table 3.5). In particular, the 1:1 and 2:2 row ratio configurations caused a significant reduction in chlorophyll a

Table 3.5 Photosynthetic pigment content of soybean leaves under different field configurations (mg dm^{-2})

Treatment	chlorophyll a	chlorophyll b	chlorophylls	Carotenoid	Total photosynthetic pigments	chlorophyll a/b
1:1	1.82 c	0.46 c	2.28 c	0.43 c	2.71 c	3.93 ab
2:2	2.84 b	0.80 b	3.64 b	0.66 b	4.29 b	3.57 b
CK	4.41 a	1.05 a	5.46 a	1.09 a	6.55 a	4.19 a

Note: 1:1, 2:2, and CK denote single maize and soybean alternating rows, two rows of maize alternating with two rows of soybean, and soybean monoculture used as a control, respectively; different lowercase letters within columns indicate significant differences at the $p < 0.05$

Table 3.6 Soybean leaf photosynthetic characteristics under different intercropping field configurations

	Pn	Gs	Tr	Ci
Treatment	mmol m^{-2} s^{-1}			mmol mol^{-1}
1:1	6.04 c	0.55 c	3.47 b	344.89 a
2:2	9.56 b	0.86 b	4.27 b	339.97 a
Soybean monoculture	15.69 a	1.06 a	6.35 a	319.56 b

Table 3.7 Soybean chlorophyll fluorescence parameters under different intercropping field configurations

Treatment	Fv/Fm	NPQ	Fv′/Fm′	qp	ΦPSII
1:1	0.77 a	2.58 b	0.52 a	0.36 a	0.24 a
2:2	0.79 a	2.84 ab	0.51 ab	0.38 a	0.20 b
Soybean monoculture	0.81 a	3.14 a	0.48 b	0.44 a	0.19 b

(50.2% and 27.9%, respectively), chlorophyll b (46.4% and 13.7%, respectively), carotenoids (49.9% and 33.6%, respectively), and the chlorophyll a/b ratio (6.9% and 16.3%, respectively), compared to soybean monoculture. The soybean photosynthetic pigment content under the traditional 1:1 field configuration of soybean under intercropping with maize was substantially reduced as a result of the more intense shading pressure from maize than under the 2:2 strip intercropping field configuration.

3.5.1.2 Soybean Photosynthetic Fluorescence Characteristics

Soybean photosynthetic fluorescence characteristics in the soybean-maize strip intercropping system were significantly affected by shading (Table 3.6). Leaf Pn was significantly ($P < 0.05$) lower under the 1:1 and 2:2 intercropping configurations than that observed in soybean plants grown as monoculture, with a notable increase recorded for 2:2 over 1:1. Additionally, g_s values significantly decreased compared to those measured on soybean leaves under monoculture, although not under 2:2 intercropping conditions. In contrast, C_i concentration experienced a significant increase of 7.3% and 5.4% under the 1:1 and 2:2 intercropping configurations, respectively, relative to soybean monoculture. Moreover, under both 1:1 and 2:2 intercropping field configurations, soybean leaves exhibited lower maximum photochemical quantum yield of PSII, photochemical fluorescence quenching coefficient, and non-photochemical fluorescence quenching coefficient (NPQ) than soybean monoculture. Furthermore, the non-photochemical fluorescence quenching coefficient of soybean under the 1:1 intercropping field configuration was significantly lower than that of soybean grown as single crop, but the difference was not significant under the 2:2 intercropping field configuration. In contrast, soybean functional leaves under both intercropping field configurations showed enhanced effective quantum yield of PSII and photochemical efficiency relative to that measured in soybean leaves grown under monoculture conditions (Table 3.7), with 1:1, but not 2:2, being significantly different.

3.5.1.3 Structural Characteristics of Soybean Leaves

Soybean leaf palisade- and spongy-mesophyll tissues and epidermal cell walls all experienced a decrease in thickness under strip intercropping relative to soybean leaves in monoculture plants, with a particularly pronounced difference observed in the leaf structure in plants grown in the 1:1 maize-soybean intercropping system (Fig. 3.19). Leaf palisade-and spongy-mesophyll tissues and upper and lower epidermal cells of monoculture soybean plants were slightly larger and tightly packed, while in strip intercropped soybean leaves, these cells were smaller and showed a somewhat looser arrangement. Consequently, there was a greater space between neighboring cells, resulting in a decrease of lower epidermal cell thickness, especially in the 1:1 maize-soybean intercropping system. Quantitative analysis of anatomical structure (Table 3.8) revealed that soybean leaf thicknesses in monoculture conditions was 2.15 and 1.69 times greater than that under 1:1 and 2:2 field configurations, respectively. Nevertheless, epidermal cell thickness and the palisade-to spongy-mesophyll ratio in soybean plants grown in the 2:2 intercropping system did not exhibit significant differences compared to that of soybean leaves from plants grown under monoculture. Additionally, the thicknesses of the upper and lower epidermis, palisade- and spongy-mesophyll tissue, leaf, and palisade- to spongy-mesophyll tissue ratio of soybeans were 20.09%, 31.91%, 64.49%, 2.73%, 30.84%, and 59.80% greater under monoculture than under the 1:1 strip intercropping system with maize.

The ultrastructure of soybean leaves grown under intercropping planting systems was analyzed and compared to monoculture soybean leaves. Results showed that the size of single cells and chloroplasts decreased while the number of chloroplasts increased significantly in the two intercropping treatments. Particularly, soybean

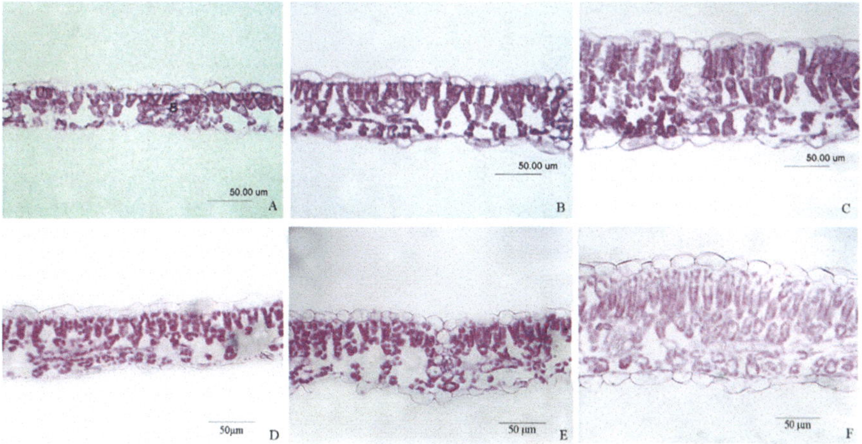

Fig. 3.19 Schematic diagram of soybean leaf anatomical structure under different field intercropping configurations. Note: **a**, **b**, and **c** represent strip intercropping configurations of one row of maize intercropped with one row of soybeans, two rows of maize intercropped with two rows of soybeans, and soybean monoculture, respectively

Table 3.8 Soybean leaf anatomical structure under different field intercropping configurations

Treatment	Thickness of upper epidermis	Thickness of lower epidermis	Thickness of palisade tissue	Thickness of sponge tissue	Leaf thickness	Palisade- to spongy-mesophyll tissue ratio
	μm					
1:1	10.50 b	8.93 b	26.36 c	26.01 b	73.26 c	1.02 b
2:2	12.61 a	11.78 a	43.36 b	26.72 b	95.85 b	1.63 a
Soybean monoculture	13.39 a	12.54 a	80.04 a	50.29 a	151.02 a	1.60 a

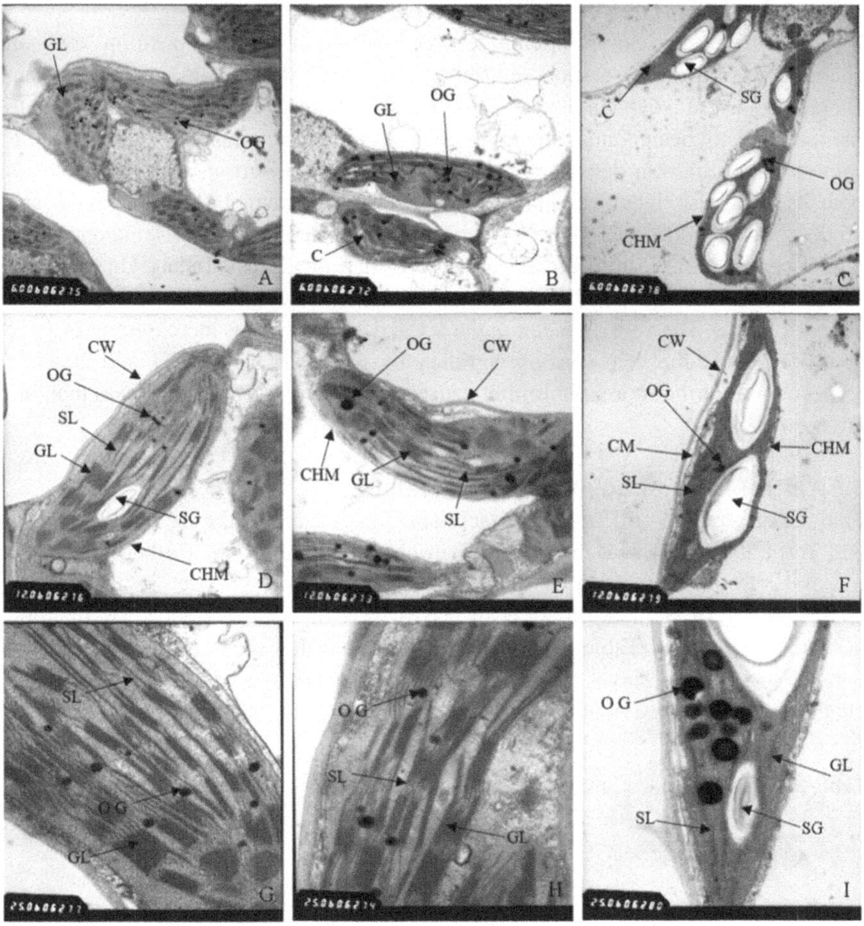

Fig. 3.20 Ultrastructure of soybean leaves under different planting configurations. Note: **a**–**c**: Chloroplast distribution (×6000); **d**–**f**: chloroplast ultrastructure (×12,000); **g**–**j**: thylakoid layer of chloroplast (×25,000); *CW* cell wall, *CHM* chloroplast membrane, *GL* grana layer, *SL* stroma layer, *SG* starch grain, *OG* osmiophilic granules, **a,d,g**: treatment A1; **b,e,h**: treatment A2; **c,f,i**: control group

leaves under the 1:1 field configuration had the largest number of chloroplasts but coupled to the smallest size. This finding indicated that the starch granules in soybean leaf chloroplasts were infrequent while eosinophilic granules increased under both 1:1 and 2:2 field configurations. Moreover, the grana and stroma lamellae were densely packed, which enhanced the absorption, transfer, and conversion of light energy and facilitated the production and accumulation of photoassimilate (Fig. 3.20).

3.5.1.4 Soybean Leaf Vein Characteristics

Maize-soybean strip intercropping led to shading of the low-position soybean canopy by the high-position maize canopy and directly affected the vein characteristics of soybean leaves (see Table 3.9 and Fig. 3.21). Compared to monoculture soybean plants, maize-soybean strip intercropping resulted in a significant reduction of vein density, vein length, and vein closure in soybean leaves in both 1:1 and 2:2 intercropping systems by 20.01% and 14.99%, 20.02% and 15.00%, and 50.00% and 48.00%, respectively. Particularly, the 1:1 intercropping model caused significantly greater reductions than the 2:2 treatment. On the other hand, strip intercropping led to significant increases in vein diameter and spacing of soybean leaves. The vein diameter in the 1:1 and 2:2 treatments increased by 24.01% and 4.59%, and the vein spacing significantly increased by 41.49% and 20.41%, respectively. Furthermore, the former caused significantly greater increases than the latter.

3.5.1.5 Features of Soybean Leaf Stomata

Leaf stomata play a critical role in facilitating gas exchange during photosynthesis and respiration under a range of environmental conditions; furthermore, they respond strongly to various environmental factors. A case in point, the characteristics of stomata on soybean leaves were directly influenced by planting system, as shown in Table 3.10. Our data revealed that the variation in stomatal density in soybean leaves shaded by the maize canopy corresponded with the changes observed in stomatal width and area. In contrast, the variations in stomatal

Table 3.9 Soybean vein characteristics under different planting configurations

Treatment	Leaf vein density (mm/mm^2)	Leaf vein length (mm)	Leaf vein diameter (μm)	Leaf vein closure (No. per mm^2)	Leaf vein spacing (μm)
1:1	47.85 c	14.33 c	32.11 a	2.36 c	188.28 a
2:2	59.25 b	17.74 b	28.11 b	3.21 b	167.92 b
Soybean monoculture	66.11 a	19.79 a	22.38 c	4.23 a	130.34 c

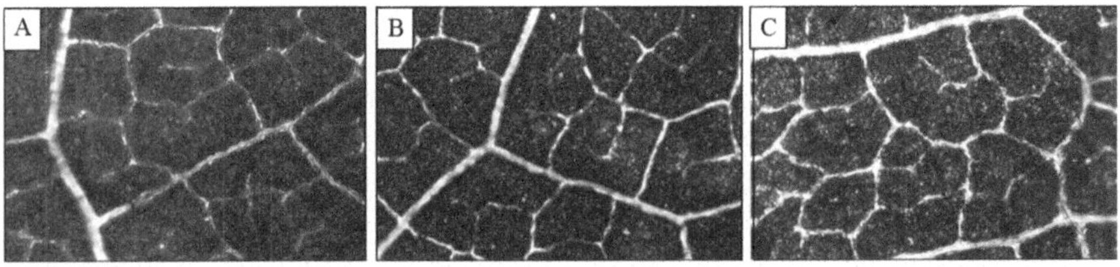

Fig. 3.21 Soybean vein characteristics under different planting configurations. **a**, **b**, and **c** represent strip intercropping configurations of one row of maize intercropped with one row of soybeans, two rows of maize intercropped with two rows of soybeans, and soybean monoculture, respectively

Table 3.10 Stomatal characteristics in soybean leaves under different planting configurations

Treatment	Stomatal density (No. per mm^2)	Stomatal length (nm)	Stomatal width (nm)	Stomatal perimeter (nm)	Stomatal area (nm^2)
1:1	122.56 b	23.90 a	2.54 c	57.08 a	98.21 b
2:2	130.78 b	22.71 b	2.98 b	56.15 a	123.42 a
Soybean monoculture	149.96 a	21.67 b	3.52 a	54.77 b	130.15 a

1:1 and 2:2 represent one row of maize intercropped with one row of soybeans and two rows of maize intercropped with two rows of soybeans, respectively

Table 3.11 Soybean light energy utilization under different intercropping configurations

Treatment	Proportion of PAR in soybean canopy	Proportion of LAI	Proportion of light interception	LUE
	%			
1:1	33 b	60 b	25 b	155 a
2:2	63 a	73 a	41 a	142 a

1:1 and 2:2 represent one row of maize intercropped with one row of soybeans and two rows of maize intercropped with two rows of soybeans, respectively

length and perimeter were inversely associated with stomatal density. Albeit the observed reductions in stomatal density by 18.27% and 12.79% under 1:1 and 2:2 treatments respectively, relative to single-crop soybean controls (CK), stomatal length and perimeter increased considerably by 10.29% and 4.80%, respectively, under 1:1, and by 4.22% and 2.52% under 2:2 planting system treatment. Notably, these degrees of decrease or increase are dissimilar, as the former is significantly greater than the latter, suggesting that the shading effect on soybeans under the 1:1 field configuration of strip intercropping was greater than that under the 2:2 configuration.

3.5.1.6 Soybean Light Energy Utilization

Photosynthetically active radiation, LAI, and light interception were significantly lower in soybean plants grown in the 1:1 or 2:2 field intercropping configurations, compared to soybean plants grown as a single crop. However, the values were larger in the 2:2 intercropping configuration than their counterparts in the 1:1 intercropping configuration, accounting for 63%, 73%, and 41%, respectively, of the corresponding values in the soybean monoculture system. Light energy utilization rates were calculated based on the light environment and light interception of soybean canopies in the 1:1 and 2:2 field configurations tested. The utilization rate was 155% for the 1:1 configuration and 142% for the 2:2 configuration, compared to soybean monoculture (Table 3.11).

3.5.2 Responses of Different Soybean Varieties to Shade

The responses of different soybean varieties to shading treatment were analyzed for comparison. Shading treatments included: no shading (100% normal light intensity, CK), mild shading (70% of normal light intensity, LS), moderate shading (50% of normal light intensity, MS), and strong shading (30% of normal light intensity, SS). Shade-tolerant Nandou 12 and shade-sensitive Guixia 7 soybean varieties were studied to compare leaf photosynthetic fluorescence characteristics, anatomical structure, and ultrastructure under the different shading treatments.

3.5.2.1 The Impact of Shading on Stomatal Density of Soybean Leaves

Stomatal density on the adaxial epidermis of the leaves of soybean variety Nandou 12 initially increased and then decreased with increasing shade (Fig. 3.22; Chen et al. 2019). Thus, under mild and moderate shading, the number of stomata on the adaxial epidermis of the leaf increased significantly by 26.9% and 18.5%, respectively, relative to the number recorded under conditions of normal light intensity. Conversely, the number of stomata in the adaxial epidermis of the leaves of soybean variety Guixia 7 was significantly reduced by 18.3%, 27.2%, and 37.8% under mild, moderate, and strong shading, in comparison with the number recorded under conditions of normal light intensity. Similarly, the number of stomata on the abaxial epidermis initially increased and then decreased with reducing light intensity,

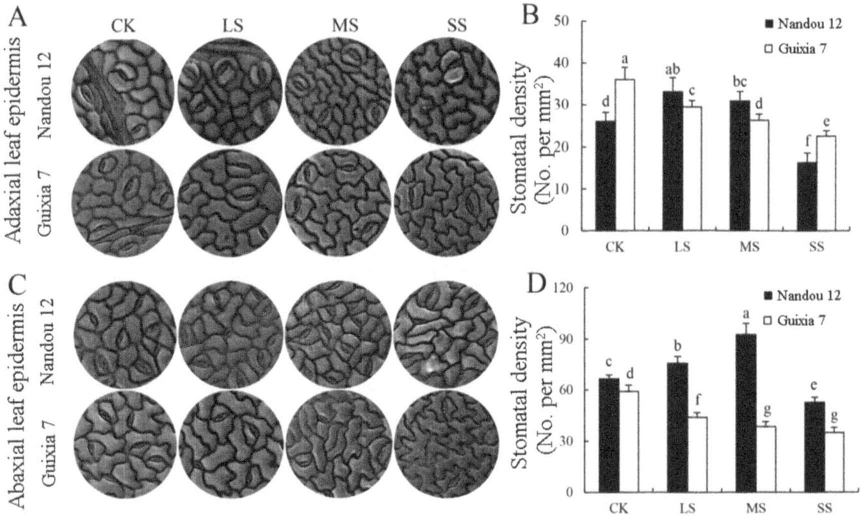

Fig. 3.22 Stomatal characteristics in different soybean varieties under shading (reprinted with permission from Scientia Agricultura Sinica). Note: **a** and **b** represent the actual stomata and stomatal density on the adaxial leaf epidermis; **c** and **d** represent the actual stomata and stomatal density on the abaxial leaf epidermis. CK, LS, MS, and SS represent normal light conditions, mild shading, moderate shading, and strong shading, respectively. Different lowercase letters within columns indicate significant differences at the 0.05 level

whereas in Nandou 12 leaves, the number of stomata on the abaxial epidermis increased significantly under mild and moderate shading (13.9% and 39.2%, respectively), although it decreased significantly by 20.8% under strong shading. In contrast, the number of stomata on the abaxial epidermis of Guixia 7 leaves significantly decreased by 25.8%, 34.9%, and 41.0% under mild, moderate, and strong shading, respectively. Thus, shade-tolerant soybean variety Nandou 12 had a significantly higher number of stomata on the abaxial epidermis than shade-sensitive Guixia 7. Furthermore, under mild and moderate shading, the number of stomata on the epidermis of Nandou 12 leaves was also significantly higher than that on leaves of Guixia 7 soybean plants.

3.5.2.2 Effects of Shading on Photosynthetic Parameters in Soybean Leaves

Mild shading improved soybean Pn, which is related to its light intensity requirement (Table 3.12). Mild shading increased Pn by 15.27% and 44.24% in Nandou 12 and Guixia 7 soybean leaves, respectively, relative to the rate measured under the normal light intensity level. However, as shading increased, Pn and g_s in Nandou 12 and Guixia 7 leaves showed a decreasing trend, with Nandou 12 registering a Pn 80.24% higher than that of Guixia 7 under severe shading. Other photosynthetic parameters exhibited similar trends. Thus, as shading increased, g_s and C_i in Nandou 12 leaves initially decreased and then increased, peaking at an average of 0.57 mm·m^{-2}·s^{-1} under mild shading. In contrast, g_s in Guixia 7 leaves gradually decreased, while C_i increased with decreasing light intensity.

3.5.2.3 Effects of Shading on Chlorophyll Fluorescence Parameters of Soybean Leaves

Although there was no significant difference in measured quantum yield in Nandou 12 leaves between mild shading and normal light treatment, the actual quantum yield significantly decreased by 20.6% and 27.1% under moderate and heavy shading treatments, respectively (Fig. 3.23). The actual quantum yield of Gui Xia 7 gradually decreased with increasing shading, and it was lower than that of Nandou 12. An increase in shading degree resulted in a decrease in maximum quantum yield for both Nandou 12 and Gui Xia 7. Further, no significant difference was found for NPQ of Nandou 12 between mild shading and normal light treatment. However, it significantly increased by 16.7% and 24.3% under moderate and heavy shading treatments, respectively. The non-photochemical quenching coefficient of Gui Xia 7 initially decreased and later increased and was 4.7% higher than that of Nandou 12 under mild shading treatment (Fan et al. 2017).

3.5.2.4 Carbohydrates in Soybean Leaves

Shading resulted in a reduction of starch content in soybean leaves (Fig. 3.24). Although Nandou 12 leaves had higher starch and soluble sugar contents than those of Guixia 7 under normal light conditions, the differences were not statistically significant. Further, when subjected to mild shading, the leaf starch content of Guixia 7 exceeded that of Nandou 12, whereas Nandou 12 leaves had significantly higher levels of soluble sugars than those of Guixia 7, demonstrating that the accumulation

Table 3.12 Photosynthetic parameters in soybean under different degrees of shading

	Pn (mmol m^{-2} s^{-1})		Gs (mmol m^{-2} s^{-1})		Ci (mmol mol^{-1})		Tr (mmol m^{-2} s^{-1})	
Treatment	Nandou12	Guixia7	Nandou12	Guixia7	Nandou12	Guixia7	Nandou12	Guixia7
CK	12.25b	10.08c	0.18 cd	0.16d	424.66d	438.21c	2.10c	1.98 cd
LS	14.12a	14.54a	0.57a	0.30bc	449.65b	425.31d	4.18a	2.92b
MS	12.29b	11.96b	0.16d	0.20c	370.37e	433.35 cd	1.98 cd	3.79a
SS	10.49c	5.82d	0.39b	0.11d	500.23a	450.47b	3.68a	1.39d

Note: Different lowercase letters within columns represent significant differences at the 0.05 level

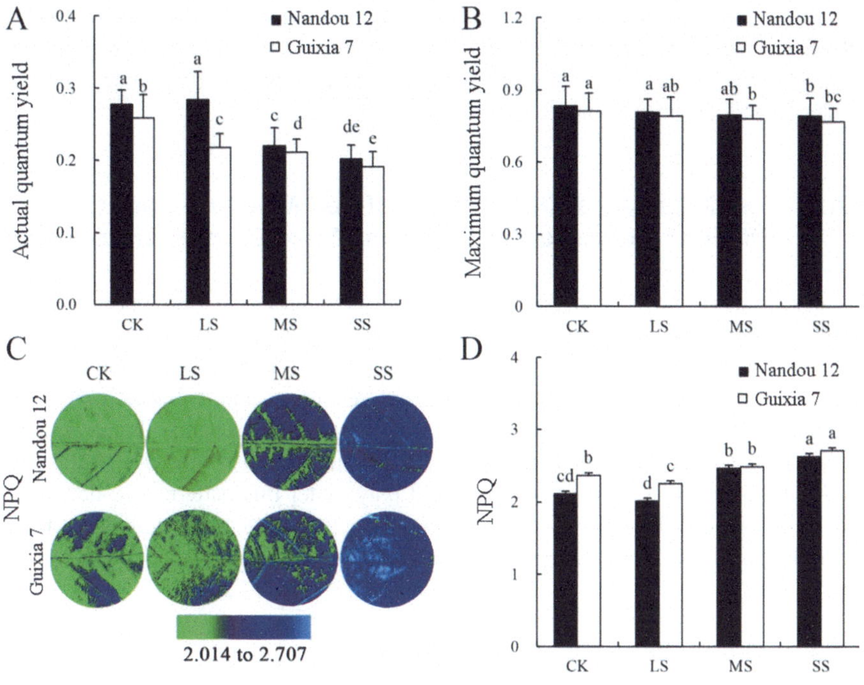

Fig. 3.23 Response of chlorophyll fluorescence parameters of different soybean varieties to shading

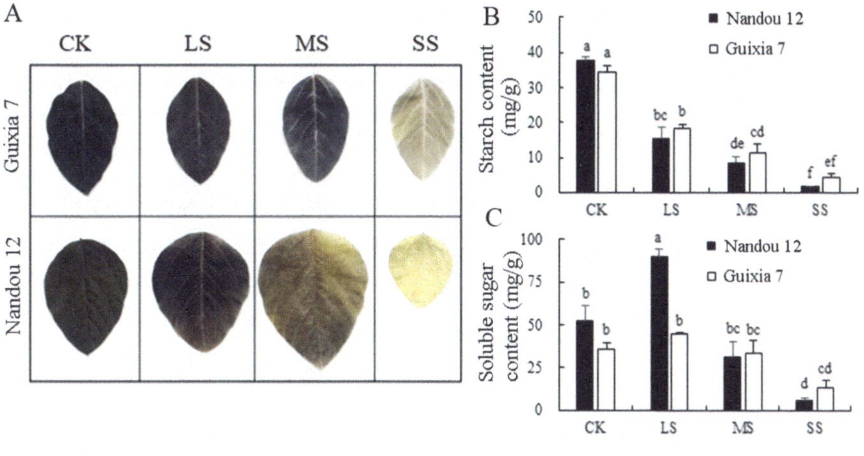

Fig. 3.24 Response of soluble sugars and starch contents to shading in different soybean varieties

of soluble carbohydrates provides benefits in improving shade tolerance of shade-tolerant soybean varieties (Li et al. 2019a). Furthermore, no significant differences were observed in starch or soluble sugar contents of the two varieties between moderate and strong shading conditions.

3.6 Spatiotemporal Distribution of Soil Moisture and its Efficient Utilization in Soybean-Maize Strip Intercropping

Water scarcity is one of the most crucial limiting factors for agricultural productivity. Strip intercropping is an intensive agricultural system that maximizes resource utilization and improves land productivity. Rational strip intercropping can increase farmland productivity, improve soil moisture retention, and enhance water use efficiency (WUE). Soybean-maize strip intercropping is currently being widely promoted in China. The elucidation of the spatiotemporal distribution of water and the high-efficiency utilization mechanism at play under this pattern provides theoretical and practical support for the development of efficient water management technologies for strip intercropping systems (Ye et al. 2015).

3.6.1 Rainfall Redistribution Patterns

The rainfall distribution patterns in strip intercropping systems were analyzed under various conditions. The production units were set to a fixed width of 200 cm, with soybean row spacing at 40 cm and maize row spacing at 20, 30, 40, 50, 60, and 70 cm. Maize and soybean monocultures served as control groups. The differences in rainfall distribution between maize inter-rows and maize-soybean inter-rows were smaller, although still significant, among the six maize row spacings, compared to the differences observed among soybean inter-rows (Table 3.13). Conversely, the average rainfall between soybean inter-rows was higher than the average rainfall in soybean monoculture, while the rainfall in each maize inter-row was lower than that in maize monoculture inter-rows. This was mainly due to the higher position above the ground of the maize canopy, which redistributed more rainfall and created heterogeneity in rainfall amounts in the maize-soybean strip intercropping system.

3.6.2 Distribution Pattern of Soil Moisture

Soil moisture distribution is closely related to rainfall redistribution and crop selection. Among the strip intercropping systems containing both maize and soybean, soil moisture content between soybean rows was noticeably higher than that between maize inter-rows or that between maize-soybean inter-rows across all six treatments. With the exception of A2 and A3, soil moisture content was also considerably higher between maize-soybean inter-rows than between maize inter-rows. As shown in Table 3.14, an increase in the spacing between maize rows was positively correlated with a corresponding increase in soil moisture content between maize inter-rows.

Table 3.13 Effect of maize row spacing on rainfall redistribution (mm)

	Treatment							
Site	A1	A2	A3	A4	A5	A6	Maize Monoculture	Soybean Monoculture
MM	13.44b(d)	17.05c(c)	21.65b(b)	25.98b(b)	36.48b(a)	36.38b(a)	37.93(a)	–
MS	15.84b	55.43b	29.94b	32.18b	62.30b	18.81c	–	–
SS	88.82a(a)	88.72a(a)	69.72a(c)	90.43a(a)	82.80a(b)	86.36a(b)	–	85.66(b)

Note: In this study, soybean and maize were strip intercropped in 200 cm width production units. Soybean rows were fixed at a spacing of 40 cm, and maize rows were tested at six different spacings: A1 (20 cm), A2 (30 cm), A3 (40 cm), A4 (50 cm), A5 (60 cm), and A6 (70 cm). Controls included maize monoculture (70 cm row spacing) and soybean monoculture (50 cm row spacing). The intercrop positions were designated as MM (maize inter-row), MS (maize-soybean inter-row), and SS (soybean inter-row). Lowercase letters outside the parentheses indicate differences in the same treatment for rainfall redistribution at different positions, and lowercase letters inside the parentheses indicate differences among treatments for rainfall redistribution at the same position

Table 3.14 Effect of maize row spacing on soil moisture distribution (%)

Treatment	A1	A2	A3	A4	A5	A6	Maize Monoculture	Soybean Monoculture
SS	19.29a(a)	21.03a(a)	19.84a(a)	21.24a(a)	21.00a(a)	19.73a(a)	–	18.12(a)
MS	18.76b	19.42b	18.98b	20.54b	20.11b	19.80a	–	–
MM	17.97c(b)	19.57b(a)	18.90b(ab)	19.78c(a)	18.62c(ab)	18.28b(ab)	20.30(a)	–

Table 3.15 Effect of maize row spacing on soil evaporation (mm/day)

	Treatment							
Site	A1	A2	A3	A4	A5	A6	Maize monoculture	Soybean monoculture
SS	2.87a(a)	2.84a(a)	2.49a(b)	2.54a(b)	2.35a(b)	2.21a(b)	–	3.03(a)
MS	2.73a	2.34b	2.07b	2.18ab	2.46a	2.22a	–	–
MM	2.03b(a)	2.01b(a)	2.17b(a)	2.09b(a)	2.19b(a)	2.11b(a)	2.05(a)	–

3.6.3 Distribution Pattern of Soil Evaporation

The role of soil moisture evaporation in field water and energy balance is significant, and it closely relates to the dynamic changes of soil moisture. Furthermore, it is of great importance to the scientific management of water and the implementation of rational irrigation and water-saving technologies. Table 3.15 shows that soil evaporation between soybean rows significantly exceeded those between maize rows and between maize and soybean rows. In addition, soil evaporation between soybean rows declined progressively with an increase in maize row spacing, and the soil evaporation between maize rows rose with increasing maize row spacing. Strip intercropping of soybeans between rows registered a lower soil evaporation rate than that in the single-crop system. On the other hand, soil evaporation between maize rows increased when the distance between them was 40–70 cm, which caused higher evaporation than that in the single-cropping system.

3.6.4 Water Transpiration Patterns

In the soybean-maize strip intercropping system, the transpiration T_r of maize leaves decreased with increasing maize row spacing, with the maximum value occurring in the treatment of 20 cm maize row spacing, while T_r of soybean leaves showed a stable increasing trend, with the maximum value in the treatment of 70 cm maize row spacing. Compared with maize or soybean monoculture, the T_r values for maize and soybean leaves in strip intercropping were significantly ($P < 0.05$) higher (Table 3.16).

3.6.5 Water Use Efficiency

The effects of maize row-spacing configurations on WUE of both maize and soybean were significant in the soybean-maize strip intercropping system, as shown in Table 3.17. Maize WUE generally increased first and then decreased as maize row spacing increased from 20 to 50 cm (in treatments A1 to A4), while soybean WUE decreased with increasing maize row spacing. Moreover, the water equivalent ratio

Table 3.16 Effect of maize row spacing on leaf transpiration rate (mmol·m^{-2}·s^{-1})

Treatment	Maize	Soybean
A1	2.07a	4.77c
A2	1.89a	4.90c
A3	1.82a	5.01c
A4	1.76a	5.75abc
A5	1.05b	5.94abc
A6	0.81b	6.80a
Maize monoculture	0.58bc	–
Soybean monoculture	–	4.75c

Table 3.17 Water use efficiency and water equivalent ratio of the soybean-maize strip intercropping system

	Water use efficiency (g/kg)		
Treatment	Maize	Soybean	Water equivalent ratio
A1	28.58e	6.14b	1.70b
A2	30.23d	6.29b	1.77a
A3	31.70c	6.13b	1.79a
A4	32.69b	5.78c	1.76a
A5	32.60b	5.24d	1.69b
A6	31.93c	4.97d	1.60c
Maize monoculture	35.12a	–	–
Soybean monoculture	–	6.93a	–

in the strip intercropping system gradually increased with increasing maize row spacing. The highest value, 1.79, was recorded in the 40 cm maize row-spacing treatment. Additionally, the combined WUE of the intercropping system was significantly greater than those observed under either monoculture cropping system.

Theory of Efficient Nutrient Utilization in Maize-Soybean Strip Intercropping

4

Taiwen Yong, Wenyu Yang, Xiaochun Wang, Chun Song, Kai Luo, Ping Lin, and Yiling Li

Abstract

Efficient nutrient utilization is the basis of high crop yield, while improved nutrient utilization efficiency is mainly achieved through the selection and breeding of high-efficiency nutrient varieties, and the application of efficient nutrient management measures. Intercropping has proved to be a high-yield, efficient, and sustainable agricultural practice, especially maize-soybean strip intercropping, through the ground underground synergy and root spatial distribution difference, the use of soybean nitrogen fixation potential, to reduce chemical nitrogen input, at the same time, and to ensure the nutrient supply needed for crop production.

Keywords

Nutrient utilization · Intercropping · Maize · Soybean · Nitrogen fixation

Efficient nutrient utilization is the basis of high crop yield, while improved nutrient utilization efficiency (NUE) is mainly achieved through the selection and breeding of high-efficiency nutrient varieties, and the application of efficient nutrient management measures. The current production process is mostly based on high-yield and high-efficiency nutrient varieties supplemented with efficient nutrient management strategies to further improve NUE.

T. Yong (✉) · W. Yang · X. Wang · K. Luo · P. Lin · Y. Li
College of Agronomy, Sichuan Agricultural University, Chengdu, China
e-mail: yongtaiwen@sicau.edu.cn; mssiyangwy@sicau.edu.cn; xchwang@sicau.edu.cn

C. Song
College of Environment, Sichuan Agricultural University, Chengdu, China
e-mail: songchun@Sichuan.edu.cn

W. Yang et al. (eds.), *Theories and Application in Maize-Soybean Strip Intercropping System*, https://doi.org/10.1007/978-981-97-8723-4_4

Intercropping is fast and widely spreading around the world to improve NUE and increase land replanting index, proving to be a high-yielding, efficient, and sustainable farming practice. Particularly, maize-soybean strip compound planting is a new intercropping system including Leguminosae, to give full play to the nitrogen (N) fixation potential of soybean, thereby allowing to reduce chemical nitrogen-fertilizer input while ensuring the nitrogen supply required for high-yielding crop production. Furthermore, maize-soybean strip compound planting provides a new way to achieve green and sustainable development of agriculture. Long-term systematic research has established the theory of efficient utilization of nitrogen and phosphorus (P) by root crown promotion and interspecific synergy (Fig. 4.1).

4.1 Regulatory Mechanism of Promotion of Crown Roots

Crop nutrient absorption and utilization is an energy-dependent process of active absorption and transport. Plants turn solar energy into chemical energy in the form of biomolecules. In turn, photomorphogenesis and development of the areal plant body directly affect the energy supply for the growth and development of underground roots and nodules. Thus, the coordinated and balanced growth of the above and belowground plant organs is the key to high crop NUE.

4.1.1 Effect of Strip Intercropping on Aboveground Growth

4.1.1.1 Leaf Green Color-Retention Characteristics

Planting system and N application level can significantly affect the intensity of the green color in maize and soybean leaves. Maize leaf color during tasseling in the absence of N application was lighter than under N application, but it was darker under strip maize-soybean intercropping (IM) than under maize monoculture (MM). Under IM and MM, leaves aged in an acropetal direction, and the leaf aging position

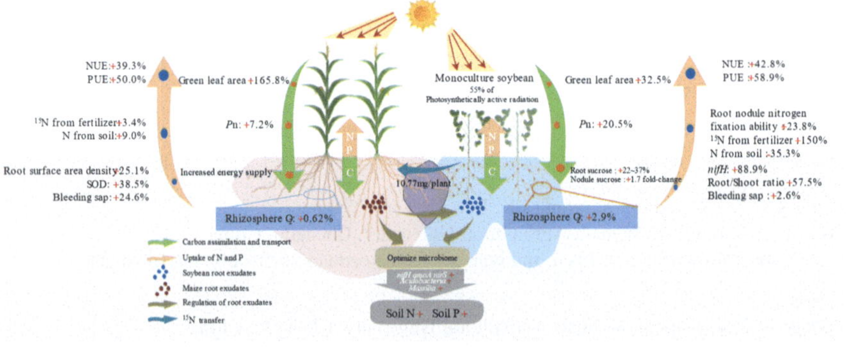

Fig. 4.1 Efficient nitrogen and phosphorus utilization for "promoting root crown and synergy between species" in maize-soybean strip compound planting

Fig. 4.2 Morphology of maize aerial plant body during male pumping (VT) and maturity (R6) in 2022

of MM at maturity (R6) was higher than that of IM (Fig. 4.2). In turn, compared to strip intercropping soybean (IS), monoculture soybean (MS) leaves were thicker, branches and leaves were sparser, and yellow leaves appeared in monoculture leaves as early as the R2 stage. The IS canopy had compact branches and internodes, and dense leaves. Further, at the R6 stage, soybean grown in IS still retained more leaves (Fig. 4.3) (Li et al. 2023).

The proportions of green leaves in both crops were consistently higher in strip intercropping than that in monoculture cropping. The overall green retention pattern observed for MM and IM leaves was NN < RN < CN. Meanwhile, the overall green retention ranking observed for MS and IS was: CN < RN < NN. Thus, N application significantly prolonged the green leaf area period (Fig. 4.4), particularly in strip intercropping plots. Indeed, 20 days after flowering (DAF), the leaf area index (LAI) under IM-NN was significantly higher than that under MM-NN by 79.4–88.5%, and at 40–50 DAF, LAI under IM-RN was significantly higher than that of MM-RN by 55.5–147.8%. In turn, peak LAI in MS was concentrated between 0–20 DAF and was significantly higher than that observed in the IS system. At 40–50 DAF, LAI in the SIS system increased significantly, whereas that in MS rapidly decreased. Furthermore, under NN, LAI in IS plots was significantly and higher than in SS

Fig. 4.3 Aboveground morphology of soybean at full blooming (R2) and beginning of seed stage (R5) in 2022. SSNN, the monocropping soybean with no N application; SSRN, the monocropping soybean with reduced N application; SSCN, the monocropping soybean with conventional N application; ISNN, the relay cropping soybean with no N application; ISRN, the relay cropping soybean with reduced N application; ISCN, the relay cropping soybean with conventional N application

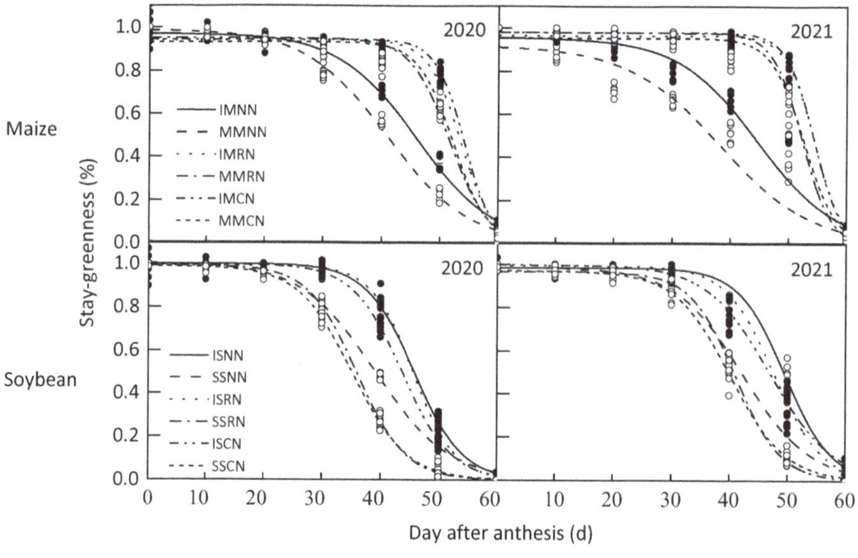

Fig. 4.4 Leaf greenness in different planting systems and nitrogen application levels

plots by 78.6–151.7%. Consistently, LAI in IS plots was significantly higher than SS by 162.6–492.3% under RN; moreover, under CN, LAI in the IS system was significantly higher than in the SS system by 125.4–747.8% (Fig. 4.5).

4.1.1.2 Leaf Photosynthesis

Data accumulated over 2 years indicated that the net photosynthesis rate (Pn) of leaves in strip intercropping after flowering was higher than that in monoculture

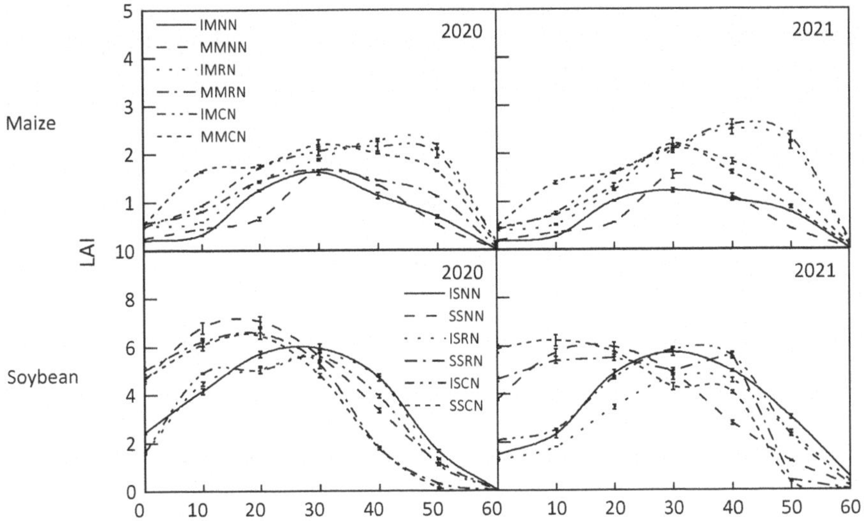

Fig. 4.5 Dynamics of LAI changes under different planting methods and nitrogen application levels

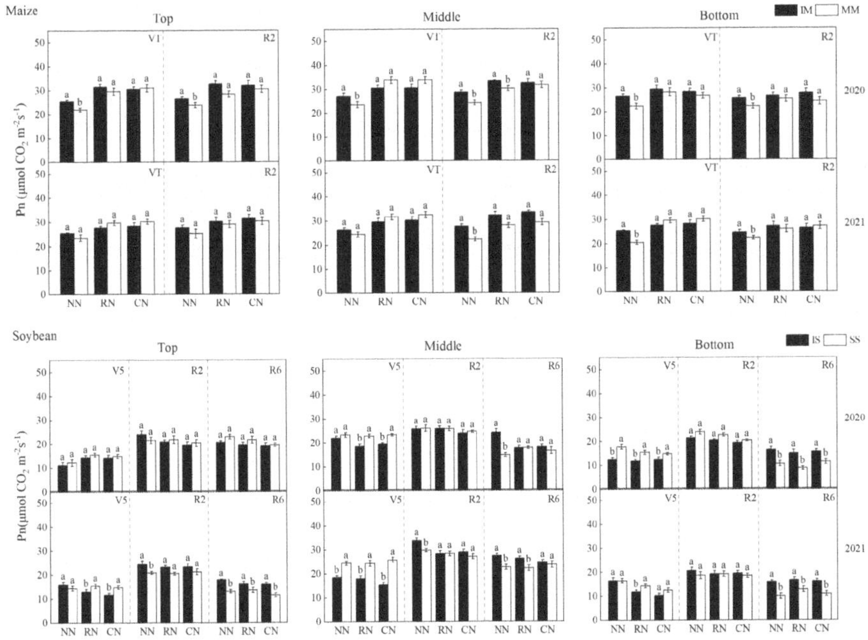

Fig. 4.6 Variation of net photosynthetic rate in maize and soybean leaves under different planting systems and nitrogen application rates

(Fig. 4.6). In maize plants, at the R2 stage, Pn in IM plots was 11.52% and 10.5% higher than that in MM under NN and CN, respectively; meanwhile, in soybean plants at the R6 stage under RN and CN conditions, Pn in IS was 11.94% and 26.47% higher than that in SS, respectively. During the VT stage of maize, Pn in the IM-RN treatment was 13.53% higher than that in the IM-NN treatment; consistently, at the R2 stage of maize, Pn in the MM-RN treatment was 13.71% higher than that in the MM-NN treatment. Meanwhile, at the R6 stage of soybean, Pn in the IS-RN treatment was 17.52% and 11.85% higher than those in the IS-NN and IS-CN treatments, respectively. Lastly, Pn in the SS-RN treatment was 12.84% and 26.37% higher than in the SS-NN and SS-CN treatments, respectively.

4.1.1.3 System Biomass

There was no significant difference in soybean biomass between strip continuous cropping and strip rotation planting under the maize-soybean strip intercropping system. In contrast, maize biomass significantly increased under strip rotation planting (Table 4.1). Thus, in 2013, maize grain yield and total plant biomass in the maize-soybean strip intercropping rotation (A2) were 7.5% and 6.0% higher than those under continuous maize-soybean strip intercropping (A1). However, in general, compared with monoculture, the biomass of maize and soybean plants grown under strip continuous or strip rotation showed a decreasing trend. Particularly, under A2, the grain yield and total plant biomass of maize decreased by 4.9% and 2.4%, respectively, and stem and total biomass of soybean plants decreased by 27.5% and 13.8%, whereas soybean seed yield significantly increased by 25.5%. When planted at equal spacing (A3), maize biomass peaked and soybean biomass was lowest. Grain yield and total biomass of soybean plants grown in A3 were 33.7% and 55.3% lower than the corresponding two-year average values recorded under MS (A5). Further analysis of the maize-soybean strip intercropping system showed that strip rotation significantly increased the biomass of the system. Thus, in 2013, the annual grain yield and total plant biomass in the A2 treatment were 6.9% and 6.8% higher than those of the A1 treatment, and 8.9% and 14.4% higher than those of the A3 treatment, respectively.

4.1.2 Effects of Strip Compound Planting on Root System

4.1.2.1 Root Growth

Monoculture maize (Fig. 4.7) and soybean (Fig. 4.8) roots showed a symmetrical horizontal distribution. However, in the strip relay intercropping model, maize and soybean roots showed a horizontally asymmetric distribution. The maize root system in the strip relay intercropping system extended to the soybean planting row and between soybean rows. Concomitantly, the soybean root system tended to grow into the maize planting row. Compared to monoculture, strip relay intercropping increased root length density of maize and soybean plants across the soil profile.

Table 4.1 Differences in maize and soybean biomass under different planting systems

Year	Plating pattern	Maize			Soybean			System		
		Stem and leaf	Grain	Total biomass	Stem and leaf	Grain	Total biomass	Stem and leaf	Grain	Total biomass
2012	A1	6262 ± 512b	6375 ± 104b	12,637 ± 408b	2159 ± 227b	1871 ± 75a	4030 ± 286a	8421 ± 578a	8246 ± 161a	16,668 ± 468a
	A2	6535 ± 213b	6464 ± 311ab	12,999 ± 364b	2118 ± 158b	1759 ± 116a	3878 ± 258a	8654 ± 125a	8223 ± 331a	16,877 ± 396a
	A3	7194 ± 198a	7126 ± 596a	14,321 ± 627a	1163 ± 112c	1011 ± 65c	2174 ± 176b	8358 ± 227a	8137 ± 660a	16,495 ± 800a
	A4	6509 ± 269b	7039 ± 298ab	13,549 ± 531ab	–	–	–	–	–	–
	A5	–	–	–	2730 ± 245a	1413 ± 202b	4143 ± 85a	–	–	–
2013	A1	6301 ± 187c	7713 ± 150b	14,015 ± 175c	3930 ± 411b	2054 ± 145a	5984 ± 556b	10,232 ± 428a	9768 ± 163b	19,999 ± 508b
	A2	6564 ± 107b	8295 ± 194a	14,860 ± 187b	4347 ± 593b	2143 ± 208a	6491 ± 776b	10,912 ± 700a	10,439 ± 226a	21,351 ± 740a
	A3	6925 ± 284a	8532 ± 130a	15,457 ± 157a	2158 ± 130c	1051 ± 84c	3210 ± 214c	9083 ± 209b	9583 ± 181b	18,667 ± 552c
	A4	6512 ± 109bc	8475 ± 184a	14,987 ± 293ab	–	–	–	–	–	–
	A5	–	–	–	6193 ± 584a	1697 ± 129b	7891 ± 483a	–	–	–

A1 maize-soybean strip intercropping continuous cropping, *A2* maize-soybean strip intercropping rotation, *A3* maize-soybean intercropping with equal row spacing for planting, *A4* maize monoculture, *A5* soybean monoculture

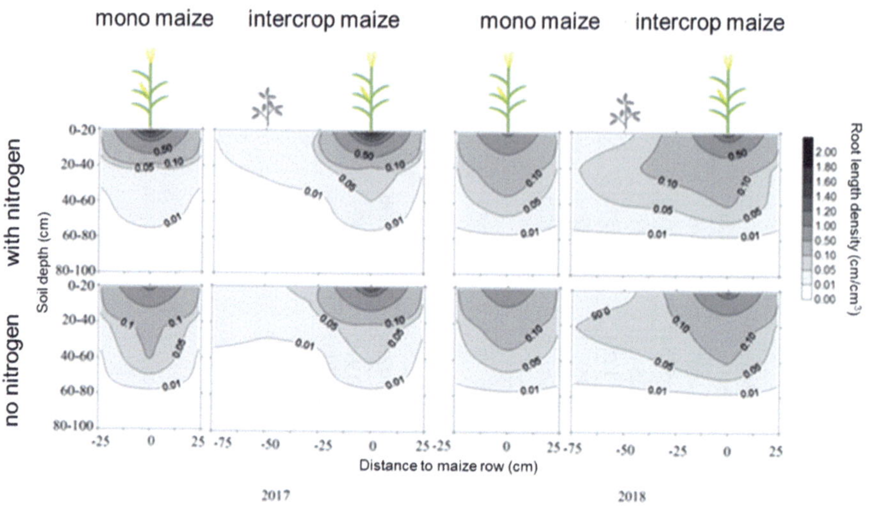

Fig. 4.7 Maize root density distribution in maize-soybean strip relay intercropping

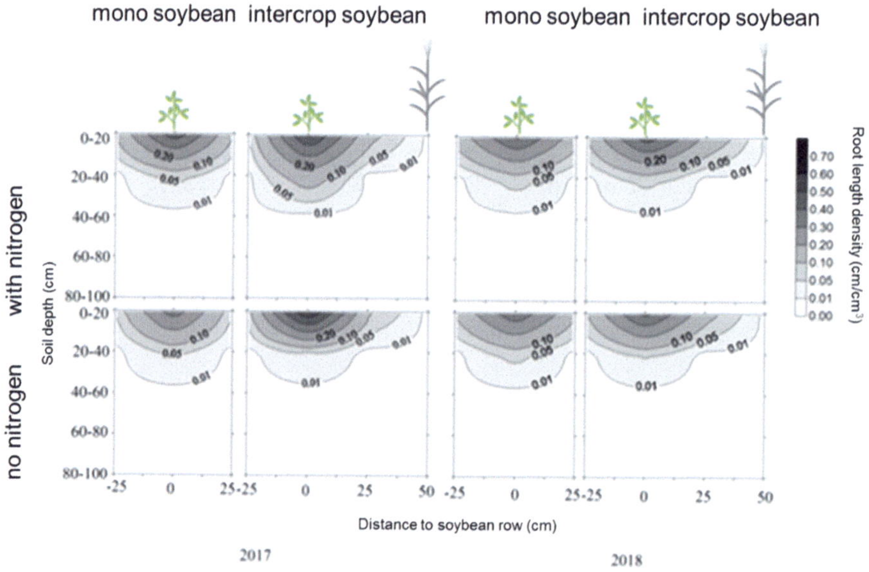

Fig. 4.8 Soybean root density distribution in maize-soybean strip relay intercropping

Therefore, maize-soybean strip relay intercropping increased root growth and the soil nutrient absorption range of the roots of the system crop members.

Compared to MM, strip relay intercropping generally increased root surface area density in maize plants across the soil profile (Fig. 4.9). Specifically, total root surface area density under IM increased significantly by 21.5% compared with

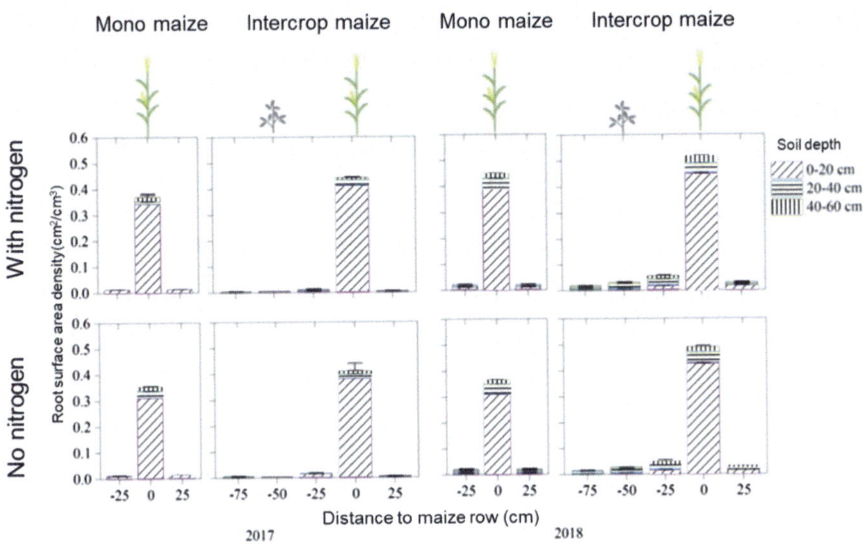

Fig. 4.9 Effects of nitrogen application rate and planting model on root surface area of maize plants

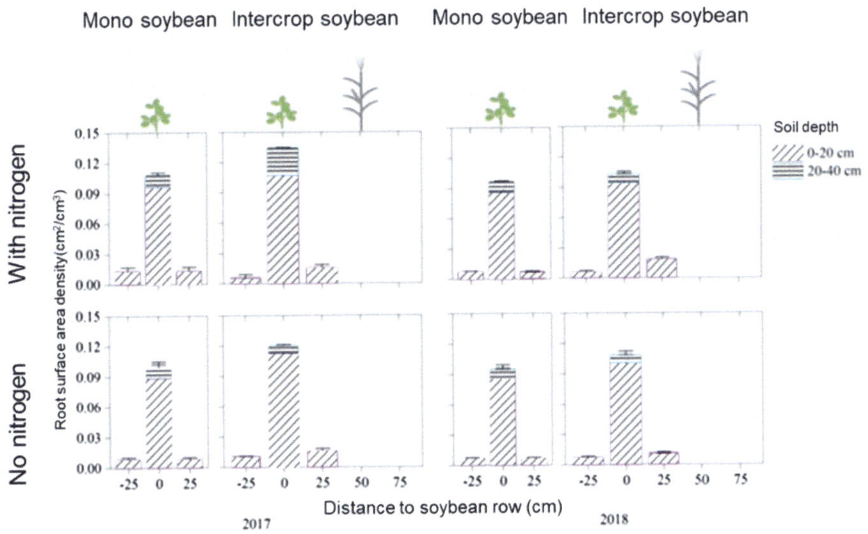

Fig. 4.10 Effects of nitrogen application rate and planting model on root surface area of soybean plants

MM (2-year average). Similarly, strip relay intercropping increased root surface area density in soybean plants across soil layers (Fig. 4.10). Further, compared with MS, total root surface area density in IS increased by 16.9% (2-year average). Therefore, compared with monocropping, maize and soybean had higher root surface area density in strip relay intercropping plots, thereby increasing root uptake area and promoting soil nutrient uptake the crop system members.

4.1.2.2 Root Activity

Nitrogen application significantly improved root activity of maize plants at the R4 growth stage (Fig. 4.11). Further, there was no significant difference in maize root activity among treatments in the absence of N application, but when N was applied, root activity in MS30 was significantly higher (32.94%) than that in MM100. Further, under the various spacing treatments in strip relay intercropping, root activity increased with increasing root interaction intensity. Thus, root activity was higher in MS30 than that in MS60 or MS75 treatments by 24.28% and 40.76%, respectively, while that in MS45 was significantly higher (27.53%) than that in the MS75 treatment.

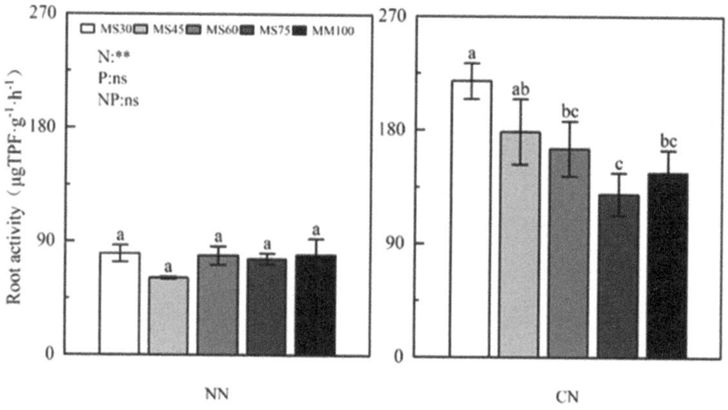

Fig. 4.11 Effects of planting system on maize root activity under different nitrogen application rates

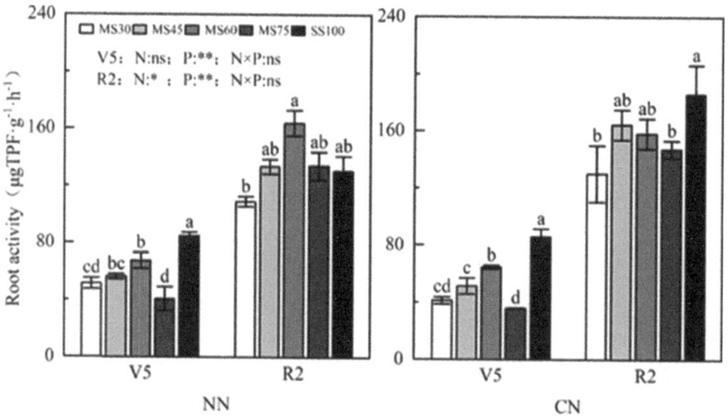

Fig. 4.12 Effects of planting system on soybean root activity at different nitrogen application rates

Similarly, N fertilizer improved soybean root activity, reaching a significant level during the R2 growth stage (Fig. 4.12). Further, planting system significantly affected soybean root activity at V5 and R2 stages, with SS100 and MS60 treatments promoting the highest soybean root activities, whereas MS75 and MS30 promoted the lowest levels recorded. Additionally, at the V5 growth stage, root activity in MS was higher than that in IS, while in MS it remained at the same level as that of IS at R2 stage, and even lower than that under IS, when N was not applied. Further, spacing effects on soybean root activity differed among growth stages. Thus, at V5 stage, root activity under MS60 was significantly higher under NN and CN treatments than those under MS30 or MS75 by 23.97% and 39.65% and by 36.13% and 44.62%, respectively. Concomitantly, under N application, root activity in MS60 was significantly higher, by 20.57%, than that in the MS45 treatment. Similarly, at R2 stage, root activity in MS60 was significantly higher than that in MS30 plots under NN treatment.

4.1.3 Effects of Strip Compound Planting on Soil Nitrogen and Phosphorus Conversion

4.1.3.1 Rhizosphere Soil Flora and Community Abundance

4.1.3.1.1 Rhizosphere Phosphorus Bacteria

The relative abundance of *Proteobacteria* and *Acidobacteria*, a bacterial phylum that grew rapidly in rhizosphere-dependent root exudates in IM, was significantly higher than that in monoculture under P application (Fig. 4.13). *Enhydrobacter*, *Dechromonas*, *Propionibacterium*, and *Massilia*, *Burkholderia*, among other genera within Proteobacteria and Acidobacteria phyla, can rapidly metabolize monosaccharides, ribose, glucose, sucrose, amino acids, and fatty acids present in root exudates. In addition, the correlation between root carbon secretion and bacterial community abundance was confirmed by redundancy analysis (Fig. 4.14). Furthermore, intercropping maize experienced better light environment to enhance root carbon secretion and alter the structure of rhizosphere bacterial communities under an appropriate P application rate.

4.1.3.1.2 Number of Soil Bacteria

The number of bacteria in the maize rhizosphere soil decreased over the period of plant growth, peaking during the V12 stage. Additionally, planting model significantly affected the number of bacteria in the maize rhizosphere soil, with MM > IMS during the pre-symbiotic stage and IMS > MM after symbiosis (Table 4.2). Compared with MM, IMS increased the number of bacteria by an average of 2.6% over the 2 years of study. Additionally, N application increased the number of bacteria in the maize rhizosphere soil. Furthermore, among N treatments, RN resulted in the highest number of bacteria, with an average increase of 9.6% and 9. 9% over the 2 years of study, compared to NN and CN treatments under the MM and IMS planting models. After symbiosis, the overall performance of soybean ranked higher

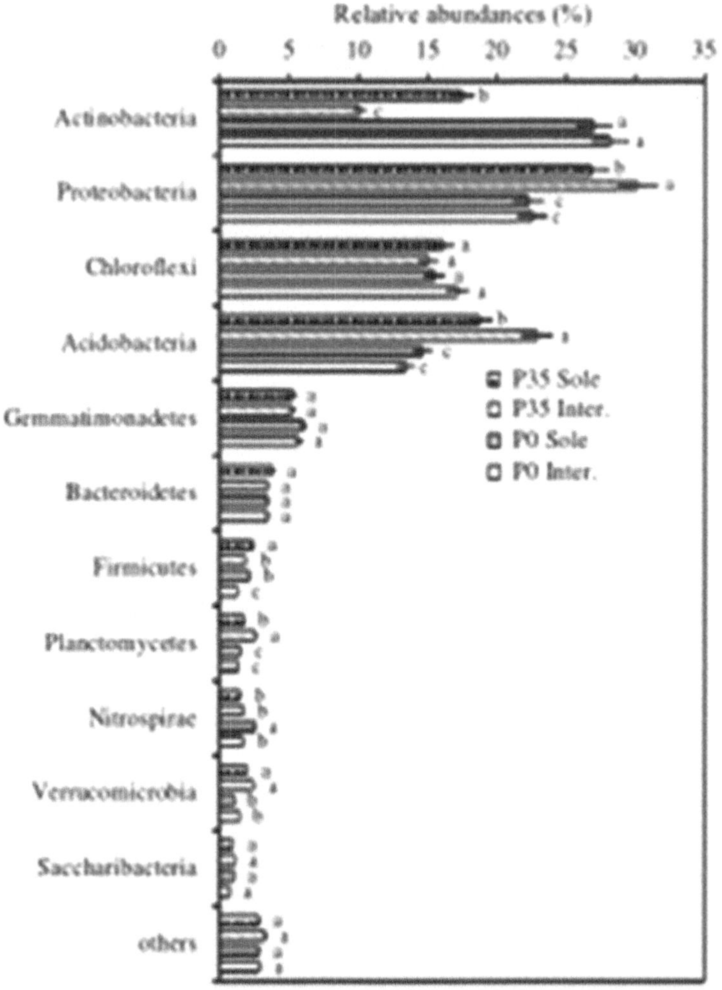

Fig. 4.13 Relative abundance of bacterial phyla in the rhizosphere soil of MM and IM cropping models under different phosphorus treatments

in the IMS than in the SS cropping scheme, with the average peaking at V5 stage. Compared to SS, IMS increased the level of soybean performance by an average of 12.9% over the 2 years of study. Lastly, the average number of soybean bacteria among N treatments in the 2 years of study reached a maximum value in the RN treatment, in which case, it was 11.7% and 11.0% higher than the number in NN or CN under SS and MS, respectively (Fu et al. 2017).

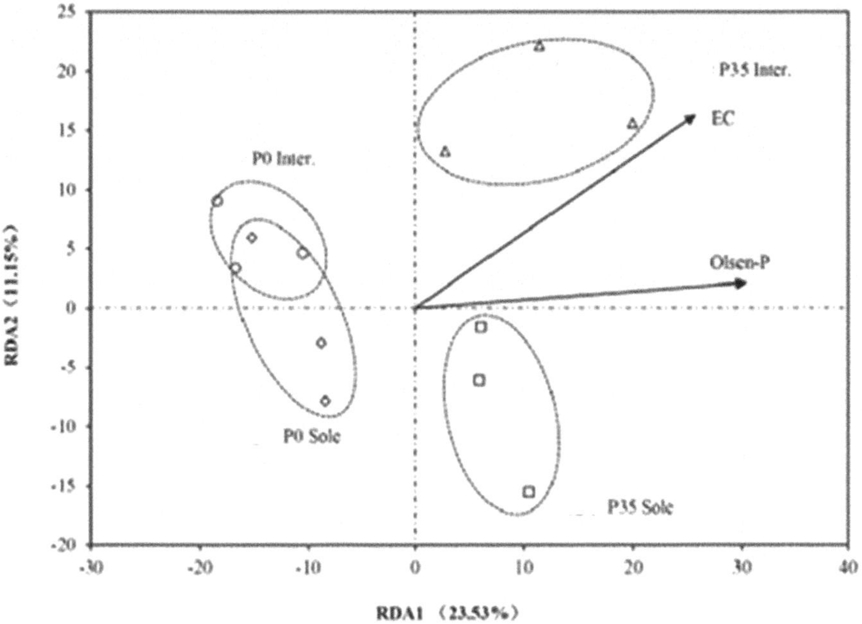

Fig. 4.14 Redundancy analysis of environmental factors and bacterial community diversity

4.1.3.1.3 Increase in Soil Microbial Richness

In maize, more than 80% of the bacterial sequences found at the grain filling stage belong to *Proteobacteria*, *Acidobacteria*, *Actinobacteria*, *Chloroflexi*, and *Gemmatimonadetes*. However, the relative abundance of *Verrucomimicrobia*, *Latescibacteria*, and other bacterial phyla with unknown or unidentifiable sequences was relatively low, accounting for less than 0.1% of the bacterial phyla in the collected soil samples.

Compared with the open field fallow treatment (CK), the percentage of soil *Proteobacteria* increased and that of *Acidobacteria* significantly decreased in maize monoculture and maize-soybean strip relay intercropping, while the trend of change in dominant bacterial phyla was inconsistent between MM and IMS under different N treatments (Fig. 4.15). Further, at soybean maturity stage, more than 75% of the bacterial sequences belonged to *Proteobacteria*, *Acidobacteria*, *Actinobacteria*, *Chloroflexi*, and *Bacteroidetes*. However, the relative abundance of *Verrucomimicrobia*, *Firmicutes*, and other bacterial phyla with unknown or unidentifiable sequences was relatively low, accounting for less than 0.1% of the bacterial phyla in the collected soil samples. Compared with the CK, the relative abundances of *Proteobacteria* and *Bacteroidetes* in MS and IMS ranked as follows: CK < MS < IMS, while the percentage of soil *Acidobacteria* and *Chloroflexi* ranked as follows: IMS < MS < CK. Therefore, the effects of N treatment on the structure of crop rhizosphere microbial communities clearly varied.

Table 4.2 Effects of planting model and nitrogen application rate on the quantity of bacteria in corn and soybean rhizosphere soils (106 cfu/g)

		Maize			Soybean			
Plant patterns	N level	V12	VT	R6	V5	R2	R4	R8
2013								
	NN	22.74 ± 2.66	14.46 ± 1.57	10.19 ± 1.34	13.39 ± 3.75	10.43 ± 1.47	7.70 ± 0.120	11.73 ± 1.63
MM/SS	RN	23.85 ± 1.46	14.99 ± 2.57	10.27 ± 1.33	16.53 ± 1.90	14.29 ± 2.99	10.27 ± 0.85	13.46 ± 6.32
	CN	19.10 ± 3.99	14.58 ± 1.02	10.19 ± 1.31	14.58 ± 2.97	11.92 ± 3.80	8.57 ± 2.000	10.42 ± 2.28
	Mean	21.89 ± 0.31	14.68 ± 2.49	10.22 ± 0.28	14.83 ± 1.58	12.21 ± 1.95	8.85 ± 1.31	11.87 ± 1.53
	NN	19.05 ± 1.21	15.83 ± 1.47	10.60 ± 1.35	13.57 ± 1.13	11.79 ± 1.85	9.64 ± 2.347	13.93 ± 2.31
IMS	RN	23.20 ± 1.79	16.49 ± 0.53	10.60 ± 1.33	17.08 ± 3.42	14.72 ± 0.12	10.48 ± 0.95	15.08 ± 3.44
	CN	20.39 ± 2.98	15.90 ± 0.96	10.48 ± 1.30	16.01 ± 8.08	12.33 ± 1.96	9.91 ± 2.13	14.06 ± 5.60
	Mean	20.88 ± 1.58	16.07 ± 2.12	10.56 ± 0.36	15.55 ± 1.80	12.95 ± 1.56	10.01 ± 0.43	14.35 ± 0.63
2014								
	NN	12.63 ± 1.21	12.20 ± 1.70	11.06 ± 1.06	13.09 ± 1.96	11.99 ± 1.34	13.67 ± 0.93	11.79 ± 1.17
MM/SS	RN	14.88 ± 1.19	13.51 ± 1.42	11.64 ± 0.86	11.92 ± 1.71	11.27 ± 1.14	12.56 ± 0.79	13.50 ± 1.23
	CN	12.77 ± 1.02	11.83 ± 1.15	11.22 ± 0.78	11.79 ± 1.15	11.64 ± 1.43	10.97 ± 1.25	11.31 ± 1.89
	Mean	13.43 ± 1.26	12.30 ± 1.05	11.31 ± 0.30	11.70 ± 0.27	11.47 ± 0.44	11.44 ± 0.97	12.20 ± 1.16
	NN	12.77 ± 1.22	12.24 ± 0.66	10.20 ± 0.61	13.82 ± 2.13	12.26 ± 0.91	11.96 ± 1.30	13.34 ± 2.07
IMS	RN	12.79 ± 0.99	14.20 ± 1.02	12.43 ± 0.90	14.87 ± 1.78	14.11 ± 1.22	11.61 ± 1.34	15.08 ± 2.02
	CN	12.28 ± 1.54	12.52 ± 1.08	11.76 ± 1.35	15.93 ± 1.57	11.97 ± 1.16	10.67 ± 1.70	13.38 ± 1.24
	Mean	12.12 ± 0.75	12.99 ± 1.06	11.46 ± 1.15	14.40 ± 0.53	12.97 ± 0.99	11.65 ± 0.42	13.93 ± 0.99

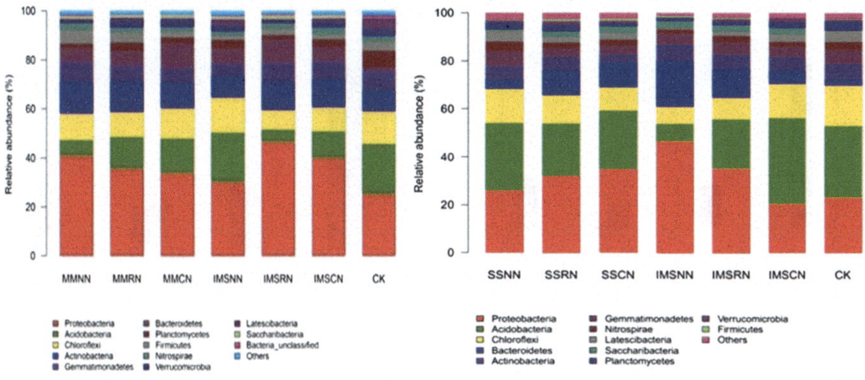

Fig. 4.15 Relative abundance of bacterial phyla under different planting systems and nitrogen application rates

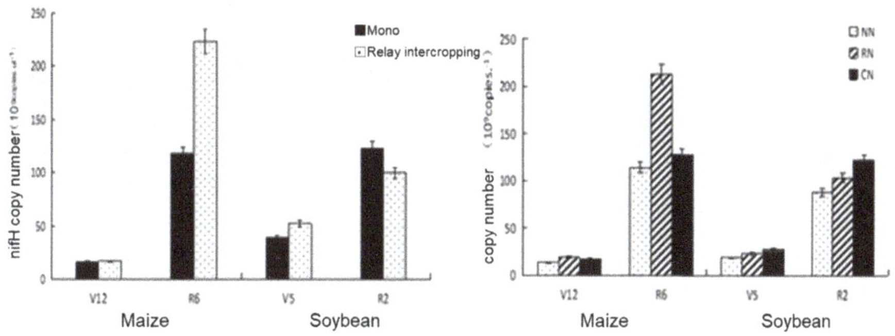

Fig. 4.16 Abundance of nifH-type nitrogen-fixing bacteria under different planting systems and nitrogen application rates

4.1.3.2 Abundance of Functional Genes from Soil Nitrogen Cycling-Related Microorganisms

The *nifH* genes of N-fixing bacteria (Fig. 4.16) in the maize rhizosphere soil were detected at higher levels in IMS than in MM in both growth stages analyzed, and there was no significant difference for *nifH* gene levels between MM and IMS at the V12 stage before symbiosis. Conversely, at the R6 stage after symbiosis, *nifH* gene abundance was significantly affected by planting scheme; specifically, compared with MM, an 88.9% increase in abundance was observed in IMS. Meanwhile, with respect to N application, *nifH* gene abundance increased with decreasing N application rate (RN) in maize rhizosphere soil, being the highest in the RN treatment before and after co-growth in both MM and IMS. As for soybean, the abundance of *nifH* genes in the rhizosphere soil increased with growth after intercropping, and with respect to N application effects, the abundance of *nifH* genes in soybean rhizosphere soil increased with increasing N application rate and was higher than without N application.

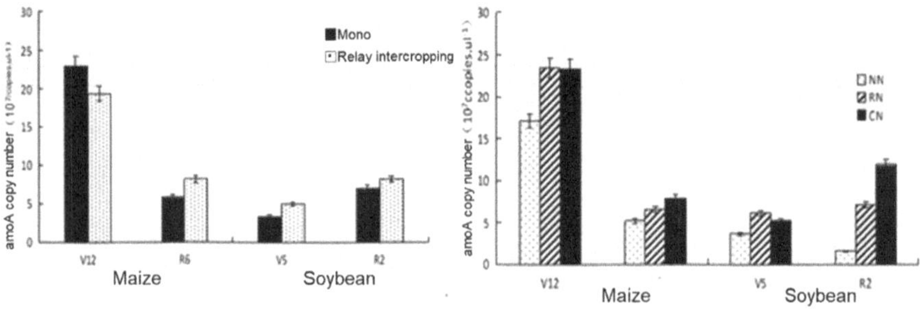

Fig. 4.17 Abundance of amoA-type nitrogen-fixing bacteria under different planting systems and nitrogen application rate

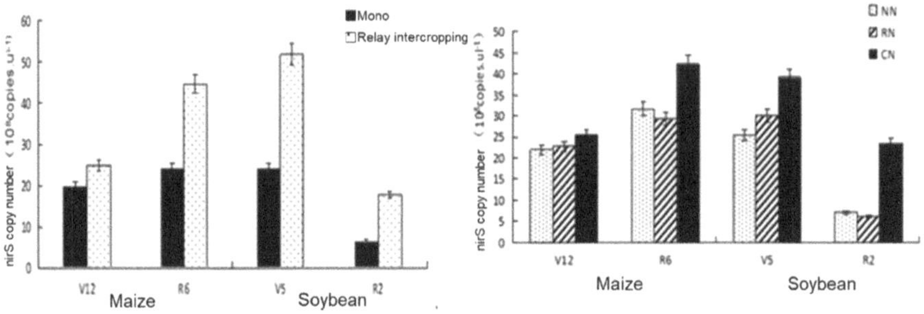

Fig. 4.18 NirS-type denitrifying bacterial community abundance under different planting systems and nitrogen application rates

Before the establishment of the maize-soybean symbiosis under the intercropping planting scheme, the abundance of the maize *amoA* gene was greater in monoculture (23.15×10^7) than that in strip intercropping (19.47×10^7); however, after symbiosis ended, the pattern was reversed and the abundance was greater (18.37×10^7) in strip intercropping than that in monoculture (5.92×10^7) (Fig. 4.17). As for N application effects, N significantly increased the abundance of the *amoA* gene in maize soil rhizosphere. Specifically, it was higher at V12, particularly in the RN treatment, in which case it was higher than in CN, while at R6, *amoA* abundance increased with increasing N application rate, being highest in the CN treatment. Similarly, *amoA* gene abundance in soybean soil rhizosphere increased with growth, and strip intercropping effectively promoted the increase of *amoA* gene abundance.

With respect to N application rate, soybean *amoA* gene reached the highest abundance in the RN treatment at V5 stage, and at R2 stage, soybean *amoA* gene abundance was significantly higher, with (108.9%), than without N application and peaked in the CN treatment.

The abundance of soil denitrification gene (*nirS*) in maize rhizosphere soil (Fig. 4.18) was higher in IMS than in MM before and after establishment of the maize-soybean symbiosis, but the increase rate in strip intercropping after symbiosis establishment was higher than that in monoculture. As for N application, the abundance of *nirS* genes among different N application treatments before maize-soybean co-growth (V12) increased with increasing N application rate, particularly in the CN treatment, where it was highest; lastly, after the maize-soybean symbiosis ended (R6), *nirS* genes abundance was lowest in the RN treatment.

With regard to soybean, *nirS* gene abundance in the rhizosphere soil was higher in IMS than in SS after maize-soybean establishment symbiosis. Further, *nirS* gene abundance showed a decreasing trend with growth both under monoculture and strip intercropping. Particularly, at V5 stage, *nirsS* gene abundance in soybean rhizosphere soil increased with increasing N application, being maximum in the CN treatment. Conversely, RN inhibited denitrification at R2 stage, whereby *nirS* gene abundance in the soil was lowest in this treatment.

4.1.4 Effect of Maize-Soybean Strip Intercropping on Nitrogen Fixation Capacity of Soybean Nodules

4.1.4.1 Photosynthetic ^{13}C Distribution in Soybean Organs

The distribution of ^{13}C in photosynthesis differs significantly between the vegetative and reproductive growth stages in soybean. Specifically, the $\delta^{13}C$ values of leaves, stems, and root nodules were higher at the V5 stage, whereas the $\delta^{13}C$ value in pods was highest at the R5 stage. Further, compared with SS, MS increased the $\delta^{13}C$ value in roots and nodules at R5 stage (Fig. 4.19), while the proportion of $\delta^{13}C$ in the nodules in MS with no nitrogen (NN) and constant nitrogen (CN) treatment was 2.2 and 1.1 times higher than that in SS. Particularly, compared to the CN treatment, the NN treatment significantly increased the $\delta^{13}C$ value of the pods at R5 stage, and the average $\delta^{13}C$ value of pods in IS under the NN treatment increased up to 49.35%

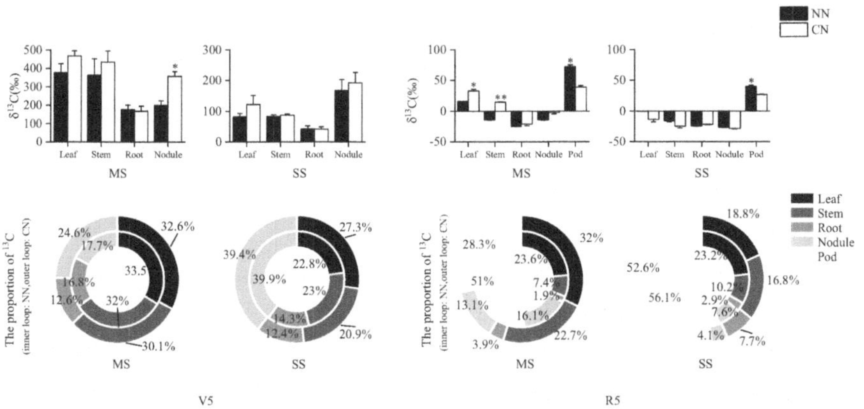

Fig. 4.19 δ13C value and 13C distribution ratio in each organ at the five-leaf stage (V5) and the first grain stage (R5) in soybean. *MS* intercropping soybean, *SS* monocropping soybean, *NN* no nitrogen application, *CN* constant nitrogen application

over that of the pods growing in the CN treatment, while the average $\delta^{13}C$ value of the pods growing in MS under NN was significantly higher (by 20.59%) than that of the pods growing under the CN treatment. Meanwhile, under the NN treatment, photosynthetic ^{13}C of pods in IS was significantly higher (by 111.84%) than that of pods in SS, and under the CN treatment, photosynthetic ^{13}C in MS was significantly higher (by 219.52%) than that observed in pods growing in SS treatment plots.

4.1.4.2 Number of Soybean Nodules and Nitrogen Fixation Capacity

Except for SS in 2019, the number of nodules per plant showed a continuous increasing trend with soybean growth and was highest at R5 stage, with an average of 233 nodules/plant (Fig. 4.20). Particularly, the number of nodules in MS treatment plots was significantly lower than that in SS at R2 stage and before, except for V5 stage in 2020. Conversely, at R5 stage, the number of nodules in MS plots was significantly higher (39.9%, on average) than that in SS plots. Further, N application rate had a significant effect on the number of nodules, which showed a decreasing trend with increasing N application rate in SS, while at R2 and later stages, the number of nodules in the MS planting scheme increased first and then decreased despite further increase in N application rate. The difference at R2 was not significant, but it was at R5 stage, in which case, the number of nodules was significantly higher in RN than those in NN or CN. Furthermore, the number of nodules was 19.7% and 94.0% higher in RN than those in NN and CN, respectively.

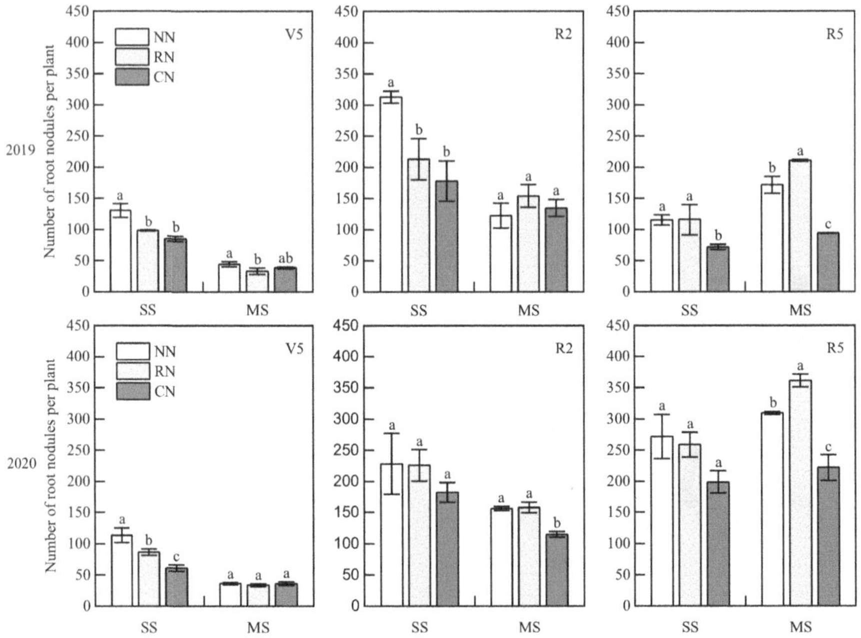

Fig. 4.20 Effects of cropping system and nitrogen application rate on the number of soybean nodules per plant

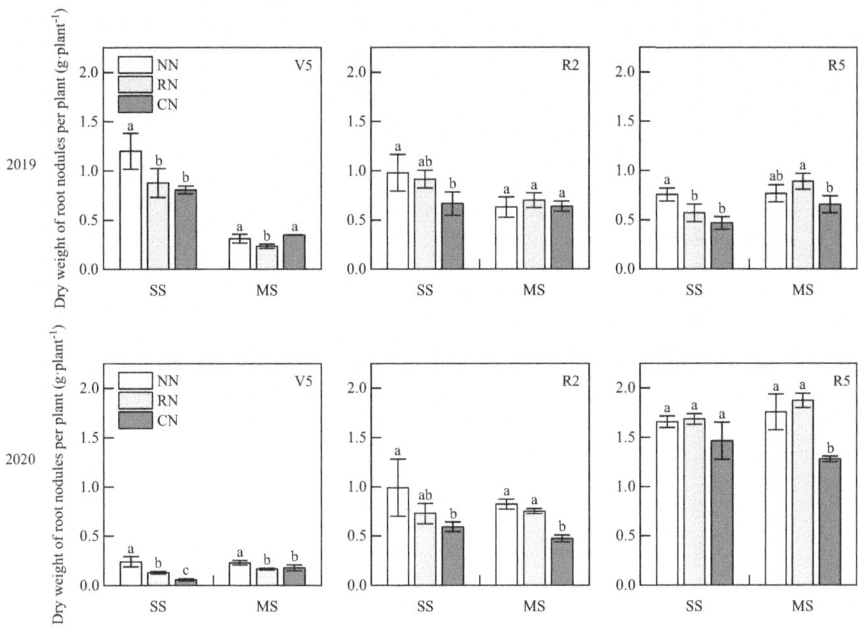

Fig. 4.21 Effects of cropping system and nitrogen application rate on soybean root nodule dry weight

As expected, the response of soybean root nodule dry weight per plant to planting system and N application rate was basically consistent with that described for the number of root nodules per plant (Fig. 4.21). Additionally, average root-nodule dry weight per plant was lower from V5 to R2 in 2020 than in 2019, and it was larger in SS that in MS plots. Meanwhile, at R5 growth stage, root nodule dry weight was larger in MS than that in SS, and in 2019, it was significantly (28.3%) higher in MS than that in SS.

Nodule dry weight showed a decreasing trend with increasing N application rate in the SS planting mode, in which case, nodule dry weight was lowest under CN treatment, with the difference relative to other treatments being significant. The effect of N application rate on nodule dry weight in the MS planting system initially decreased and then increased at V5 stage, but in the later soybean growth stages, it first increased and then decreased. This pattern was clearer at R5 stage. Nodule dry weight was highest in the RN treatment and, although there was no significant difference between RN and NN, their nodule dry weight was significantly higher (by 35.8–46.3%) than those of CN across the 2 years, respectively.

The activity of nitrogenase per unit mass of soybean increased first and then decreased with growth, peaking at R2 stage (Table 4.3). Planting model had a significant effect on nitrogenase activity of soybean nodules (except at V5 stage), and in MS planting, it was 142.1% higher than that in SS at R5 stage. The N application rate had a significant effect on the nitrogenase activity of nodules but

Table 4.3 Effect of planting system and nitrogen application rate on nitrogenase activity per unit mass of soybean and nitrogen fixation potential of single nodules, in 2020

		V5		R2	R5		
Cropping system	Application	Nitrogenase activity (ul·g^{-1}·h^{-1})	Nitrogen potential (μl·h^{-1}·plant^{-1})	Nitrogenase activity (ul·g^{-1}·h^{-1})	Nitrogen potential (μl·h^{-1}·plant^{-1})	Nitrogenase activity (ul·g^{-1}·h^{-1})	Nitrogen potential (μl·h^{-1}·plant^{-1})
SS	NN	295.21a	72.03a	631.17a	615.63a	23.80b	39.38b
	RN	118.41b	15.67b	463.25b	336.76b	66.10a	111.28a
	CN	42.20c	2.42b	63.54c	37.18c	32.40b	46.31b
	Mean	151.94	30.04	385.99	329.86	40.77	65.66
MS	NN	153.46c	35.43a	232.11ab	188.46a	75.85b	133.35b
	RN	240.71a	39.91a	312.91a	235.48a	136.35a	255.04a
	CN	106.46b	19.16b	137.35b	64.28b	83.88b	107.09b
	Mean	166.88	31.5	227.46	162.74	98.69	165.16
				ANOVA (F-value)			
Cropping pattern (*A*)		2.63^{ns}	0.10^{ns}	26.58^{**}	19.32^{**}	116.98^{**}	127.39^{**}
N application (*B*)		93.36^{**}	29.80^{**}	45.67^{**}	29.55^{**}	35.39^{**}	59.46^{**}
A × *B*		75.78^{**}	17.53^{**}	19.73^{**}	12.64^{**}	1.32^{ns}	7.48^{**}

the specific trend was affected by planting system. Furthermore, V5 and R2 stages showed a decreasing trend in nitrogenase activity with increasing N application rate in SS plots, while in MS plots, it first increased and then decreased, while RN treatment promoted the highest nitrogenase activity at R5 stage, which was significantly higher, by 79.8% and 62.6%, respectively, than those in NN and CN plots.

As growth proceeded, the nitrogen fixation potential of individual root nodules showed a continuous increasing trend in the MS planting system and reached a maximum value at the R5 stage. Compared with the SS mode, the nitrogen fixation potential of individual root nodules in the MS mode increased by 151.6% at R5 stage. The effect of N application rate on the nitrogen fixation potential of individual root nodules was consistent with nitrogenase activity, and treatment differences reached a highly significant level. Specifically, compared with CN, nitrogen fixation potential in the RN treatment in the MS planting scheme significantly increased by 108.3%, 266.4%, and 138.1% at V5, R2, and R5 stages, respectively.

4.1.5 Effects of Maize-Soybean Strip Intercropping on Nutrient Uptake

4.1.5.1 Nitrogen Uptake in Different Organs

Pool planting $^{15}NH_4NO_3$ and $NH_4{}^{15}NO_3$ labeling experiments (Table 4.4), revealed that, compared with MM, nitrogen uptake and ^{15}N uptake in the IM planting system increased significantly, with stems and leaves increasing by 13.06% and 21.60%, grains increasing by 5.15% and 25.07%, and total N uptake and total ^{15}N uptake increasing by 9.48% and 23.18%, respectively (Chen et al. 2017).

Compared with RN-$NH_4{}^{15}NO_3$, N uptake by maize roots under RN-$^{15}NH_4NO_3$ treatment showed no significant change, but ^{15}N uptake increased significantly. Meanwhile, in the MM planting system, the total ^{15}N increased by 12.81%, 12.03%, and 12.45%, respectively, while in IM plots, they increased by 16.01%, 15.56%, and 15.79%, respectively.

Further analysis of soybean N uptake under different isotope-labeling treatments (Table 4.5) showed that, compared with the SS system, N uptake and ^{15}N uptake by soybean stems and leaves under the IS system were significantly reduced but the increase in pods and grains resulted in a significant increase of 10.42%, 21.11%, and 41.83% in grain N uptake, pod ^{15}N uptake, and grain ^{15}N uptake; furthermore, total N uptake showed no significant change, while total ^{15}N uptake increased significantly by 21.26%. Meanwhile, compared with RN-$NH_4{}^{15}NO_3$, N uptake of soybean under RN-$^{15}NH_4NO_3$ treatment showed no significant change, while ^{15}N uptake increased significantly. Additionally, stem, leaf, pod, seed, and total soybean ^{15}N uptake in SS plots increased by 37.63%, 57.09%, 18.91%, 45.97%, and 44.02%, respectively. Meanwhile, in the IS planting system, soybean ^{15}N uptake increased by 45.54%, 20.98%, 17.97%, 21.63%, and 23.40%, respectively. In addition, although soybean N uptake in IS plots was 11.61% lower than that in SS plots, the variation

Table 4.4 N uptake and 15 N uptake by maize plants under different planting systems

	N uptake (g·plant^{-1})				^{15}N uptake (mg·plant^{-1})			
	Stem		Grain		Stem		Grain	
N treatments	MM	IM	MM	IM	MM	IM	MM	IM
RN-$NH_4{}^{15}NO_3$	1.449a	1.714a	1.253a	1.246a	245.20b	293.76b	204.47b	251.53b
RN-$^{15}NH_4NO_3$	1.568a	1.696a	1.194a	1.326a	276.60a	340.78a	229.06a	290.67a
Mean	1.508	1.705*	1.223	1.286	260.9	317.27*	216.76	271.1*

MM monoculture maize, *IM* maize-soybean strip relay intercropping

Table 4.5 N and 15 N uptake by soybean plants under different planting systems

N treatments		Stem		Leaf		pod		Grain	
		SS	IS	SS	IS	SS	IS	SS	IS
N uptake (g·plant^{-1})	RN-$NH_4{}^{15}NO_3$	0.162a	0.135a	0.363a	0.240a	0.151a	0.165a	0.975a	1.011a
	RN-$^{15}NH_4NO_3$	0.146a	0.139a	0.357a	0.259a	0.161a	0.168a	0.925a	1.088a
	Mean	0.154	0.137*	0.360	0.249*	0.156	0.166	0.950	1.049*
^{15}N uptake (mg·plant^{-1})	RN-$NH_4{}^{15}NO_3$	8.53a	7.29a	11.70b	12.15a	6.40a	7.79a	32.11b	50.58b
	RN-$^{15}NH_4NO_3$	11.74a	10.61a	18.38a	14.70a	7.61a	9.19a	46.87a	61.52a
	Mean	10.13	8.95*	15.04	13.42*	7.01	8.49*	39.49	56.01*

SS monocropping soybean, *IS* intercropping soybean

Table 4.6 Biological nitrogen fixation in soybean plants under different planting systems

Plant patterns	N uptake (g/plant)	%Ndfa	Ndfa (g/plant)
SS	1.404	51.39	0.722
IS	1.241*	63.64*	0.790*

pattern of N uptake in SS and IS, without N application, was consistent (Table 4.6); however, compared with SS plots, the N fixation ratio and N fixation amount in IS increased significantly by 23.83% and 9.42%, respectively.

4.1.5.2 Nitrogen and Phosphorus Uptake in Maize-Soybean Strip Intercropping

Compared with monoculture, there was minor change in N or P uptake by maize plants under strip continuous cropping or strip rotation (Table 4.7) but the latter significantly reduced N and P uptake by soybean stems and leaves, as well as total P and K uptake by soybean plants. Soybean seeds increased N and P uptake and plant total N uptake. Among them, N uptake by soybean seeds, total N uptake by plants, and P uptake by seeds under maize-soybean intercropping rotation (A2) were 24.4%, 9.8%, and 9.6% higher than those of soybean monoculture (A5) treatment, respectively. Further, compared with monoculture, in maize-soybean intercropping at equal row spacing (A3), N and P uptake by maize plants showed an increasing trend, while those of soybean plants significantly decreased. Nitrogen uptake, total N uptake, total P uptake, and total P uptake by soybean seeds in A3 were 35.7%, 43.1%, 38.2%, and 48.8% lower than the average values recorded in A5, respectively. Under the maize-soybean relay strip intercropping system, the strip intercropping continuous system and the strip intercropping rotation system did not significantly affect N or P uptake by soybean plants. However, after strip rotation, N and P uptake by maize plants increased significantly. In 2013, N uptake, total N uptake, P uptake, and total P uptake in A2-treated corn were 20.4%, 18.5%, 8.5%, and 9.1% higher than those in A1 (strip intercropping continuous), respectively. Further, strip intercropping rotation increased plant N and P uptake in the maize-soybean intercropping system. Specifically, in 2013, grain N uptake, plant total N uptake, grain P uptake, and plant total P uptake in treatment A2 were 11.2%, 11.6%, 4.5%, and 5% higher than those in the A1 treatment, respectively, and 46.9%, 39.3%, 14.5%, and 14.6% higher than those in treatment A3.

4.1.5.3 Nutrient Harvest Index in a Maize-Soybean Strip Intercropping System

Compared with monoculture, there was no significant difference in the N and P harvest index values in the IM planting system, but those in the IS system did show a significant increase. Particularly, the N harvest index values of soybean plants in A1 and A2 treatments were 13.6% and 11.9% higher than the average values observed

Table 4.7 Effects of planting system on N and P uptake by maize and soybean plants

Nutrient	Year	Plant patterns	Maize			Soybean			System		
			Stem and leaf	Grain	Total uptake	Stem and leaf	Grain	Total uptake	Stem and leaf	Grain	Total uptake
N	2012	A1	52.6 ± 4.3b	71.8 ± 2.0a	124.5 ± 2.6a	18.5 ± 2.4b	127.3 ± 6.2a	145.8 ± 7.4a	71.1 ± 5.4a	199.1 ± 8.1a	270.3 ± 7.1a
		A2	55.3 ± 1.3ab	72.9 ± 5.5a	128.1 ± 6.1a	20.0 ± 2.3ab	120.2 ± 6.8a	140.3 ± 8.8a	75.3 ± 3.6a	193.1 ± 5.4a	268.4 ± 8.9a
		A3	61.7 ± 1.4a	76.1 ± 9.1a	137.7 ± 7.7a	9.6 ± 0.8c	66.4 ± 6.0c	76.1 ± 6.6c	71.3 ± 0.8a	142.5 ± 15.1b	213.8 ± 14.3b
		A4	48.0 ± 5.4b	78.2 ± 3.4a	126.3 ± 8.6a	–	–	–	–	–	–
		A5	–	–	–	24.3 ± 2.2a	98.1 ± 11.4b	122.4 ± 9.2b	–	–	–
	2013	A1	49.0 ± 13.8a	104.7 ± 6.7b	153.7 ± 17.4b	37.2 ± 3.9b	158.0 ± 11.2a	195.2 ± 11.6a	86.3 ± 17.5a	262.7 ± 4.7a	348.9 ± 15.2a
		A2	56.0 ± 2.3a	126.1 ± 7.7a	182.1 ± 7.4a	41.3 ± 3.8b	166.2 ± 18.2a	207.4 ± 17.4a	97.3 ± 5.7a	292.2 ± 24.3a	389.5 ± 22.7a
		A3	58.3 ± 1.7a	117.3 ± 7.8ab	175.6 ± 8.8ab	22.4 ± 1.3c	81.6 ± 9.4c	104.1 ± 10.7b	80.8 ± 2.6a	198.9 ± 17.2b	279.7 ± 19.5b
		A4	52.1 ± 3.4a	116.2 ± 2.5ab	168.3 ± 5.6ab	–	–	–	–	–	–
		A5	–	–	–	61.9 ± 5.8a	132.2 ± 10.1b	194.2 ± 6.2a	—	–	—
P	2012	A1	3.6 ± 1.2a	13.4 ± 0.2a	17.0 ± 1.2a	4.7 ± 0.5b	13.3 ± 0.6a	18.1 ± 1.0a	8.3 ± 0.9a	26.7 ± 0.7a	35.0 ± 0.3a
		A2	3.3 ± 0.5a	12.3 ± 0.6a	15.6 ± 1.1a	5.2 ± 0.3ab	12.3 ± 1.1a	17.5 ± 1.4a	8.6 ± 0.8a	24.6 ± 1.3a	33.2 ± 2.1a
		A3	4.3 ± 0.9a	12.6 ± 0.9a	16.8 ± 0.7a	2.7 ± 0.2c	7.5 ± 0.4c	10.2 ± 0.5b	7.0 ± 0.7a	20.1 ± 0.9b	27.1 ± 0.5b

(continued)

Table 4.7 (continued)

Nutrient	Year	Plant patterns	Maize			Soybean			System		
			Stem and leaf	Grain	Total uptake	Stem and leaf	Grain	Total uptake	Stem and leaf	Grain	Total uptake
		A4	3.7 ± 0.7a	12.6 ± 0.7a	16.3 ± 1.3a	–	–	–	–	–	–
		A5	–	–	–	6.5 ± 1.4a	10.2 ± 0.9b	16.7 ± 0.45a	–	–	–
	2013	A1	7.2 ± 0.8a	24.6 ± 2.0a	31.8 ± 1.5b	5.3 ± 0.7b	15.5 ± 0.4a	20.7 ± 0.9b	12.5 ± 1.4a	40.1 ± 1.9a	52.6 ± 0.8a
		A2	7.9 ± 0.3a	26.7 ± 3.9a	34.7 ± 3.8ab	5.9 ± 0.6b	15.2 ± 1.8a	21.1 ± 2.4b	13.8 ± 0.4a	41.9 ± 5.6a	55.7 ± 6.0a
		A3	8.2 ± 0.2a	28.7 ± 1.8a	36.9 ± 2.0a	3.7 ± 0.4c	8.0 ± 0.7b	11.7 ± 1.2c	12.0 ± 0.6a	36.6 ± 2.4a	48.6 ± 2.9a
		A4	8.5 ± 2.3a	27.9 ± 0.6a	36.4 ± 2.1a	–	–	–	–	–	–
		A5	–	–	–	11.2 ± 1.1a	14.9 ± 0.4a	26.1 ± 0.9a	–	–	–

A1 maize-soybean strip intercropping continuous cropping, *A2* maize-soybean strip intercropping rotation, *A3* maize-soybean single-row intercropping, *A4* maize monocropping, *A5* soybean monocropping

Table 4.8 Effects of planting system on nutrient harvest index of maize and soybean

		2012			2013		
Nutrient	Plant patterns	Maize	Soybean	System	Maize	Soybean	System
N	A1	57.8ab	87.3a	73.7a	68.4a	80.9a	75.4a
	A2	56.8ab	85.7a	71.9a	69.2a	80.0a	75.0a
	A3	55.1b	87.3a	66.5b	66.8a	78.4a	71.1a
	A4	62.0a	–	–	69.1a	–	–
	A5	–	80.0b	–	–	68.1b	–
P	A1	79.0a	73.8a	76.2a	77.2a	74.6a	76.2a
	A2	78.6a	70.3a	74.3a	76.9a	72.1a	75.1a
	A3	74.7a	73.8a	74.3a	77.7a	68.1b	75.4a
	A4	77.4a	–	–	76.8a	–	–
	A5	–	61.1b	–	–	57.0c	–

A1 maize-soybean strip intercropping continuous cropping, *A2* maize-soybean strip intercropping rotation, *A3* maize-soybean single-row intercropping, *A4* maize monocropping, *A5* soybean monocropping

in plants growing in A5 treatment plots, respectively, while the P harvest index values recorded in A1 and A2 plots were 25.7% and 20.6% higher than those in A5 plots (Table 4.8). Conversely, there was no significant difference in the N and P harvest indices between A1, A2, and A3 treatments for maize and soybean plants grown in the maize-soybean relay intercropping system. Further, N and P harvest index values for plants grown at equal spacing in the strip intercropping continuous cropping and strip intercropping rotation systems showed an increasing trend. Specifically, N harvest index values of plants in A1 and A2 treatment plots were 8.4% and 6.8% higher than those of plants growing in A3 plots.

4.2 Interspecific Collaborative Regulation Mechanism

The existence of interspecific competition and compensation effects in crop intercropping and interplanting is well known. Furthermore, an adequate interspecific layout and field configuration can give full play to interspecific compensation and reduce interspecific disharmony caused by excessive competition. Intercropping between leguminous and gramineous species can improve nutrient utilization through ecological niche complementation, changing nutrient competition into compensation, reshaping root architecture, and activating soil N and P nutrients, among other mechanisms.

4.2.1 Effect of Maize-Soybean Strip Intercropping on Soil Aeration

4.2.1.1 Dynamic Changes in O_2 Content in the Soybean Rhizosphere

Soil O_2 content in the IS planting system was higher than that in the MS system under reduced N application treatment. Specifically, soil O_2 content in IS was 3.4%,

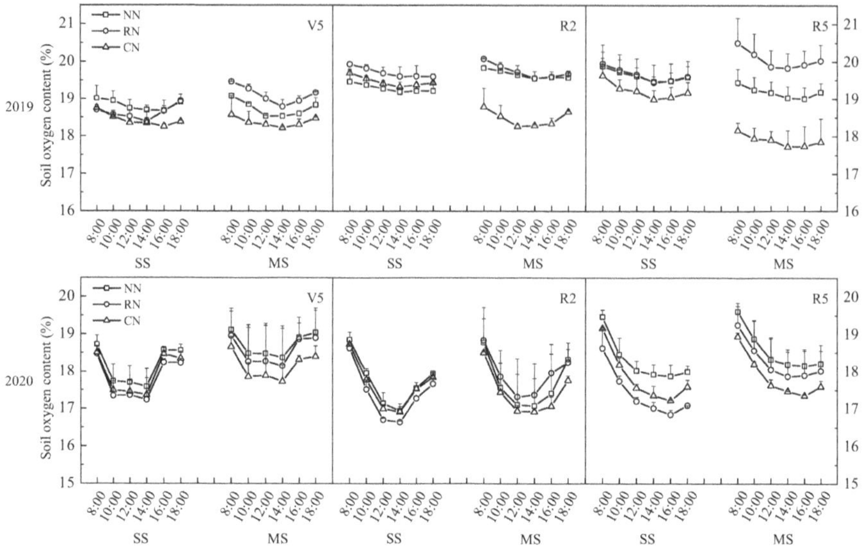

Fig. 4.22 Dynamic changes in soil O2 content at V5, R2, and R5 soybean growth stages

1.6%, and 2.4% higher than the corresponding levels in MS, respectively, at V5, R2, and R5 growth stages (Fig. 4.22). In 2019, the daily average soil O_2 content under N application treatment was highest in RN followed by those in NN and CN treatments; in turn, the soil O_2 content in the MS system under RN treatment was higher than that for any other treatment, especially at R5 stage, when it was 4.3% and 10.5% significantly higher than the O_2 content in NN and CN, respectively. The specific impact of N application rate on soil O_2 content in the MS planting model ranked in the order of RN > NN > CN at R2 stage in 2020, but the difference was not significant (Peng et al. 2022).

4.2.1.2 Variation in Soybean Soil Respiration Rate

There were significant differences in the changes of soybean soil respiration rate under different planting schemes. Specifically, in the MS model, soil respiration rate was significantly lower than that in SS (i.e., 24.9% and 27.6% lower in 2019 and 2020, respectively), at V5 soybean growth stage in the symbiotic phase. However, at R2 and R5 stages (i.e., after corn harvest), the average soil respiration rate in MS was significantly higher than that in SS by 52.0% and 137.8%, respectively. Further, the effect of N application on soybean soil respiration rate was significant. Particularly, soil respiration in MS mainly showed a decreasing trend with increasing N application rate, while in IS, it first increased and then decreased with increasing N application rate. Especially at the R2 stage, soil respiration rate in the RN treatment was 58.4% and 19.9% higher than the corresponding rates in treatments NN and CN, respectively (Fig. 4.23).

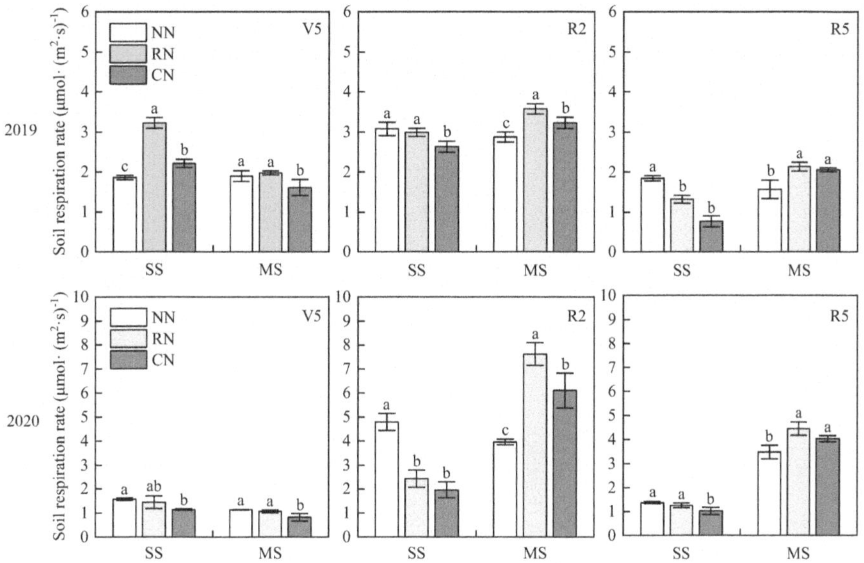

Fig. 4.23 Effects of planting system and nitrogen application rate on soybean soil respiration rate

4.2.1.3 Distribution Characteristics of Water-Stable Soil Aggregates in Soybean Intercropping Systems

Planting system had a significant impact on the content of water-stable soil aggregates of various particle sizes (Table 4.9). Compared with SS, the MS system significantly increased the percent content of >5 mm soil particles, with an average increase of 31.3%. Particularly, the percentage of 2–5 mm particles increased significantly in 2019, whereas the percentage of particles in the <0.25, 0.25–0.5, and 0.5-1 mm particle fractions decreased significantly, with average reduction rates of 6.8%, 17.8%, and 25.8%, respectively. The effect of N application rate on water-stable soil aggregates with different particle sizes was significant, but the variation pattern was inconsistent. The percentage of water-stable soil aggregates with particle sizes of 0.5 ~ 1 mm and 0.25 ~ 0.5 mm increased first and then decreased with increasing N application rate, with the RN treatment having the highest percentage and differences being significant. In contrast, the variation pattern for >5 mm particle size aggregates were opposite to that of the 0.5 ~ 1 mm and 0.25 ~ 0.5 mm particle size fractions, with the percentage of water-stable soil aggregates being lowest in the RN treatment. Meanwhile, in the SS planting model, water-stable soil aggregates in treatment RN were 4.3% and 14.4% lower than those in the NN and CN treatments, respectively. Similarly, in the MS planting mode, water-stable soil aggregates in RN were 1.1% and 1.7% lower than those in the NN and CN treatments, respectively. Moreover, N application reduced the percentage of water-stable soil aggregates with particle size of <0.25 mm. Finally, in 2020, the percentage of water-stable soil aggregates under RN in the MS planting system was significantly lower (by 26.5%) than that in the NN treatment.

Table 4.9 Mass percentage content of water-stable soil aggregates under different planting systems and nitrogen application rates

Year	Cropping pattern	N application rate	Particle size distribution of soil water-stable aggregates (%)					
			>5 mm	2 ~ 5 mm	1 ~ 2 mm	0.5 ~ 1 mm	0.25 ~ 0.5 mm	<0.25 mm
2019	SS	NN	20.43b	13.17b	8.48b	16.95a	17.27b	23.71a
		RN	19.87b	13.72ab	9.71a	17.05a	19.28a	20.36b
		CN	25.59a	14.61a	9.40ab	17.05a	16.46b	16.89c
		Mean	21.96	13.83	9.20	17.02	17.67	20.32
	MS	NN	29.15b	18.16a	9.91a	10.31b	11.44c	21.03a
		RN	28.733b	14.17b	8.99a	14.19a	14.22a	19.69b
		CN	31.92a	16.66a	9.20a	13.08a	12.85b	16.29c
		Mean	29.94	16.33	9.37	12.53	12.84	19.01
					ANOVA (F-value)			
	Cropping pattern (A)		153.00**	58.53^{**}	0.49^{ns}	342.11^{**}	377.73^{**}	15.94^{**}
	N application rate (B)		19.12**	12.12^{**}	0.15^{ns}	24.01^{**}	36.758^{**}	104.04^{**}
	A × B		1.62^{ns}	16.63^{**}	7.142^{**}	21.33^{**}	6.81^{*}	4.28^{*}
2020	SS	NN	28.03a	15.69a	9.69a	16.11b	14.61a	15.87b
		RN	26.41a	14.54b	8.61b	18.90a	15.30a	16.23b
		CN	28.22a	16.55a	8.45b	15.96b	11.85b	18.98a
		Mean	27.56	15.59	8.92	16.99	13.92	17.03
	MS	NN	31.59b	16.65a	8.29b	11.85b	12.75b	18.88a
		RN	31.37b	16.22a	9.41a	14.52a	14.61a	13.87b
		CN	41.50a	14.76b	6.39c	11.69b	10.97c	14.69b
		Mean	34.82	15.88	8.03	12.69	12.78	15.81
					ANOVA (F-value)			
	Cropping system (A)		121.18^{**}	0.91^{ns}	18.01^{**}	411.11^{**}	24.65^{**}	16.39^{**}
	N application rate (B)		31.63^{**}	2.40^{ns}	25.47^{**}	77.96^{**}	80.95^{**}	21.91^{**}

MS maize-soybean relay strip intercropping, *NN* no nitrogen application, *RN* reduced nitrogen application, *CN* constant nitrogen application

Different lowercase letters within columns indicate significant ($P < 0.05$) differences among different nitrogen application rates and planting system. * indicates significant, and * * indicates highly significant

Table 4.10 Effects of row spacing on morphological characteristics of the maize root system

Year	Treatment	Root length (cm)	Root surface area (cmm^2)	Root volume (cm^3)	Root dry weight (g)	Average root diameter (mm)
2019	MS30	3035.54 ± 253.74b	1509.25 ± 23.45b	25.03 ± 1.16b	4.3 ± 0.20b	0.57 ± 0.08a
	MS45	3376.46 ± 200.6 1ab	1553.81 ± 129.07b	26.71 ± 4.15b	4.9± 0.27b	0.54 ± 0.09a
	MS60	2392.95 ± 11.91c	1341.97上 67.88b	23.15 ± 3.14b	4.48 ± 0.49b	0.64 ± 0.06a
	MM100	3675.02 ± 25.57a	1945.77± 4.99a	31.23 ± 3.07a	8.53 ± 0.23a	0.68 ± 0.02a
2020	MS30	3893.95 ± 79.34bc	1154.58 ± 20.41ab	34.8 ± 3.59a	6.11 ± 0.46ab	0.59 ± 0.01a
	MS45	5187.74 ± 290.73ab	1267.58 ± 110.99ab	38.5 ± 2.52a	6.74 ± 083ab	0.55 ± 0.02b
	MS60	3632.25 ± 37.03c	1037.8 ± 14.45b	36.8 ± 0.44a	5.21 ± 0.01b	0.56 ± 0.01ab
	MM100	5845.21 ± 772.39a	1315.66 ± 125.2a	38.86 ± 2.16a	7.54 ± 0.80a	0.50 ± 0.01c
F-value						
Years (*Y*)		46.53**	49.17**	43.75**	5.88*	2.84 ns
Interspecific distance (*I*)		12.50**	10.76**	3.27*	16.76*	0.50 ns
Y × *I*		1.72 ns	2.03 ns	1.12 ns	3.60*	1.80 ns

4.2.2 Spatial Distribution and Root Nitrogen Uptake

4.2.2.1 Root Morphological Characteristics of Corn and Soybean

Except for average root diameter, significant differences in root morphology parameters were observed among years and treatments at R4 stage of maize growth (Table 4.10). Thus, root length, root surface area, root volume, and root dry weight were higher in MM than the corresponding values in IM planting. The average increase rates in these parameters, relative to MS45 over the 2 years of study, were 11.16%, 15.60%, 7.48%, and 38%, respectively, and differences were significant.

Root length, root surface area, root volume, and root dry weight in IM first increased and then decreased with increasing distance between maize and soybean plant rows. Compared with MS30, root length, root surface area, root volume, and root dry weight of MS45 increased by an average of 23.59%, 5.91%, 8.99%, and 11.82% over the 2 years of study, respectively (Ren et al. 2022).

Table 4.11 Effects of interspecific distance on the morphological characteristics of the soybean root system

Year	Treatment	Root length (cm)		Root surface area (cmm2)		Root volume (cm3)		Root dry weight (g)		Average root diameter (mm)	
		V5	R5	V5	R5	V5	R5	V5	R5	V5	R5
2019	MS30	184.32 ± 25.89b	1763.96 ± 16.46c	123.86 ± 2.71b	731.08 ± 34.28a	3.95 ± 0.26b	20.38 ± 1.56b	0.1 ± 0.01b	4.53 ± 21a	0.41 ± 0.04a	0.57 ± 0.03a
	MS45	199.59 ± 3.03b	1888.87 ± 40.3b	131.21 ± 7.98b	929.41 ± 259.04a	4.24 ± 0.36b	30.18 ± 3.45a	0.15 ± 0.02b	4.89 ± 0.04a	0.42 ± 0.03a	0.62 ± 0.03a
	MS60	283.76 ± 25.48ab	1921.8 ± 29.01b	149.87 ± 15.78b	887.77 ± 10.94a	4.74 ± 0.38ab	27.15 ± 0.12ab	0.16 ± 0.03b	4.63 ± 0.04a	0.44 ± 0.02a	0.46 ± 0.01b
	MM100	618.77 ± 223.59a	2509.44 ± 7.38a	259.97 ± 52.19a	976.92 ± 55.63a	6.31 ± 0.98a	27.34 ± 3.05ab	0.48 ± 0.05a	4.79 ± 0.02a	0.37 ± 0.01a	0.54 ± 0.04ab
2020	MS30	233.42 ± 42.44b	3451.57 ± 33.06ab	139.99 ± 8.48c	735.73 ± 18.43a	3.39 ± 0.11c	28.72 ± 0.44b	0.23 ± 0.05c	5.26 ± 0.27b	0.33 ± 0.02a	0.48 ± 0.01b
	MS45	347.75 ± 49.28b	3699.95 ± 112.31ab	193.35 ± 8.65b	780.95 ± 7.14a	3.97 ± 0.11b	37.32 ± 1.31a	0.55 ± 0.04b	7.47 ± 0.44a	0.36 ± 0.04a	0.47 ± 0.05b
	MS60	373.8 ± 128.96b	3162.52 ± 249.54b	227.58 ± 5.93b	722.74 ± 6.23a	4.37 ± 0.06b	29.59 ± 0.29ab	0.31 ± 0.14bc	6.9 ± 0.37ab	0.4 ± 0.03a	0.55 ± 0.01ab
	MM100	915.45 ± 166.89a	3918.37 ± 225.69a	303.66 ± 25.46a	833.37 ± 84.86a	5.69 ± 0.30a	35.58 ± 4.53ab	1.28 ± 0.03a	7.65 ± 1.00a	0.39 ± 0.01a	0.57 ± 0.01a
F-value											
Years (*Y*)		3.41 ns	291.45**	10.29*8	2.59 ns	2.32 ns	14.780**	84.30**	49.55**	3.67 ns	2.20 ns
Interspecific distance (*I*)		10.25**	11.28**	17.30**	1.09 ns	11.53**	5.51**	65.14**	4.21*	0.99 ns	1.26 ns
Y × *I*		0..47 ns	2.08 ns	0.72 ns	0.32 ns	0.07 ns	0.67 ns	14.71**	2.51 ns	1.13 ns	7.633*

From V5 to R5 stages of soybean growth (Table 4.11), except for average root diameter, root morphology parameters significantly increased, and there were differences between years and treatments. Particularly, at V5 stage, root length, root surface area, root volume, and root dry weight in IS plots were significantly lower than those in SS. Specifically, compared with monoculture, the average values for the parameters above in MS45 significantly decreased in 2 years by 64.32%, 42.42%, 31.58%, and 60.32%, respectively. At R5 soybean growth stage, compared with SS, the IS planting system reduced root length and root surface area relative to the values recorded in SS plots. Thus, average root length in the MS45 planting mode/interspecific distance treatment combination significantly decreased by 13.05% in 2 years, but root volume was higher than in SS, with an average increase of 7.28% for the two-year period. Meanwhile, in the relay strip intercropping treatment, at V5 stage, except for root dry weight, all root morphology parameters increased with increasing interspecific distance, being largest in MS60 and smallest in MS30; in turn, at R5 stage, root volume and root dry weight first increased and then decreased with increasing interspecific distance, average values being largest in MS45 and smallest in MS30.

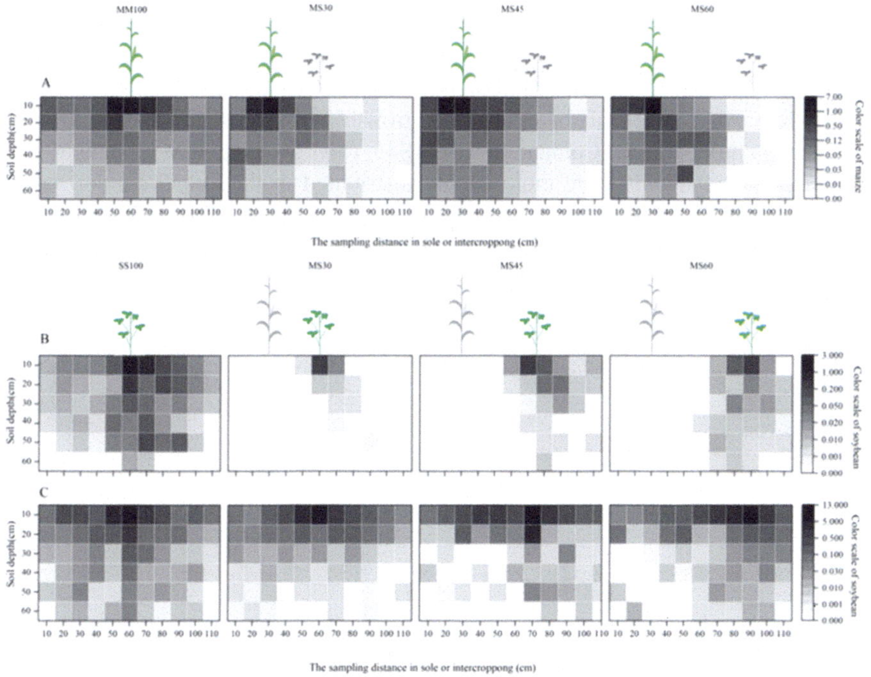

Fig. 4.24 Spatial distribution of maize and soybean root systems at different interspecific planting distances in 2020. *A* maize dough stage, R4, *B* soybean fifth leaf stage V5, *C* soybean initial grain stage R5. Each square represents the dry weight of roots within a volume of 1800 cm^3 of soil, green plants represent the root distribution of the crop and gray plants represent the corresponding positions of symbiotic crops

4.2.2.2 Spatial Distribution of Maize and Soybean Root Systems

At R4 stage of maize growth (Fig. 4.24), the distribution of maize roots was mainly in the horizontal direction, within a radius of 15 cm at the base of the stem, and in the vertical direction ranging from 0 to 40 cm. Compared with the IM planting system, the root system of in MM was more evenly distributed in the soil, with a symmetrical distribution on both sides; furthermore, the distribution space was larger than that in the IM system. In the horizontal direction, the root distribution in IM first increased and then decreased with increasing interspecific distance, and all roots grew below the soybean row space, with the MS45 combination treatment showing the widest root distribution. In the vertical direction, compared to the MM system, IM increased the root distribution in intercropping by 40 ~ 60 cm soil, with the increase in MS45 being particularly significant.

At V5 stage of soybean growth (Fig. 4.23b), the root system in SS showed a symmetrical distribution due to underground intraspecies competition. Compared with SS, the root growth space in IS was significantly reduced. Further, in the relay strip intercropping system, roots in MS30 and MS45 combination treatments were more inclined towards soybean strip growth, showing a significant skewness, but the root distribution in both horizontal and vertical directions in the IS system increased as the interspecific distance increased. Thus, compared with MS30, combination treatments MS45 and MS60 increased root distribution in both horizontal and vertical directions. Later, at R5 stage of soybean growth (Fig. 4.23c), the root system in IS extended in the horizontal direction to the maize belt in the 0–20 cm soil layer; meanwhile, in the vertical direction, root growth in MS60 returned to the level observed in monoculture. Compared with MS45, roots growing in MS30 and MS60 combination treatments were more evenly distributed in the 20–50 cm soil layer towards the corn belt, but the root distribution in the 0–20 cm soil layer increased in the MS45 combination treatment.

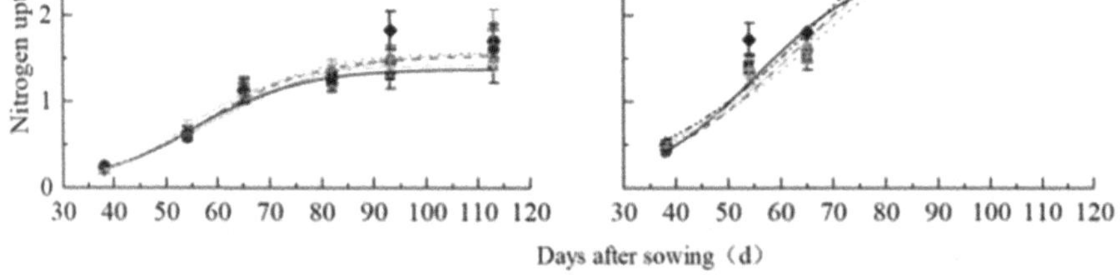

Fig. 4.25 Changes in nitrogen absorption in maize plants under different nitrogen application rates and interspecific distances

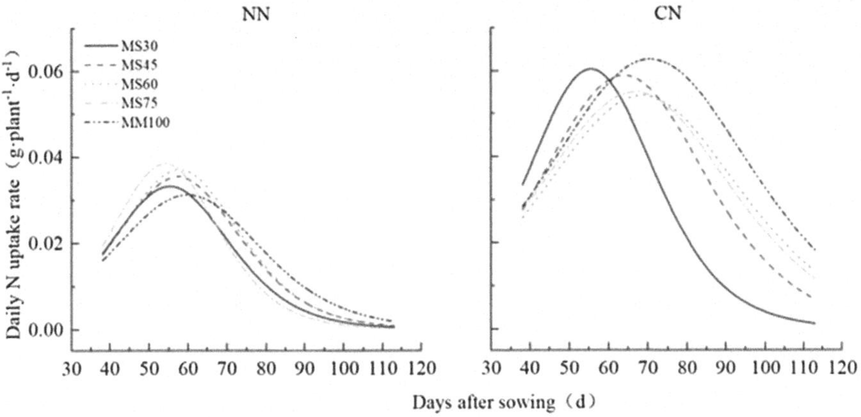

Fig. 4.26 Changes in daily nitrogen absorption rate in maize under different nitrogen application rates and intervals

Table 4.12 Characteristic parameters of nitrogen absorption in maize under different nitrogen application rates and spacing

N application	Interspecific distance	$T_{\text{max-N}}$ (d)	$W_{\text{max-N}}$ (g)	$V_{\text{max-N}}$ (×10^{-2} g·plant^{-1}·d^{-1})	$V_{\text{mean-N}}$ (×10^{-2} g·plant^{-1}·d^{-1})
NN	MS30	55.29 ± 3.48a	0.69 ± 0.08b	3.33 ± 0.48a	1.33 ± 0.04a
	MS45	57.3 ± 5.75a	0.77 ± 0.1a	3.55 ± 0.76a	1.42 ± 0.15a
	MS60	57.69 ± 6.12a	0.79 ± 0.36a	3.72 ± 0.56a	1.47 ± 0.1a
	MS75	54.12 ± 2.69a	0.71 ± 0.01ab	3.86 ± 2.58a	1.46 ± 0.27a
	MM100	59.87 ± 3.89a	0.79 ± 0.2a	3.14 ± 4a	1.33 ± 0.22a
CN	MS30	55.58 ± 4.79c	1.31 ± 0.16b	6.04 ± 0.36a	2.46 ± 0.07b
	MS45	63.95 ± 4.28ab	1.66 ± 0.17ab	5.89 ± 0.44ab	2.55 ± 0.12b
	MS60	68.56 ± 2.46ab	1.81 ± 0.1ab	5.41 ± 0.31b	2.47 ± 0.07b
	MS75	66.83 ± 5.34ab	1.8 ± 0.25ab	5.5 ± 0.22b	2.51 ± 0.1b
	MM100	70.68 ± 2.97a	2.16 ± 0.07a	6.27 ± 0.06a	2.86 ± 0.04a
F-value					
N application (N)		3.11*	60.92**	6.79*	145.16**
Interspecific distance (I)		4.26*	5.81*	1.39 ns	5.11*
$N \times I$		1.37 ns	1.05 ns	0.27 ns	1.05 ns

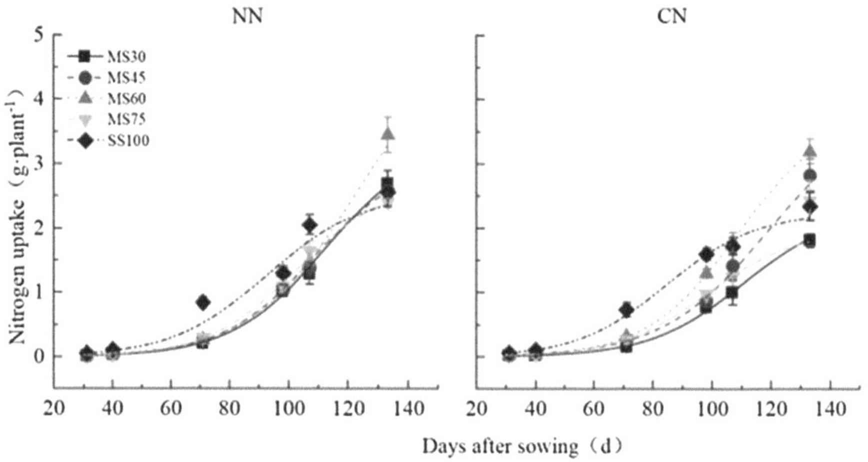

Fig. 4.27 Changes in nitrogen absorption in soybean plants under different nitrogen application rates and row spacings

4.2.2.3 Nitrogen Absorption in Maize and Soybean Plants

The logistic equation fit best the N absorption process in maize (Figs. 4.25 and 4.26), and the adjusted value for R^2 was above 0.97. The N absorption process and daily N absorption rate in maize plants were significantly affected by N application (Table 4.12). Compared with NN, the CN treatment significantly increased N absorption in maize. Particularly, the MS45 combination treatment significantly increased N accumulation at maximum N absorption rate (Wmax-N), maximum N absorption rate (Vmax-N), and average N absorption rate (Vmean-N) by 115.56%, 65.92%, and 79.58%, respectively. In turn, the MS60 combination treatment significantly increased the same parameters by 129.11%, 45.43%, and 68.03%, respectively. Furthermore, CN treatment delayed the time to reach maximum N uptake rate in maize (Tmax-N), and the CN treatment in MS45 and MS60 delayed it by 6.65 and 10.87 days, respectively, compared with NN. Row spacing had a significant impact on the N absorption process and the daily N absorption rate in maize. Under CN, compared with MM, IM significantly reduced the accumulation of Wmax-N, Vmax-N, and Vmean-N, while MS60 significantly decreased the values for these parameters by 16.20%, 13.72%, and 13.64%, respectively. The time to reach Tmax-N was shortened by IM planting relative to monoculture planting. Further, Wmax-N and Vmean-N values first increased and then decreased with increasing row spacing. When the N absorption rate was at its maximum, Wmax-N was uniformly highest in MS60 and lowest in MS30.

The logistic equation fit best the N absorption process of soybean (Figs. 4.27 and 4.28), and the adjusted value for R^2 was above 0.96. Nitrogen application reduced the N absorption rate in soybean plants (Table 4.12). Among treatments, compared with NN, the values for Wmax-N and Vmax-N in MS60 decreased by 17.30%, 14.93%, and 12.03% under CN, respectively, and the Tmax-N in MS60 was attained

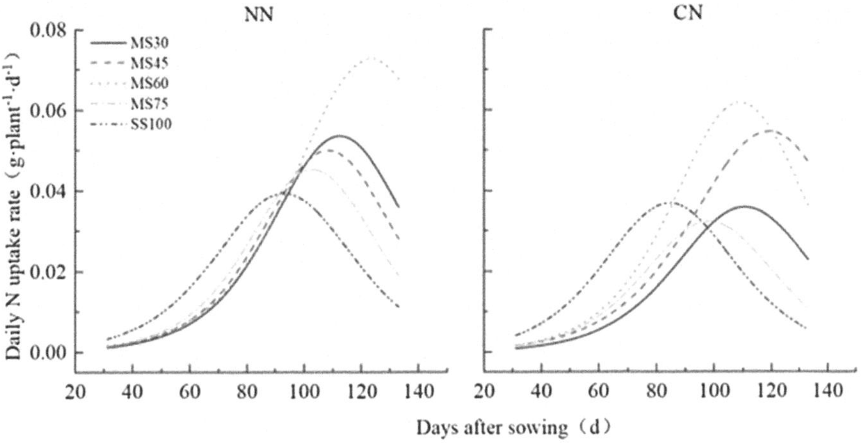

Fig. 4.28 Changes in daily nitrogen uptake rate in soybean plants under different nitrogen application rates and row spacings

9.80 days earlier. Interspecific distance had a significant impact on the N absorption process and daily N absorption rate in soybean plants. Compared with SS, N uptake in the MS60 treatment combination increased in the later stages of growth, and the amount of N accumulated at Wmax-N, Vmax-N, and Vmean-N significantly increased by 60.32%, 62.34%, and 43.32% under NN, respectively, which in turn increased by 53.36%, 31.39%, and 24.23% in CN-treated plots, respectively. Compared with SS, the Tmax-N in the IS planting system was delayed 21.83 and 22.9 days under NN and CN treatments, respectively, in the MS60 combination treatment. Further, as row spacing increased, the values for Wmax-N, Vmax-N, and Vmean-N first increased and then decreased. Except for the NN treatment in 2021, the highest and lowest N absorption rates were recorded in the MS60 and MS30 combination treatments, respectively (Table 4.13).

4.2.3 Soybean Root Exudates and Their Functions in Maize-Soybean Strip Intercropping

4.2.3.1 Effects on the Secretion of Flavonoids in Soybean Roots

Flavonoids are signaling substances that are active in the process of induction of nodule formation. The genistein and daidzein therein are key substances in soybean nodule induction. The secretion of flavonoids from soybean roots at different growth stages indicated that relay strip intercropping effectively increased flavonoid secretion from roots of various soybean varieties, and the secretion of some flavonoid components showed an upward trend (Fig. 4.29). Further, in the relay strip intercropping planting system, the secretion of daidzein and genistein in variety Nandou 12 significantly increased by 85.0% and 43.9%, respectively, compared to soybean monoculture. Similarly, genistein and quercetin contents in the root

Table 4.13 Characteristic parameters of nitrogen absorption in soybean plants under different nitrogen application rates and row spacings

N application	Interspecific distance	$T_{max\text{-}N}$ (d)	$W_{max\text{-}N}$ (g)	$V_{max\text{-}N}$ (×10^{-2} g·plant^{-1}·d^{-1})	$V_{mean\text{-}N}$ (×10^{-2} g·plant^{-1}·d^{-1})
NN	MS30	112.35 ± 4.58b	1.7 ± 0.29b	5.37 ± 1.08a	1.82 ± 0.43a
	MS45	108.31 ± 7.97b	1.55 ± 0.1bc	5.01 ± 0.21a	1.71 ± 0.13a
	MS60	123.26 ± 8.38a	2.61 ± 0.26a	7.28 ± 2.08a	2.51 ± 0.65a
	MS75	102.79 ± 6.19bc	1.37 ± 0.19bc	4.54 ± 0.87a	1.58 ± 0.06a
	MM100	92.84 ± 4.21c	1.28 ± 0.24c	3.95 ± 0.82a	1.52 ± 0.27a
CN	MS30	110.55 ± 5.63a	1.15 ± 0.2b	3.6 ± 1.34ab	1.24 ± 0.51a
	MS45	118.85 ± 3.89a	1.97 ± 0.25a	5.46 ± 1.11a	1.93 ± 0.78a
	MS60	108.96 ± 1.8a	1.94 ± 0.1a	6.18 ± 0.44ab	2.12 ± 0.13a
	MS75	98.17 ± 8.16b	0.99 ± 0.05b	3.25 ± 1.83ab	1.17 ± 0.1a
	MM100	84.43 ± 6.38c	1.13 ± 0.23b	3.7 ± 0.37b	1.44 ± 0.12a
F-value					
N application (*N*)		2.81 ns	12.56**	0.33 ns	0.10 ns
Interspecific distance (*I*)		21.45**	29.85**	1.31 ns	0.76 ns
N × *I*		3.48*	6.50**	0.93 ns	0.77 ns

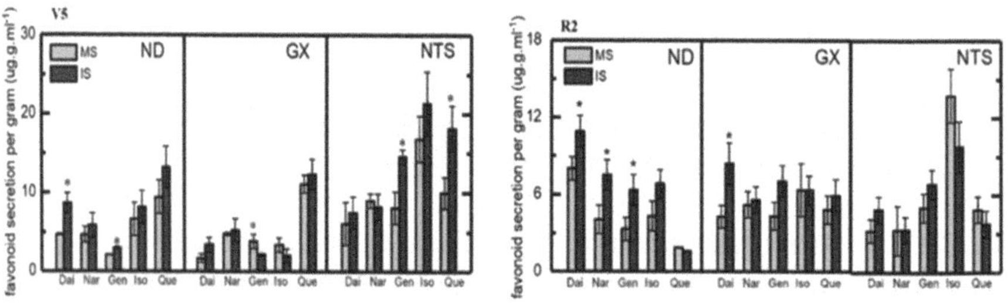

Fig. 4.29 Effect of planting system on flavonoids secreted by soybean roots. *MS* monocropping soybean, *IS* maize-soybean relay strip intercropping

exudates of variety NTS1007 significantly increased by 80.1% and 82.6%, respectively, in relay strip intercropping compared to monoculture. Conversely, the secretion of genistein from the roots of variety Guixia 3 was significantly reduced by 42.8% compared to that observed in soybean monoculture.

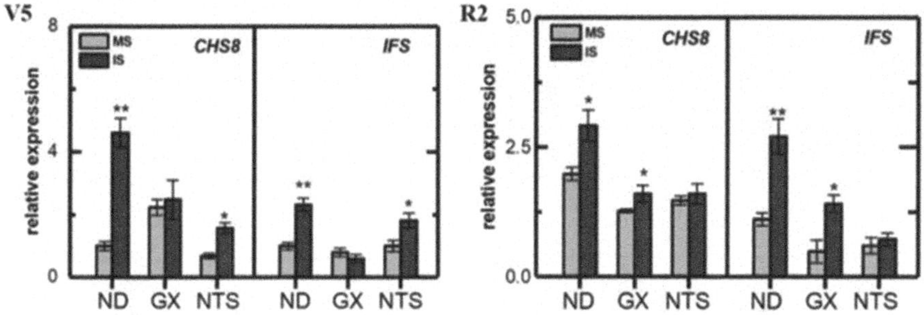

Fig. 4.30 Expression level of key genes involved in the flavonoid biosynthesis pathway in soybean roots (2019)

During the full bloom (R2) stage, the content of various flavonoid components in the root exudates of varieties Nandou 12 and Guixia 3 generally increased in relay strip intercropping compared to the monoculture system. Among such flavonoids, the levels of daidzein, naringin, and genistein in the root exudates from Nandou 12 significantly increased by 36.2%, 85.9%, and 90.5% in relay strip intercropping, respectively, compared with monoculture. Similarly, the amount of daidzein secreted by the roots of Guixia 3 increased significantly by 95.8%. Meanwhile, there was no significant difference in the secretion of various flavonoid components from the roots of variety NTS1007 between relay strip intercropping and monoculture.

A comparison of the level of expression of genes encoding for key enzymes in the flavonoid biosynthesis pathway in soybean roots at different growth stages (Fig. 4.30) revealed that relay strip intercropping up-regulated the expression of key enzyme genes *CHS* (chalcone synthase) and *IFS* (isoflavone synthase) in Nandou12 and NTS1007, thereby enhancing the secretion of flavonoids. Further, compared with monoculture, the expression levels of *CHS* and *IFS* genes in Nandou12 under relay strip intercropping were significantly increased 3.60 and 1.32 times, respectively. Similarly, the same gene expression levels in NTS1007 showed a 1.30-fold and 0.83-fold up-regulation, respectively. Additionally, at R2 stage, *CHS* and *IFS* gene expression in three soybean varieties showed up-regulation under relay strip intercropping. Compared with monoculture, the *CHS* expression level in Nandou12 under relay strip intercropping significantly increased 1.47 times and the *IFS* expression level increased highly significantly 1.88 times. Overall, gene expression levels in Nandou12 and Guixia 3 were up-regulated 1.26 and 1.43 times, respectively, under the strip intercropping planting system.

4.2.3.2 Up-Regulation of Gene Expression Related to Root Exudate-Mediated Nodule Formation in Soybean

The expression of the key gene *GmIFS1* in the flavonoid biosynthesis pathway in roots of soybean variety Nandou12 was significantly up-regulated by treatment with maize root exudate, during the maize-soybean symbiotic period. Concomitantly, the expression level of *GmEXPB2*, a crucial root nodule development-related gene, was

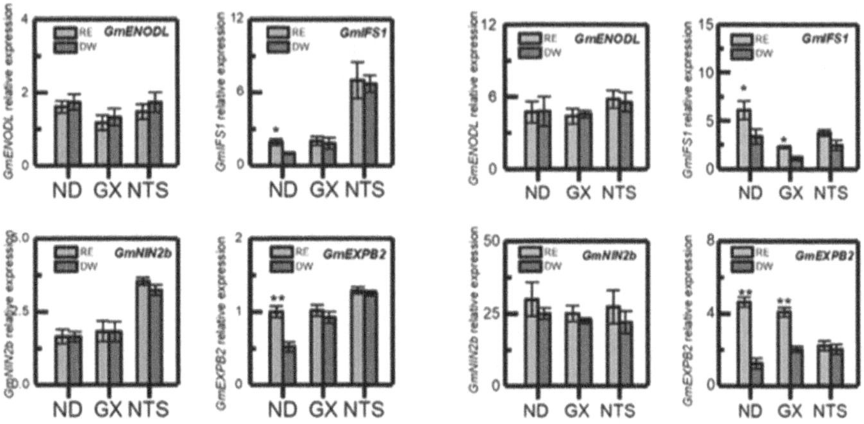

Fig. 4.31 Effect of maize root exudates on the expression of genes related to soybean nodule formation

also significantly up-regulated (Fig. 4.31), with an increase of 37.0% compared to the control level, which was beneficial for root formation and nodule development. Consistently, the expression levels of the early nodulation genes *NIN2b* and *GmIFS1* in NTS1007 were significantly higher than those in Nandou 12 and Guixia 3, suggesting their strong effect on the nodulation process. At R2 stage (Fig. 4.31), the expression of *GmEXPB2* genes in Nandou 12 and Guixia 3 were significantly up-regulated 2.72 and 1.02 times upon treatment with maize root exudate, respectively, while *GmIFS1* genes were significantly up-regulated by 79.9% and 107.8%, respectively, thereby promoting lateral growth of root nodules and facilitating nitrogenase formation.

After spraying soybean roots with maize root exudates (RE), the number of root nodules in each soybean variety showed no significant change compared to the control distilled water (DW) treatment (Fig. 4.32). The number of nodules formed in each soybean variety differed with growth stage. Thus, at V5 stage, the soybean varieties tested ranked as follows: NTS1007 > Nandou 12 > Guixia 3; which at R2 changed to: Guixia 3 > Nandou 12 > NTS1007; and at R4 to: Nandou 12 > Guixia 3 > NTS1007. Further, nodule growth rate varied among soybean varieties, with the maximum number of nodules in Nan Dou 12 occurring in the late stage of growth (full pod stage), which was more conducive to N fixation at that stage. Although NTS1007 is a high-nodulating variety, its nodule growth and decay rates at R2 stage were remarkably high, which was not conducive to sustained N fixation. Meanwhile, root treatment with RE increased root nodule dry weight. Specifically, at R2, nodule dry weight in RE-treated Nandou 12 roots was 81.5% higher than that of control roots, with the difference being significant. This increasing trend continued and at R4, when root nodule dry weight further and significantly increased by 37.6%. Furthermore, at V5, RE treatment increased root nodule dry weight by 36.8% and 44.4% in Nandou 12 and Guixia 3.

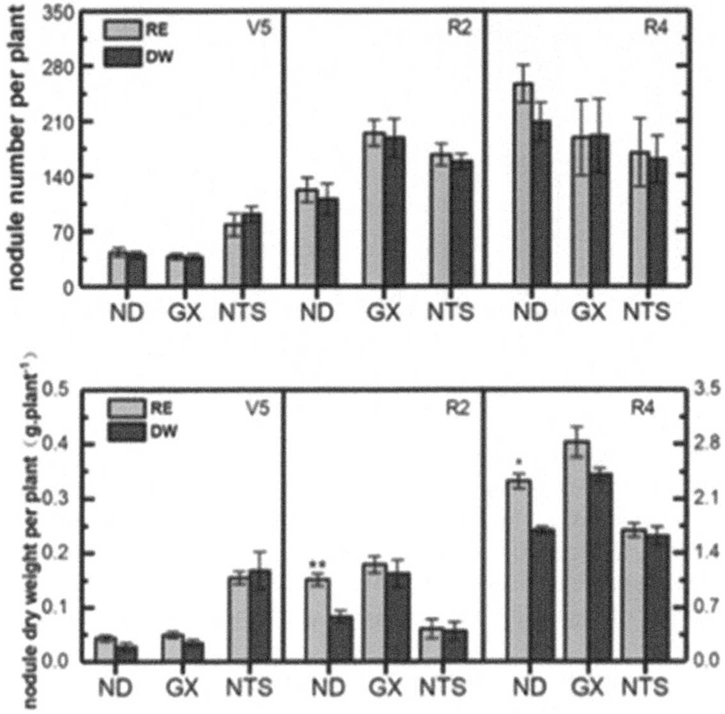

Fig. 4.32 Effect of maize root secretions on the number and dry weight of root nodules

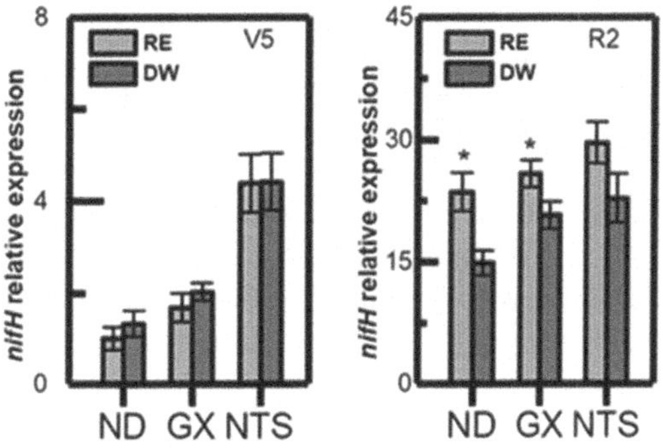

Fig. 4.33 Effect of maize root secretions on the expression of the nitrogenase gene

4.2.3.3 Root Exudate-Mediated Up-Regulation of Nitrogen Fixation-Related Gene Expression in Soybean Root Nodules

Nitrogenase is the key enzyme for developing effective N-fixing nodules and unleashing N fixation ability. In this study, there was no significant difference in the expression of the nitrogenase gene *nifH* among various soybean varieties between monoculture and relay strip intercropping at V5 stage. However, at R2 stage (full bloom stage), the expression of the gene *nifH* in Nandou 12 was

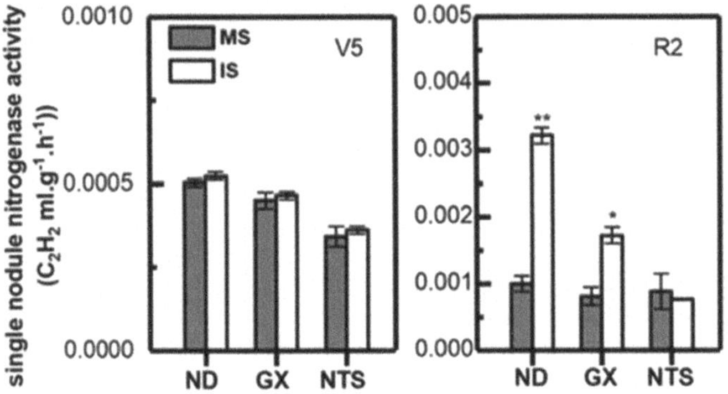

Fig. 4.34 Effect of planting pattern on nitrogenase activity in root nodules

significantly up-regulated in relay strip intercropping, showing an increase of 107.6% relative to monoculture (Fig. 4.33). Similarly, the *nifH* gene was also significantly up-regulated by 70.8% in Guixia 3 but no significant difference was detected for *nifH* gene expression in NTS1007. Therefore, the light distribution environment in the relay intercropping planting system effectively improved nitrogenase activity of Nandou 12 and Guixia 3, and increased their nitrogen fixation ability.

Nitrogenase activity per unit mass was significantly increased from V5 to R2 stage in soybean varieties Nandou 12, Guixia 3, and NTS1007. Further, there was no significant difference in nitrogenase activity per unit mass between MS and IS planting systems. Nonetheless, single-root nodule nitrogenase activity differed significantly under relay strip intercropping conditions (Fig. 4.34). Moreover, at R2, compared with monoculture, single-root nodule nitrogenase activity significantly increased 2.22 and 1.12 times in Nandou 12 and Guixia 3, respectively, but there was no significant difference in NTS 1007.

4.2.3.4 Root Exudate-Mediated Up-Regulation of Root Auxin-Related Gene Expression

The expression level of the *GmYUCCA14* gene in relay strip intercropping was significantly higher than that in monoculture at both V5 and R3 stages, especially under low P treatment; except for V5 stage, no significant difference in expression was detected between the two P treatments in the monoculture planting system (Fig. 4.35a,b). The expression pattern of *GmTIR1C* in P and light treatments was similar to that of *GmYUCCA14*, but relay strip intercropping had a significant advantage over monoculture only without P treatment, and there was no significant difference between planting systems under P treatment (Figure 4.40c,d). Further, at V5 and R3 stages, the expression level of *GmARF05* in the intercropping treatment was significantly higher than that under monoculture, except for the P20 treatment at V5 stage, but the expression level at R3 was less affected by P treatment. Lastly, there was no significant difference in *GmARF05* expression between the two P

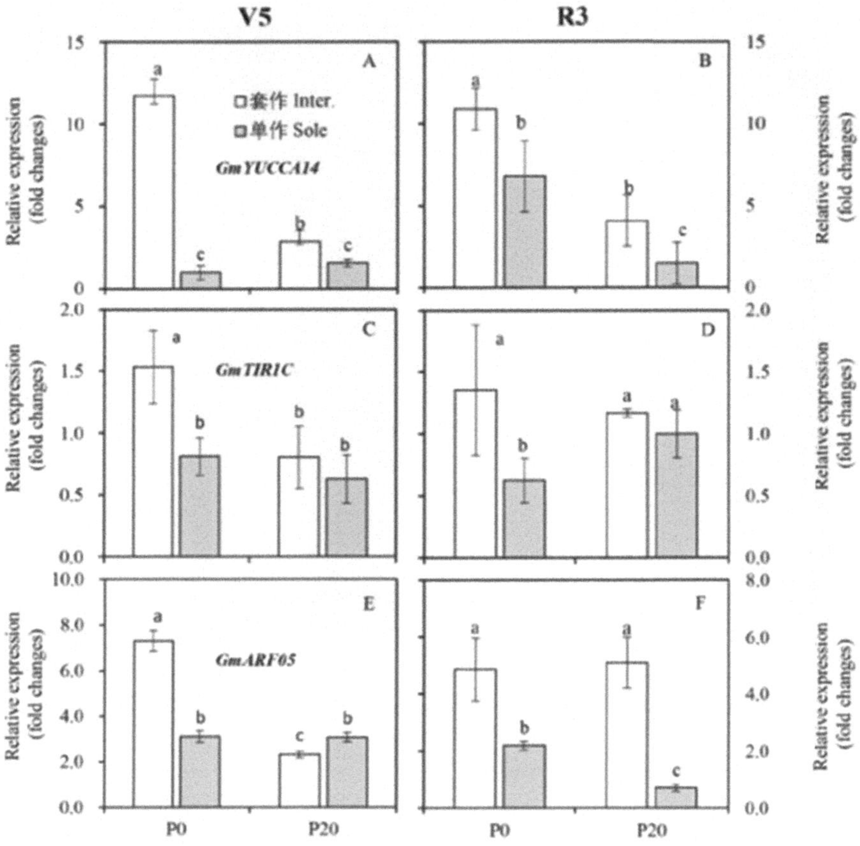

Fig. 4.35 Expression levels of genes related to auxin biosynthesis and their response to phosphorus treatment in the soybean root system in strip intercropping

treatments under relay strip intercropping, and it was higher under monoculture when no P was applied than when it was (Fig. 4.35e,f).

4.2.4 Soybean Root Exudates and their Role in Maize-Soybean Strip Intercropping

4.2.4.1 Characteristics of Soybean Root Exudates

A total of 32 significantly differential metabolites were screened between SS100 and MS30 (Fig. 4.36a), and 13 between SS100 and MS60 (Fig. 4.36b), respectively. These differential metabolites include mainly flavonoids, flavonols, isoflavones, and flavanones. Among them, two metabolites in soybean root exudates, namely, tangeretin and nobiletin, significantly accumulated under SS100 and MS30

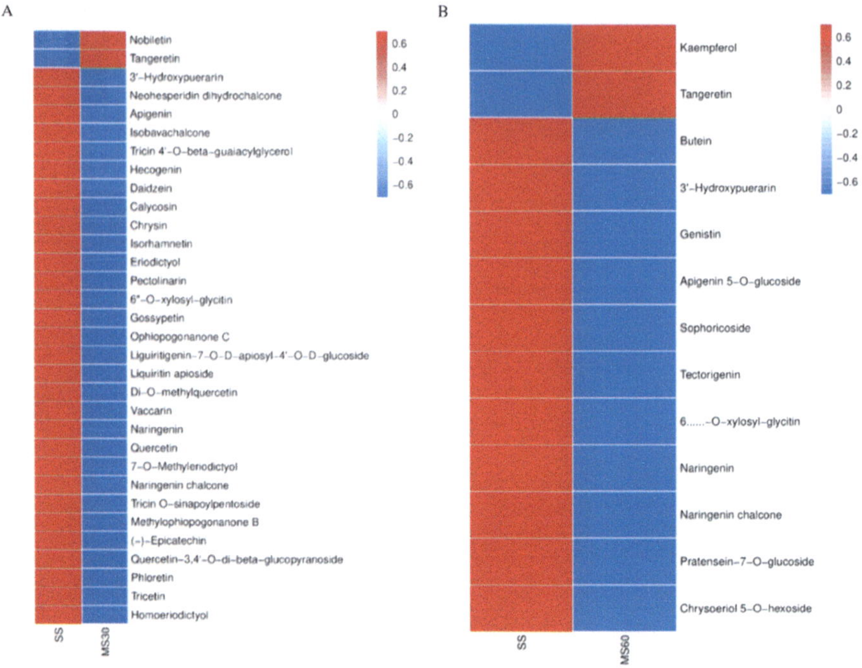

Fig. 4.36 Heat map of differential metabolites among SS100, MS30, and MS60 treatment groups

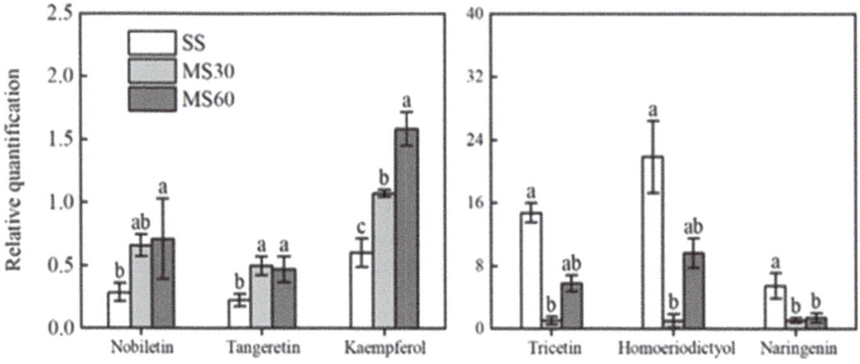

Fig. 4.37 Changes in flavonoids in soybean root exudates under different planting systems

combination treatments; additionally, two metabolites, kaempferol and tangeretin, significantly accumulated in soybean root exudates under SS100 and MS60, respectively.

Analysis of significantly differential metabolites revealed that, compared with monoculture, under MS60 conditions, nobiletin, tangeretin, and kaempferol contents increased 2.47, 2.09, and 2.62 times, respectively. Similarly, under MS30 treatment, nobiletin, tangeretin, and kaempferol contents increased 2.29, 2.20, and 1.77 times, respectively. Concomitantly, under the intercropping planting system, the secretion of tricetin, homoeodictyl, and naringenin significantly decreased compared with monoculture (Fig. 4.37). These findings indicated that the secretion of flavonoids and flavonols in soybean root exudates increased, while the secretion of flavanones decreased after intercropping with maize.

4.2.4.2 Effects of Soybean Root Exudates on Hormone Secretion and Response Factor Expression in Maize Roots

Maize root hormone (IAA, JA, and 5DS) contents were higher after treatment with soybean RE than those in the control groups (Fig. 4.38). Moreover, changes in JA content were more significant than those recorded for other hormones, and significantly ($p < 0.05$) and highly significantly ($p < 0.01$) higher than those of controls at 10D and 30D, by 64.21% and 75.05%, respectively. In turn, 5DS content at 20D was significantly higher (44.62%) than that of controls. Meanwhile, after treatment with soybean RE, ACC and ABA contents were lower than those in controls, although the difference was not significant.

The relative expression of maize root auxin-response factors *ZmARF1* and *ZmARF27* increased with soybean RE treatment (Figs. 4.39 and 4.40). Compared

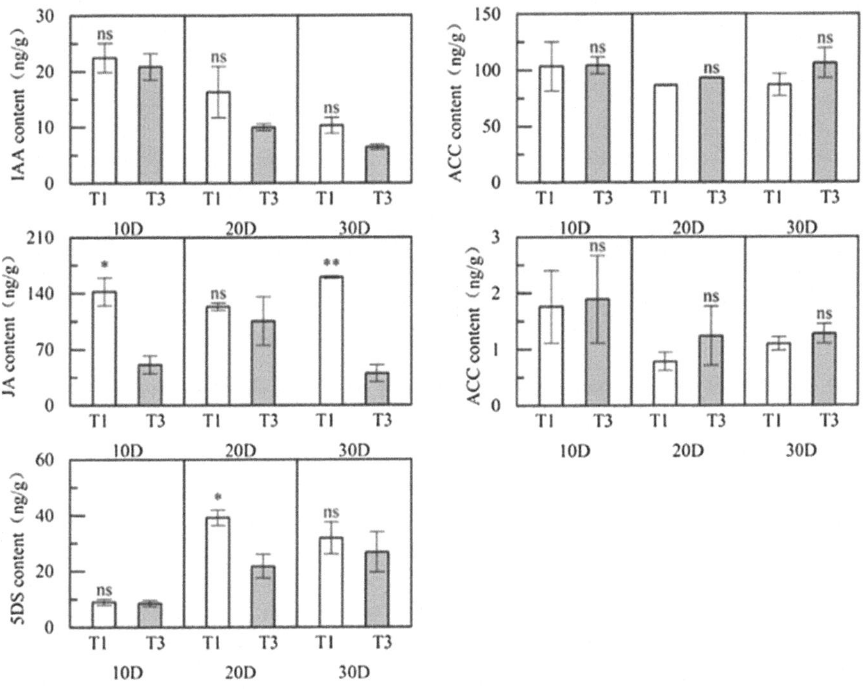

Fig. 4.38 Effect of soybean root exudates on hormone or precursor contents in maize roots. *IAA* Indole-3-acetic acid, *JA* jasmonic acid, *ACC* 1-Aminocyclopropane carboxylic acid, *ABA* Abscisic acid, *5DS* 5-Deoxystrigol

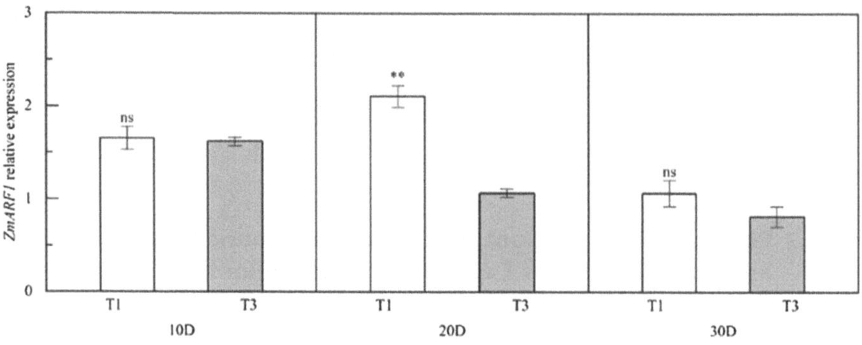

Fig. 4.39 Effect of soybean root exudates on the relative expression of ZmARF1 in maize roots

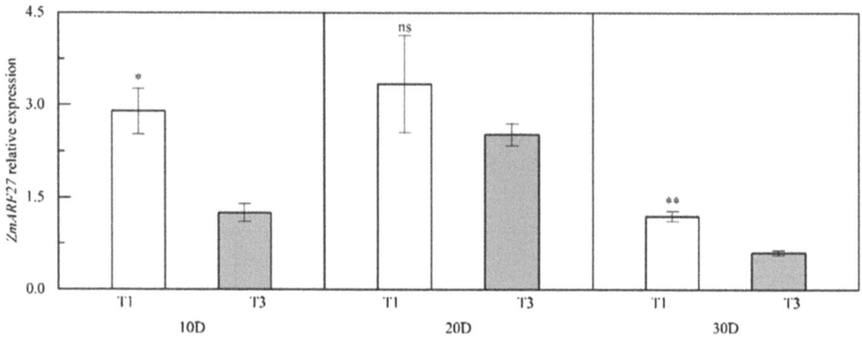

Fig. 4.40 Effect of soybean root exudates on the relative expression of ZmARF27 in maize roots

with the control treatment, treatment with soybean RE significantly increased the expression of *ZmARF1* by 49.32% at 20D; similarly, that of *ZmARF27* at 10D was significantly increased by 56.96%, while that of *ZmARF27* at 30D was highly significantly increased by 49.66%.

4.2.4.3 Effect of Soybean Root Exudates on Crop Root Growth and Development

Root growth and development of were significantly enhanced as time of soybean RE treatment increased, relative to that observed in controls (Table 4.14). Thus, at 10D, the first-order lateral root length and root biomass were significantly higher than those of controls by 23.58% and 30.68%, respectively. Then, at 20D, there was no significant difference between the two but at 30D, root exudate treatment significantly increased root length, primary lateral root length, root surface area, root volume, and root biomass by 26.27%, 39.30%, 29.68%, 32.67%, and 40.82%, respectively, relative to the control plants.

The effect of lateral root order on number of roots for different soybean RE treatment times differed significantly (Fig. 4.41), with secondary lateral roots contributing the most at 10D and 20D, while, at 30D, the contribution of third-

Table 4.14 Effects of soybean root exudates on maize root growth

Treatment		Root length (cm)	First lateral root (c)	Root surface (cm^2)	Root volume (cm^3)	Root height (g)
10D	T1	116.22a	55.04a	29.81a	0.54a	41.07a
	T3	105.13a	42.07b	38.83a	0.84a	28.47b
Mean		110.68	48.56	34.32	0.69	34.77
20D	T1	1070.00a	304.95a	147.74a	1.4a	165.53a
	T3	1043.65a	306.1a	137.41a	1.19a	158.83a
Mean		1056.83	305.53	142.58	1.3	162.18
30D	T1	5398.47a	362.1a	737.6a	8.02a	802.6a
	T3	3980.46b	219.78b	518.66b	5.4b	475b
Mean		4689.47	290.94	628.13	6.71	638.8

10D, 20D, and 30D: 10 days, 20 days, and 30 days of mist cultivation; T1: soybean root exudates and mist-cultured maize, T3: nutrient solution mist-cultured maize

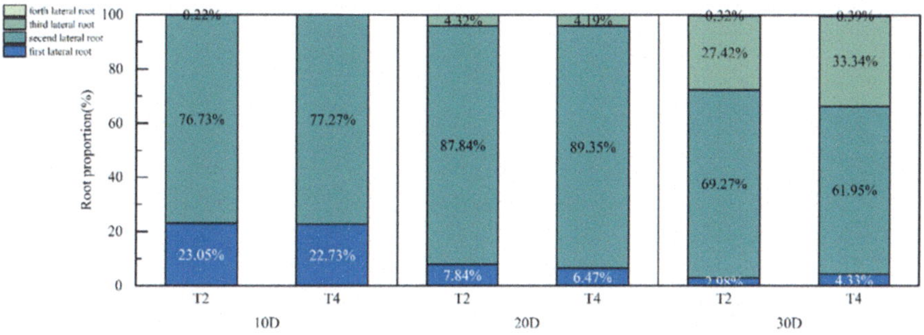

Fig. 4.41 Effect of soybean root exudates on maize lateral root grading

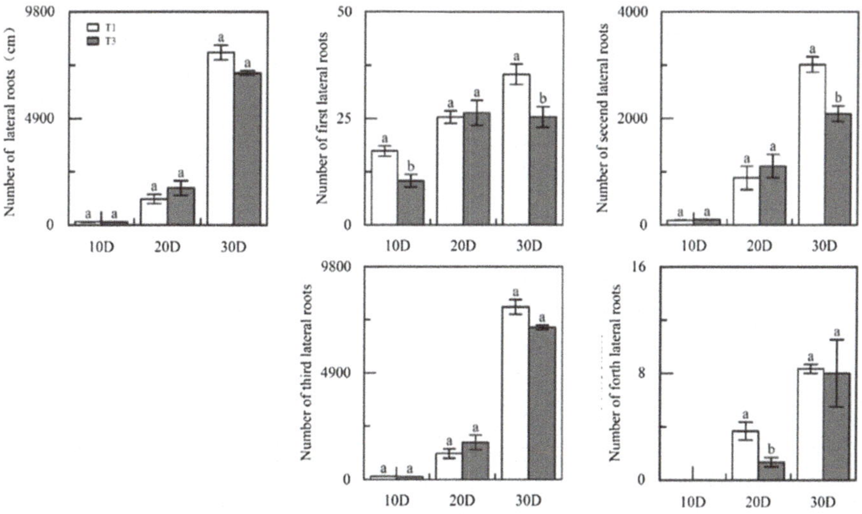

Fig. 4.42 Effect of soybean root exudates on the number of lateral roots in maize

order lateral roots was the largest (Fig. 4.42). Additionally, the effect of soybean RE on the number of maize lateral roots increased with increasing treatment time, including the emergence of third- and fourth-order lateral roots. However, the effect of soybean RE on the total number of maize roots was not significant. Specifically, at 10D and 30D, the number of maize primary lateral roots after soybean RE treatment was significantly higher than that of plants under the control treatment by 40.39% and 28.30%, respectively; additionally, at 30D, there was a significant difference in the number of secondary lateral roots, and at 20D, there was a significant difference in the number of fourth level lateral roots.

Theory of Soybean Canopy Architecture Regulation in Maize-Soybean Strip Intercropping

5

Weiguo Liu, Xiaoling Wu, and Yushan Wu

Abstract

This chapter investigates the regulatory mechanisms of canopy architecture in shorter-height crops within maize-soybean intercropping (IC) systems, focusing on soybean adaptation to shade conditions imposed by taller maize plants. It examines soybean canopy responses to various light environments, especially under strip intercropping, including the effects on leaf size, epidermal cells, gene expression, and hormone levels due to shading. Additionally, the study assesses the impact of the light environment on soybean petioles and stems, emphasizing physiological, biochemical, and molecular responses to optimize light capture and enhance yield in intercropping systems. These results will provide a theoretical basis and technical support for the canopy architecture regulation and yield improvement in short-height crops within maize-soybean strip intercropping.

Keywords

Canopy architecture · Maize · Soybean · Intercropping · Light environments · Shade adaptation · Physiological responses · Yield

As the shorter-height crop in maize-soybean intercropping (IC) systems, soybean is inevitably shaded by the taller maize plants during the early stages of growth, or during the intermediate and late growth stages in the case of maize-soybean strip intercropping systems. Shading leads to a series of adaptive changes in plant architecture, such as excessive internode elongation, thinner stems and leaves, and smaller leaflet angles, among others, as well as reduced accumulation and

W. Liu (✉) · X. Wu · Y. Wu
College of Agronomy, Sichuan Agricultural University, Chengdu, China
e-mail: lwgsy@sicau.edu.cn; wulx@sicau.edu.cn; yushan.wu@sicau.edu.cn

W. Yang et al. (eds.), *Theories and Application in Maize-Soybean Strip Intercropping System*, https://doi.org/10.1007/978-981-97-8723-4_5

imbalanced distribution of photoassimilate. Overall, such responses result in greater lodging, fewer pods, and ultimately, lower yield. Hence, to date, the improvement of the light environment for short-height crops in intercropping systems still relies mainly on crop field configuration, while research on shade tolerance mechanisms and screening for shade-tolerant varieties of short-height crops remains scarce at best. Furthermore, the corresponding theoretical support for canopy architecture regulation is still lacking, making it very difficult to increase the yield of short-height crops in intercropping systems and to improve the productivity of such systems.

5.1 Response Mechanism of Soybean Canopy Architecture to Strip Intercropping Light Environment

Effective photosynthetic radiation in strip intercropping systems is weakened due to the shading effect of corn, leading to an increase in the ratio of red light to far-red light (R:FR) and a decrease in light intensity, thus conforming a typical shaded environment. However, the degree of shading of short-height crops varies greatly depending on corn canopy architecture, field layout, and crop growth stage. Particularly, from flowering initiation onward, the light environment for soybean in a strip intercropping system may in fact be better after corn harvest than that in soybean monoculture. Therefore, it is extremely important to clarify the process of dynamic adaptation to the changing light environment at play in soybean under strip intercropping, such as to be able to cultivate those varieties that are particularly well suited for strip intercropping and to develop the corresponding management practices.

5.1.1 The Influence of Light Environment on Soybean Leaves

5.1.1.1 The Effect of Shading on Soybean Leaf Size

The impact of light environment on soybean leaves differed with developmental stage. Thus, for example, the results of a pot experiment conducted to test two treatments, i.e., shading (400 μmol m^{-2} s^{-1}) and full sunlight (1100 μmol m^{-2} s^{-1}) intensity, showed that the total leaf area of soybean plants grown under the shading treatment was significantly lower than that of plants grown under full sunlight intensity. Furthermore, the size of young, intermediate, and fully mature leaves was 0.7–0.8 times those corresponding to leaves grown under full sunlight intensity, with the difference being significant. In addition, the lengths of the leaf petioles at different positions were significantly greater than those of leaf petioles in the full sunlight-intensity treatment (Table 5.1).

Table 5.1 Effects of light environment on soybean total leaf area, leaflet area, and petiole length

	Leaf area (cm^2)				Lobular area (cm^2)			Petiole length (cm)		
Treatment	Young leaf	Intermediate leaf	Mature leaf	Total leaf area	Young leaf	Intermediate leaf	Mature leaf	Young leaf	Intermediate leaf	Mature leaf
Shaded	40.93b	66.88b	58.88b	199.65b	13.64b	22.29b	19.63b	6.72a	12.78a	11.43a
Full sunlight	55.60a	83.97a	73.60a	239.73a	18.53a	27.98a	24.53a	5.58b	11.02b	10.67b

Note: Different lowercase letters within columns indicate significant differences at the 0.05 level

5.1.1.2 The Effects of Shading on the Number and Size of Epidermal Cells in Soybean Leaves

Shading treatment (400 μmol m^{-2} s^{-1}) significantly reduced the number and size of epidermal cells in young, intermediate, and mature soybean leaves (Fig. 5.1). Further, the effect of shading on the number of epidermal cells in young leaves (0.79) was significantly lower than that in mature leaves (0.84). Similarly, the effect of shading on the size of epidermal cells in young leaves (0.84) was also significantly lower than that in mature leaves (0.96). Furthermore, the thickness and cell length of epidermal cells at different leaf-development stages also decreased significantly under shading.

5.1.1.3 Effect of Shading on the Expression of Cell Division- and Cell Expansion-Related Genes in Soybean Leaves

The extent to which light environment affects cell division and expansion of leaf cells at different developmental stages varies. Specifically, the expression of nine genes involved in soybean leaf-cell division (i.e., ANT, AN3, GRF5, KLUH, UBP15, CYCD3, JAG, ROT4, and ARGOS), four genes involved in cell expansion (EXP10, TOR, ROT3, and SAUR19), five genes involved in both cell division and expansion (ARF2, EBP1, RGA, DA1, and EOD1), and two genes involved in leaf primordia (SWP) and shoot apical meristem (PPD2) development were detected under different light conditions (Fig. 5.2). The results showed that, compared with full sunlight-intensity treatment (1100 μmol m^{-2} s^{-1}), in axillary buds under shade

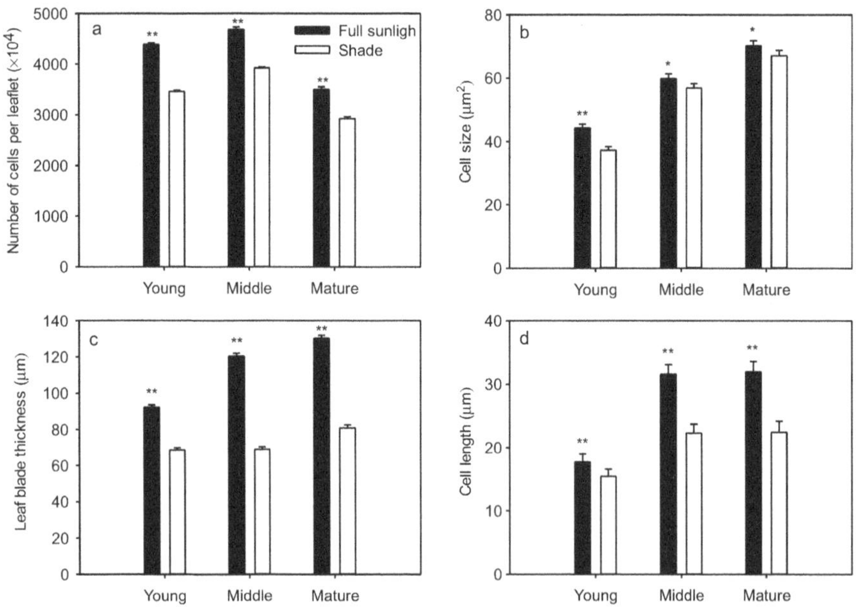

Fig. 5.1 The Effects of light environment on the number and size of palisade mesophyll cells in soybean leaves

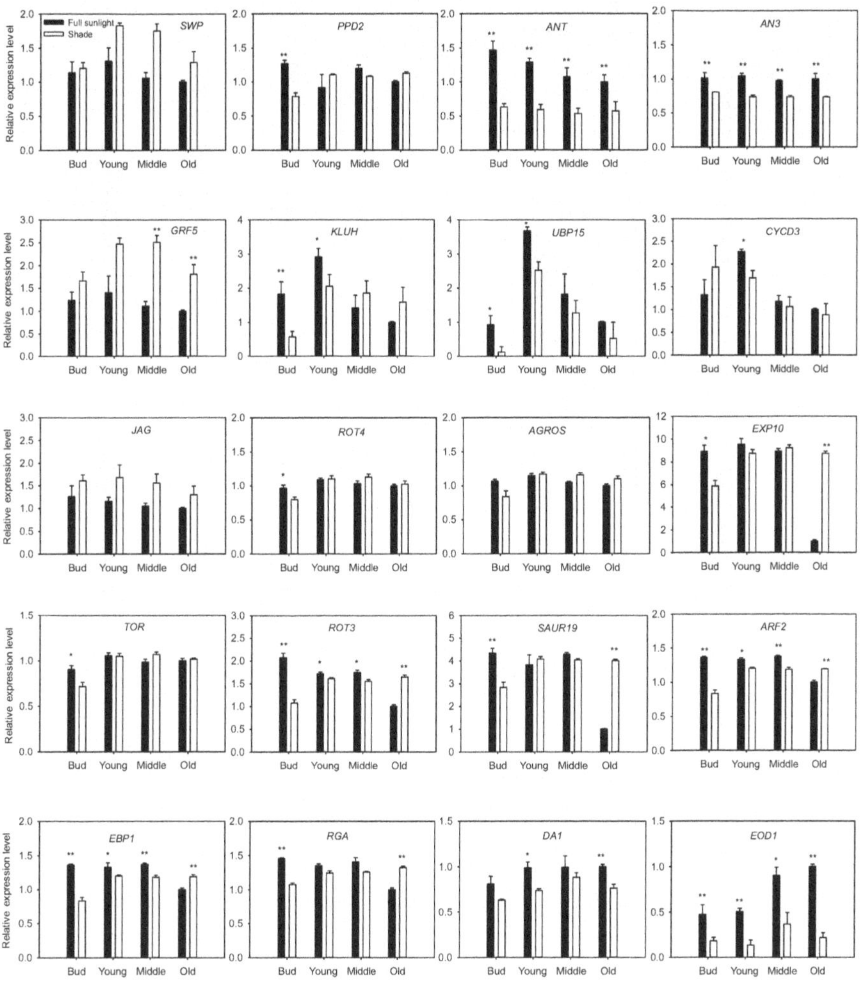

Fig. 5.2 Effects of light environment on the expression levels of genes related to cell division and expansion in soybean leaves

treatment (400 μmol m^{-2} s^{-1}), the expression of one shoot apical-meristem gene (PPD2), five cell division genes (ANT, AN3, KLUH, UBP15, and ROT4), four cell expansion genes (EXP10, TOR, ROT3, and SAUR19), and four genes involved in both cell division and expansion (ARF2, EBP1, RGA, and EOD1) were significantly downregulated. In turn, four cell division genes (ANT, AN3, KLUH, UBP15, and CYCD3), one cell expansion gene (ROT3), and four genes involved in both cell division and expansion (ARF2, EBP1, DA1, and EOD1) were significantly downregulated in young leaves. Meanwhile, two cell division genes (ANT and AN3), one cell expansion gene (ROT3), and three genes involved in both cell division and expansion (ARF2, EBP1, and EOD1) were significantly downregulated

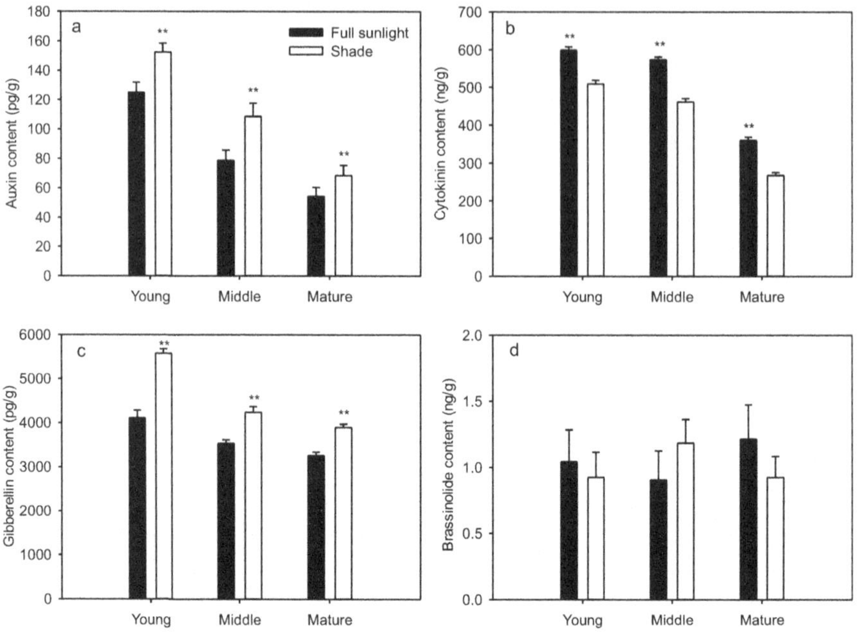

Fig. 5.3 Effects of the environment on hormone contents in soybean leaves

in intermediate leaves. Finally, in mature leaves, two cell division genes (ANT and AN3) were significantly downregulated, and three cell expansion genes (EXP10, ROT3, and SAUR19) and three genes involved in both cell division and expansion (ARF2, EBP1, and RGA) were significantly upregulated.

5.1.1.4 Effect of Shading on Hormone Contents in Soybean Leaves

Plant hormones are important determinants of leaf size. As shown in Fig. 5.3, shading (400 μmol m^{-2} s^{-1}) significantly increased leaf auxin and gibberellin (GA) contents, significantly decreased cytokinin content, and had no significant effect on brassinosteroid contents at the different leaf-development stages under study.

5.1.2 The Effect of Light Environment on Soybean Leaf Petioles and Stems

Supporting the leaf blades to capture light energy and maintaining leaf posture for maximum photosynthesis, leaf petioles are an important determinant of soybean plant structure. However, the response mechanism of petiole length and angle to changes in light intensity and quality in intercropping systems has been rarely reported. The mechanism whereby shading regulates soybean hormone metabolism and affects petiole growth remains unclear, resulting in a lack of a sound theoretical basis for shaping an ideal soybean plant structure for maize-soybean intercropping systems and making it difficult to improve high-yield cultivation techniques. Therefore, using field and pot experiments, as well as methods and techniques from plant physiology, biochemistry, molecular biology, and transcriptomics, we studied the mechanism responsible for the response of soybean petioles to changes in the intercropping light environment.

5.1.2.1 The Influence of Different Planting Systems on the Petiole Angle of Soybean Leaves

Analysis of 155 soybean materials showed that the petiole angle under maize-soybean intercropping (IC) was smaller than that under monoculture cropping (MC). Additionally, there was no significant difference in mean petiole angle under different planting systems or growth periods (Fig. 5.4a,b). In particular, at the V2 stage, the petiole angle under SC was 61.2°, whereas under RC, it was 29.7°. Similarly, during the V5 period, the petiole angle under SC was 62.1°, whereas under RC, it was 30.2°. On the other hand, the mean petiole angle under SC was 2.1 times that under RC during both the V2 and V5 stages. The variation range of petiole angles was larger under SC than under RC, and larger at V2 than at V5. Overall, the petiole angle was negatively correlated with the amount of light captured. Further, at V2, the value for the linear correlation coefficient (R^2) for the correlation between the petiole angle under SC and the amount of light captured was the smallest, at 0.38. In addition, the value of R^2 for the correlation between the petiole angle under RC and the amount of light captured was greater than that under SC, while during V5, the R^2 value was similar. Finally, at the V5 stage, the value of R^2 for the correlation between the petiole angle and the amount of light captured under SC reached its maximum, at 0.80 (Fig. 5.4c–f).

5.1.2.2 Effects of Light Intensity and Quality on Soybean Stems

Soybean plants respond to shaded light environments (i.e., weaker light intensity) and a decreased ratio of red to far-red light (R:FR) by reducing the stem angle, increasing stem diameter, and elongating the stem. Among these responses, decreasing the stem angle and elongating the stem are the primary characteristics of stem responses to a decreased R:FR ratio, particularly in the tender stem tissues (Fig. 5.5).

5.1.2.3 The Effects of Light Intensity and Quality on Hormone Contents in Soybean Petioles

Regardless of light intensity, treatment with a low R:FR ratio significantly increased indole-3-acetic acid (IAA) and indole-3-butyric acid (IBA) contents in soybean leaves, especially IAA, which was 2.67 and 2.94 times that of the treatment with high R:FR ratio (Fig. 5.6). The content of the intermediate product of IAA synthesis, indole-3-pyruvic acid (IPA), was also significantly increased under treatment with low R:FR ratio, especially under low light intensity, reaching 4.73 times that of the treatment with high R:FR ratio. Additionally, the decrease in light intensity significantly induced an increase in the content of two types of gibberellins, namely, GA3

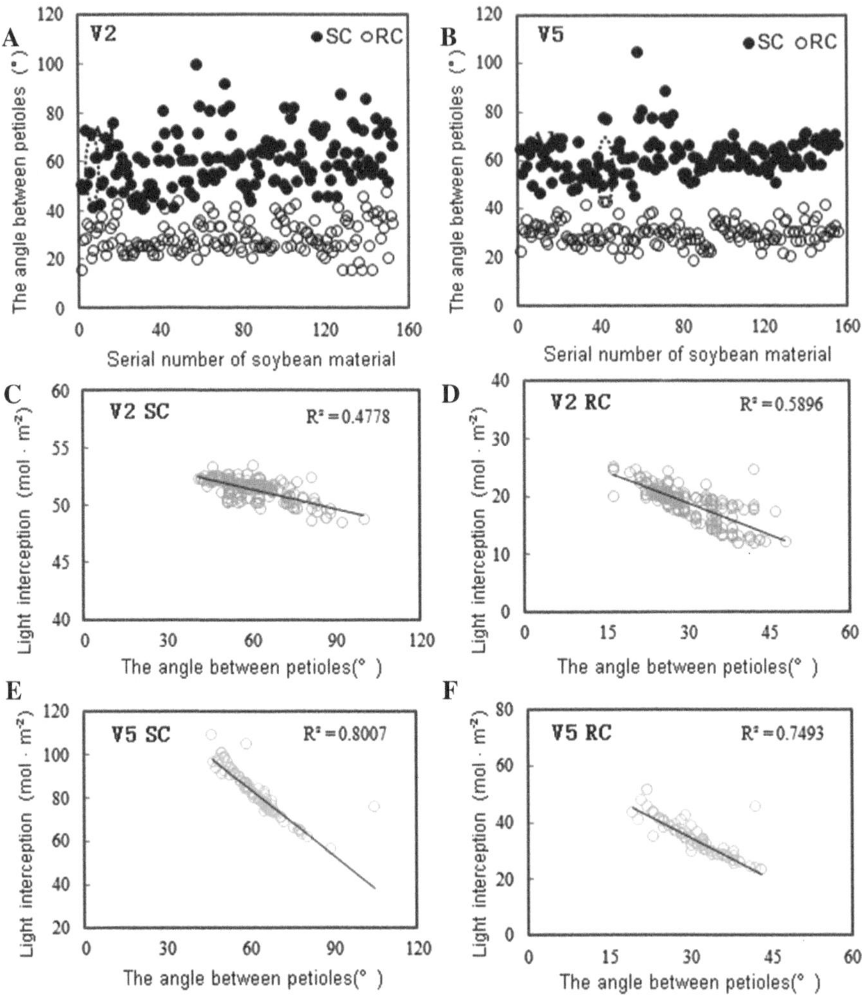

Fig. 5.4 Effect of planting system on soybean stem angle at the seedling stage and its relationship with light interception. Note: MC stands for monoculture and IC stands for intercropping. **a** and **b** show the stem angle of soybean at V2 and V5 stages, respectively. **c**, **d** and **e**, **f** show the linear relationship between stem angle and light interception at V2 and V5 soybean growth stages

and GA4, and treatment with low R:FR ratio significantly increased their contents, compared with the treatment with high R:FR ratio, but this increase was not as great as that caused by the decrease in light intensity.

The trends in hormone content changes, including auxin, gibberellin, and cytokinin, in the leaf cushions were similar to those in the leaf blades, but the effect of light intensity on IBA was not significant. After the weakening of light intensity, IAA and gibberellin GA3 and GA4 contents increased, and the change in gibberellin GA4 was the largest, whereas the change in IAA was greater under treatment with low R:FR light ratio.

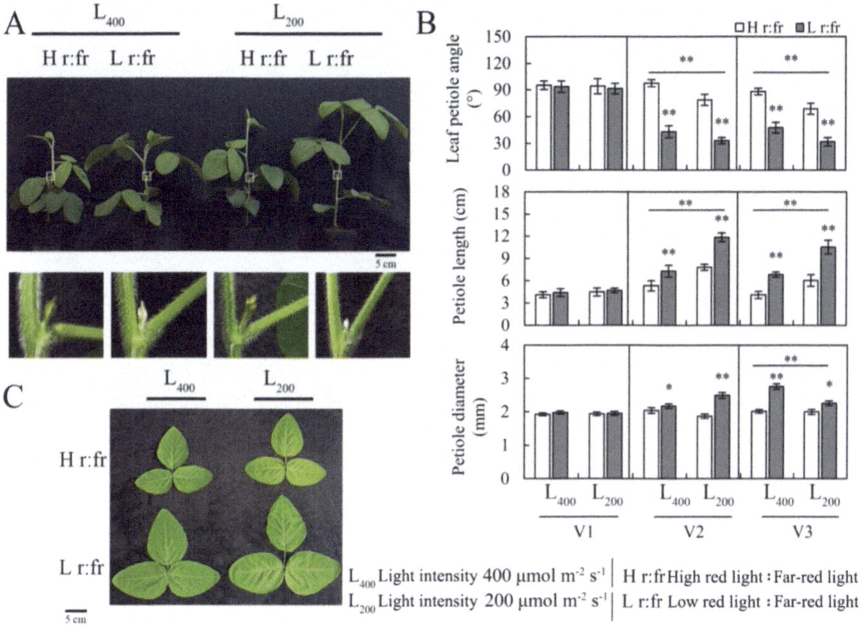

Fig. 5.5 Response of soybean stem and other related traits to light intensity and R:FR ratio. Note: Plants were grown under two light intensities and two R:FR ratios (light intensity: photon flux density of 400 μmol m^{-2} s^{-1}, L400, and 200 μmol m^{-2} s^{-1}, L200; R:FR represents the ratio of red light to far-red light, H R:FR = 2.08, L F:FR = 0.52. Asterisks indicate significant differences between treatments: $(0.001 \leq p \leq 0.01)$*, **($p \leq 0.001$). (**a**) Whole plant morphology; (**b**) V2 trifoliate leaf area; (**c**) Leaf angle; (**d**) Whole plant leaf area; (**e**) Leaf angle, stem length, and diameter; (**f**) Internode length; (**g**) Plant height

5.1.2.4 Effects of Light Intensity and Quality on the Expression of Genes Related to Auxin Synthesis and Signaling in the Pulvinus

Except for GmIAA19a, genes related to auxin synthesis and signaling were significantly regulated by light intensity and quality. Genes involved in auxin synthesis, including GmTAA1d and Gmyucca4/Gmyucca11, which encode tryptophan aminotransferase and flavin monooxygenase, respectively, were significantly upregulated under the low R:FR ratio treatment, whereas their relative expression levels significantly decreased under low light intensity. Genes related to auxin signaling, such as GmTIR1a, GmARF11, GmARF8, GmSAUR19, and GmGH3, showed similar expression patterns to that of GmTAA1d under low R:FR ratio and low light intensity treatments. However, the expression of GmIAA19a, a gene encoding an auxin response protein, was significantly suppressed under the low R: FR ratio treatment (Fig. 5.7).

The GA 20-oxidase gene GmGA20ox1 and GA 3-oxidase gene GmGA3ox1 are key genes in GA synthesis. Decreased light intensity significantly promoted their upregulation. High R:FR ratio values upregulated their expression compared with low R:FR ratio treatment and the same change was observed for the GA receptor

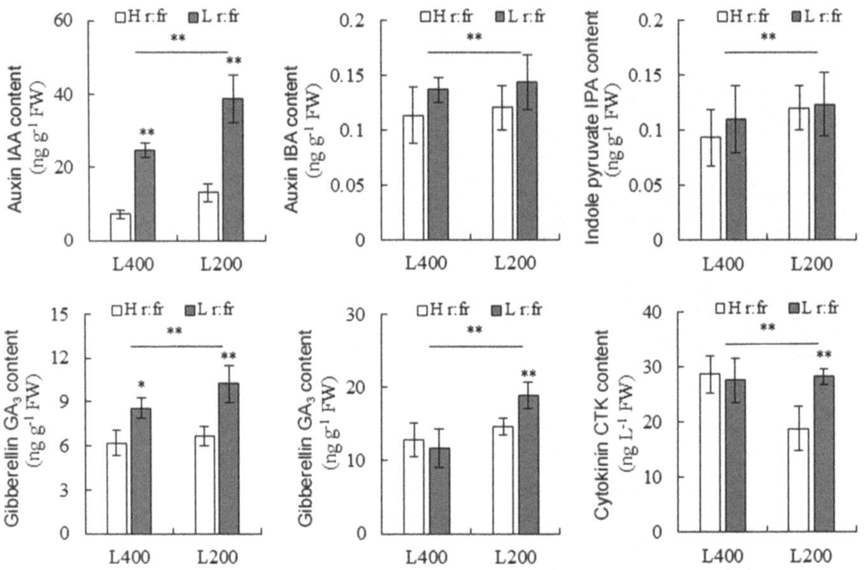

Fig. 5.6 Effects of light intensity and quality on hormone contents in the leaf axil. Note: * indicates significant differences between treatments ($0.01 \leq p \leq 0.05$), and ** indicates highly significant differences between treatments ($p \leq 0.01$)

gene, GmGID1. In turn, GA 2-oxidase gene GmGA2ox1 has the function of inhibiting active GA, and its expression was downregulated under low R:FR ratio treatment, as was also the expression of GmRGA1 and GmGAI1, which encode the DELLAs protein family (Fig. 5.8).

The expression of the cytokinin synthase gene GmIPT1, the receptor gene GmCRE1, and the cytokinin-phosphotransfer protein gene GmAHP1 were significantly inhibited under weakened light intensity and low R:FR ratio treatment. In contrast, the expression of the response regulator gene GmARR1 was upregulated (Fig. 5.9).

5.1.3 Effects of Light Environment on Soybean Stem Growth and Lodging Resistance

In maize-soybean strip intercropping systems, soybeans are shaded, which increases plant height in an attempt to capture more light but decreases stem strength, leading to stem sprawling and lodging, which in turn results in reduced soybean yield and mechanical harvest efficiency. To clarify the physiological mechanism of soybean stem resistance to lodging in response to shading stress, in this study we selected two soybean varieties, ND12 and NN99-6, and set up four shading treatment levels: S0 (no shading, normal light), S1 (light shading, 43% decrease in light intensity), S2 (moderate shading, 58% decrease in light intensity), and S3 (strong shading, 73% decrease in light intensity).

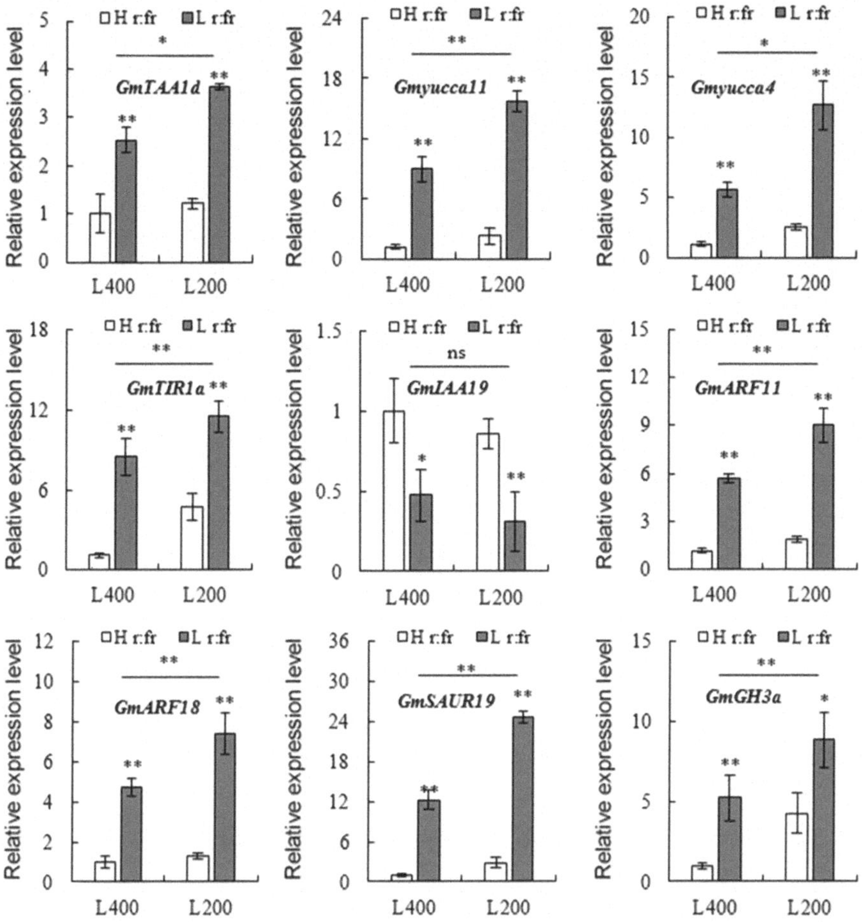

Fig. 5.7 Expression of genes involved in auxin synthesis and signaling in the pulvinus. Note: The asterisk (*) indicates a significant difference between treatments ($0.01 \le p \le 0.05$), and the double asterisk (**) indicates a highly significant difference between treatments ($p \le 0.01$)

5.1.3.1 Effects of Shading on Soybean Stem Lodging Resistance Phenotype

Soybean plant height (i.e., main stem length) is most sensitive to light environment. When subjected to shading, the plant height of both soybean varieties at the early flowering stage increased significantly with increasing shading level (Fig. 5.10), and neither stem thickness nor bending resistance changed significantly under shading conditions. However, under moderate to strong shading, stems of both varieties became thinner, stem thickness and bending resistance decreased significantly, and the value of the lodging resistance index decreased, thereby increasing the risk of lodging.

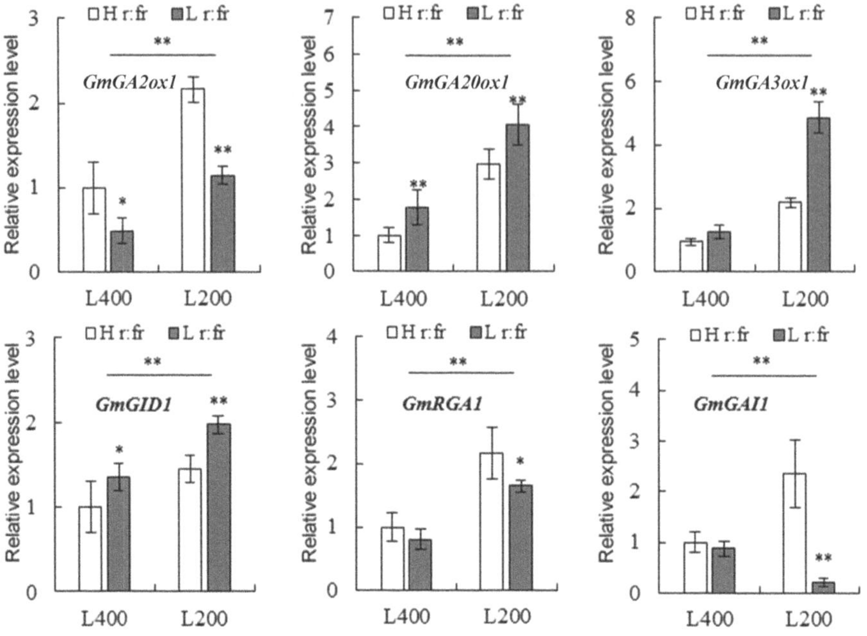

Fig. 5.8 Effects of light intensity and quality on the synthesis of indole-3-acetic acid and the expression of signaling genes in leaf nodes. Note: The asterisk (*) indicates a significant difference between treatments ($0.01 \leq p \leq 0.05$), and the double asterisk (**) indicates a highly significant difference between treatments ($p \leq 0.01$)

5.1.3.2 Effects of Shading on the Anatomical Structure of Soybean Stems

The area of lignified (xylem) tissue and pith tissue in soybean stems significantly decreased with increasing shading intensity (Table 5.2). Regarding the proportions of lignified and pith tissues in each case, mild shading had no significant effect, whereas moderate and strong shading significantly increased the proportion of pith tissue. Further, under mild shading, the proportion of lignified tissue in the two varieties tested increased by 3.44% and 1.04%, respectively, compared with full sunlight-intensity treatment, but as shading intensity increased, the proportion of lignified tissue significantly decreased.

5.1.3.3 Effects of Shading on Nonstructural Carbohydrates in Soybean Stems

Nonstructural carbohydrate contents in soybean stems were significantly higher under moderate and strong shading than those under full sunlight (Fig. 5.11). Similarly, soluble sugars also increased with increasing shading intensity. Particularly, sucrose contents in both varieties were 36.10% and 38.16% higher under strong shading (S3) than under full sunlight (S0), whereas the differences were not significant under S1 or S2, except in variety E93. In turn, starch accumulation in the

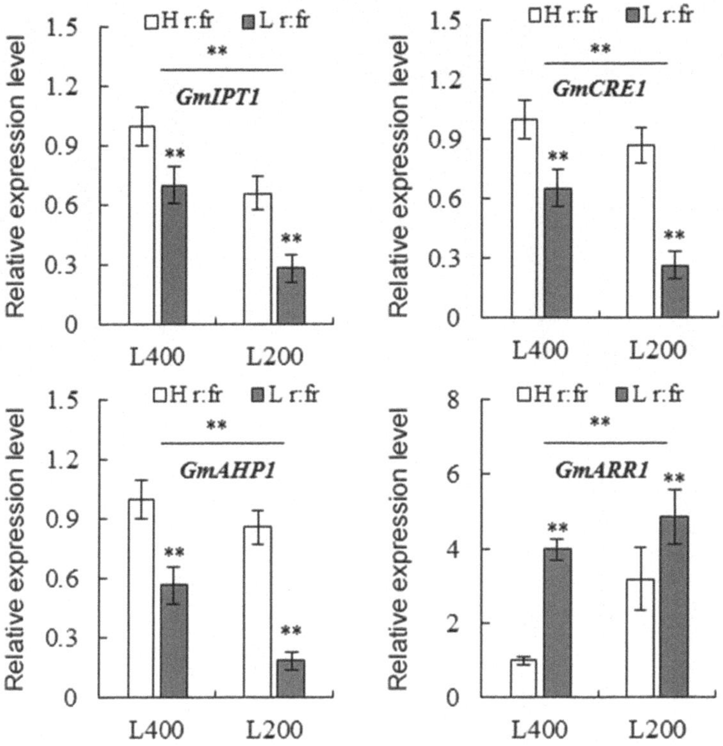

Fig. 5.9 Effects of light intensity and quality on the expression of cytokinin-signaling genes in leaf nodes

stems ranked as follows: S3 > S2 > S1. Under mild shading, soluble sugar and sucrose contents in the stems were not significantly different from those in full sunlight, whereas they were significantly lower than those in the control group, indicating that the rate of conversion of soluble sugar and sucrose to structural carbohydrates was higher.

5.1.3.4 The Effect of Shading on Structural Carbohydrates in Soybean Stems

Lignin content in the stems of both soybean varieties tested was highest under the light shading treatment and decreased linearly with further increasing shading intensity. Additionally, cellulose was more sensitive to shading than lignin, and its content in the stems decreased significantly under shading, although this decrease was not significant with further shading (Fig. 5.12).

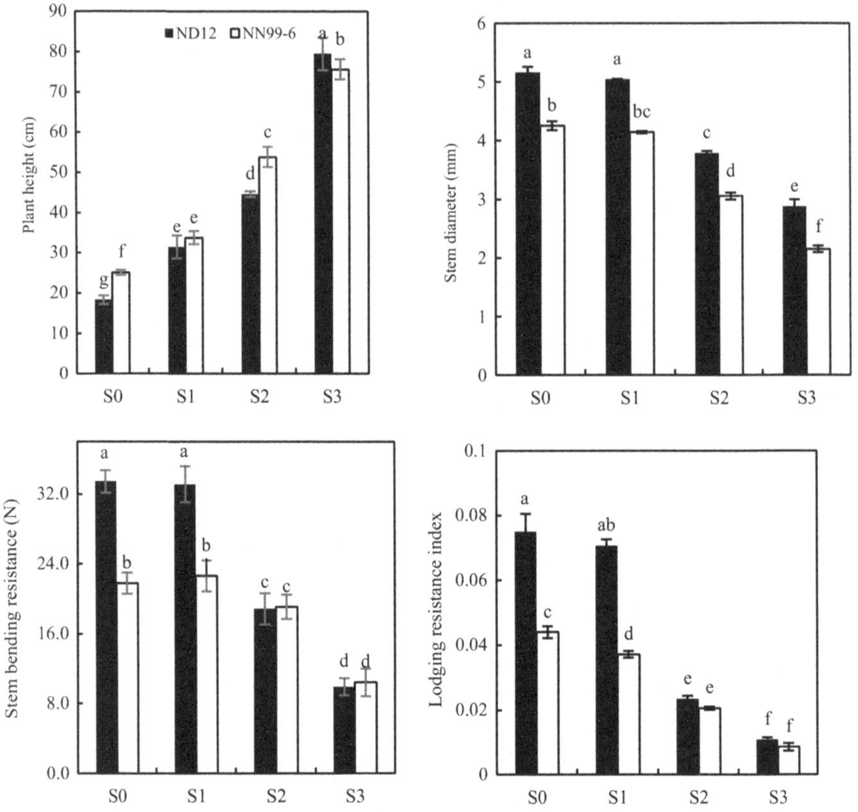

Fig. 5.10 Effects of shading on soybean stem lodging resistance phenotype at the seedling stage

Table 5.2 Effect of shading on lignified and pith areas in soybean stems

Variety	Handle	Xylem area (mm^2)	Pith area (mm^2)
ND12	S0	13.493 ± 0.013 a	5.587 ± 0.008 c
	S1	12.293 ± 0.009 b	4.681 ± 0.005 b
	S2	6.389 ± 0.013 e	4.035 ± 0.007 e
	S3	3.614 ± 0.005 g	2.584 ± 0.004 g
NN99-6	S0	10.576 ± 0.020 c	6.252 ± 0.12 a
	S1	8.457 ± 0.005 d	5.115 ± 0.008 d
	S2	4.363 ± 0.019 f	2.769 ± 0.002 f
	S3	1.816 ± 0.011 h	1.599 ± 0.003 h

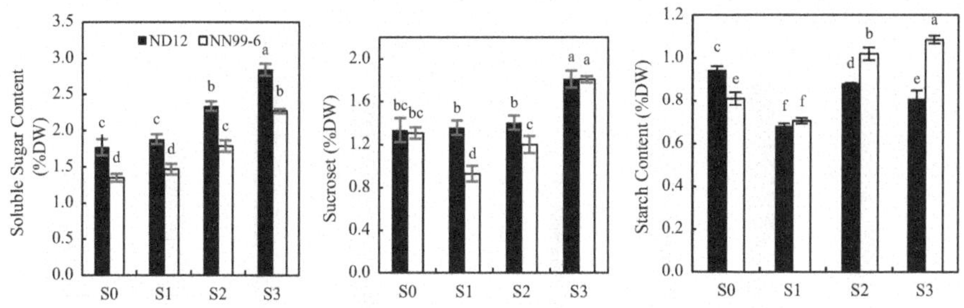

Fig. 5.11 Nonstructural carbohydrate contents in soybean stems under different levels of shading

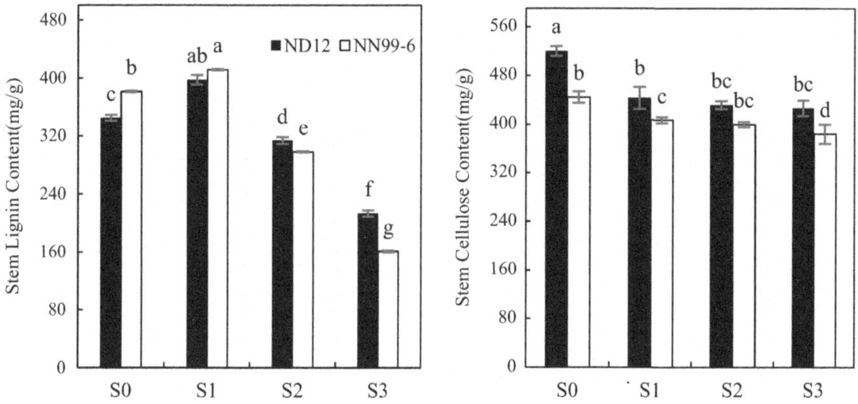

Fig. 5.12 Cellulose and lignin contents in the stems of two soybean varieties under different shading levels

5.1.3.5 The Impact of Shading on the Activity of Key Enzymes Involved in Lignin Synthesis in Soybean Stems

No significant difference in PAL activity was found between the two soybean varieties or between light shading (S1) and full sunlight. However, as shading increased, PAL activity decreased significantly, ranking in the following order: S1 > S2 > S3. Further, light shading (S1) significantly increased 4CL activity in the stem of both soybean varieties by 6.89% and 14.12%, respectively, although this activity decreased significantly with further shading. In turn, peroxidase (POD) activity differed between the two varieties, as shading significantly reduced POD activity in the stem of ND12, although there was no significant difference between light and moderate shading treatments. Light shading increased POD activity in NN99-6 but further shading lowered it. Lastly, except for strong shading, there was no significant difference in cinnamyl alcohol dehydrogenase (CAD) activity among shade treatments for ND12. In contrast, CAD activity in NN9906 was significantly higher under light and moderate shading than under full sunlight (Fig. 5.13).

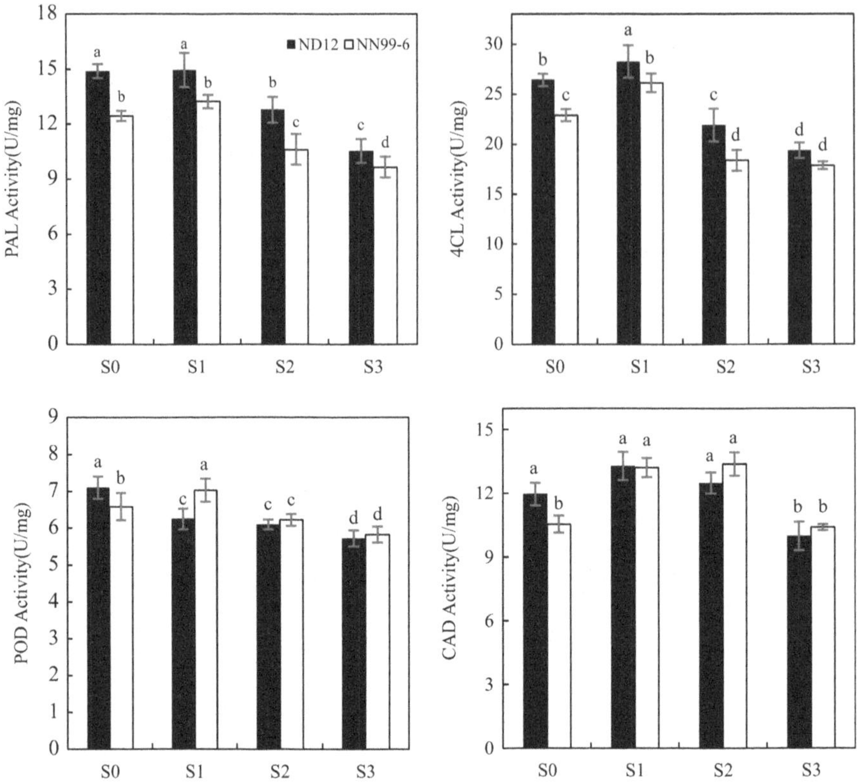

Fig. 5.13 Shading effects on the activities of key lignin synthesis-related enzymes in soybean stems

5.1.3.6 Shading Effects on the Expression of Key Lignin Biosynthesis-Related Genes in Soybean Stems

The expression levels of the LAC, C3H, and 4CL genes in the ND12 soybean variety showed an increasing followed by a decreasing trend as shading increased. Specifically, under mild shading (S1), the three genes were upregulated by 23.53%, 31%, and 2.8%, respectively, compared with the control group. Moreover, under the moderate shade treatment (S2), the LAC and C3H genes continued to be upregulated, but to a lesser extent than that under S1. Further, under S1, the C3H gene of the NN99-6 variety was significantly upregulated, whereas the LAC gene did not show any significant changes. However, further shading (S2 and S3) led to the downregulation of the expression levels of LAC, C3H, and 4CL genes in both varieties compared with those under full sunlight conditions.

On the other hand, the expression levels of the PAL and C4H genes in both soybean varieties under mild shading (S1) were not significantly different from those under full sunlight (S0). However, as shading increased to moderate (S2) and strong

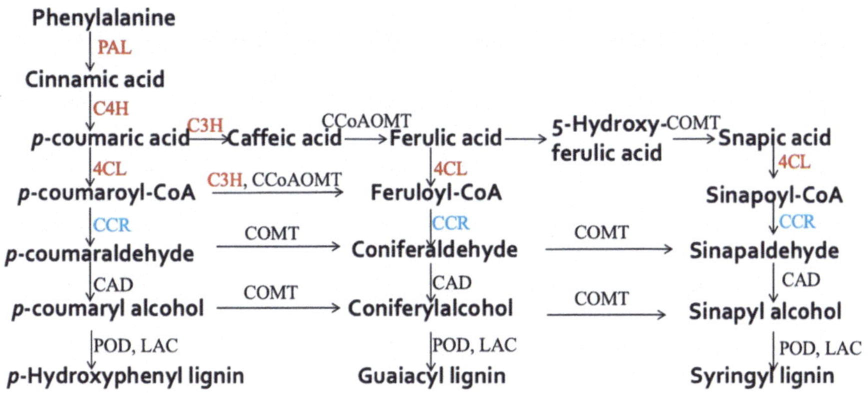

Fig. 5.14 Mild shading effects on the expression of genes involved in lignin biosynthesis in soybean stems

(S3), the expression of the PAL and C4H genes in the ND12 variety gradually decreased, whereas, in the NN99-6 variety, a sudden decrease was observed, suggesting that NN99-6 was more sensitive to the shading treatment. Additionally, the changing trend of the PAL gene expression level was consistent with the corresponding enzyme activity (Fig. 5.14).

The changing trends in the expression of the CAD and COMT genes in both varieties were similar under different shading intensities. Specifically, as shading increased, the expression of both genes gradually decreased. However, the expression level of the COMT gene in ND12 was higher than that in NN99-6 across treatments. Interestingly there was no significant difference in the expression levels of these two genes between S1 and S0 in ND12.

The expression of the CCoAOMT gene in both varieties was downregulated under shading treatments. Specifically, the expression of the CCoAOMT gene in ND12 gradually decreased with increasing shading, whereas in NN99-6, it first decreased, then increased, and then decreased again. Moreover, the expression level of this gene in NN99-6 under S1 was significantly lower than that in ND12. Lastly, the expression of the CCR gene showed a similar trend in both varieties, decreasing with increasing shading. However, unlike those of other genes, the expression level of the CCR gene in NN99-6 was higher than that in ND12 across treatments (Table 5.3).

Note: Red, blue, and black colors indicate upregulation, downregulation, and no significant change under mild shading, respectively.

Table 5.3 Z-scores of the expression levels of lignin biosynthesis-related genes in two soybean varieties under mild (S1), moderate (S2), and strong (S3) shading treatments

	ND12			NN99-6		
Gene	S1	S2	S3	S1	S2	S3
PAL	1.16	0.63	−0.86	1.11	−0.69	−1.34
C4H	1.35	0.35	−0.46	1.07	−1.10	−1.22
C3H	1.10	1.01	−1.16	0.75	−0.40	−1.30
4CL	1.05	0.55	−0.92	0.79	0.27	−1.75
CCoAOMT	1.21	0.49	−1.60	0.29	0.69	−1.08
COMT	1.59	−0.06	−0.71	0.86	−0.20	−1.47
CCR	0.32	−0.34	−1.61	1.36	0.94	−0.68
CAD	1.50	0.15	−0.88	0.93	−0.29	−1.41
LAC	1.55	0.88	−1.24	0.22	−0.37	−1.04

Note: Z-scores were calculated using Microsoft Excel 2016 software. $z = (X-\mu)/\sigma$, where $X-\mu$ is the deviation from the mean and σ is the standard deviation

5.2 Mechanism of Adaptation in Shade-Tolerant and Lodging-Resistant Soybean Genotypes to the Light Environment in the Maize-Soybean Strip Intercropping System

5.2.1 The Ideal Plant Type and Screening for Shade Tolerance and Lodging Resistance in Soybean

Due to the shading effect of the taller maize plants on soybean during their co-growth period in the strip intercropping system, the light environment is directly changed, affecting the accumulation and distribution of photosynthetic products in soybean leaves and inhibiting plant growth. In turn, soybean plants try to adapt to such changes by altering their morphology to capture more light energy and continue normal growth. Thus, under weak light conditions, soybean stems become slender and weak as internodes elongate excessively and become thinner, changing from an upright to a semi-upright or trailing position leading to lodging, and ultimately, affecting yield.

5.2.1.1 Requirements for the Agronomic Traits of Soybean in Strip Intercropping

Under strip intercropping, maize mainly shades soybean at the seedling stage, causing its internodes to elongate, plant height to increase, and the stems to become weak and slender. In severe cases, soybean plants may become sprawling, leading to serious lodging and reduction of grain yield and quality.

Using the shade tolerance coefficient for each trait, principal component analysis and cluster analysis were performed on 17 traits of 139 soybean families and, as a result, the soybean germplasm tested was classified into four types: strongly shade tolerant, moderately shade tolerant, weakly tolerant, and non-shade tolerant. Strongly shade-tolerant genotypes showed the lowest increases in plant height, stem thinning, and decreased branching under shading at the seedling stage. They also exhibited good recovery upon full sunlight restoration, with the greatest decrease in plant height and the least decrease in effective branching, and the greatest increase in branching-pod number, single plant-pod number, single plant-grain number, and single plant-grain weight at maturity. In contrast, non-shade-tolerant genotypes showed the greatest increase in plant height and stem thinning, and decreased branching at the seedling stage; furthermore, they showed the greatest increase in plant height and the greatest decrease in effective branching at maturity. Ultimately, trait analysis showed that shade-tolerant soybeans were characterized by greater plant height at the seedling stage but shorter at maturity, thicker stems, a larger number of branches, effective branches, branching pods, and pods per plant, and a higher grain weight per plant.

Analysis of variance was performed on plant height, main stem node number, and internode length of 51 soybean genotypes under full sunlight and shading. Plant height and internode length under full sunlight were significantly or highly significantly lower than under shading, whereas the main stem-node number was significantly higher under full sunlight than under shading. Further, the number of main stem nodes in all materials was significantly lower under shading than under full sunlight, whereas plant height increased significantly, indicating that soybean genotypes mainly increased their plant height by increasing internode length under shading conditions (Fig. 5.15). An analysis of the ratio of internode length to plant height under different light conditions showed that the trends of change in the values of this ratio were consistent under both light conditions (Fig. 5.16). Specifically, the ratio of first internode length to plant height was the highest, and there was no significant difference in the ratio values of the second to fifth internode length to plant height ($P < 0.05$). However, the ratio values of the sixth to eleventh internode length to plant height were significantly higher under shading than under full

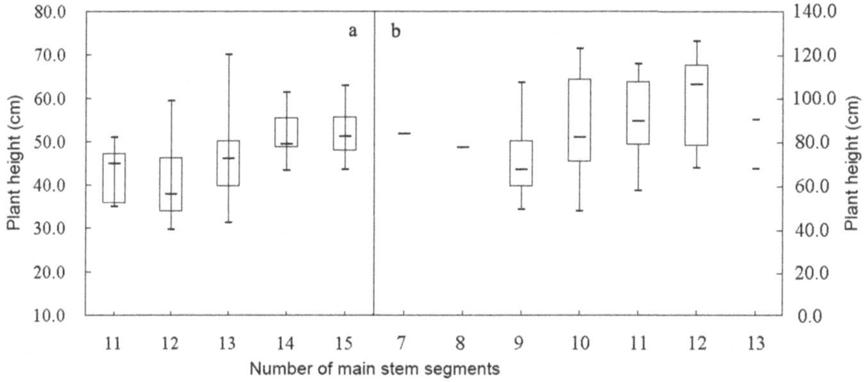

Fig. 5.15 Box plots showing the relationship between main stem nodes and plant height under different light intensity conditions: (**a**) full sunlight; (**b**) shaded environment

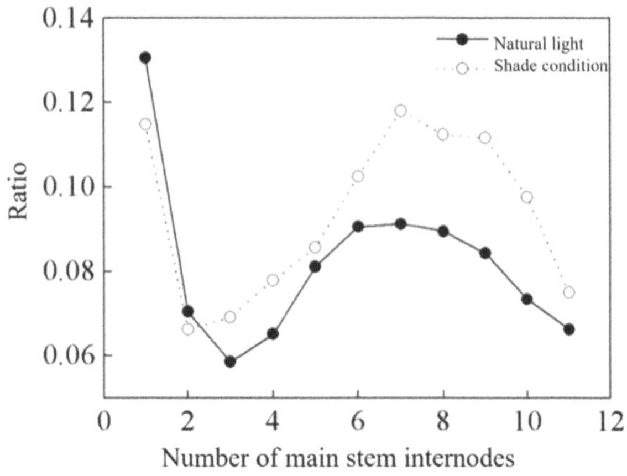

Fig. 5.16 Changes in the ratio of soybean stem-internode length to plant height under different light conditions

sunlight ($P < 0.001$), indicating that internode length had a greater impact on plant height in the later stage of soybean growth under shading compared with full sunlight.

Altogether, our data clearly indicated that the suitable shade-tolerant soybean phenotype for maize-soybean strip intercropping systems should have the following morphological characteristics: under shade conditions, the reduction in the number of main stem nodes, elongation of internodes, increase in plant height, decrease in stem thickness, and decrease in branching number at the seedling stage should be the lowest possible. Additionally, at maturity, the reduction in plant height should be the greatest possible, whereas the reduction in effective branching number should be as low as possible, and the increase in the number of branching pods, pods per plant, grains per plant, and grain weight per plant should be as large as possible.

5.2.1.2 Requirements for Agronomic Traits of Soybean for Strip Intercropping

Under strip intercropping, soybean plants grow normally at the early stages of growth but usually suffer from shading stress caused by maize shading after the beginning of flowering, resulting in lodging and a relatively high rate of flower and pod shedding, which may severely affect soybean yield. By studying the effects of strip intercropping on soybean through two planting methods, namely monoculture planting and strip intercropping, we found that, compared with monoculture, strip intercropping significantly increased soybean plant height, while reducing stem diameter, the number of main stem nodes, branches, and leaves, while concomitantly shortening the duration of maximum leaf area. Meanwhile, above and belowground dry matter accumulation decreased, and the number, weight, and quantity of soybean pods on the effective branches were all significantly reduced.

Under strip intercropping conditions, soybean yield per plant was significantly and positively correlated with stem thickness and growth period, followed by plant height, number of main stems, number of branches, height of the lowest pod, nutritional growth period, 100-grain weight, number of pods per plant, and number of seeds per plant. The gray correlation ranking of yield with other major traits was the following: stem thickness > growth period > nutritional growth period > number of pods per plant >100-grain weight > height of the lowest pod > number of seeds per plant > number of branches > number of main stems > plant height > reproductive growth period. Trait contribution to yield varied in importance; therefore, when selecting soybean varieties for strip intercropping, it is important to focus on materials with relatively thick stems, longer early and vegetative growth periods, larger number of pods per plant, and larger 100-grain weight (Table 5.4).

Using principal component and cluster analysis on 15 traits of 55 soybean varieties under strip intercropping, the tested soybean materials were classified into three types. The first type mainly included 15 varieties that performed better among all varieties tested in terms of yield components, such as pod number, single seed number, seed weight per plant, growth period and stem thickness, and higher yield under strip intercropping. For example, varieties such as Qihuang 35, Fendou 79, and Jidou 15, which grew thick stems, had high pod and seed numbers, high seed weight, showed a moderate growth period, and were suitable for planting in the Huang-Huai-Hai region. The second type comprised 32 varieties that did not significantly differ in stem thickness, main stem number, branch number, lowest branch height, pod height, or average growth period, compared with the first type, but had higher plant height, which makes them more prone to lodging and indirectly affects yield. The average values of single plant pod number, single plant seed number, and single plant seed weight also differed significantly from those of the first type. Lastly, the third type mainly included eight varieties with medium plant height, thin stem, short growth period, were early maturing varieties, and showed relatively low yield components, such as pod number, seed number, and seed weight per plant. These varieties are not suitable for intercropping in the Huang-Huai-Hai region.

5.2.1.3 Screening for Shade Tolerance and Lodging Resistance in Soybean

As a short-height crop, soybeans are shaded by high-stalk crops, and various morphological traits, yield, and quality undergo significant changes. The variation coefficient of most traits under shading was higher than that under normal light conditions, and different varieties responded differently to shading. Through correlation analysis of 24 morphological and physiological indicators of 19 soybean materials, we found that there were different degrees of correlation between each indicator, and the correlation between some traits reached a significant or highly significant level. In addition, the role of each individual indicator in shade tolerance of different soybean materials also differed, indicating that soybean shade tolerance is a complex comprehensive trait, and no single indicator alone can accurately estimate it. Hence, the selection of correlated indicator traits is crucial for the reliable evaluation of shade tolerance.

Table 5.4 Analysis of grey correlation degree between soybean yield and major agronomic traits

	X1	X2	X3	X4	X5	X6	X7	X8	X9	X10	X11
Correlation	0.4065	0.4910	0.4176	0.4241	0.4281	0.4825	0.4459	0.3333	0.4308	0.4376	0.4277
Ranking	10	1	9	8	6	2	3	11	5	4	7

Note: Plant height (X1), stem diameter (X2), number of main stem nodes (X3), branch number (X4), bottom pod height (X5), growth period (X6), vegetative growth period (X7), reproductive growth period (X8), 100-grain weight (X9), number of pods per plant (X10), and number of grains per plant (X11)

Table 5.5 Coefficients and contribution rates of each composite index

	Main ingredient					
	CI_1	CI_2	CI_3	CI_4	CI_5	CI_6
Eigenvalues	7.319	5.097	2.880	2.571	1.925	1.244
Contribution rate	30.495	21.239	12.000	10.713	8.021	5.184
Cumulative contribution rate	30.495	51.734	63.734	74.447	82.468	87.652

Table 5.6 Performance of different traits in the clustering results of different shade-tolerant types

	Full sunlight				Shade			
Type	*LDW*	g_s	*PH*	F_m	*LDW*	g_s	*PH*	F_m
Strong shade tolerance	0.46	0.08	12.53	1486.08	0.43	0.12	42.01	1769.04
Moderate shade tolerance	0.60	0.16	11.74	1594.58	0.49	0.13	38.59	1856.78
Shade intolerance	0.54	0.15	13.39	1487.80	0.25	0.07	34.89	1747.98

Note: Leaf dry weight (LDW), Stomatal conductance (g_s), Plant height (PH), Maximum fluorescence yield in the dark (Fm)

Principal component analysis can transform several originally correlated individual indicators into a few new and independent comprehensive indicators. Thus, by conducting principal component analysis on the shade tolerance coefficients of 24 individual indicators, the cumulative contribution rate of the first six comprehensive evaluation indicators reached 87.65%, representing most of the information carried by the original indicators (Table 5.5). Then, using the membership function method to calculate the comprehensive shade tolerance-evaluation value (D) and conducting cluster analysis on it, soybean materials were classified into strongly tolerant, moderately tolerant, and intolerant types (Table 5.6).

To clarify the relationship between the indicators and the shade tolerance of the test materials, it is convenient to select reliable shade tolerance-identification indicators and establish a mathematical model that can be used to evaluate soybean seedling shade tolerance. Therefore, taking the shade tolerance comprehensive D value as the dependent variable and the shade tolerance coefficient of each individual indicator as the independent variable, the optimal regression equation was established using the stepwise regression method: $D = -0.301 + 0.419\text{LDW} + 0.169\text{Gs} + 0.031\text{PH} + 0.255\text{Fm}$ ($R^2 = 0.959$, $P = 0.0001$). This regression equation can predict the shade tolerance of soybean seedlings with good effect and high accuracy.

Table 5.7 Effect of corn shading on photosynthetic parameters of soybean leaves

Variety	Treatment	Pn ($\mu mol \cdot m^{-2} \cdot s^{-1}$)	gs ($\mu mol \cdot m^{-2} \cdot s^{-1}$)	Ci ($\mu mol \cdot m^{-2}$)	Tr ($mmol \cdot m^{-2} \cdot s^{-1}$)
Nandou 12	A1	15.27 ± 0.52c	0.27 ± 0.02c	251.67 ± 20.17a	3.06 ± 0.22c
	A2	17.03 ± 0.76b	0.33 ± 0.01b	248.67 ± 8.06a	4.69 ± 0.14b
	CK	27.91 ± 0.78a	0.46 ± 0.00a	238.67 ± 3.30a	5.17 ± 0.06a
GuiXia 3	A1	14.97 ± 0.39b	0.24 ± 0.01c	258.33 ± 11.44a	3.08 ± 0.36c
	A2	16.10 ± 2.19b	0.29 ± 0.03b	239.00 ± 12.08a	4.87 ± 0.14b
	CK	26.40 ± 1.70a	0.44 ± 0.02a	235.00 ± 1.41a	5.50 ± 0.02a

Note: A1, A2, and CK represent 1 row of corn intercropped with 1 row of soybean, 2 rows of corn intercropped with 2 rows of soybean, and soybean monoculture, respectively. Pn, net photosynthetic rate, g_s, stomatal conductance, Tr, transpiration rate. Different lowercase letters in the same column indicate significant differences ($P < 0.05$)

5.2.2 Physiological Mechanisms of Shade Tolerance in Soybean Leaves of Different Genotypes

5.2.2.1 Photosynthetic Characteristics of Soybean Leaves with Different Shade Tolerance

Compared with monoculture, intercropping with shading significantly reduced Pn, gs, and Tr of soybean leaves, but had no significant effect on intercellular CO_2 concentration. Under the two intercropping treatments, Pn of the shade-tolerant variety Nandou 12 was reduced by 45.2% and 38.9%, respectively, compared with monoculture cropping, whereas that of the non-shade-tolerant variety Guixia 3 was reduced by 43.4% and 39%, respectively. However, Pn and gs of Nandou 12 were significantly higher than those of Guixia 3 (Table 5.7).

5.2.2.2 Characteristics of Chlorophyll Fluorescence in Soybean Genotypes Differing in Shade Tolerance

The potential activity of PSII and the initial light energy conversion efficiency of PSII in soybean leaves under shading were significantly higher than those in monoculture, whereas the actual photochemical efficiency of PSII was significantly lower than that in monoculture (Table 5.8). Additionally, there was no significant difference in the value of the non-photochemical fluorescence quenching coefficient and the effective photochemical quantum yield of PSII among treatments. Furthermore, the potential activity of PSII, the initial light energy conversion efficiency of PSII, and the actual photochemical efficiency of PSII in the leaves of Nan 12 were all higher than those in Gui Xia 3 under the same treatment conditions.

Table 5.8 The effect of maize shading on chlorophyll fluorescence parameters in soybean leaves

Variety	Treatment	Potential activity of PS II	PSII primary light energy conversion efficiency	Non-photochemical fluorescence quenching coefficient	Effective photochemical quantum yield of PS II	Actual photochemical efficiency of PS II
NanDou 12	A1	4.53 ± 0.01a	0.82 ± 0.01a	2.34 ± 0.06a	0.58 ± 0.00a	0.26 ± 0.01b
	A2	4.32 ± 0.20a	0.81 ± 0.01a	2.43 ± 0.25a	0.56 ± 0.02a	0.24 ± 0.01b
	CK	3.70 ± 0.30b	0.78 ± 0.01b	2.21 ± 0.15a	0.54 ± 0.03a	0.29 ± 0.01a
GuiXia 3	A1	4.03 ± 0.45a	0.80 ± 0.02a	2.46 ± 0.19a	0.54 ± 0.02a	0.25 ± 0.01b
	A2	4.06 ± 0.38a	0.80 ± 0.01a	2.44 ± 0.21a	0.55 ± 0.02a	0.23 ± 0.01c
	CK	3.15 ± 0.14b	0.77 ± 0.01b	2.43 ± 0.24a	0.51 ± 0.01a	0.27 ± 0.01a

Note: A1, A2, and CK represent, one corn row intercropped with one soybean row, two corn rows intercropped with two soybean rows, and soybean monoculture, respectively. Different lowercase letters within columns indicate significant differences ($P < 0.05$)

Table 5.9 Effects of maize shading on the anatomical structure of soybean leaves

Variety	Treatment	Blade thickness (μm)	Upper skin thickness (μm)	Lower skin thickness (μm)	Palisade tissue thickness (μm)	Spongy tissue thickness (μm)
NanDou12	A1	73.37 ± 5. 35c	11.93 ± 1.08bc	6.97 ± 1.03c	21.48 ± 1.29b	32.70 ± 2.55a
	A2	93.21 ± 4.13b	12.42 ± 1.24ab	8.75 ± 0.69b	25.76 ± 1.58b	37.02 ± 2.37a
	CK	118.52 ± 5.28a	13.65 ± 1.32a	12.67 ± 0.96a	46.74 ± 2.33a	38.92 ± 3.02a
GuiXia3	A1	65.99 ± 3.25c	10.65 ± 1.22c	6.32 ± 0.71c	20.89 ± 1.42b	24.36 ± 2.63b
	A2	86.49 ± 4.23b	11.43 ± 0.99bc	7.92 ± 0.84b	23.64 ± 1.36b	43.70 ± 2.34a
	CK	127.26 ± 5.76a	12.44 ± 1.02ab	12.93 ± 1.23a	57.36 ± 2.36a	46.32 ± 1.66a

Note: A1, A2, and CK represent one corn row intercropped with one soybean row, two corn rows intercropped with two soybean rows, and soybean monoculture, respectively. Different lowercase letters within columns indicate significant differences ($P < 0.05$)

5.2.2.3 Leaf Anatomical Characteristics in Soybean Genotypes Differing for Shade Tolerance

Shading significantly reduced soybean leaf, and mesophilic palisade and spongy tissues; furthermore, the arrangement of spongy and palisade tissues became sparse compared with those of plants grown under the monoculture cropping system (Table 5.9). In treatments A1 and A2, leaf thickness, lower epidermis thickness, and palisade tissue thickness of Nan Dou 12 were significantly reduced by 38.09%, 21.36%, and 45.24%, respectively, relative to those of controls, and the thickness of the mesophilic spongy tissue was reduced by 30.94%, 54.04%, and 44.89%. Similarly, Guixia 3 was significantly reduced by 48.15%, 32.04%, and 51.12%, respectively, and the thickness of the spongy tissue was reduced by 38.75%, 63.58%, and 58.79%. Under maize shading, leaf thickness, upper and lower epidermis thickness, and mesophilic palisade and spongy tissue thickness of Nan Dou 12 were higher than those of Guixia 3 under the same treatment.

5.2.2.4 Chloroplast Ultrastructure in Leaves of Soybean Genotypes Differing for Shade Tolerance

The thickness of thylakoid lamellae and the ratio of thylakoid area to chloroplast area of the two soybean varieties significantly increased compared with the monoculture cropping treatment, under shading (Table 5.10). The starch granule area and the ratio of starch granule area to chloroplast area increased significantly compared with monoculture cropping of Gui Xia 3, whereas the starch granule area and the ratio

Table 5.10 Effects of maize shading on the ultrastructure of chloroplasts in soybean leaves

Variety	Treatment	Base particle thickness (μm)	Number of basal particles	Grana area/ chloroplast area	Starch particle area (μm^2)	Starch grain area/ chloroplast area
NanDou12	A1	0.28 ± 0.01a	34.33 ± 1.25a	0.28 ± 0.01a	9.14 ± 0.47a	0.40 ± 0.03a
	A2	0.21 ± 0.01b	31.33 ± 2.05a	0.21 ± 0.01b	4.33 ± 0.94b	0.22 ± 0.03b
	CK	0.18 ± 0.01c	22.67 ± 0.94b	0.18 ± 0.01c	4.24 ± 0.47b	0.21 ± 0.03b
GuiXia3	A1	0.27 ± 0.01a	24.00 ± 0.82b	0.28 ± 0.00a	6.73 ± 0.60a	0.28 ± 0.04a
	A2	0.18 ± 0.00b	31.33 ± 2.49a	0.18 ± 0.01b	1.58 ± 0.15b	0.13 ± 0.01b
	CK	0.16 ± 0.01c	32.00 ± 0.82a	0.16 ± 0.01c	0.52 ± 0.04c	0.09 ± 0.03c

Note: A1, A2, and CK represent one corn row intercropped with one soybean row, two corn rows intercropped with soybean rows, and soybean monoculture, respectively. Different lowercase letters within columns indicate significant differences ($P < 0.05$)

of starch granule area to chloroplast area of NanDou 12 in treatment A1 increased significantly compared with the monoculture cropping treatment, but there was no significant difference in the starch granule area and the ratio of starch granule area to chloroplast area in treatment A2, compared with CK. Changes in thylakoid number in the two soybean varieties differed. Specifically, the number of thylakoids in NanDou 12 increased under intercropping, whereas that in Gui Xia 3 decreased compared with the monoculture cropping. In addition, the thickness of thylakoid lamellae, the ratio of thylakoid area to chloroplast area, and the ratio of starch granule area to chloroplast area of the two soybean varieties in A1 were significantly higher than those in A2, and the thickness of thylakoid lamellae, the starch granule area, and the ratio of starch granule area to chloroplast area of NanDou 12 were higher than those of Gui Xia 3.

5.2.2.5 Leaf Vein Characteristics in Leaves of Soybean Genotypes Differing for Shade Tolerance

Table 5.11 shows that, compared with monoculture cropping, vein density, vein length, and vein closure of NanDou12 significantly decreased by 14.99% and 20.01%, 15.00% and 20.02%, and 48.00% and 50.00%, in A2 and A1 treatments, respectively, whereas vein diameter significantly increased by 4.59% and 24.01%, respectively. Meanwhile, vein density and vein length in GuiXia3 were significantly reduced by 10.38% and 27.62%, 10.38% and 27.62%, respectively, and vein closure was significantly reduced by 24.14% and 44.37%, whereas vein diameter increased by 25.60% and 43.48%, respectively. The change trend in vein characteristics of GuiXia3 and NanDou12 was consistent, but the changes in vein density, vein length, vein diameter, and vein spacing in GuiXia3 were greater than those of NanDou12

Table 5.11 Effects of maize shading on the characteristics of soybean leaf veins

Variety	Treatment	Leaf vein density ($mm \cdot mm^{-2}$)	Leaf vein length (mm)	Leaf vein diameter (μm)	Leaf vein closure (no./mm^2)	Leaf vein spacing (μm)
NanDou12	CK	66.11 ± 4.88a	19.79 ± 2.46a	22.38 ± 1.65c	4.23 ± 1.52a	130.34 ± 5.72c
	A2	59.25 ± 3.66b	17.74 ± 1.38b	28.11 ± 1.86b	3.21 ± 0.93b	167.92 ± 7.77b
	A1	47.85 ± 3.05c	14.33 ± 1.47c	32.11 ± 2.39a	2.36 ± 0.79c	188.28 ± 6.44a
GuiXia3	CK	61.84 ± 4.82a	18.51 ± 2.27a	26.80 ± 1.68b	3.85 ± 1.27a	133.48 ± 5.81c
	A2	52.57 ± 4.33b	15.74 ± 1.36b	28.03 ± 1.72b	2.00 ± 1.45b	160.72 ± 8.04b
	A1	49.46 ± 3.97c	14.81 ± 1.43c	33.26 ± 2.17a	1.92 ± 0.84b	188.86 ± 7.96a
F						
Variety		24.406**	24.406**	3.731	11.641**	10.853**
Shade treatment		2.677	2.677	24.259**	3.470*	2.387
Variety × shade treatment		6.012**	6.012**	4.969*	19.198**	8.194**

Note: A1, A2, and CK represent the intercropping of one corn row with one soybean row, the intercropping of two con rows with two soybean rows, and the monoculture cropping of soybean, respectively. Different lowercase letters within columns indicate significant differences ($P < 0.05$)

Table 5.12 Effect of maize shading on soybean stomatal features

Variety	Treatment	Stomatal density (No./mm^2)	Stomata length (nm)	Stomata width (nm)	Stomatal circumference (nm)	Stomatal area (nm^2)
NanDou12	CK	149.96 ± 11.56a	21.67 ± 0.41b	3.52 ± 0.18a	54.77 ± 0.82b	130.15 ± 6.93a
	A2	130.78 ± 2.82b	22.71 ± 1.49b	2.98 ± 0.09b	56.15 ± 0.37a	123.42 ± 1.06a
	A1	122.56 ± 6.68b	23.90 ± 1.38a	2.54 ± 0.28c	57.08 ± 1.01a	98.21 ± 4.06b
GuiXia3	CK	151.12 ± 0.63a	21.50 ± 0.64c	2.83 ± 0.05a	55.06 ± 3.28b	130.91 ± 3.78a
	A2	127.29 ± 0.55ab	23.19 ± 1.42b	2.08 ± 0.12b	56.78 ± 0.93a	91.56 ± 2.30c
	A1	117.18 ± 1.99b	24.495 ± 0.28a	1.85 ± 0.01b	57.93 ± 2.25a	120.00 ± 3.28b
F						
Variety		0.17	0.92	138.89**	2.96	1.57
Shade treatment		9.33*	25.05**	86.65**	21.11**	67.66**
Variety × shade treatment		0.11	0.59	1.05	0.25	0.7

Note: A1, A2, and CK represent one corn row intercropped with one soybean row, two corn rows intercropped with two soybean rows, and soybean monoculture cropping, respectively. Different lowercase letters within columns indicate significant differences ($P < 0.05$)

Table 5.13 Effects of light intensity on photosynthetic pigment contents in soybean leaves

Variety	Treatment	Chlorophyll a content ($mgdm^{-2}$)	Chlorophyll b content ($mgdm^{-2}$)	Carrot-like content ($mgdm^{-2}$)
NanDou12	CK	3.746b	1.032a	0.715 b
	A1	5.771a	1.048a	0.899 a
	A2	3.414b	1.245 a	0.685 b
GuiXia3	CK	3.758b	1.034 a	0.603 a
	A1	4.808 a	1.045 a	0.656 a
	A2	3.428 b	1.135 a	0.574 a

Note: CK: normal light conditions; A1: shading rate 10%; A2: shading rate 36%. Lowercase letters within columns indicate significant differences ($P < 0.05$)

under A2, whereas the change in vein closure was smaller than that of NanDou12. Moreover, there were highly significant differences in vein density, vein length, closure ring number, and vein spacing between varieties

5.2.2.6 Stomatal Characteristics of Soybean Varieties Differing for Shade Tolerance

Stomatal density in Nandou 12 leaves decreased significantly by 12.79% and 18.27% under the A2 and A1 treatments, respectively, relative to monoculture (Table 5.12). In turn, stomatal length increased significantly by 4.80% and 10.29%, and the stomatal perimeter increased significantly by 2.52% and 4.22%, respectively. Meanwhile, stomatal density in Gui Xia 3 leaves decreased significantly by 15.77% and 22.46%, and the stomatal length increased significantly by 7.86% and 13.93%. The stomatal perimeter increased significantly by 3.12% and 5.21%. The above data indicate that the shade-tolerant variety, NanDou 12, was affected to a lesser extent by shading than the non-shade-tolerant variety, GuiXia 3.

5.2.2.7 Photosynthetic Pigment Contents in Soybean Varieties Differing for Shade Tolerance

Increasing shading caused chlorophyll a, carotenoids, and the value of chlorophyll a/b in both soybean varieties to increase first and then decrease (Table 5.13). Compared with CK, the chlorophyll a content of NanDou 12 and GuiXia 3 under A1 increased significantly by 54.06% and 27.94%, respectively, and the carotenoid content increased by 25.73% and 8.79%, respectively.

5.2.2.8 Characteristics of Photosynthetic Assimilation in Soybean Varieties Differing for Shade Tolerance

Sucrose and starch contents in soybean leaves decreased first and then increased with increasing shading (Fig. 5.17). Compared with the peak value under the CK treatment, under A1, maximum sucrose content in NanDou 12 leaves decreased by 3.92% and 6.29%, respectively, whereas that in GuiXia 3 increased by 0.13% and 4.33%, respectively. Further, compared with CK, under A2, starch accumulation in

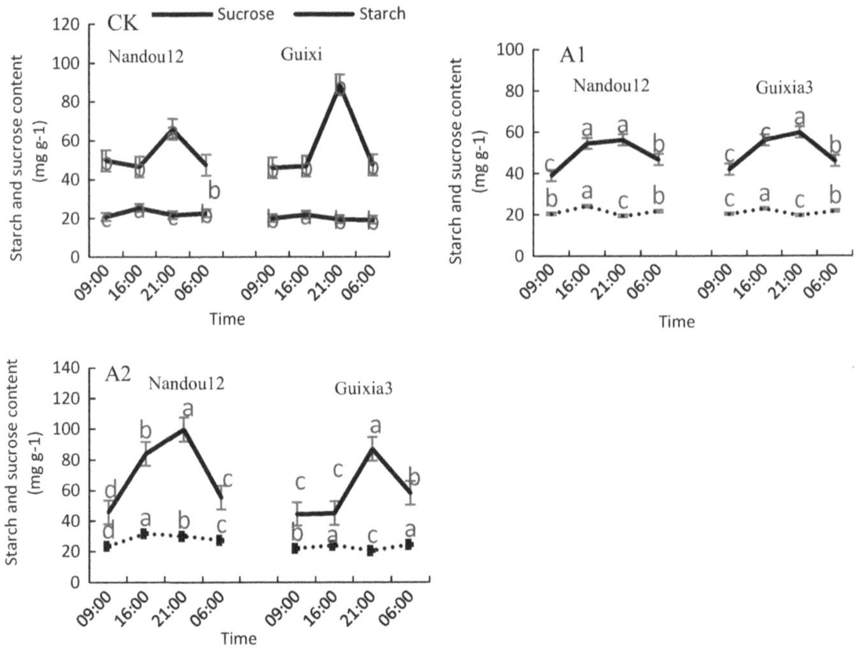

Fig. 5.17 Effects of shading on diurnal changes in sucrose and starch accumulation in soybean leaves. Note: CK represents full sunlight conditions; A1 represents 10% shading; A2 represents 36% shading

NanDou 12 and GuiXia 3 increased by 52.62% and 86.44%, respectively, and it was significantly higher in NanDou 12 than in GuiXia 3.

5.2.3 Response of Soybean Varieties with Differing Petiole Angles to the Light Environment in a Strip Intercropping System

The relationship between soybean petiole angle and shade tolerance was studied under the conditions of maize-soybean strip intercropping by a two-factor, split-plot design, with shading as main factor with three levels represented by three different maize varieties: M1 compact dwarf maize Denghai 605, light shade; M2 semi-compact tall maize Rongyu 1210, moderate shade; M3 semi-compact medium-tall maize Zhenghong 505, strong shade. The secondary factor was soybean variety-petiole angle: Nandou 25 (ND, slightly shade-susceptible genotype) and Rongxian Winter Bean (RD, moderately susceptible genotype). The results are described below.

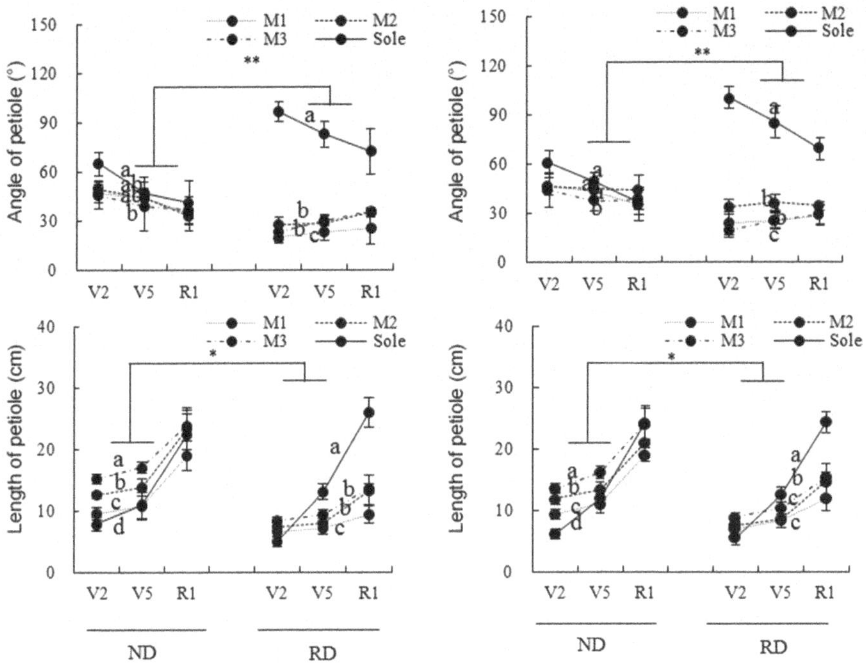

Fig. 5.18 Soybean leaf petiole angle and petiole length

5.2.3.1 Stem Characteristics of Soybean Genotypes in Response to Shading

Stem angle and length differed significantly between the two soybean genotypes studied under maize shading. Overall, the stem angles of ND soybean plants were greater than those of RD soybean plants, and shading effects on stem angle within the same genotype ranked in the following order: M2 > M1 > M3. Further, stem length in ND soybean plants was greater than that of RD soybean plants, and in this case, the effect of shading on stem length ranked in the following order within the same genotype: M3 > M2 > M1 (Fig. 3.8). Furthermore, the analysis of stem angle and length in different spatial positions showed that a certain degree of shading under strip intercropping would change the spatial layout of soybean, adjusting the stem angle and length to adapt to the changing light environment. However, there were no significant differences in stem angle or length in soybean plants at the upper, middle, or lower canopy levels under net cropping for any of the 2 years of study (Fig. 5.18). Conversely, there were significant differences in soybean stem angle and length between the upper and middle/lower levels for 2 years in strip intercropping under maize shading in M1. Specifically, the stem angles of ND soybean plants were greater in the upper than in the middle/lower canopy levels, whereas the differences in stem angles in RD soybean plants among the upper, middle, and lower canopy levels were not significant. Overall, the stem lengths of the two soybean genotypes were greater in the upper than in the middle/lower canopy levels, whereas there were

Table 5.14 Fresh weight of soybean leaves with different petiole angles

Soybean	Treatment	2019			2020		
Variety	SL	V2	V5	R1	V2	V5	R1
ND	S	9.51a	18.32 a	82.66 a	8.31 a	15.46 a	86.65 a
	M1	3.49 b	11.18 b	61.41 b	3.41 b	9.63 b	57.77 b
	M2	2.34 c	8.68 c	43.39 c	2.70 bc	7.10 c	42.55 c
	M3	2.10 c	7.44 c	31.41 d	1.94 c	6.32 c	31.21 d
RD	S	6.63 a	9.13 a	79.20 a	7.04 a	11.28 a	75.95 a
	M1	3.02 b	4.71 b	44.51 b	3.19 b	4.86 b	48.69 b
	M2	1.98 c	5.00 b	35.32 bc	2.02 b	4.53 b	35.32 c
	M3	1.63 c	3.81 c	23.22 c	1.77 c	3.95 c	26.45 d
F value	Variety	7.19*	57.46**	13.94**	6.94*	81.06**	13.31**
	SL	54.67**	22.42**	88.33**	143.26**	94.94**	109.65**
	Var. × SL	2.46 ^{ns}	3.08 ^{ns}	1.31 ^{ns}	1.35 ^{ns}	2.36 ^{ns}	0.34 ^{ns}

Table 5.15 Fresh weight of soybean stems with different petiole angles

Soybean	Treatment	2019			2020		
Variety	SL	V2	V5	R1	V2	V5	R1
ND	S	4.06 a	6.48 a	44.58 a	4.00 a	6.74 a	46.05 a
	M1	1.14 b	3.57 b	20.83 b	1.12 b	4.37 b	25.19 b
	M2	0.79 c	2.07 c	16.21 bc	0.95 bc	2.75 c	18.10 bc
	M3	0.664 c	1.59 c	11.98 b	0.71 c	2.23 c	12.48 c
RD	S	2.90 a	3.13 a	44.42 a	3.09 a	3.95 a	36.59 a
	M1	0.64 b	1.00 b	15.26 b	0.68 b	1.24 b	14.7 b
	M2	0.36 b	0.94 b	12.02 bc	0.39 b	1.05 b	12.51 b
	M3	0.28 b	0.94 b	8.85 c	0.35 b	0.76 c	9.37 c
F value	Variety	11.76**	49.51**	2.62	61.17**	174.27**	23.37**
	SL	63.35**	35.08**	58.84**	396.79**	100.73**	88.38**
	Var. × SL	1.03 ^{ns}	5.27*	0.32 ^{ns}	2.84 ^{ns}	5.53**	1.52 ^{ns}

no significant differences in stem angle or length among the three maize types under shading.

5.2.3.2 Accumulation and Distribution of Aboveground Dry Matter in Soybean Genotypes Differing for Leaf-Petiole Angle

Different leaf petiole angles affect the accumulation and partitioning of aboveground materials in soybean. Hence, different shading intensities of corn over soybean plants had significant effects on the fresh weight of aboveground organs such as leaves, petioles, and stems, but there was no significant interaction between soybean

Table 5.16 Fresh weight of stems of soybean plants with different leaf petiole angles

Soybean	Treatment	2019			2020		
Variety	SL	V2	V5	R1	V2	V5	R1
ND	S	4.83 a	11.36 a	85.56 a	4.94 a	12.79 a	91.96 a
	M1	2.58 b	10.19 a	43.54 b	2.26 b	9.26 b	47.45 b
	M2	2.01 c	8.37 b	31.77 c	2.43 b	8.02 c	30.27 c
	M3	1.98 c	8.36 b	17.87 d	1.89 c	8.32 c	18.77 d
RD	S	3.27 a	6.95 a	105.82 a	3.38 a	8.04 a	91.84 a
	M1	2.25 b	4.83 b	36.93 b	2.44 b	4.97 b	38.04 b
	M2	1.34 c	4.73 b	25.13 c	1.56 c	4.87 b	26.43 bc
	M3	1.26 c	3.66 c	20.71 c	1.70 c	4.66 b	18.15 c
F value	Variety	11.48**	35.61**	0.95	14.88**	101.47**	5.08*
	SL	21.80**	3.31*	183.56**	46.02**	23.12**	354.23**
	Var. × SL	1.45 ^{ns}	0.22 ^{ns}	6.28**	5.78**	0.81 ^{ns}	1.6 ^{ns}

genotype and corn shading (Tables 5.14, 5.15, 5.16). The fresh weight of aboveground organs followed the order of S > M1 > M2 > M3 under net plot and different corn shading levels in both years. Except for the soybean leaves in the R1 stage in 2019, the fresh weight of aboveground organs in other years and measuring periods was larger in ND than in GD. Similarly, the average fresh weight of ND leaves, petioles, and stems was 1.24, 1.79, and 1.19 times that of GD at V2, V5, and R1 stages, respectively, in the two years. Additionally, for the same growth stages, the corresponding values were 1.54, 2.29, and 1.28; and 1.34, 1.80, and 1.01 times that of GD, respectively. Total dry matter accumulation in ND was significantly greater than that in RD at all growth stages in both years, and the trend of change in total dry matter accumulation was consistent with that of the fresh weight of each organ. Further, under different corn shading levels, M1 was closer to the net plot, and the average dry weight of ND over the 2 years (mean of each measuring period) was 57.9% that in the net plot, which was 15.9% higher than that in RD. In addition, during the recovery period (R1 stage), ND grew faster, and the growth rates under M1, M2, and M3 shading were 29.70%, 45.17%, and 30.15% higher than those under RD (Fig. 5.19a,b). During the cocultivation period of corn and soybean (V2, V5), the interannual variation in dry matter allocation between ND and RD was consistent, and the proportion of ND leaves and petioles gradually decreased with increasing corn shading under strip intercropping, whereas that of stems increased. Under M1, M2, and M3 shading, the average proportion of leaves and petioles decreased by 4.97% and 20.03%; 10.06% and 30.69%; and 14.13% and 37.45% over the two years, respectively. Meanwhile, the proportion of RD leaves remained basically unchanged but that of petioles decreased by an average of 45.21%, 50.15%, and 50.91%, respectively, over the two years. During the R1 stage of strip intercropping, the proportion of soybean leaves increased significantly, and under M1, M2, and M3 treatments, ND and RD increased by 20.27%, 21.80%, and

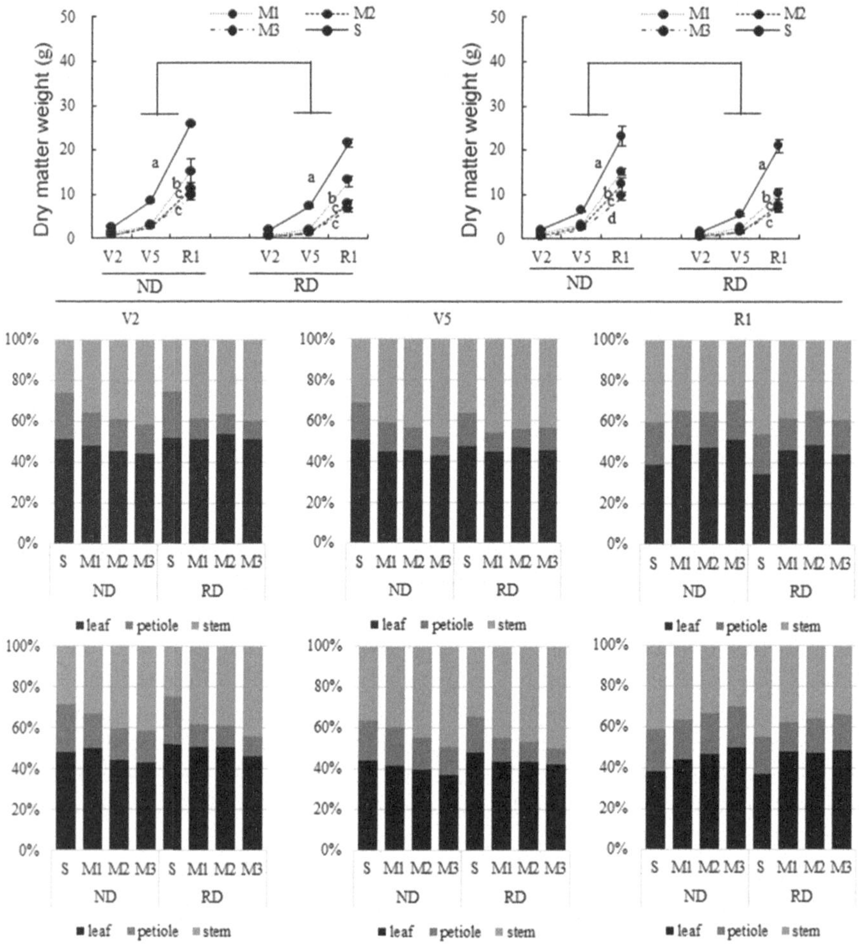

Fig. 5.19 Dry matter accumulation and distribution in soybean genotypes with different leaf petiole angles

30.77%; and by 31.24%, 34.6%, and 29.67% (two-year average), respectively, while the change in petioles was relatively small.

5.2.3.3 Light Interception and Utilization in Soybean Plants with Different Leaf Angles

There was a significant difference in light interception between the two soybean cultivars studied. On average, over two years, ND showed a 29.15% higher light interception rate than RD. The variation in light interception under different maize shading levels was smaller for ND and larger for RD. Moreover, the difference between the two soybean cultivars was significant at all shading levels, and both performed in the order of M1 > M2 > M3 over the two-year period (Fig. 5.20a,b).

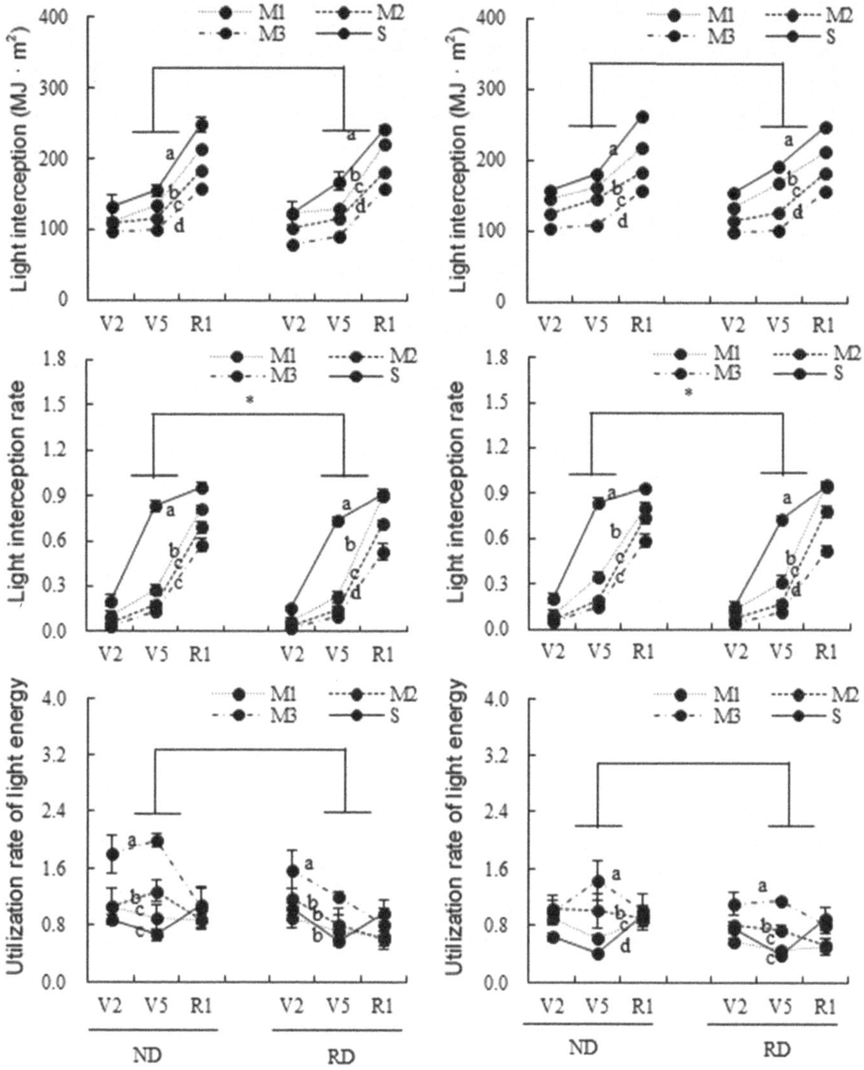

Fig. 5.20 Light interception and utilization efficiency in soybean plants with different petiole angles

The change in light interception rate was basically consistent with that of light interception. Additionally, the light interception rate increased rapidly from V2 to V5 stages and slowly from V5 to R1 stage in net intercropping, whereas in strip intercropping, the increase was slow from V2 to V5 stages and rapid from V5 to R1 stage. The two types of intercropping showed opposite trends. In addition, there was no significant difference between M2 and M3 under different shading levels in strip intercropping (Fig. 5.20c,d). Further, the two soybean cultivars differed significantly

Table 5.17 Yield components in soybean genotypes varying in petiole angles

Year	Soybean variety	Treatment SL	100-grain weight (g)	Number of grains per plant	Number of pods per plant	Grain weight per plant (g)
2019	ND	S	25.38 b	88.48 a	48.40 a	22.46 a
		M1	25.95 a	80.39 b	46.24 a	20.86 a
		M2	25.33 b	72.82 c	41.65 b	18.45 b
		M3	25.22 b	65.43 d	38.23 b	16.50 c
	RD	S	9.39 a	168.34 a	86.13 a	15.81 a
		M1	9.72 a	155.28 b	83.22 a	15.09 a
		M2	9.18 b	132.36 c	73.23 b	12.15 b
		M3	9.01 b	96.80 d	65.27 c	8.72 c
2020	ND	S	25.14 a	86.11 a	50.00 a	21.65 a
		M1	25.53 a	74.43 b	50.18 a	19.00 b
		M2	24.98 ab	68.49 c	42.16 b	17.11 c
		M3	25.12 b	62.15 c	38.33 b	15.61 d
	RD	S	9.59 a	130.26 a	101.46 a	12.49 a
		M1	9.68 a	120.73 b	92.75 b	11.69 b
		M2	9.37 a	112.56 bc	82.34 c	10.55 c
		M3	8.98 b	94.23 c	77.36 d	8.46 d
F value	Variety		58.63**	26.37**	16.83**	36.25**
	SL		6.83 ns	47.37**	28.46**	10.26**
	Var. × SL		21.36 ns	5.89*	3.28*	17.28*

in light use efficiency, with the overall performance ranking in the order of M3 > M2 > M1 > S. Except for RD in 2019, the two soybean cultivars in strip intercropping showed an increase followed by a decrease over 2 years for M2 and M3, whereas S showed a decrease followed by an increase, and M1 showed a slow decrease. The net light use efficiency of ND and RD was the same, at 0.77, over the two years. Lastly, under strip intercropping, ND was 1.09, which was 30.54% higher than the 0.84 value for RD, and the corresponding values under M1, M2, and M3 were 37.90%, 32.90%, and 24.69% higher than those recorded for RD under maize shading, respectively (Fig. 5.20e,f).

5.2.3.4 Yield Components in Soybean Plants with Different Petiole Angles

The soybean yield of ND was greater than that of RD in both years, and the main reason for the difference in yield between these two soybean genotypes was due to the large difference in seed weight, which was mainly variety dependent, while shading had no significant effect on it. Further, increasing maize shading reduced yield per plant of both ND and RD, and yield components including seed weight, number of grains per plant, and number of pods per plant with the minimum values registered under the M3 maize shading treatment. The differences in the number of grains and pods per plant under different shading treatments were significant. Unlike seed weight, these yield components were influenced by both variety and shading degree, and a significant interaction between the two factors. The two-year average yield per plant for ND and RD under strip intercropping in M1, M2, and M3 were 22.05, 19.93, and 17.78 g, 14.15, 13.39, and 11.35 g, and 8.59 g, respectively. Importantly, the relative change in yield under different shading treatments was different for strip intercropping compared with monoculture cropping. The yield of RD, a strongly shade-susceptible material in relation to leaf angle was 94.52%, 80.65%, and 61.46% of the yield observed in monoculture cropping under M1, M2, and M3 shading, respectively. On the other hand, the yield of ND, a slightly shade-susceptible material in relation to leaf angle was 90.34%, 80.59%, and 72.80% of the yield in the monoculture cropping under M1, M2, and M3 shading, respectively. The variation range of RD was 5.48%, 19.35%, and 38.54%, whereas that of ND was 9.66%, 19.41%, and 27.20%. Under relatively light shading, the yield reduction in RD was smaller, whereas under strong shading, the yield reduction in ND was smaller (Table 5.17).

5.2.4 The Physiological Mechanism of Shade Tolerance and Lodging Resistance in Different Soybean Genotypes

Through years of field screening experiments, soybean varieties with different levels of lodging resistance in the maize-soybean strip intercropping system have been obtained. Further, by combining field and indoor experiments, the physiological mechanism of shade tolerance and lodging resistance of soybean was studied under different light environments, from a perspective of stem morphology, anatomical structure, and content and metabolism of structural and nonstructural carbohydrates.

5.2.4.1 Field Lodging Rate

Soybean plants are susceptible to shading stress caused by taller maize plants and, therefore, prone to lodging under strip intercropping. Soybean lodging rate gradually increased as plant growth proceeded and reached a maximum value at 58 days after emergence (i.e., after maize harvest and at the end of the intercropping period), exhibiting severe lodging or even heavy lodging, while no lodging occurred in soybean monoculture at the seedling stage (Table 5.18). Further, field lodging rates varied significantly among soybean varieties. The slightly lodging-tolerant soybean variety B1 (Nandou 032–4) had the highest lodging rate 30 days after sowing and maintained a relatively high growth level throughout the seedling stage, with its lodging rate significantly higher than that of B2 (September Yellow) and B3 (Nandou 12). Based on two-year field data, at the end of the intercropping period of maize and soybean, the lodging rates of B1, B2, and B3 soybean varieties were 89.81%, 83.05%, and 55.93%, respectively, indicating that there is a significant difference in the lodging resistance of soybean varieties under the intercropping environment, such that the more shade tolerant a variety is, the less severe the lodging will be.

Table 5.18 Field lodging rates (%) of different shade-tolerant soybean varieties

Cultivars	Days after emergence (2013年)					Days after emergence (2014年)				
	30d	37d	44d	51d	58d	30d	37d	44d	51d	58d
B1	25.9 a	65.3 a	74.1 a	83.6 a	94.2 a	27.9 a	64.3 a	71.2 a	79.2 a	91.2 a
B2	26.5 c	42.6 b	55.4 b	62.2 b	85.9 b	26.8 c	50.4 b	62.9 b	72.5 b	81.3 b
B3	22.4 b	31.8 c	32.5 c	44.1 c	51.3 c	21.4 b	32.8 c	41.3 c	47.5 c	53.1 c

5.2.4.2 Stem Strength

There were significant differences ($P < 0.01$) in the bending resistance of soybean stems under different planting methods. For the same soybean variety, the bending resistance of the stem gradually increased with growth from 30 to 58 days after emergence (Table 5.19). Compared with monoculture, intercropping significantly reduced the bending resistance of soybean stems, but the degree of reduction varied among varieties. Based on 2 years of data, under intercropping, the bending resistance of the stems of B1, B2, and B3 decreased by 71.37%, 63.97%, and 55.10%, respectively, ultimately ranking in the order of B3 > B2 > B1. This indicates that the strong shade-tolerant variety B3 was less affected by shading under intercropping and has better mechanical properties in terms of lodging resistance.

5.2.4.3 Stem Anatomical Structure

Anatomical structure responses of soybean stems were as follows: intercropping significantly reduced the cross-sectional area, pith area, and lignified area of soybean stems, as well as the proportion of lignified to the cross-sectional area, while significantly increasing the proportion of pith to the cross-sectional area (Table 5.20). Further, as growth and development under intercropping proceeded, the cross-sectional area, pith area, lignified area, and the proportion of lignified area gradually increased, whereas the proportion of pith decreased. The strongly shade-tolerant soybean variety B3 had significantly higher cross-sectional area, pith area, lignified area, and proportion of lignified tissue than the weakly shade-tolerant soybean varieties B1 and B2 throughout the seedling stage, with a significantly lower proportion of pith than B1 and B2. At the end of the symbiotic period, compared with monoculture, the lignified area of B1, B2, and B3 varieties decreased by 92.04%, 87.27%, and 83.05%, respectively, and it decreased by 18.57%, 19.61%, and 6.82%, respectively, whereas the proportion of pith increased 3.73, 3.20, and 3.09 times. This indicates that the strong shade-tolerant soybean variety was less affected by intercropping shading, and its stem became lignified earlier, resulting in a larger lignified area, a higher proportion of lignified tissue, and a lower proportion of pith, thereby improving stem mechanical strength and enhancing plant lodging resistance.

5.2.4.4 Stem Cellulose Content

As shown in Fig. 5.21, from 30 to 58 days after sowing, soybean and corn plants were in a symbiotic stage. The stem cellulose content of soybeans gradually increased with growth, and the stem cellulose content in intercropping was significantly lower than that in monoculture for the same soybean variety. Further, soybean varieties differed significantly in stem cellulose content under the same planting system, but it was higher in the strongly shade-resistant soybean variety B3 than that of B1 or B2 regardless of the planting system.

The accumulation of sucrose in the stem increased continuously with growth under monoculture conditions, and it showed the order of B3 > B2 > B1, with the peak values appearing at approximately 58 days after sowing. Further, the accumulation of sucrose in the stem showed a trend of “increase-decrease-increase” under

Table 5.19 Stem bending resistance of different shade-tolerant soybean varieties

Planting system		Days after emergence (2013年)					Days after emergence (2014年)				
	Cultivar	30d	37d	44d	51d	58d	30d	37d	44d	51d	58d
Monoculture	B1	16.250 b	39.167 c	50.323 c	102.867 c	166.460 c	10.580 d	25.083 c	46.673 c	98.527 c	158.147 c
	B2	15.773 b	42.393 b	72.953 b	125.233 b	190.520 b	15.680 b	42.593 b	73.020 b	123.080 b	192.447 b
	B3	31.870 a	62.960 a	91.733 a	166.207 a	203.060 a	38.840 a	66.267 a	113.020 a	195.460 a	225.393 a
Strip intercropping	B1	6.227 e	8.453 f	16.460 d	22.750 e	32.473 e	5.613 f	10.087 f	16.167 e	20.540 f	32.887 e
	B2	8.317 d	13.020 e	21.380 d	27.230 e	35.707 e	8.100 e	13.573 e	19.657 e	26.067 e	35.773 e
	B3	12.093 c	17.233 d	26.203 d	36.207 d	47.190 d	11.953 c	18.020 d	27.787 d	37.447 d	46.917 d

Table 5.20 Anatomical structure of stems (transverse section) in soybean varieties with different shade tolerance and lodging resistance

Days after emergence	Planting system	Cultivar	Medullary (mm^2)	Xylem (mm^2)	Transverse section (mm^2)	Proportion of medullary (%)	Xylem percentage (%)
20	Strip intercropping	B1	1.09 d	0.23 e	1.97 e	55.60 a	11.42 e
		B2	1.13 d	0.28 e	2.27 e	49.74 b	12.27 d
		B3	2.08 c	0.62 d	4.54 d	45.81 c	13.71 c
	Monoculture	B1	2.17 c	1.26 c	5.14 c	42.11 d	24.79 b
		B2	2.58 b	1.60 b	6.40 b	40.33 e	25.07 b
		B3	3.54 a	2.44 a	8.95 a	39.52 e	27.30 a
34	Strip intercropping	B1	1.07 d	1.65 f	4.05 f	26.50 a	40.73 f
		B2	1.14 d	1.85 e	4.47 e	25.45 b	41.46 e
		B3	1.73 c	3.07 d	7.10 d	24.38 c	43.17 d
	Monoculture	B1	2.75 b	8.44 c	17.83 c	15.45 e	47.31 c
		B2	4.13 a	11.52 b	23.53 b	17.54 d	48.99 b
		B3	4.18 a	14.01 a	28.30 a	14.75 e	49.51 a
48	Strip intercropping	B1	1.38 f	2.53 f	5.80 f	24.04 a	43.93 f
		B2	1.78 e	3.64 e	7.91 e	22.52 b	46.07 e
		B3	2.47 d	6.15 d	11.84 d	20.99 c	52.14 d
	Monoculture	B1	2.99 b	31.75 b	58.86 b	5.08 e	53.94 c
		B2	2.68 c	28.62 c	49.94 c	5.36 d	57.30 a
		B3	3.32 a	36.30 a	64.87 a	5.12 e	55.95 b

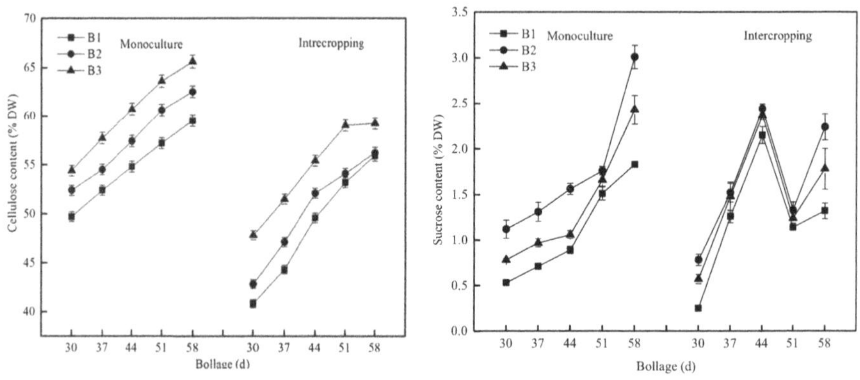

Fig. 5.21 Stem cellulose and sucrose content of different shade-tolerant soybean varieties

intercropping conditions, reaching the highest level at approximately 44 days after sowing and then decreased before increasing again. This indicates that before 44 days after soybean sowing, the shading by corn on soybeans was strong, and a large amount of sucrose in soybean stems was used to maintain rapid stem elongation, thereby reducing the sucrose content converted into cellulose. Therefore, the sucrose content in the stem was higher. After 44 days, as corn matured, the shading on soybeans gradually decreased, and the rate of sucrose synthesis into cellulose increased. However, the transport of photosynthetic products from the leaves to the stems was insufficient, resulting in a decrease in sucrose content. After 51 days, with further improvement in the light environment and the resulting increase in photosynthetic rate, the sucrose content in the stem was restored. Under the shade in intercropping, the stem sucrose content of the strong shade-resistant soybean variety B3 changed the least. The more photosynthetic products were allocated to the stem and quickly converted into cell wall substances,cellulose, the stronger the stem became, and the greater its resistance to lodging.

5.2.4.5 Activity of Stem Cellulose Synthesis-Related Enzymes

The activity of sucrose-phosphate synthase (SPS) in soybean stems varied with growth in an unimodal pattern (Fig. 5.22) regardless of monoculture or intercropping, and peaking at approximately 44 days after sowing. As growth proceeded, intercropped soybean stems showed more marked changes in sucrose phosphate synthase (SPS) activity compared with monoculture. Within 30 days after sowing, SPS activity in intercropped soybean stems was lower than that in monoculture, but then rapidly increased, and by 44 days after sowing, it was significantly higher than in monoculture. However, it then rapidly decreased, and by 51 days after sowing, it was significantly lower than in monoculture. This indicated that, before 44 days after sowing, shading from maize was more severe, and soybean stems increased SPS activity to synthesize more sucrose for their rapid elongation, thus showing higher enzyme activity. After 44 days, shading from maize gradually decreased, and sucrose content continuously increased, whereby soybean stems

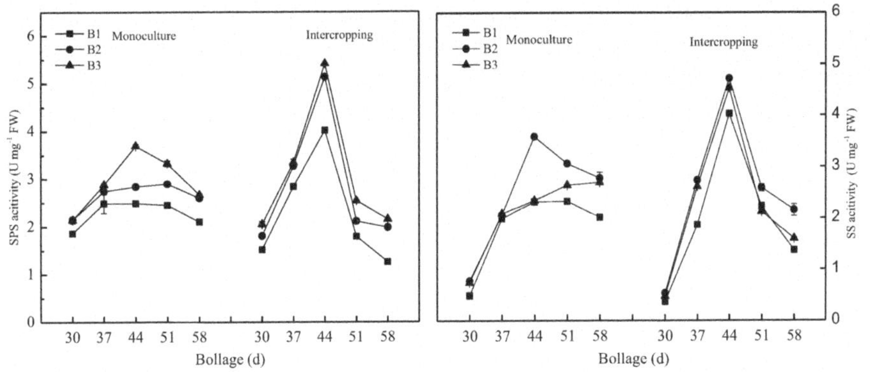

Fig. 5.22 Changes in SPS and SS activities in different soybean genotypes

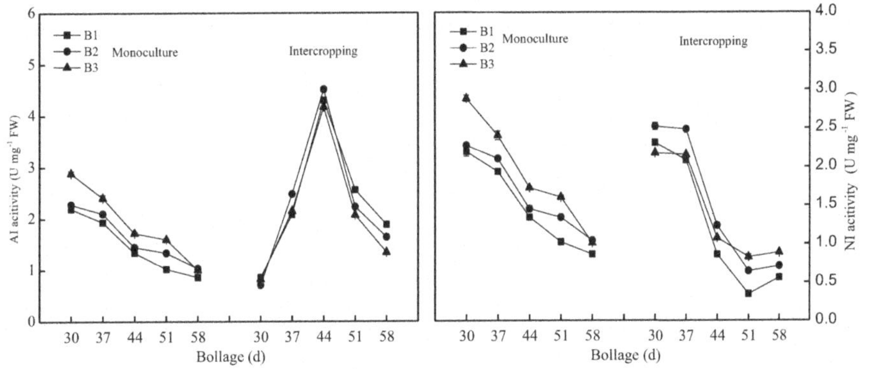

Fig. 5.23 Changes in AI and NI activities of different soybean genotypes

lowered SPS activity to suppress the excessive increase of sucrose content, resulting in lower enzyme activity. Comparing different varieties, we found that the strong shade-tolerant soybean variety B3 had the smallest variation in stem sucrose content, and its SPS activity was consistently higher than that in monoculture soybean plots, indicating that it was advantageous for sucrose accumulation, thereby increasing the material basis and energy for cellulose synthesis.

Sucrose synthase (SS) activity in soybean stems during development followed a similar trend to that of SPS activity (Fig. 5.23), with the difference being mainly in the level of enzyme activity throughout the growth period. Regardless of the planting system, the peak activity occurred approximately 44 days after sowing. Under monoculture conditions, SS activity changed slowly; conversely, under intercropping conditions, before 44 days after sowing, shading from maize was more severe, whereby SS activity in soybean stems increased to synthesize more Uridine Diphosphate Glucose(UDPG), which provided material support for cellulose synthesis and led to higher enzyme activity. After 44 days, shading from maize gradually decreased, and the resulting large amount of sucrose inhibited the reversible enzyme SS activity, resulting in a gradual decrease in enzyme activity. Comparing different varieties, we found that the strong shade-tolerant soybean variety B3 effectively maintained a higher level of stem SS activity under intercropping conditions, which was advantageous to sucrose breakdown, thereby providing more UDPG as a substrate for cellulose synthesis.

The activity of acid invertase (AI) in soybean stems varied with growth stage under different planting systems (Fig. 5.23). Thus, under monoculture conditions, AI activity in stems gradually decreased, whereas under strip intercropping conditions, it showed an inverted V-shape trend, and the stem AI activity of each soybean variety reached a maximum 44 days after sowing. The comparison among varieties revealed that the stem AI of the shade-tolerant variety B3 under monoculture planting was significantly higher than those in the other two varieties, but under strip intercropping conditions, there was no significant difference between B3 and B2 before 44 days after sowing, after which, the speed of AI activity resumption was higher than those of the other two varieties.

Meanwhile, the activity of the neutral invertase (NI) in soybean stems gradually decreased under monoculture (Fig. 5.23), whereas it showed a V-shape trend under strip intercropping conditions. The comparison among varieties showed that under strip intercropping conditions, before 44 days after sowing, the NI activity of the three soybean varieties differed significantly, following the order: B3 > B2 > B1. This indicated that, under severe shading conditions by corn on soybean, the shade-tolerant variety B3 maintained a higher NI activity to metabolize sucrose and provide more carbon sources and energy for rapid growth. Later, due to the improvement of the light environment, there was no significant difference in NI activity among soybean varieties, and finally, they ranked in the following order: B1 > B2 > B3. This indicated that shade-tolerant cultivar B3 reduced stem NI activity and sucrose decomposition, which was conducive to sucrose accumulation.

5.2.4.6 Stem Lignin Content

Soybean stem lignin content showed a gradual increasing trend with growth and development during the seedling stage. Further, stem lignin content was significantly lower for the same soybean variety in strip intercropping than that in monoculture cropping. Furthermore, there were significant differences in stem lignin content among soybean varieties in both strip intercropping and monoculture cropping (Table 5.21). Additionally, stem lignin content in each variety grown in strip intercropping or monoculture cropping systems at the same growth stage ranked as B3 > B2 > B1, which was basically consistent with the variation trend of varietal stem bending resistance. At the end of the symbiotic period, stem lignin contents of B1, B2, and B3 in strip intercropping were 32.20%, 33.99%, and 25.94% lower than those in monoculture cropping, respectively. Additionally, analysis of variance showed that planting system and variety both had a significant impact on stem lignin content, while their interaction effect was also significant. Varieties with strong shade tolerance and lodging resistance maintained a higher level of lignin content in the strip intercropping environment, effectively improving lodging resistance.

Table 5.21 Variation in stem lignin content among different soybean genotypes

Planting system	Cultivar	Days after emergence				
		20	27	34	41	48
Strip intercropping	B1	0.577 d	0.970 e	1.810 f	2.101 f	2.234 f
	B2	0.667 cd	1.102 d	1.954 e	2.253 e	2.414 e
	B3	0.740 c	1.382 c	2.266 d	2.569 d	2.868 d
Monoculture	B1	0.983 b	1.814 b	2.531 c	3.009 c	3.295 c
	B2	1.082 ab	1.936 a	2.853 b	3.386 b	3.657 b
	B3	1.185 a	2.020 a	2.987 a	3.529 a	3.849 a
*F*值						
C		127.565**	686.417**	317.963**	625.479**	401.098**
G		24.856**	60.668**	171.947**	95.305**	257.250**
C × G		0.305 ns	8.570**	8.702**	5.458*	13.094**

The F value is the result of a two-factor analysis of variance. C represents the planting system, G represents variety, and C × G represents the interaction effect of the planting system and variety. “ns” indicates no significant difference at the 0.05 level

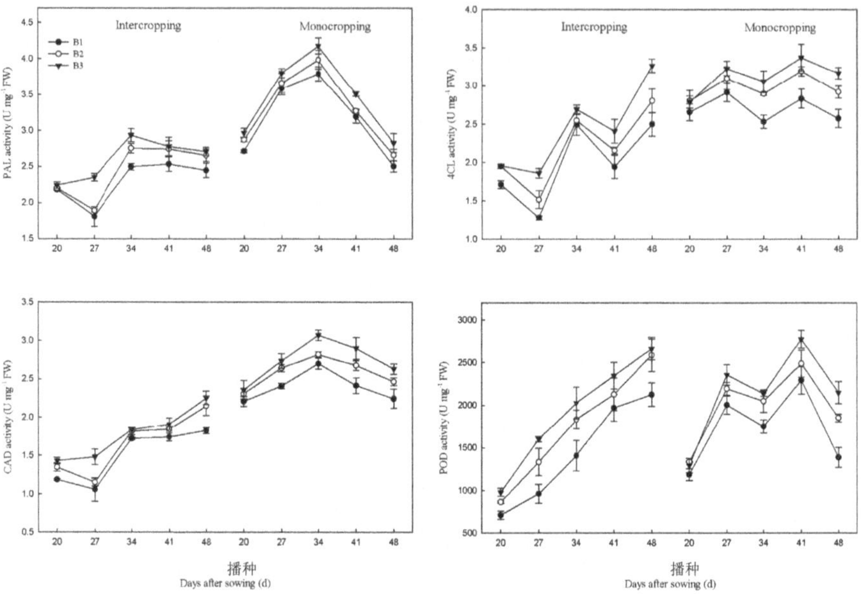

Fig. 5.24 Variation in PAL, 4CL, CAD, and POD activities in different soybean genotypes

5.2.4.7 Activities of Enzymes Related to Stem Lignin Synthesis

Phenylalanine ammonia-lyase (PAL) activity: The trend in PAL activity in soybean stems varied between strip intercropping and monoculture, showing an overall lower level in the former than in the latter system (Fig. 5.24). Further, PAL activity showed an upward trend followed by a downward trend during stem growth under monoculture, reaching a maximum value at 34 days after sowing. Among the varieties, PAL activity generally ranked in the order of B3 > B2 > B1. On the other hand, under strip intercropping, PAL activity in B1 and B2 showed a downward trend followed by an upward trend and then a downward trend again, whereas PAL activity in B3 showed a slow upward trend followed by a downward trend. At 27 days after sowing, PAL activity in B1 and B2 reached a minimum level, whereas in B3 it was significantly higher than those in B1 or B2. It can be speculated that 27 days after sowing is a critical period for the differences in lignin synthesis among different shade-tolerant and lodging-resistant varieties and that maintaining a higher PAL activity in the low-light environment caused by maize shading can effectively increase the stem lignin content.

4-Coumaroyl-CoA ligase (4CL) activity: The overall 4CL activity in soybean stems in intercropping was generally lower than that in monoculture, and the trend of change was inconsistent between intercropping and monoculture (Fig. 3.4). Thus, in the latter case, 4CL activity showed an "M"-shaped bimodal trend as stem growth proceeded, ranking in the order of B3 > B2 > B1 among different varieties. Conversely, as stem growth proceeded under intercropping, 4CL activity showed a "W"-shaped change trend, and varieties performed basically the same as in monoculture, i.e., B3 > B2 > B1. It is worth noting that at 27 days after sowing, both varieties B1 and B2 showed a significant decrease in 4CL activity, while the corresponding decrease in B3 variety was smaller, indicating that shade-tolerant and lodging-resistant varieties are less affected by shading stress, and their higher 4CL activity is conducive to lignin synthesis in stems, thereby improving plant lodging resistance.

Cinnamyl alcohol dehydrogenase (CAD) activity: Compared with monoculture, CAD activity in soybean stems was significantly lower under intercropping at all stages (Fig. 5.24). Under monoculture, CAD activity first increased and then decreased with stem growth and development, and the activity of the different varieties ranked in the order of B3 > B2 > B1. In contrast, under intercropping, CAD activity in B1 and B2 significantly decreased at 27 days after sowing and slowly increased thereafter, whereas the activity in B3 showed a slow but steady upward trend throughout the seedling stage; then, at 27 days, CAD activity was significantly higher in B3 than in B1 or B2. As with PAL and 4CL activities, the difference in CAD activity may be the reason for the difference in lignin content in the stems of different shade-tolerant and lodging-resistant varieties.

Peroxidase (POD) activity: POD activity in soybean stems gradually increased with growth and development under intercropping, reaching a maximum value at 48 days after sowing; whereas under monoculture, it showed a bimodal trend with peaks at 27 days and 41 days after sowing (Fig. 5.24). Overall, before 41 days after sowing, POD activity in intercropping soybean was significantly lower than that in monoculture, whereas at 48 days, intercropping showed significantly higher POD activity than monoculture. Lastly, with regard to variety, POD activity ranked as follows: B3 > B2 > B1 regardless of planting system.

5.2.4.8 Expression of Stem Lignin Synthesis-Related Genes

Light environment significantly affected the expression of key enzyme genes in the stem lignin synthesis pathway of soybean. The change patterns of the expression levels of each gene under different light environment conditions were basically consistent as the duration of shading increased. At any sampling time point, almost all genes in the shade-tolerant and lodging-resistant soybean variety Nan 12 were downregulated, while shading treatment suppressed the expression of four genes (C3H, CCR, CCoAOMT, and POD) in the weakly shade-tolerant and lodging-resistant soybean variety Nan 032-4 (Fig. 5.25). With the increase in shading time, most genes in Nan 12 showed larger differences in expression compared with Nan 032-4 with increasing duration of shading. After shading treatment for 8.5 h, except for the PAL, C4H, and COMT genes, the expression levels of all other genes in Nan 12 were higher than those in Nan 032-4. Analysis of variance of the expression levels of each gene showed that the interaction between light environment and variety had a highly significant impact on the expression of eight genes, PAL, C4H, C3H, 4CL, COMT, CCR, CAD, and LAC. Meanwhile, PAL, 4CL, CAD, C3H, COMT, CCoAOMT, CCR, and POD were significantly affected by the light environment, and except for POD, the expression levels of all other genes were significantly affected by variety.

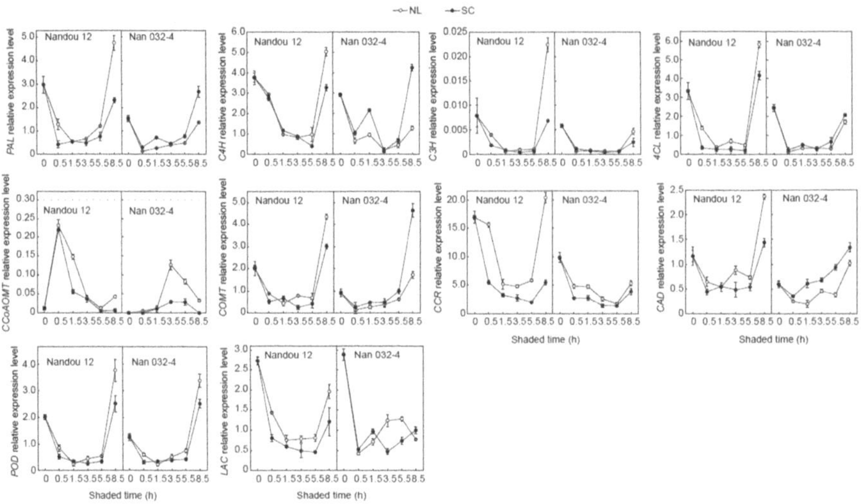

Fig. 5.25 Expression of lignin biosynthesis-related genes in the stem of soybean varieties with different shade tolerance and lodging resistance

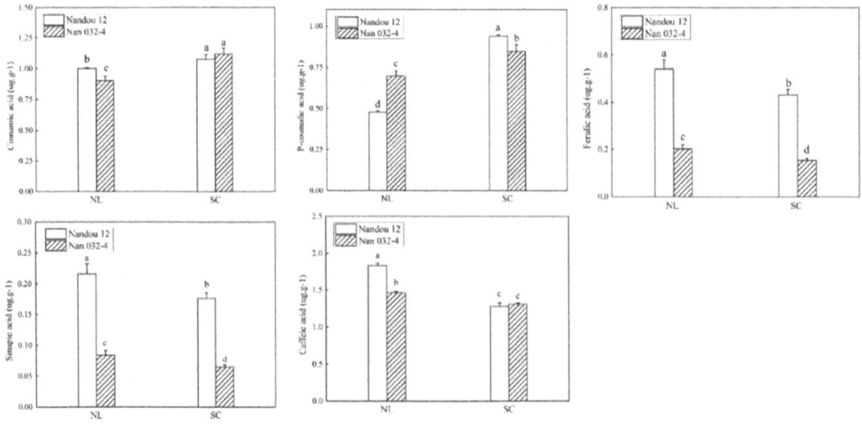

Fig. 5.26 Lignin synthesis-metabolite contents in soybean stems

5.2.4.9 Metabolite Content in the Lignin-Biosynthesis Pathway in Soybean Stems

Shading treatment significantly increased cinnamic and ferulic acid contents in the lignin biosynthesis pathway of the stem and reduced those of caffeic, syringic, and sinapic acids, compared with full sunlight conditions (Fig. 5.26). Particularly, in the strong shade-tolerant and lodging-resistant variety ND12, cinnamic and ferulic acids increased by 7.72% and 96.39%, respectively, whereas caffeic, syringic, and sinapic acids decreased by 30.04%, 20.11%, and 20.24%, respectively. Similarly, in the weakly shade-tolerant and lodging-resistant variety ND032-4, cinnamic and ferulic acids increased by 23.99% and 21.21%, respectively, whereas caffeic, syringic, and sinapic acids decreased by 10.7%, 24.13%, and 24.04%, respectively. Furthermore, under shading conditions, sinapic, ferulic, and syringic acids in ND12 were 64.53%, 9.79%, and 64.48%, respectively, higher than those in ND032-4, respectively.

5.3 Plant Regulator-Mediated Regulation of Plant Architecture in Intercropped Soybean

5.3.1 Regulation on Soybean Yield and Yield Components by Uniconazole

Uniconazole treatment increased the yield of intercropped soybean, and the effect varied with the spraying period and concentration used (Table 5.22). The highest yield was obtained with the 30 mg L^{-1} treatment at the V5 stage, followed by the 60 and 90 mg L^{-1} treatments. All uniconazole resulted in significantly higher yields than that of the control treatment. Specifically, the 30 mg L^{-1} treatment increased yield by 35.5% compared with the control. In contrast, the 120 and 150 mg L^{-1} treatments resulted in lower yields than those of the controls. The differences in yield among treatments at the R1 stage were small, with only the 150 mg L^{-1} treatment significantly increasing yield over the control level. Specifically, the differences in the number of effective pods and 100-grain weight among treatments at each growth stage were consistent with the overall yield performance. Thus, for example, at the V5 stage, treatment with 30 or 60 mg L^{-1} uniconazole resulted in a significantly higher number of effective pods and 100-grain weight than those in the control treatment, whereas the corresponding averages under the 120 and 150 mg L^{-1}

Table 5.22 Effect of foliar spraying of uniconazole on yield and yield components of intercropped soybean

Application growth stage	Concentration (mg L^{-1})	Number of effective pods	Number of pods	100-grain weight (g)	Yield per plant (kg/hm^2)
V5	0	60.0BCc	1.63ABb	23.88Cc	2335.5CDd
	30	75.3Aa	1.50Cc	28.05Aa	3164.1Aa
	60	65.3Bb	1.58BCb	26.69ABb	2759.3Bb
	90	59.5BCc	1.63ABb	26.08Bb	2535.9BCc
	120	53.9CDd	1.72Aa	23.86Cc	2211.8Dde
	150	50.5Dd	1.72Aa	23.70Cc	2061.7De
R1	0	60.0Bb	1.63Aa	23.85Cc	2332.3Bbc
	30	60.5ABb	1.58ABb	23.89Cc	2283.6Bc
	60	60.7ABb	1.56Bbc	24.42BCc	2312.4Bbc
	90	60.8ABb	1.55Bbc	24.42BCc	2301.3Bbc
	120	61.0ABb	1.54Bc	25.40ABb	2386.1Bb
	150	62.2Aa	1.54Bc	26.37Aa	2530.6Aa

treatments were lower than those of the controls. Lastly, at the R1 stage, the average values under the 150 mg L^{-1} treatment were significantly higher than those for the controls, whereas the order of performance with respect to pod number among the treatments was the opposite of that recorded for yield.

5.3.2 Regulatory Mechanism of Soybean Leaf Morphology and Function by Uniconazole

5.3.2.1 Effects of Uniconazole on Soybean Leaf Morphology

Uniconazole treatment significantly reduced the leaf area of intercropped soybean plants at the seedling stage (Table 5.23). Subsequently, at 46 days after flowering, the leaf area index (LAI) value under treatments B1, B2, and B3 decreased by 18.6%, 20.3%, and 25.3%, respectively, relative to control values. However, at 61 days after flowering, the LAI value of each treatment exceeded that of the control, indicating that uniconazole treatment delayed soybean leaf senescence (Fig. 5.27).

Table 5.23 Effects of dry seed treatment with uniconazole on leaf area of intercropped soybeans (cm^2 $plant^{-1}$)

	Days after seeding			
Treatment	21	28	35	42
B0	149.4 Aa	291.8 Aa	409.9 Aa	1030.6 Aa
B1	69.9 ABb	99.3 Bb	325.1Bb	601.2 Bb
B2	44.1 ABb	93.3 Bb	290.9 BCbc	565.0 Bb
B3	29.7 Bb	91.7 Bb	256.4 Cc	429.4 Cc

Note: B0, B1, B2, and B3 represent 0, 2, 4, and 8 mg kg^{-1} enoxazole dry mixing. Different lowercase and uppercase letters within columns indicate significant differences at the 0.05 and 0.01 levels, respectively

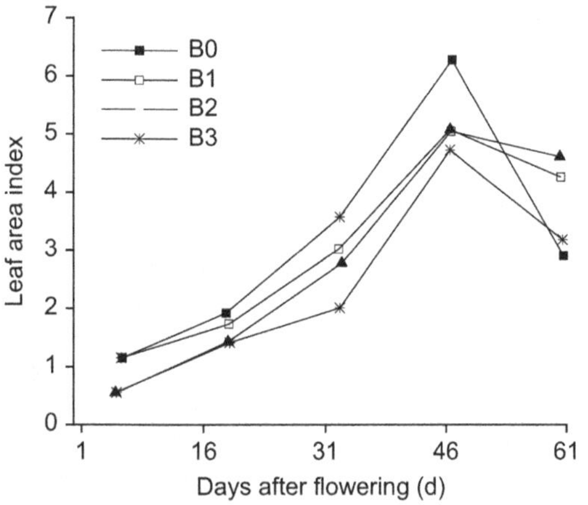

Fig. 5.27 Effects of dry seed treatment with uniconazole on leaf area index of strip intercropped soybeans. Note: B0-B3 represents different concentrations of uniconazole seed treatment: 0, 2, 4, and 8 mg/kg

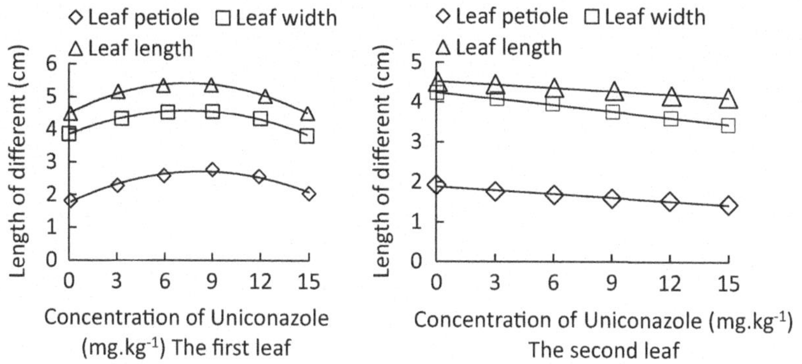

Fig. 5.28 The effects of dry seed treatment with uniconazole on soybean leaf morphology

Uniconazole treatment on dry seeds inhibited the growth in length and width, and the length-width ratio of the first trifoliate petioles. The normal increase of these parameters with growth showed a linear downward trend with the increase of uniconazole concentration, with the lowest decrease ranging from 0.5% to 14% compared with the control and the highest decrease ranging from 8.1% to 25% (Fig. 5.28). The elongation of the petioles, and the increase in length, and width of the second trifoliate showed a promotion effect. The elongation length first increased and then decreased with increasing uniconazole concentration. Thus, at 9 mg·kg^{-1}, the length, width, and length-width ratio of the second trifoliate reached maximum values that were 51%, 25.5%, and 26.4% higher than those in controls, respectively. The second trifoliate also showed higher values for these parameters at a concentration of 12 mg·kg^{-1}, with an average increase of 39.5% compared with the control.

5.3.2.2 Effects of Uniconazole on Chlorophyll in Soybean Leaves

Uniconazole treatment on dry seeds significantly increased the chlorophyll content of soybean leaves. As shown in Fig. 5.29, from 1 to 61 days after flowering, total chlorophyll and chlorophyll a contents in soybean plants treated with different concentrations of uniconazole showed a single-peak curve with an initial increase followed by a decrease, reaching a maximum value at 46 days after flowering. Treatment B3 resulted in the highest total chlorophyll content, and treatment B2 resulted in the highest content of chlorophyll a.

5.3.2.3 The Effect of uniconazole on Soybean Leaf Photosynthesis

Uniconazole dry mixing significantly affected the photosynthesis of soybean leaves. As shown in Table 5.24, treatment with different concentrations of uniconazole significantly increased Pn, g_s, and Tr of soybean leaves. Compared with controls, the B2 treatment performed the best, increasing Pn by 12.75% after 31 days of flowering, whereas the B1 and B3 treatments increased it by 11.65% and 8.8%, respectively.

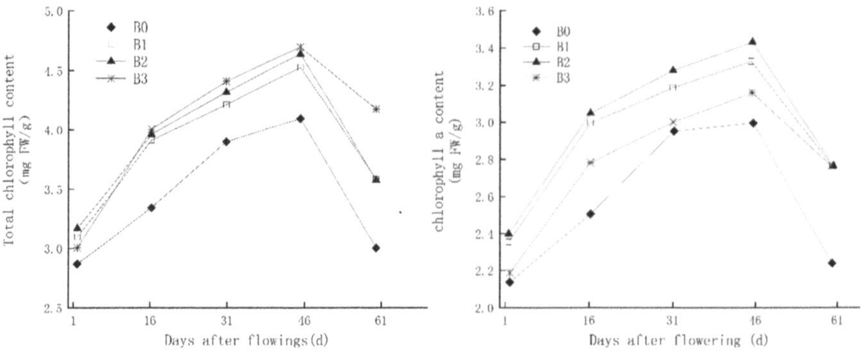

Fig. 5.29 The effect of uniconazole on leaf chlorophyll content of soybeans under strip intercropping. Note: B0, B1, B2, and B3 represent 0, 2, 4, and 8 mg kg^{-1} of uniconazole in dry mixing treatment

5.3.2.4 The Effect of Uniconazole on Antioxidant Enzyme Activity in Soybean Leaves

The activities of SOD and POD in soybean leaves at the seedling stage showed an increasing trend followed by a decrease, reaching a maximum value 35 days after sowing. Table 5.25 shows that SOD and POD activities in the B2 treatment were the highest, followed by B1, B3, and B0. Subsequently, at 42 days after sowing, SOD activity in the B2-treated leaves was 50.1%, 88.4%, and 274.4% higher than that in the B1, B3, and B0 treatments, respectively. Lastly, POD activity under the B2 treatment was also significantly higher than that in the B1, B3, or B0 treatments.

Malondialdehyde (MDA) content in leaves of soybean seedlings showed an increasing trend with time after sowing, but uniconazole treatment significantly reduced it. Further, MDA content was lowest in the B2 treatment from 21 to 42 days after sowing (Table 5.26). Then, at 42 days, MDA contents in the B2, B1, and B3 treatments were 27.6%, 17.1%, and 7.4% lower than that in the control treatment, B0, respectively. Conversely, proline (Pro) decreased with time after sowing, but uniconazole treatment significantly increased it. Both B2 and B1 treatments significantly promoted Pro content over control levels, ranking in the following order: B2 > B1 > B3 > B0 (Table 5.26).

5.3.3 Regulation of Petiole Angle by uniconazole in Soybean Plants

Shading significantly reduced petiole angle and increased petiole length. Exogenous application of 100 μmol gibberellin (GA) further reduced petiole angle, but exogenous application of 10 μmol GA synthesis-inhibitor uniconazole significantly

Table 5.24 The effect of uniconazole dry mixing on photosynthesis of soybeans under strip intercropping

Uniconazole treatment	Pn (μmol m^{-2} s^{-1})			g_s (mol m^{-2} s^{-1})			Tr (mmol m^{-2} s^{-1})		
	Days after flowering			Days after flowering			Days after flowering		
	1	31	61	1	31	61	1	31	61
B0	12.9Cd	18.1Bb	7.5Bc	0.35Cc	0.44Dd	0.18Dd	3.69Bc	4.29Bb	2.51 Bd
B1	17.6Abb	20.2Aa	11.4Aab	0.38Bb	0.60Bb	0.31Bb	3.94Aab	4.74Aa	3.50Ab
B2	19.0Aa	20.4Aa	12.4Aa	0.41Aa	0.64Aa	0.34Aa	4.06 Aa	4.82Aa	3.78Aa
B3	16.4Bc	19.7Aa	11.0Ab	0.38Bb	0.50Cc	0.21Cc	3.87Abb	4.42Abc	2.74Bc

Note: B0, B1, B2, and B3 represent 0, 2, 4, and 8 mg kg^{-1} uniconazole in dry mixing treatment. Different lowercase and uppercase letters within columns indicate significant differences at the 0.05 and 0.01 levels

Table 5.25 The effect of uniconazole on SOD and POD activities in soybean leaves under strip intercropping

Uniconazole treatment	SOD				POD			
	Days after seeding				Days after seeding			
	21	28	35	42	21	28	35	42
B0	50.8Cc	132.0 Cc	220.0 Dd	156.4 Dd	1116.5 Cd	1360.7 Dd	1436.1 Cc	1385.0 Cc
B1	94.4 Bb	204.9 Bb	433.7 Bb	390.3 Bb	1605.0 Bb	1941.4 Bb	2562.5 Aa	2049.8 ABb
B2	130.9 Aa	240.9 Aa	606.4 Aa	585.7 Aa	1704.9 Aa	2150.0 Aa	2672.5 Aa	2196.2 Aa
B3	87.7 Bb	181.9 Bb	359.5 Cc	310.8 Cc	1171.3 Cc	1628.4 Cc	2027.5 Bb	1968.6 Bb

Note: B0, B1, B2, and B3 represent 0, 2, 4, and 8 mg kg^{-1} of uniconazole

Table 5.26 Effects of seed treatment with uniconazole on MDA and Pro contents in leaves of soybean seedlings

Uniconazole treatment	MDA				Pro			
	Days after seeding				Days after seeding			
	21	28	35	42	21	28	35	42
B0	0.0216 aa	0.0352 aa	0.0354 aa	0.0637 aa	516.3 Bc	345.9 Bc	204.1 cc	67.6 Dd
B1	0.0131 cc	0.0215 bb	0.0241 cc	0.0528 cc	643.0 ABb	435.4 bb	425.0 bb	286.3 bb
B2	0.0123 cc	0.0141 cc	0.0240 cc	0.0461 Dd	760.0 aa	632.9 aa	529.5 aa	475.1 aa
B3	0.0170 bb	0.0218 bb	0.0307 bb	0.0590 bb	641.5 ABb	390.4 Bbc	368.3 bb	143.1 cc

Note: B0, B1, B2, and B3 represent 0, 2, 4, and 8 mg kg^{-1} of uniconazole

Table 5.27 Effects of uniconazole on soybean petiole angle

	Petiole angle(°)			Petiole length(cm)		
	Gibberellin	Uniconazole	CK	Gibberellin	Uniconazole	CK
Full sunlight	29.99	86.76	83.49	15.97	4.30	6.21
Shade	25.41	60.99	33.98	20.76	4.90	14.37

increased it, indicating that uniconazole inhibited the shading-induced reduction of soybean petiole angle under strip intercropping planting system (Table 5.27).

5.3.4 Regulation of Uniconazole on Lodging Resistance of Soybean Stem

5.3.4.1 Effects of Uniconazole on Soybean Internode Length and Lodging Resistance

Foliar application of Uniconazole under shading conditions significantly reduced soybean internode length soybean compared with that under full sunlight. Concomitant with the shortening of each internode, plant height was significantly reduced (Fig. 5.30), thus significantly improving the lodging resistance of soybean under shading.

5.3.4.2 The Effect of Uniconazole on Cell Length in Soybean Stems

Shading significantly promoted the elongation of lignified cells in soybean stems. Compared with the control group under full sunlight, the lengths of the lignified stem cells in the two soybean varieties under shading increased by 72.0% and 131.2%, respectively. However, after foliar application of uniconazole under the same conditions, the lengths of the stem lignified cells recovered to the level of the control group under normal light in the two varieties (Fig. 5.31).

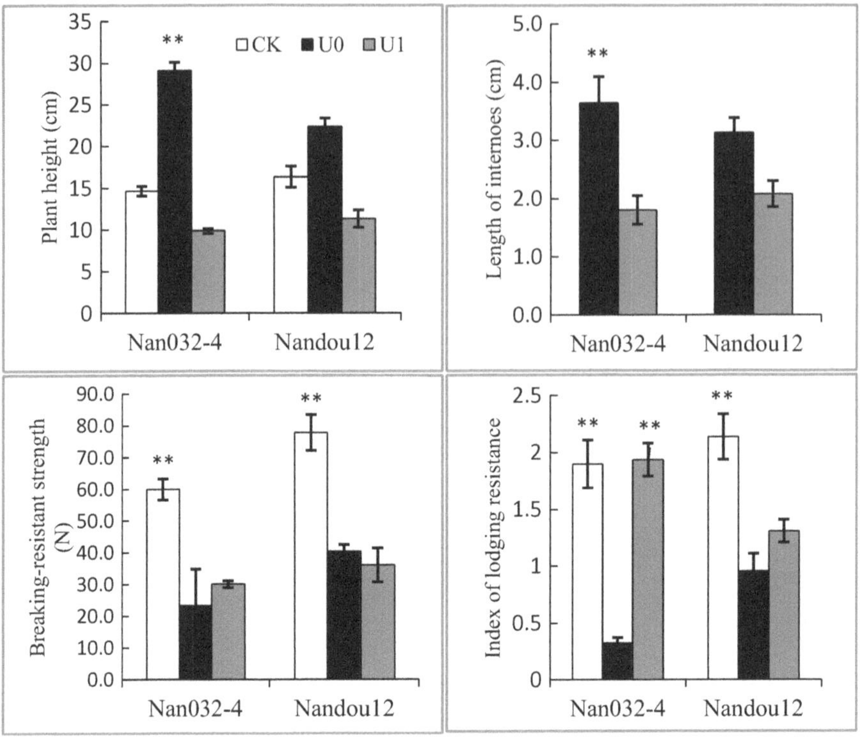

Fig. 5.30 Effects of uniconazole on internode length and lodging resistance in soybean

Fig. 5.31 Effect of uniconazole on stem cell length in soybean

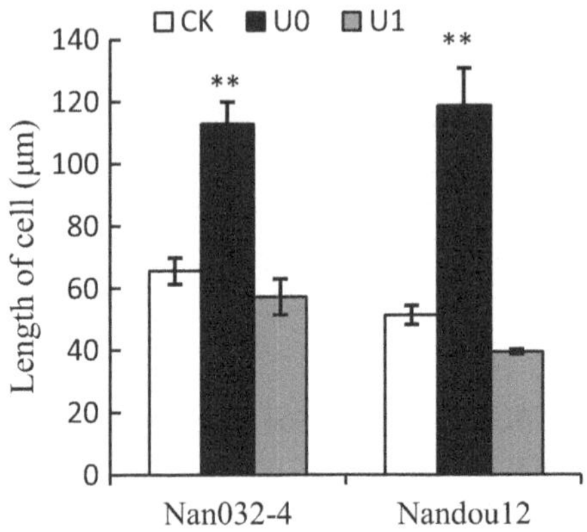

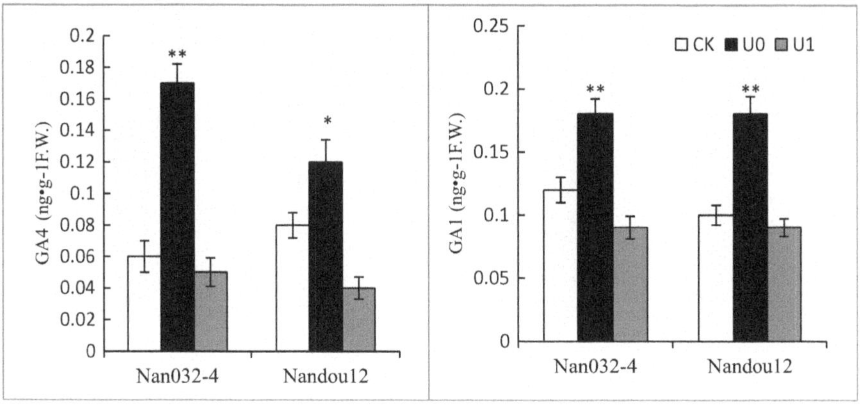

Fig. 5.32 Effects of uniconazole on gibberellin content in soybean stems

5.3.4.3 The Effect of Uniconazole on the Gibberellin Content in Soybean Stems

Shading significantly increased GA1 and GA4 contents in the stems of the two soybean varieties (Fig. 5.32). Compared with full sunlight treatment, shading increased GA1 content by 50% and 80% and GA4 content by 183.3% and 50% in the Nan032–4 and Nandou12 soybean varieties, respectively. However, when uniconazole was sprayed on the leaves, Stem GA1 and GA4 contents in both varieties decreased to the level recorded under the full sunlight treatment.

5.3.4.4 Effects of Uniconazole on GA Metabolism in Soybean Stems

Gibberellin metabolism-related gene expression was analyzed by using qPCR. The results showed that, compared with full sunlight, shading significantly upregulated the expression levels of three GA metabolism-related genes, namely, GmGA3ox1, GmGA20ox2, and GmGA2ox4, in the stems of the two soybean varieties, and downregulated the expression of GmGA20ox6 (Fig. 5.33). Further, treatment with uniconazole significantly downregulated the expression of gibberellin synthesis genes, GmGA3ox1 and GmGA20ox2, thereby inhibiting the synthesis of GA1 and GA4. Conversely, uniconazole treatment upregulated the expression of GA breakdown genes, GmGA2ox4 and GmGA2ox6, thereby increasing GA oxidase activity, catalyzing the transformation of GAs from the active to the inactivate types GA8 and GA34, respectively, and reducing the levels of the two active gibberellins GA1 and GA4 in soybean stems under shading conditions.

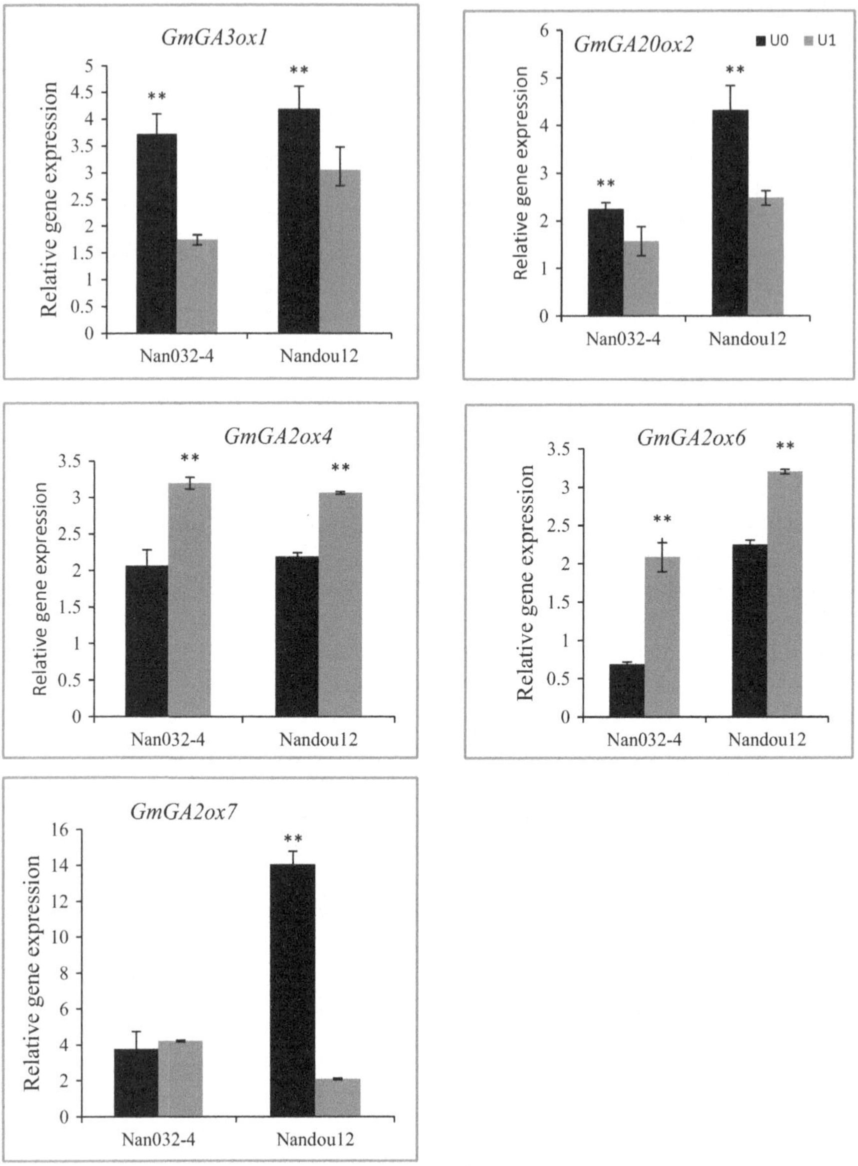

Fig. 5.33 Effects of uniconazole on gibberellin metabolism in soybean stems

6 Crop Quality Formation and Its Environmental Regulation in Maize-Soybean Strip Intercropping

Jiang Liu, Jianhua Chen, Yuanyuan He, Juncai Deng, Caiqiong Yang, Yanhong Yan, and Weiguo Liu

Abstract

This chapter examines the impact of maize-soybean intercropping on the quality of soybeans. The presence of corn in the intercropping system modifies various environmental factors, including light, moisture, and temperature, which depend on the specific field configurations and planting times. Maize-soybean intercropping significantly influences the chemical composition of soybean kernels, such as protein, fat, isoflavones, anthocyanins, and carotenoids. It also affects the sensory, nutritional, and visual qualities of fresh soybeans. Furthermore, this intercropping system has implications for the fermentation characteristics and nutritive value of corn-soybean silage, expanding its use as forage.

Keywords

Light environment · Quality · Isoflavones · Protein · Fat

Under field conditions, changes in light, water, temperature, and other microclimate factors directly or indirectly regulate crop metabolic processes and ultimately affect crop quality in maize-soybean strip intercropping systems. Such systems have realized a variety of benefits, including maize and soybean grain, fresh food, and silage, thus expanding the market. Therefore, the study and understanding of the regulation of crop quality formation by the environment in strip intercropping systems for improving crop stress resistance and crop output values is of paramount importance for the successful development of maize-soybean intercropping.

J. Liu (✉) · J. Chen · Y. He · J. Deng · C. Yang · Y. Yan · W. Liu
College of Agronomy, Sichuan Agricultural University, Chengdu, China
e-mail: jiangliu@sicau.edu.cn; chenjianhua@stu.sicau.edu.cn; 2020201031@stu.sicau.edu.cn

W. Yang et al. (eds.), *Theories and Application in Maize-Soybean Strip Intercropping System*, https://doi.org/10.1007/978-981-97-8723-4_6

6.1 Regulation of Soybean Grain Quality Formation in Maize-Soybean Intercropping Systems

Soybean products obtained from maize-soybean strip intercropping systems have multiple uses, among which grain production is the most widespread, because soybean kernels are rich in protein, fat, and other nutrients, making them an important source of human dietary protein and oil. Additionally, soybean kernels contain functional components such as isoflavones, anthocyanins, and carotenoids, which have important medicinal and nutritional values. Altogether, these metabolites determine the chemical quality of soybean grain crops. The changes in the light environment within the soybean canopy in intercropping systems differ from those that occur under monoculture conditions. Further, they are specific and complex in terms of time and space. In particular, due to the influence of the taller maize plants in a maize-soybean intercropping system, various chemical components in the developing soybean grains are modified. Therefore, soybean quality improvement requires a thorough understanding of the regulatory mechanisms of soybean grain chemical composition under the changing light environment of the maize-soybean strip intercropping system.

6.1.1 Effects of the Maize-Soybean Strip Intercropping System on Protein and Fat of the Soybean Grain

6.1.1.1 Effects of Field Configuration on Grain Protein and Fat in Intercropped Soybean

Proper field configuration can optimize crop population structure and make the best use of crop fine traits. A field configuration experiment on a maize-soybean intercropping system laid out in a 2:3 row ratio showed that different production unit widths (W1: 2.0 m, W2: 2.2 m, W3: 2.4 m) and different planting densities (D1: 52500 plants/hm^2, D2: 65000 plants/hm^2, D3: 82500 plants/hm^2, D4: 97500 plant/hm^2) had important effects on field microclimate and soybean grain quality. Specifically, soybean canopy transmittance of side and middle rows showed an increasing trend with increasing unit width (W3 > W2 > W1). Compared with W1 and W2, soybean canopy transmittance of side and middle rows in the W3 unit-width treatment significantly increased by 16.8% and 13.1%, and 16.8% and 8.6%, respectively. The soybean canopy received more light with increasing production unit width (Table 6.1). Similarly, planting density had a significant effect on soybean canopy transmittance ranked as follows: D4 > D3 > D2 > D1. Soybean canopy transmittance in side and middle rows in the D4 treatment increased by 12.4%, 2.9%, and 1.5%, and 10.3%, 2.7%, and 1.4%, relative to treatments D3, D2, and D1, respectively. Furthermore, when the production unit width was 2.4 m, D1 showed the lowest canopy transmittance among all planting density treatments, although the differences among other treatments were not significant. Soybean canopy transmittance in the side and middle rows in the D4 treatment significantly increased by 3.1% and 2.6%, respectively, compared with treatment D1. The light transmittance of the

Table 6.1 Soybean canopy transmittance under different field configurations

	Side rows				Middle rows			
Treatments	W_1	W_2	W_3	Average	W_1	W_2	W_3	Average
D_1	48.3d	51.3b	62.2b	53.9c	63.7d	71.7b	82.3b	72.6d
D_2	54.3c	58.0a	64.6a	58.9b	72.0c	78.4a	83.5ab	78.0c
D_3	56.1b	58.5a	64.3a	59.7b	73.5b	79.3a	84.2a	79.0b
D_4	59.9a	57.7a	64.13a	60.6a	77.5a	78.5a	84.4a	80.1a
Average	54.6c	56.4b	63.8a		71.6c	77.0b	83.6a	

Note: W1: Width of production unit 2.0 m; W2: Width of production unit 2.2 m; W3: Width of production unit 2.4 m. D1: Soybean population density: 52500 plants/hm^2; D2: Soybean population density 67,500 plants/hm^2; D3: Soybean population density 82,500 plants/hm^2; D4: Soybean population density 97,500 plants/hm^2. Means within columns followed by different lowercase letters are significantly different at the 0.05 level

soybean canopy in the middle row was higher than that in the side row. Furthermore, the light transmittance of the soybean canopy in the middle rows in treatments W1, W2, and W3 increased by 31.1%, 36.5%, and 31.0%, respectively, compared with that in the side rows. The canopy light transmittance in the middle rows in treatments D1, D2, D3, and D4 increased by 34.7%, 32.4%, 32.3%, and 32.2%, respectively, compared with that in the side rows, indicating that plants in the middle rows were better illuminated (Table 6.1).

Soybean grain protein content increased with increasing production unit width; specifically, that of the side-row plants in production units with widths of 2.4 m significantly increased by 1.5% and 1.4%, respectively, compared with that of production units with widths of 2.0 and 2.2 m, respectively. Population density also had a significant effect on soybean grain protein content. Thus, the grain protein content of side- and middle-row plants increased first and then decreased as population density increased, and that in the D2 and D3 treatments was higher. Furthermore, compared with treatments D1 and D4, grain protein content in the side and middle rows under treatment D3 significantly increased to 0.7% and 1.6%, and 2.5% and 2.5%, respectively. In contrast, only a minor difference was observed in grain protein content among density treatments at a bandwidth of 2.4 m, whereas that of side-row plants was marginally higher than that of middle-row plants (Table 6.2). These findings indicated that an adequate field configuration can increase soybean grain protein content.

Soybean grain fat content in side rows increased first and then decreased with increasing width of production units, ranking as follows: BW1 > BW3 > BW2 (Table 6.3). However, no significant differences were observed among treatments. Contrary to the pattern followed by protein content in soybean grain, soybean grain fat content decreased first and then increased as soybean planting density increased, with that in plants grown under treatments D1 and D4 being higher and those in the side and middle rows of D4-treated plants significantly increased to 2.0% and 3.3%, and 1.8% and 3.9% compared with D2 and D3, respectively. Further, under BW3, the ranking of the grain fat content of plants in the side and middle rows was D1 > D4 > D2 > D3, and that in plants growing in the side and middle rows in

Table 6.2 Soybean grain protein content under different field configurations

Treatments	Side row				Middle row			
	W_1	W_2	W_3	Average	BW_1	BW_2	BW_3	Average
D_1	39.70b	40.19a	40.46a	40.12ab	39.28b	39.73b	39.95b	39.65b
D_2	40.33a	40.42a	40.66a	40.47a	40.54a	40.13b	40.74a	40.47a
D_3	40.54a	40.01a	40.70a	40.42a	40.69a	40.66a	40.63a	40.66a
D_4	39.39b	39.42b	40.52a	39.78b	39.11b	39.85b	40.06b	39.68b
Average	39.99b	40.01b	40.59a		39.91a	40.09a	40.35a	

Note: W1: Width of production unit 2.0 m; W2: Width of production unit 2.2 m; W3: Width of production unit 2.4 m. D1: Soybean population density: 52500 plants/hm^2; D2: Soybean population density 67,500 plants/hm^2; D3: Soybean population density 82,500 plants/hm^2; D4: Soybean population density 97,500 plants/hm^2. Means within columns followed by different lowercase letters are significantly different at the 0.05 level

Table 6.3 Soybean fat content under different field configurations

Treatments	Side row				Middle row			
	W_1	W_2	BW_3	Average	BW_1	BW_2	BW_3	Average
D_1	17.45a	17.71ab	17.40a	17.52ab	17.70ab	17.50a	18.01a	17.74a
D_2	17.35a	17.33b	17.10ab	17.26bc	17.47bc	17.45a	17.44ab	17.45a
D_3	16.77b	17.63ab	16.73b	17.04c	17.37c	16.96b	17.02b	17.11b
D_4	17.78a	17.91a	17.11ab	17.61a	18.04a	17.47a	17.8a	17.77a
Average	17.34a	17.65a	17.10b		17.64a	17.35a	17.57a	

Note: W1: Width of production unit 2.0 m; W2: Width of production unit 2.2 m; W3: Width of production unit 2.4 m. D1: Soybean population density: 52500 plants/hm^2; D2: Soybean population density 67,500 plants/hm^2; D3: Soybean population density 82,500 plants/hm^2; D4: Soybean population density 97,500 plants/hm^2. Means within columns followed by different lowercase letters are significantly different at the 0.05 level treatment D1 significantly increased by 1.8% and 4.0%, and by 3.3% and 5.8% compared with D2 and D3, respectively. Lastly, contrary to the variation pattern of grain protein content, the grain fat content of middle-row soybean plants was slightly higher than that of side-row plants.

The increase in production unit width in maize-soybean intercropping systems led to the increase of soybean canopy transmittance and the improvement of the microclimate within soybean bands. When the intercropping strip width was 2.4 m (W3), the protein content of soybean grain increased significantly, whereas fat content concomitantly decreased to some extent. Further, protein content increased first and then decreased with increasing planting density, peaking at 67500 (D2) and 82,500 plants/hm^2 (D3). Meanwhile, in the south, high protein contents greatly contributed to soybean quality and, with respect to zonal maize-soybean intercropping, at 2.4 m for production unit width and 67,500 to 82,500 plants/hm^2, soybean canopy light transmittance was higher, the field microclimate environment was more suitable, and soybean quality was greater than the corresponding values observed under any other unit width-planting density combination treatment.

6.1.1.2 Field Configuration Effects on Protein and Fat Contents in Strip Intercropping Soybean Grain

Compared with soybean-maize strip intercropping, the scope of application of relay strip intercropping is relatively narrow and is concentrated mainly in southwest China. Most available studies have focused on yield traits, whereas little attention has been paid to grain quality, and research results are controversial. Table 6.4 summarizes the protein and fat contents of soybean grains under different field configurations of maize-soybean relay strip intercropping over the past 10 years. As can be seen, soybean grain protein content under relay intercropping initially increased from 1.66% to 6.15% and then decreased from 0.20% to 3.14% compared with the values observed in monoculture soybean. Conversely, fat content in the same test materials decreased from 1.63% to 5.97% and increased from 0.45% to 2.83%. Thus, protein and fat contents in soybean grain under relay intercropping did not only increase or decrease but were closely related to the specific field configuration.

Correlation analysis of soybean grain protein and fat contents with the main field configuration parameters, such as maize-soybean spacing, soybean row spacing, plant spacing, and population density, showed that soybean grain protein content decreased significantly as maize row spacing and soybean row spacing increased (Fig. 6.1a,b). Specifically, soybean protein content increased significantly with increasing plant spacing (Fig. 6.1c). However, it also increased with increasing soybean planting density, although the correlation was not significant (Fig. 6.1d). In fact, within a certain range of maize-soybean and soybean row spacing (maize-soybean spacing <57.2 cm, Fig. 6.1a intercept; soybean row spacing <41.4 cm, B intercept in Fig. 6.1), grain protein content of banded intercropping soybean increased compared with that of net intercropping soybean, while that in relay strip intercropping soybean showed a negative growth trend when the spacing between maize and soybean rows continued to increase. However, the effects of

Table 6.4 Comparison of soybean grain quality under different field configurations of maize-soybean relay strip intercropping

Row ratio	Maize-soybean spacing (cm)	Row spacing (cm)	Plant spacing (cm)	Density (ten thousand plant/hm^2)	Protein content (%)			Fat content (%)			
					Monoculture	Strip intercropping	Amplitude	Monoculture	Strip intercropping	Amplitude	
2:3	40.0	35.0	15.0	11.4	44.17	45.83	3.76%	18.07	17.49	−3.21%	Yu (2009)
2:3	40.0	35.0	15.0	11.4	44.17	45.31	2.58%	17.76	17.47	−1.63%	Yu (2009)
2:3	45.0	30.0	15.0	13.3	42.25	44.85	6.15%	19.81	19.42	−1.97%	Song (2009)
2:3	45.0	30.0	15.0	13.3	41.98	44.26	5.43%	20.35	19.78	−2.80%	Song (2009)
2:2	58.5	33.0	15.0	10.1	45.05	46.17	2.49%	17.43	16.39	−5.97%	Ye (2010)
2:2	60.0	30.0	10.0	16.7	40.47	41.14	1.66%	17.68	17.38	−1.70%	Chen (2011a, b)
2:2	72.5	33.5	11.2	13.3	47.7	46.2	−3.14%	20.00	20.09	0.45%	Xiang (2012)
2:2	60.0	40.0	8.3	15.0	48.95	48.26	−1.41%	17.12	17.61	2.83%	Tao (2013)
2:2	50.0	50.0	10.0	10.0	50.7	50.6	−0.20%	/	/	/	Cai et al. (2016)
2:2	55.0	50.0	10.0	10.0	44.05	43.075	−2.21%	/	/	/	Liu (2019)
2:2	60.0	40.0	10.0	12.5	49.85	48.74	−2.23%	17.89	18.04	0.84%	Xiao (2020)

Note: “row ratio” refers to the ratio of the number of maize rows to that of soybean rows; “maize-soybean spacing” is the spacing between maize and soybean rows; “amplitude” expresses the ratio of the difference in protein or fat contents between relay strip intercropping and monoculture, to the corresponding content observed in monoculture soybean

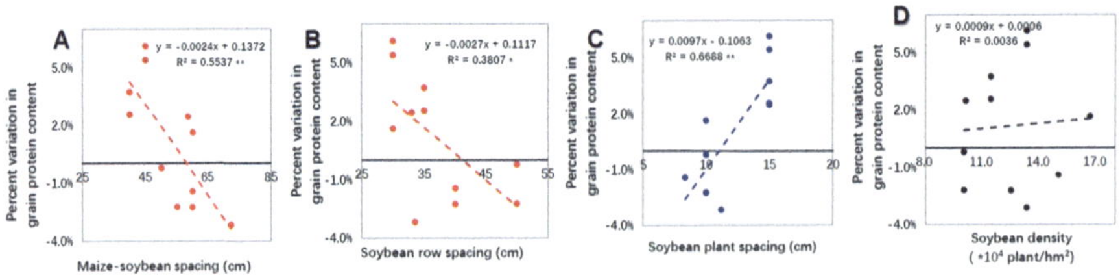

Fig. 6.1 Correlation analysis between main field configuration parameters and soybean grain protein content in relay strip intercropping

maize-soybean plant spacing and planting density on protein content showed an opposite trend. Particularly, when plant spacing was less than 11.0 cm (Fig. 6.1c-intercept), the grain protein content of relay strip intercropping soybean was lower than that of net cropping soybean and showed a positive growth trend as plant spacing increased and exceeded the threshold value.

Studies have confirmed that, during the drum developmental stage, shading leads to protein accumulation in the soybean grain; thus, in intercropping systems, the shading of the soybean canopy by maize leaves results in the improvement of soybean grain quality by indirectly influencing grain protein content. Further, in the relay intercropping system, shading of the short-height soybean canopy by the taller maize plants occurs mainly during vegetative growth of the soybean plants, mainly affecting the soybean canopy structure and leading to a decrease in the number of leaves and their size. Subsequently, when soybean plants enter the reproductive growth period, the light environment is restored, hence reducing intraspecies shading-mediated protein accumulation. However, when the spacing between soybean rows was below a certain threshold, inter-grain shading among soybean plants increased with increasing planting density, thereby resulting in an increase in soybean grain protein content (Fig. 6.1d). Therefore, interspecific shading at the grain drum stage caused by strip intercropping, and intraspecific shading caused by reduced row spacing in relay strip intercropping may be key factors to manipulate to increase grain protein content in intercropped soybean.

Numerous studies have also confirmed that the synthesis and accumulation of protein and fat in soybean grains are under opposite regulatory mechanisms. Based on the comparative data of maize and soybean grain quality recorded under different configurations of relay strip intercropping fields (Table 6.4), correlation analysis of grain protein and fat contents further confirmed those results. Additionally, compared with monoculture soybean, the variation in protein content in relay strip intercropping soybean grain was significantly and negatively correlated with the variation in fat content (Fig. 6.2).

Therefore, the relationship between field configuration parameters and soybean grain fat content was similar to that of protein content but the correlation trend was the opposite, namely, soybean grain fat content increased as maize-soybean spacing and soybean row spacing increased. When the threshold values of these parameters reached 72.4 cm (Fig. 6.3a intercept) and 37.9 cm (Fig. 6.3b intercept), respectively,

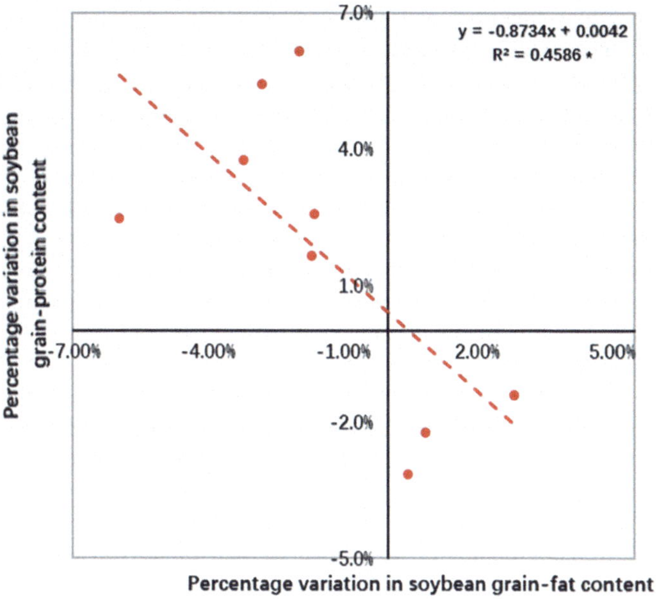

Fig. 6.2 Correlation analysis of grain protein and fat contents in relay strip intercropping soybean

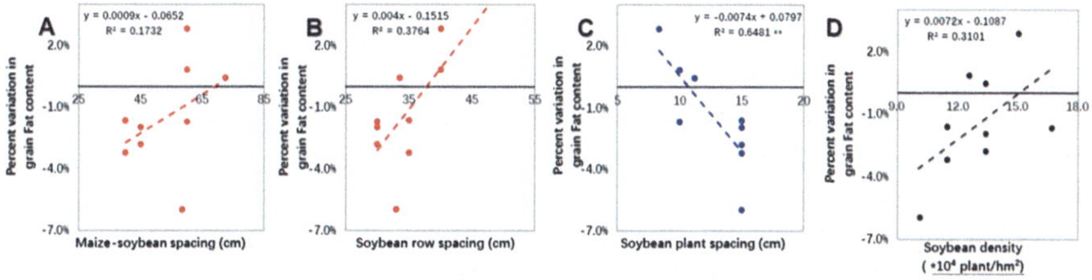

Fig. 6.3 Correlation between main field configuration parameters and soybean grain fat content in relay strip intercropping

soybean grain fat content increased. However, as the distance between plants increased, soybean grain fat content decreased, and when the threshold value was greater than 10.8 cm (Fig. 6.3c intercept), soybean grain fat content decreased. Further, fat content increased with increasing planting density. When the planting density was lower than 151,000 plants/hm^2 (d intercept of Fig. 6.3), the grain fat content of belt intercropping soybean was lower than that of monoculture soybean. When population density exceeded this threshold, grain fat content increased. However, the correlation between protein and fat content and planting density was not significant, indicating that the synthesis and accumulation of protein and fat in relay strip intercropping soybean was mainly determined by the distance between maize and soybean rows, the distance between soybean rows, and between plants along rows, compared with monoculture soybean at the same planting density.

To ensure a high yield per unit area, concomitantly achieving the goal of "no reduction of maize yield and more soybean harvest in one season," the maize-soybean strip intercropping system adopted the strategy of expanding the width of maize-soybean strips to ensure high light intensity for the soybean canopy and a marginal effect of maize-soybean shading. Similarly, the strategy aimed to reduce the distance between plants to ensure an effective number of plants per unit area and overall crop yield. According to the above results, the increase of maize-soybean spacing and soybean row spacing and the reduction of distance between plants were not conducive to protein accumulation, although they did enhance fat accumulation. Therefore, by optimizing field configuration and fitting the above parameters to an adequate interval, the total protein and lipid content in relay strip intercropping soybean can be increased.

6.1.1.3 Regulation Patterns of Soybean Grain Protein and Fat Contents by Spatiotemporally Differing Shading Treatments

Shading by the maize canopy is an important environmental factor regulating the formation of soybean quality. Under the maize-soybean strip intercropping system, the shade due to intercropping or relay intercropping showed spatial and temporal differences. In relay intercropping soybean, shading affected plant growth from the time of seedling emergence to the end of the first or full flowering stage. In contrast, in the strip intercropping system, the maximum extent of shading began at the initial or full flowering stage and continued through the end of the full ripening stage (Wang et al. 2016). These observations clearly underline the enormous theoretical significance and practical value of exploring the mechanisms underlying the control of soybean quality formation in field experiments under controlled shading conditions at different growth stages. This would further optimize the field configuration of mixed planting.

Compared with the control treatment, shading increased protein content and reduced the crude fat content of soybean grains, but different shading treatments had different effects (Table 6.5). The protein content at VEV6, VER1, and VER2 in the early shading treatment was 0.22%, 3.10%, and 2.12% higher than that of the control, respectively. There was no significant difference between VEV6 and the control, indicating that the protein content of soybean grains in the early shading to V6 stage was not significant and that the protein content of soybean grains would increase significantly after shading in the R1 stage. Meanwhile, the crude lipid content at VEV6, VER1, and VER2 in the early shading treatment was 0.31%, 1.55%, and 3.91% lower than that of the control, respectively. There was no significant difference between the contents at VEV6 and the control, indicating that the influence of shading at the V6 stage was not significant and that the crude lipid content in soybean grains would decrease significantly after shading to the R1 stage. The protein content at R1R8, R2R8, and R5R8 in the late shading treatment was 9.20%, 8.98%, and 6.23% higher than that of the control, respectively, indicating that the influence of late shading on soybean protein was greater than that observed for early shading. Furthermore, the longer the shading time during late shading, the greater the increase in protein content. The crude lipid content at R1R8,

Table 6.5 Effects of different shading treatments on protein and crude fat contents in soybean grains

		Protein (%)		Crude fat (%)	
Year	Treatment	Guixia 2	Nandou 12	Guixia 2	Nandou 12
2011	VER8	44.15% a	45.04% a	18.5% d	18.70% c
	VEV6	40.67% d	40.48% d	20.26% a	20.53% a
	VER1	41.48% c	42.28% c	19.78% b	20.37% a
	R1R8	43.78% a	44.72% a	18.73% d	18.60% c
	R5R8	42.67% b	43.66% b	19.37% c	19.62% b
	CK	40.32% d	40.79% d	20.2% a	20.47% a
2012	VER8	44.49% a	45.31% a	18.48% c	18.00% d
	VER5	43.46% b	44.13% b	18.71% c	18.90% c
	VER2	41.54% c	42.17% c	19.71% b	19.48% b
	R2R8	44.24% a	45.16% a	18.11% d	18.14% d
	CK	40.58% d	41.66% d	20.45% a	20.71% a

Note: Shading from emergence to ripening stage (VER8); shading from emergence to sixth festival stage (VEV6); shading from emergence to flowering stage (VER1); shading from the beginning of flowering to the ripening stage (R1R8); shading from the beginning of grain development to the ripening stage (R5R8); no shading at any time during growth (CK); shading from emergence to the beginning grain development (VER5); shading from emergence to full flowering stage (VER2); shading from full flowering to full ripening stage (R2R8). LSD was used for comparison among different treatments; different lowercase letters indicate significant differences among shading treatments for the same variety at the P = 0.05 level

R2R8, and R5R8 during the late shading treatment was 9.31%, 10.74%, and 4.28% lower than that of the control, respectively, indicating that the influence of late shading on crude lipid content in soybean grains was greater than that recorded upon early shading and that the longer the shading time, the greater the reduction in the crude lipid content.

6.1.2 Effects of Maize-Soybean Strip Intercropping on Functional Components of Grain Soybean

6.1.2.1 Comparison of Grain Isoflavone and Carotenoid Contents in Monoculture and Intercropping Soybean

The results of long-term continuous field experiments have shown that the accumulation of isoflavones in soybean grains increased with grain growth, whereas the increasing trend in relay intercropping soybean began to be higher than that in monoculture at the R7 stage and peaked after natural air-drying. The isoflavone contents in soybean grains grown under relay strip intercropping were significantly higher (18.85%) than those in monoculture soybean (Fig. 6.4a). In contrast, grain carotenoid content showed a decreasing trend with growth, particularly in relay strip intercropping soybean grain, with a maximum decrease of 43%. As grains gradually matured, their carotenoid content did not decrease significantly and did not change significantly after harvest. However, during the entire process from podding to the

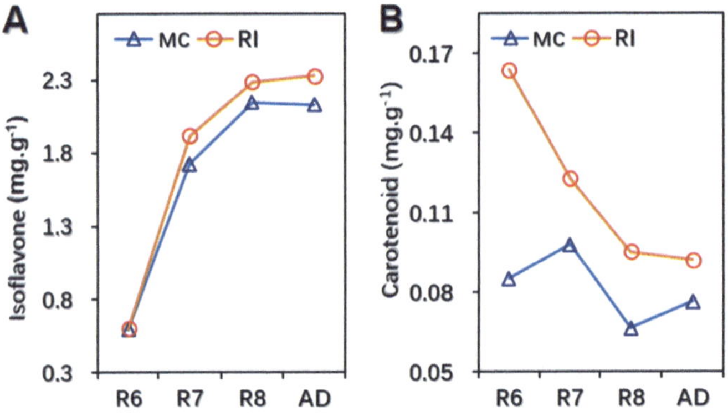

Fig. 6.4 Comparison of soybean-grain isoflavone and carotenoid accumulation in monoculture and relay strip intercropping systems. Note: *MC* monoculture, *IC* relay strip intercropping, *R6* grain-filling stage, *R7* early maturity, *R8* full maturity, *AD* air-drying period

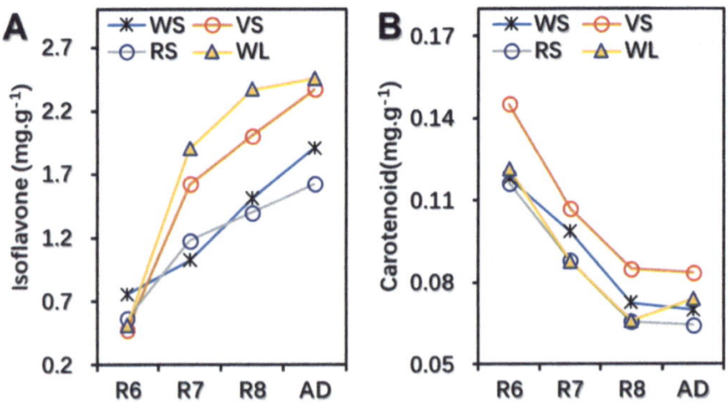

Fig. 6.5 Effects of shading at different growth stages on isoflavone and carotenoid accumulation in soybean grains

air-drying stage, the carotenoid content of soybean grains in relay strip intercropping was always higher than that in monoculture soybean. Overall, the average increase rate of carotenoid content in soybean grains was approximately 34.1% (Fig. 6.4b).

6.1.2.2 Effects of Shading at Different Growth Stages on Isoflavone and Carotenoid Accumulation in Soybean Grains

Years of shading experiments at different growth stages have established that grain-isoflavone contents are highest under shading during vegetative growth (VS) or under normal light (WL) during the whole growth period. Conversely, they are typically lowest under shading treatment during reproductive growth (RS) or throughout growth (WS) (Fig. 6.5a). Specifically, compared with the WL treatment, grain-isoflavone content decreased by 22%, 8%, and 33% under WS, VS, and RS, respectively. Thus, shading does not favor the synthesis nor the accumulation of isoflavones during grain development. Whereas, if applied during vegetative growth, shading may influence the synthesis and accumulation of isoflavones in soybean grains during reproductive growth by affecting soybean canopy structure.

Soybean-grain carotenoid contents under shading at different growth stages are shown in Fig. 6.5a. Particularly when treated with VS or WL, grain carotenoid contents were higher, whereas lower levels were observed with RS and WL treatments. Overall, compared with the WL control treatment, carotenoids in shaded grains decreased by 11%. Specifically, a 20% decrease was observed in the RS relative to the WL treatment, whereas a 15.5% increase was registered in the VS relative to the WL treatment. Further, as grains matured, carotenoid content decreased significantly. In contrast, shading treatment seemingly benefitted the accumulation of carotenoids in soybean grain, especially when applied during vegetative growth.

The light environment during reproductive growth is crucial for the quality of soybean grains. Thus, compared with the equal row-spacing treatment, the light environment during the reproductive growth period in strip intercropping soybean improved, leading to the accumulation of isoflavones and carotenoids in the developing grain.

6.2 Effects of Strip Intercropping on the Quality of Fresh Edible Soybean

Fresh edible soybean is one of the most important final value-added products of the maize-soybean strip intercropping system. The quality of fresh edible soybean is complex, primarily including appearance, palatability/digestibility, and nutritional quality. Hence, overall superior quality involves coordinating all quality components. Further, ambient variables such as light, temperature, and rainfall affect the quality of fresh edible soybean. As these variables differ with the cropping system, studying the differences in appearance and nutrient accumulation in fresh edible soybean between monoculture and strip intercropping is of great significance for quality improvement of fresh edible soybean produced by intercropping and, ultimately, for the enhancement of the economic benefits of soybean intercropping.

6.2.1 Effects of Strip Intercropping on the Phenotype of Fresh Edible Soybean

The phenotypic traits such as green pod size, pod color, number of grains per pod, and grain size are considered important quality traits of fresh edible soybean. Generally, people prefer large grains and large pod varieties with long, wide, and thick pods. Compared with monoculture, 100-pod fresh weight, 100-pod grain weight, the number of pods with one and two grains, and the rate of two-grain pods were significantly reduced in maize strip intercropping with different varieties of fresh edible soybean. The 100-grain fresh weight and the number of shriveled pods of intercropped Zhexiandou 4 and Qingsu 5 varieties were higher than those of the corresponding monoculture, whereas those of Shenxian 6 were lower. In turn, 100-pod fresh weight, 100-pod grain weight, 100-grain fresh weight, and the number of pods with three grains were affected by planting method, variety, and the interaction between them. Except for the number of shrunken pods, the other appearance quality indexes were affected by the planting patterns, and the number of pods was mainly affected by the interaction between the planting system and variety. Among quality indexes, the 100-pod fresh weight and 100-pod grain weight of Zhexiandou 4 were the highest, whereas the 100-grain fresh weight and two-grain pod rate were lower than those in other varieties. Under monoculture, the 100-pod fresh weight of Zhexiandou 4 was 39.6 and 13.4 g higher than those of Qingsu 5 and Shenxian 6, respectively, the 100-pod grain weight was 11.8 and 2.11 g higher than those of Qingsu 5 and Shenxian 6, respectively, the fresh weight of 100-pods was 10.7 and 9.1 g lower than those of Qingsu 5 and Shenxian 6, respectively, and the two-grain pod rate was 6.0% and 4.5% lower than those of the same two varieties, respectively. Furthermore, under strip intercropping, the 100-pod fresh weight of Zhexiandou 4 was 24.9 and 28.8 g higher than those of Qingsu 5 and Shenxian 6, respectively, the 100-pod grain weight was 11.7 and 24.6 g higher than those of Qingsu 5 and Shenxian 6, respectively, the 100-grain fresh weight was 18.3 and 3.0 g lower than those of Qingsu 5 and Shenxian 6′, respectively, and the two-grain pod rate was 2.0% and 6.5% lower than those of Qingsu 5 and Shenxian 6, respectively (Table 6.6).

6.2.2 Effects of Strip Intercropping on the Chemical Quality of Fresh Edible Soybean

Nutritional and edible quality determine the market price of fresh edible soybean. The umami of fresh soybean mainly comes from a large number of free amino acids. Compared with ordinary soybean, the metabolic rate of free amino acids in fresh soybean differs with grain development, and the content is nearly twice as high as that of ordinary soybean. Fresh soybean can provide more essential amino acids than ordinary soybean. Additionally, starch and soluble sugar contents are directly related to the edible quality of fresh soybean. Varieties with high soluble sugar content have a strong sweetness, those with a high starch content are strongly waxy, and those with a high-fat content have a soft texture.

The protein content of fresh edible soybean in maize-soybean strip intercropping was significantly higher than that in monoculture, ranging from 360.4 to 390.4 mg·g^{-1} in the former and from 353.7 to 377.8 mg·g^{-1} in the latter (Table 6.7). Meanwhile, crude fat content decreased, but the difference was not significant. Among different varieties, Shenxian 6 had the best protein and fat contents in both planting systems. Further, free amino acid levels differed with variety and planting system. Among them, umami, sweet, bitter, aromatic, and total free amino acids increased in Zhexiandou 4 under strip intercropping, whereas they decreased in the Qingsu 5 variety. Meanwhile, the difference in amino acids in

Table 6.6 Phenotypic quality of fresh soybean under different planting systems (from Fang et al. 2016)

Planting system	Variety	100-pod fresh mass /g	100-pod grains mass/g	100-grain fresh mass /g	Number of shrunken pods	Number of pods with one grain	Number of pods with two grains	Number of pods with three grains	Two-grain pod rate (%)
Monoculture	Zhexiandou 4	244.977a	130.014a	61.172c	1.62ab	8.67c	11.22b	4.78a	42.68a
	Qingsu 5	205.377d	118.183b	71.857b	1.28b	11.97a	14.77a	2.17b	48.68a
	Shenxian 6	231.552b	127.903a	70.254b	2.10a	10.48b	16.11a	5.38a	47.20a
Strip intercropping	Zhexiandou 4	216.571c	129.269a	64.615c	1.68ab	3.58d	3.98c	1.92b	35.43b
	Qingsu 5	191.657e	117.526b	83.311a	1.98ab	5.03d	4.74c	1.29b	36.38b
	Shenxian 6	187.738e	104.634c	67.646b	1.61ab	4.13d	5.03c	1.21b	41.96a
F value	*F* value								
	P	99.495**	36.275**	13.933**	0.241 ns	278.753**	405.716**	131.104**	39.111**
	C	43.389**	38.166**	60.352**	0.96 ns	13.943**	14.255**	21.218**	5.956*
	P × C	9.152**	30.336**	13.771**	7.417**	2.236 ns	5.99*	17.255**	2.52 ns

Note: P represents the planting system, C represents vegetable soybean varieties, and P × C represents the interaction between the planting system and vegetable soybean variety. Different lowercase letters within columns indicate significant differences at the 0.05 level. The F value was the result of a two-factor analysis of variance. * and * * indicate significant effects at the 0.05 and 0.01 levels, respectively; ns represents no significant difference at the 0.05 level

Table 6.7 Amino acid content in fresh edible soybean under different planting systems

System	Variety	Free amino acids				
		Umami amino acids	Sweet amino acids	Bitter amino acids	Aromatic amino acids	Total free amino acids
Monoculture	Zhexiandou 4	3.550c	7.006b	2.377c	0.582bc	13.515c
	Qingsu 5	5.615a	8.775a	3.760a	1.683a	19.833a
	Shenxian 6	1.772d	2.632d	1.664d	0.202c	6.270d
Strip intercropping	Zhexiandou 4	4.595b	7.738ab	3.203b	0.678bc	16.215b
	Qingsu 5	3.272c	5.699c	2.424c	0.740b	12.136c
	Shenxian 6	1.870d	2.574d	1.610d	0.403bc	6.457d

Note: Different lowercase letters indicate significant differences among shading treatments for the same variety at the P = 0.05 level

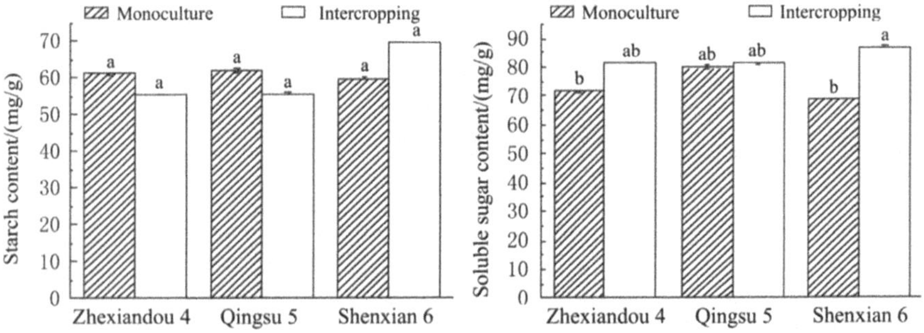

Fig. 6.6 Starch and soluble sugar contents in fresh edible soybean under different planting systems. Note: Different lowercase letters indicate significant differences among shading treatments for the same variety at the $P = 0.05$ level

Shenxian 6 between the two planting systems was not significant. The total free amino acid content in Zhexiandou 4 under strip intercropping was significantly higher than that in monoculture, whereas it was significantly lower in Qingsu 5, and it increased in Shenxian 6 (Table 6.7).

There was no significant difference in starch content among different planting systems or varieties; although in this case, the trend change was different. Thus, in Shenxian 6, starch increased in strip intercropping and decreased in the other two varieties. Moreover, soluble sugar content also increased in intercropping, and the difference relative to monoculture was significant. Although starch and soluble sugar contents in Shenxian 6 were lower than those of the other two varieties, they were found at optimal levels in strip intercropping (Fig. 6.6).

6.2.3 Effects of the Planting System and Variety on Vegetable Soybean Quality

Analysis of variance showed that planting systems had significant effects on protein, soluble sugar, umami, sweet, aromatic, and total free amino acid contents in fresh soybean, whereas crude fat, starch, and bitter amino acid contents were not affected by the planting system. Further, variety had significant effects on protein, crude fat, and various free amino acid types, not on soluble sugar or starch. Finally, the interaction between the planting system and variety had significant effects on protein and various amino acid types but not on crude fat, soluble sugar, or starch (Table 6.8).

The protein content of fresh edible soybean increased significantly under strip intercropping, while soluble sugar increased and crude fat decreased. Further, sweet, umami, bitter, and total free amino acids and water content were highest in Zhexiandou 4, whereas crude fat, soluble sugar, and starch content were the lowest. The protein content in Qingsu 5 was the lowest, while other indexes were intermediate. Although in Shenxian 6, protein, crude fat, soluble sugar, and starch were highest, various amino acid types and water content were the lowest. Moreover, soluble sugar, starch, free amino acids, and protein were higher than those in monoculture; therefore, Shenxian 6 showed the best nutritional quality among the three varieties, followed by Zhexiandou 4 and Qingsu 5.

In contrast with chemical quality, planting systems, and varieties had significant effects on phenotypic quality (except for the number of shrunken pods). Planting systems had significant effects on umami amino acids, proteins, and soluble sugars considering nutritional quality. Meanwhile, starch content was stable and was not affected by planting system, variety, or the interaction between them.

Compared with fresh edible soybean produced in maize-soybean strip intercropping, although its phenotypic quality was reduced, except for free amino acids, chemical quality traits including protein, crude fat, starch, and soluble sugar showed better levels, and variety were major sources of the observed differences. Further study of the mechanism of formation of fresh edible soybean quality under strip intercropping will allow screening for high-quality vegetable soybean varieties to further develop maize-soybean strip intercropping systems aimed at producing fresh edible vegetable soybean.

6.3 Quality Evaluation of Mixed Silage Produced by Strip Intercropping

Maize is rich in fermentable water-soluble carbohydrates that are readily ensiled. Indeed, it is the main silage raw material; however, its crude protein content is low and cannot meet livestock protein needs. Whereas soybean has a high crude protein content, which makes it of great nutritional value, although its buffering energy is large and water-soluble carbohydrate content is low, whereby single silage cannot meet the needs of lactic acid bacteria for fermentation substrates, which can easily

Table 6.8 Analysis of variance for factors affecting fresh edible soybean quality

F value	Protein content	Crude fat	Soluble sugar	Starch	Umami amino acids	Sweet amino acids	Bitter amino acids	Aromatic amino acids	Total free amino acids
P	164.039**	1.576 ns	9.294**	4.369 ns	4.799**	6.823*	1.866 ns	5.162*	6.134*
C	268.97**	11.911**	0.527 ns	1.354 ns	80.581**	104.698**	41.612**	31.439**	88.049**
P × C	10.556**	0.361 ns	2.190 ns	0.117 ns	305.65**	14.339**	20.856**	14.799**	23.4**

Note: P represents the planting system, C represents vegetable soybean varieties, and P × C represents the interaction between the planting system and vegetable soybean variety. * and * * indicate significant effects at the 0.05 and 0.01 levels lead to silage fermentation failure, large nutrient loss, and difficulty in successful silage production. Alternatively, mixed silage expands the feed source, improves silage fermentation quality and nutritional value, and is an effective method to remedy the high nutritional value silage shortage due to the lack of high-quality forage. Proper selection of cropping practices, such as maize and soybean varieties, and adjustment of crop growth period can make simultaneous harvesting of maize and soybean for mixed silage feasible.

6.3.1 Analysis of the Nutritional Quality of Maize and Soybean Silage under Different Row Ratios

The effects of strip intercropping at different row ratios on the quality of mixed silage using soybean and maize varied markedly. We hypothesized that an appropriate strip intercropping row ratio should maximize the economic benefits derived from simultaneous maize and soybean cropping. To test this hypothesis, an experiment was designed to evaluate two contrasting soybean varieties (S1: Nandou No.25, S2: Fendou Mulv No. 2) and three kinds of strip intercropping row ratio configurations of maize (Fig. 6.7), with maize and soybean monocultures as control treatments. When maize kernels were at the milky stage in a field planted using a maize-to-soybean ratio of 1:2, plants were randomly mowed, kneaded, vacuum packaged, and ensiled at room temperature (25–30 °C). Then, after 60 days of silage, nutritional and fermentation quality were analyzed.

The results showed that the dry matter ratio recorded for S1 and B1 treatments was lower than that of monoculture maize, but the dry matter quality of other strip intercropping treatments was higher than that of monoculture maize (Table 6.9). Soybean variety, row ratio, and the interaction between them had significant effects on dry matter quality and neutral-detergent fiber (NDF) content. Similarly, soluble carbohydrate content was significantly affected by soybean variety-row ratio interaction but not row ratio. Meanwhile, crude protein content was significantly affected by row ratio, soybean variety, and the interaction between the two. Overall, the soluble carbohydrate content of monoculture maize was significantly higher than that recorded for other treatment groups, whereas crude protein content was significantly lower than that of other treatment groups. In turn, NDF and acid-detergent fiber (ADF) contents were significantly affected by soybean variety in strip intercropping fields. Furthermore, row ratio and soybean variety had a significant effect on NDF but not on ADF content. The relative feeding value of maize strip intercropping with different soybean varieties at a row ratio of 2:3 was higher than those for other row ratio treatments. Further, compared with strip intercropping and monoculture maize, NDF and ADF contents were higher in monoculture soybean but the relative feeding value was lower (Table 6.9).

Analysis of silage nutritional quality at 60 days after treatment initiation showed that, in addition to dry matter mass and ADF content, row ratio had a significant effect on crude protein, soluble carbohydrates, and NDF contents (Table 6.10). The soybean variety-row ratio interaction had a significant effect on dry matter mass,

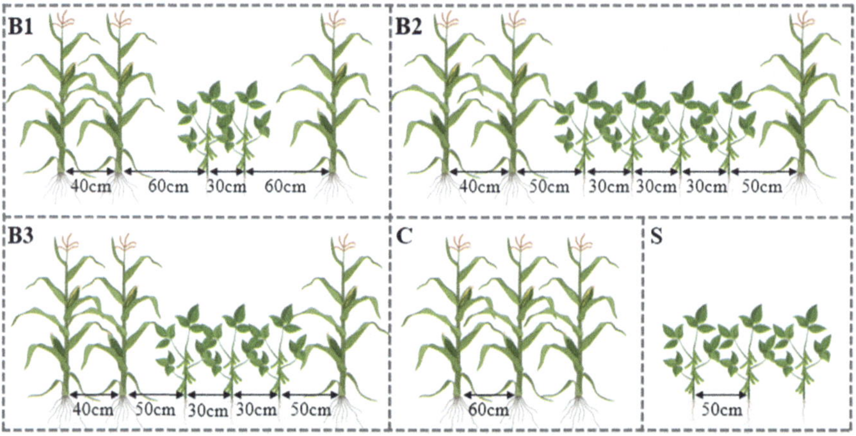

Treatment	Row ratio	Maize row spacing (cm)	Maize-soybean spacing (cm)	Soybean row spacing (cm)	Strip width(cm)	Maize plant spacing (cm)	Soybean plant spacing (cm)
B1	2:2	40	60	30	190	14.03	7.01
B2	2:3	40	55	30	210	12.70	9.52
B3	2:4	40	50	30	230	11.59	11.59
C	1:0	60	-	-	-	23.00	-
S	0:1	-	-	50	-	-	13.33

Fig. 6.7 Field configurations used to test the effects of maize-soybean strip intercropping on the nutritional quality of maize-soybean mixed silage

crude protein, and NDF contents, but not on soluble carbohydrate content. Overall, dry matter quality in the S2B2 treatment was significantly higher than that observed in other treatments; further, soluble carbohydrate content in maize and different soybean varieties decreased with increasing number of soybean rows. The crude protein content of each row ratio treatment was lower than that of monoculture soybean and higher than that of monoculture maize. After silage, NDF content in the S2B1 treatment was the highest. Additionally, the highest relative feeding value of maize was 146.45%, and the lowest relative feeding value of intercropping, as in the S2B1 treatment plots, was 126.72%.

6.3.2 Analysis of Silage Fermentation Quality of Maize and Soybean in Intercropping at Different Row Ratios

Analysis of silage fermentation quality after 60 days in the different silage treatments showed that in the intercropping treatment of maize and Nandou 25 soybean, the pH value and the ammonia nitrogen/total nitrogen ratio (NH3-N/TN) for the S1B2

Table 6.9 Nutritional quality of different silage materials (from Zeng et al. 2020)

Treatment	DM%	WSC/% DM	CP/% DM	NDF/% DM	ADF/% DM	Relative feed value/%
S1B1	35.40c	15.51b	8.10d	42.78b	22.39c	155.38
S1B2	36.32b	13.67bc	8.69c	41,.72b	21.89c	160.20
S1B3	35.77bc	13.69bc	8.85c	42.10b	22.09c	158.41
S2B1	35.52c	14.56b	7.98d	47.75a	25.81b	134.02
S2B2	37.09a	16.22ab	8.67c	41.67b	23.86bc	156.97
S2B3	36.44b	15.28b	9.63b	43.40b	23.13c	151.93
S1	31.67e	7.81d	14.79a	47.94a	31.39a	125.05
S2	30.34e	8.99c	14.60a	47.41a	29.67a	129.08
C	34.38d	17.63a	6.85e	42.94b	21.03c	157.10
Soybean variety (V)	**	**	*	**	**	–
Cropping pattern (P)	**	NS	**	**	NS	–
V*P	**	**	**	**	NS	–

Note: *DM* dry matter, *WSC* soluble carbohydrate, *CP* crude protein, *NDF* neutral-detergent fiber, *ADF* acid-detergent fiber

Table 6.10 Nutritional quality of silage after 60 days of silage

Treatment	DM%	WSC/% DM	CP/% DM	NDF/% DM	ADF/% DM	Relative feed value/%
S1B1	34.92c	4.98a	7.81 cd	45.79b	27.08bcd	137.75
S1B2	35.15c	3.64bc	7.96c	44.81bc	25.17 cd	143.85
S1B3	34,.38 cd	2.3de	7.46e	49.08a	27.21bcd	128.32
S2B1	35.17c	4,.41b	an7.40e	49.35a	27.82bc	126.72
S2B2	36.43a	3.25 cd	7.73 cd	45.04bc	25.35 cd	142.82
S2B3	36.15b	2.47de	7.58de	47.14ab	25,.64 cd	136,0.02
S1	31.27e	1.37e	13.07b	42.47c	29,41ab	144.54
S2	29.75e	1.69e	13.90a	42.26c	30.76a	142.94
C	33.82d	2.98 cd	6.44f	44.28bc	24,63d	146.45
V	NS	NS	**	NS	NS	–
P	NS	**	**	**	NS	–
V*P	*	NS	**	**	*	–

Note: P represents the planting system, C represents vegetable soybean varieties, and P × C represents the interaction between the planting system and vegetable soybean variety. Different lowercase letters within columns indicate significant differences at the 0.05 level. The F value was the result of a two-factor analysis of variance. * and * * indicate significant effects at the 0.05 and 0.01 levels, respectively; ns represents no significant difference at the 0.05 level

treatment were the lowest, 3.81 and 7.03%, respectively, and the corresponding values recorded in the S2B2 treatment were the lowest, 3.88 and 8.69%, respectively (Table 6.11). The pH values for silage produced with the two soybean varieties were greater than 4.20, and the NH3-N/TN was greater than 12.00%. Additionally, the lactic acid content in maize was significantly higher than that recorded for other

Table 6.11 Silage fermentation quality after 60 days of silage in different treatments

Treatment	pH value	Ammonia nitrogen/Total nitrogen (%)	Lactic acid (mg·g^{-1} DM)	Acetic acid (mg·g^{-1} DM)	Propionic acid (mg·g^{-1} DM)	Butyric acid (mg·g^{-1} DM)	V-score
S1B1	3.91 cd	9.54bc	20.02e	2.18f	0.94c	ND	90.92
S1B2	3.81d	7.03c	30.95c	4.19e	1.95b	ND	95.93
S1B3	4.11bc	10.51bc	24.19de	5.11de	1.14c	ND	87.95
S2B1	3.92 cd	8.86bc	27.39 cd	7.65c	2.50ab	ND	91.68
S2B2	3.88 cd	8.69c	36.73b	6.42 cd	2.46ab	ND	92.26
S2B3	3.99 cd	9.15bc	36.13b	10.27b	2.77a	ND	90.77
S1	4.43a	15.52a	6.41f	20.70a	ND	ND	65.84
S2	4.31ab	12.23b	6.23f	19.90a	ND	ND	79.18
C	3.90 cd	9.27bc	45.77a	5.33de	2.25ab	ND	91.45
V	NS	NS	**	**	**	–	–
P	**	*	**	**	NS	–	–
V*P	NS	NS	NS	*	*	–	–

Note: ND means not detected. P represents the planting system, C represents vegetable soybean varieties, and P × C represents the interaction between the planting system and vegetable soybean variety. Different lowercase letters within columns indicate significant differences at the 0.05 level. The F value was the result of a two-factor analysis of variance. * and * * indicate significant effects at the 0.05 and 0.01 levels, respectively; ns represents no significant difference at the 0.05 level

treatments. The lactic acid content of strip intercropping mixed silage was between those of maize and soybean in monoculture. Whereas the acetic acid content of monoculture soybean silage was higher than that of any other treatment. Specifically, acetic and propionic acid contents in maize and S1 soybean intercropping silage were lower than those observed in maize and S2 soybean intercropping silage. Additionally, in the maize and Nandou 25 soybean strip intercropping treatment, the V-Score value of the S1B2 treatment was the highest; similarly, in the maize and "Fendoumulv 2" soybean strip intercropping silage treatment, the V-Score value of the S2B2 treatment was the highest. The V-Score value of strip intercropping silage was higher than that of soybean monoculture (S1 and S2) silage. Additionally, under maize and soybean strip intercropping, soybean varieties had significant effects on lactic, acetic, and propionic acid contents, but not on pH or NH3-N/TN. Consistently, row ratio had a significant effect on NH3-N/TN, pH, lactic, and acetic acids. Except for acetic and propionic acid contents, which were significantly affected by the interaction of soybean variety and row ratio, other fermentation quality parameters were not affected by the interaction between the two (Table 6.11).

In short, water-soluble carbohydrate content in silage raw materials in strip intercropping was higher than that of soybean, and crude protein content was 16.50–40.44% higher than that of maize. Further, the crude protein content of strip intercropping mixed silage was 14.91–23.60% higher than that of monoculture maize silage. Further, after 60 days of silage, the pH values of strip intercropping and maize single storage were less than 4.2, whereas the pH value of monoculture soybean silage was greater than 4.2. The lactic acid content of maize single storage was significantly higher than that of other treatments, which was 45.77 mg·g^{-1}. The lactic acid content of strip intercropping mixed silage was intermediate between that of monoculture maize and monoculture soybean silage. Based on these findings, we concluded that mixed silage after soybean-maize strip intercropping improved silage fermentation quality, increased silage crude protein content, and made it easier to obtain high-protein and high-quality silage. Furthermore, the maize-soybean row ratio can be manipulated to improve soybean biological yield while ensuring that of maize. Thus, intercropping effectively allowed for high-quality mixed silage production. Lastly, among the intercropping treatments tested, strip intercropping of two rows of maize and three rows of soybean (2:3) provided the best results with different soybean varieties and is, therefore, worthy of promotion and recommendation.

Patterns of Pest, Disease, and Weed Incidence in Strip Intercropping Systems

7

Jing Shang, Xiaoli Chang, and Jizhi Yang

Abstract

Crop economic losses greater than 30% can result from damage caused by pests, diseases, and weeds. Therefore, effective pest and disease control is essential for ensuring high crop yields and economic benefits. Crop pest and disease incidence comprehensively reflect the interaction between harmful organisms, crops, and environmental conditions. The strip intercropping system for soybeans and maize comprises a closely spaced maize strip alternating with a soybean strip consisting of two to six rows. Changes in field configuration can lead to significant changes in microclimate, which, in turn, may alter the pattern of pest and disease incidence. We systematically and comprehensively studied the patterns of pest incidence in this intercropping system and found that major pests, such as maize borers, bean pod borers, and soybean aphids, were suppressed, whereas root rot, bacterial and fungal leaf spots, and viral disease incidence were reduced. Furthermore, beneficial predators and insect pest-parasitoids were enhanced. Lastly, a row ratio of 3:2 or 4:2 was found to be optimal for pest and disease control.

Keywords

Intercropping · Disease incidence · Pest incidence

J. Shang (✉) · X. Chang · J. Yang
College of Agronomy, Sichuan Agricultural University, Chengdu, China
e-mail: xl_changkit@sicau.edu.cn

W. Yang et al. (eds.), *Theories and Application in Maize-Soybean Strip Intercropping System*, https://doi.org/10.1007/978-981-97-8723-4_7

7.1 Incidence Patterns of Major Diseases

7.1.1 Incidence Patterns and Rhizosphere Regulatory Mechanism of Soybean Root Rot

Root rot, a worldwide soil-borne fungal disease, has become a significant limiting factor that hinders soybean production and reduces yield in continuous soybean monoculture cropping. Through four-year field location experiments of soybean-maize strip intercropping and soybean monoculture, studies have revealed varying incidence patterns of root rot disease in different years. Strip intercropping significantly reduced the incidence rate and disease index of soybean root rot. The analysis of pathogen population and identification of the causal agents of root rot showed a higher diversity of *Fusarium* species was observed in strip intercropping than that in monoculture cropping. Furthermore, *Fusarium oxysporum* showed the stronger pathogenicity and higher isolation frequency in monoculture fields than those in strip intercropping fields. Further, analysis of microbial diversity in the soybean rhizosphere indicated a shift in the population structure of pathogenic and beneficial microbes in strip intercropping.

7.1.1.1 Effects of Crop Cropping Patterns on the Incidence of Soybean Root Rot

The disease incidence and disease index of soybean root rot were investigated from field location experiments in soybean-maize strip intercropping and soybean monoculture over four consecutive years, and results revealed that the incidence rate and disease index of soybean root rot increased gradually as the cropping continued. Thus, in 2018, the incidence rate of soybean root rot in monoculture cropping was 61.02%, whereas that observed in strip intercropping was 24.28% (Fig. 7.1), indicating that continuous cropping significantly affected the incidence rate of soybean root rot. Furthermore, compared with monoculture, strip intercropping

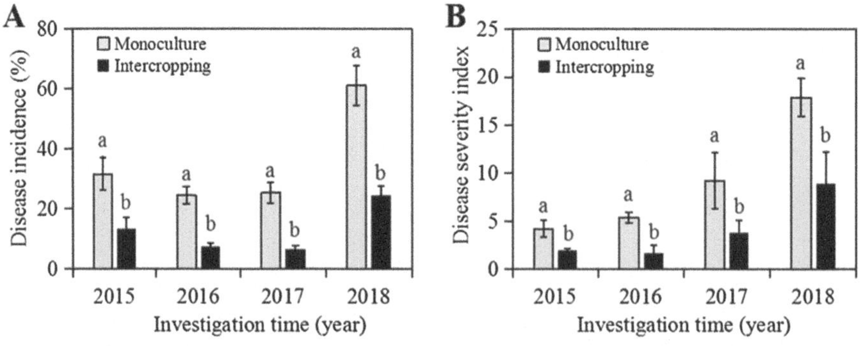

Fig. 7.1 Incidence and disease index of soybean root rot under strip intercropping and monoculture from 2015 to 2018

reduced the incidence rate and disease index of root rot, thus showing better resistance to this disease.

(A) Disease incidence of soybean root rot; (B) Disease severity index of soybean root rot. Statistical analysis was performed using SPSS. Different lowercase letters indicate significant differences between the cropping patterns as per Duncan's test with a P value greater than 0.05.

7.1.1.2 Effects of Strip Intercropping on Population Diversity and Isolation Frequency of Pathogenic *Fusarium* in Soybean

7.1.1.2.1 Effects of Strip Intercropping on the Diversity of Pathogenic *Fusarium* Species in Soybean Root Rot

Healthy and diseased soybean plants were collected from intercropping and monoculture fields, and *F. oxysporum* was isolated from infected roots using a tissue separation method. The *EF-1α* and *RPB2* nucleotide sequences of each fungal isolate were amplified using PCR, sequenced, and identified by BLASTn analysis based on the databases of *Fusarium* MLST (http://www.wi.knaw.nl/Fusarium/Biolomics.aspx) and *FUSARIUM*-ID (http://isolate.fusariumdb.org/guide.php). Subsequently, a multigene phylogenetic tree was constructed using the maximum parsimony method in the MEGA6.0 software, with NRRL52709 *Nectriaceae* sp. and NRRL 52754 *Nectriaceae* sp. as outgroups. The obtained isolates of the *F. solani* species complex (FSSC), *F. oxysporum*, *F. commune*, *F. fujikuroi*, *F. proliferatum*, *F. verticillioides*, *F. graminearum*, *F. asiaticum*, and *F. meridionale* were clustered to one branch with their respective reference strains, and no apparent genetic differentiation was observed among isolates of the same *Fusarium* species from different cropping systems (Fig. 7.2).

A double-gene phylogenetic tree of the *EF1-α* and *RPB2* sequences was constructed using the maximum parsimony method, and the bootstrap support values were calculated from 1000 repeats. Strains NRRL 52709 (*Nectriaceae* sp.) and NRRL 52754 (*Nectriaceae* sp.) were used as outgroups. The related parameters for the phylogenetic tree were as follows: tree length (TL) was 1053, consistency index (CI) was 0.762, retention index (RI) was 0.975, and rescale consistency index (RCI) was 0.734.

The *Fusarium* population responsible for soybean root rot disease in strip intercropping fields consisted of FSSC, *F. verticillioides*, *F. oxysporum*, *F. fujikuroi*, *F. graminearum*, *F. asiaticum*, and *F. proliferatum*, whereas those in monoculture fields included FSSC, *F. oxysporum*, *F. commune*, *F. asiaticum*, and *F. meridionale*. It is noteworthy that *F. verticillioides*, *F. fujikuroi*, and *F. proliferatum* were found only in intercropping fields. Relevant information on *Fusarium* species associated with soybean root rot in strip intercropping and monoculture systems is summarized in Table 7.1.

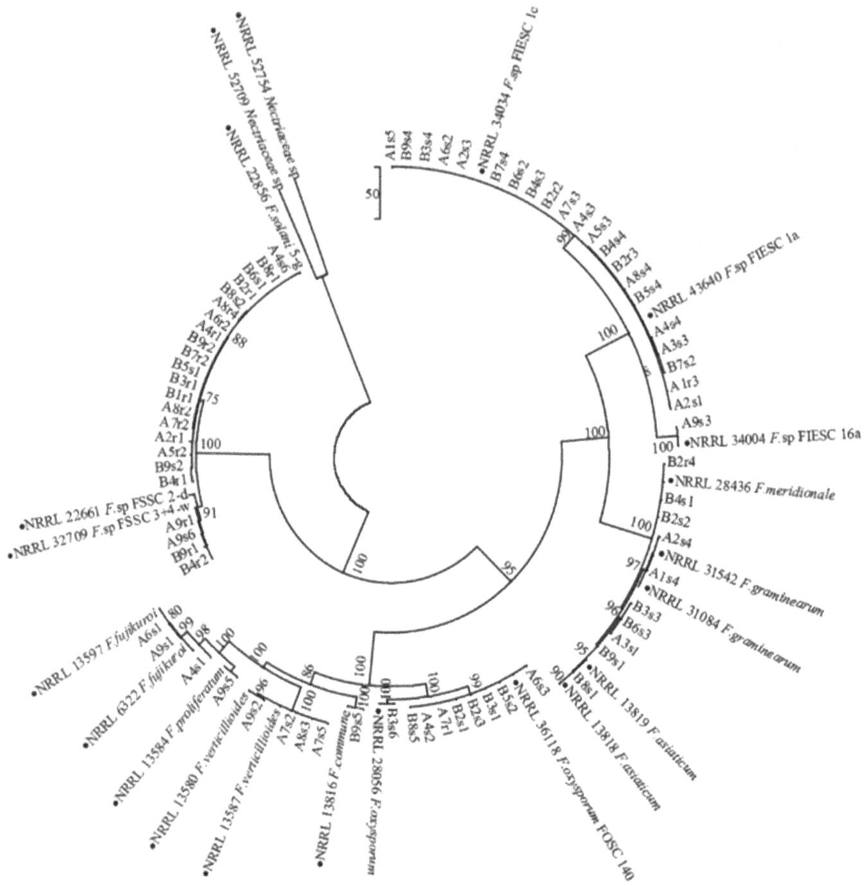

Fig. 7.2 Clustering tree of pathogenic *Fusarium* in soybean root rot based on the combined analysis of the *EF-1α* and *RPB2* gene sequences under strip intercropping and monoculture

7.1.1.2.2 Effects of Cropping Patterns on the Isolation Frequency of Soybean Pathogenic *Fusarium* Species

There was a significant difference in the isolation frequency of *Fusarium* causing soybean root rot disease between strip intercropping and monoculture. The isolation frequency of *F. solani* species complex (FSSC) was higher than those of other *Fusarium* species in both cropping patterns. However, the isolation frequency of highly pathogenic *F. oxysporum* was significantly higher in monoculture (16.22%) than that in strip intercropping plots. Furthermore, *F. verticillioides*, *F. proliferatum*, and *F. fujikuroi*, which belong to the *F. fujikuroi* species complex (FFSC), were the specific pathogens in intercropping plots, with an isolation frequency of 11.11%, 2.78%, and 8.33%, respectively, whereas *F. commune* was specific to monoculture with an isolation frequency of 2.7%. Therefore, FSSC was the primary pathogen causing soybean root rot disease in both cropping systems, although the isolation

Table 7.1 Information of *Fusarium* species associated with soybean root rot in strip intercropping and monoculture and their respective gene accession numbers in GenBank for the *EF-1α* and *RPB2* sequences

Strain code	Planting patterns	GenBank accession number		Species name	Species complex
		EF-1α	*RPB2*		
A1s4	Intercropping	MK560306	MN892318	*F. graminearum*	*Fusarium graminearum* species complex, FGSC
A1s5	Intercropping	MK560320	MN892327	FIESC	*Fusarium incarnatum-equiseti* species complex, FIESC
A1r3	Intercropping	MK560319	MN892328	FIESC	FIESC
A2s1	Intercropping	MK560321	MN892326	FIESC	FIESC
A2s3	Intercropping	MK560322	MN892325	FIESC	FIESC
A2s4	Intercropping	MK560307	MN892317	*F. graminearum*	FGSC
A2r1	Intercropping	MK560280	MN892289	FSSC	*Fusarium solani* species complex, FSSC [40]
A3s1	Intercropping	MK560335	MN892344	*F. asiaticum*	FGSC
A3s3	Intercropping	MN892351	MN892338	FSSC	FSSC
A4s1	Intercropping	MK560308	MN892321	*F. fujikuroi*	*Fusarium fujikuroi* species complex, FFSC
A4s2	Intercropping	MK560300	MN892307	*F. oxysporum*	*Fusarium oxysporum* species complex, FOSC
A4s3	Intercropping	MK560323	MN892336	FIESC	FIESC
A4s4	Intercropping	MK560324	MN892341	FIESC	FIESC
A4s6	Intercropping	MK560283	MN892295	*F. solani*	FSSC
A4r1	Intercropping	MK560282	MN892296	FSSC	FSSC
A5s3	Intercropping	MK560325	MN892342	FIESC	FIESC
A5r2	Intercropping	MK560284	MN892284	FSSC	FSSC
A6s1	Intercropping	MK560309	MN892320	*F. fujikuroi*	FFSC
A6s2	Intercropping	MK560326	MN892324	FIESC	FIESC
A6s3	Intercropping	MK560301	MN892306	*F. oxysporum*	FOSC
A6r2	Intercropping	MK560285	MN892294	*F. solani*	FSSC
A7s2	Intercropping	MK560261	MN892281	*F. verticillioides*	FFSC
A7s3	Intercropping	MK560327	MN892323	FIESC	FIESC
A7s5	Intercropping	MK560262	MN892280	*F. verticillioides*	FFSC
A7r1	Intercropping	MK560302	MN892305	*F. oxysporum*	FOSC
A7r2	Intercropping	MK560286	MN892293	*F. solani*	FSSC
A8s3	Intercropping	MK560263	MN892279	*F. verticillioides*	FFSC
A8s4	Intercropping	MK560328	MN892343	FIESC	FIESC
A8r2	Intercropping	MK560288	MN892291	FSSC	FSSC

(continued)

Table 7.1 (continued)

Strain code	Planting patterns	GenBank accession number		Species name	Species complex
		EF-1α	*RPB2*		
A8r4	Intercropping	MK560287	MN892292	*F. solani*	FSSC
A9s1	Intercropping	MK560310	MN892319	*F. fujikuroi*	FFSC
A9s2	Intercropping	MK560264	MN892278	*F. verticillioides*	FFSC
A9s3	Intercropping	MK560329	MN892322	FIESC	FIESC
A9s5	Intercropping	MK560292	MN892349	*F. proliferatum*	FFSC
A9s6	Intercropping	MK560290	MN892285	FSSC	FSSC
A9r1	Intercropping	MK560291	MN892290	FSSC	FSSC
B1r1	Monoculture	MK560265	MN892304	FSSC	FSSC
B2s1	Monoculture	MK560293	MN892313	*F. oxysporum*	FOSC
B2s2	Monoculture	MK560303	MN892316	*F. meridionale*	FGSC
B2s3	Monoculture	MK560294	MN892312	*F. oxysporum*	FOSC
B2r1	Monoculture	MK560266	MN892288	FSSC	FSSC
B2r2	Monoculture	MK560311	MN892335	FIESC	FIESC
B2r3	Monoculture	MK560312	MN892337	FIESC	FIESC
B2r4	Monoculture	MK560304	MN892315	*F. meridionale*	FGSC
B3s1	Monoculture	MK560295	MN892311	FIESC	FOSC
B3s3	Monoculture	MK560331	MN892348	*F. asiaticum*	FGSC
B3s4	Monoculture	MK560313	MN892334	FIESC	FIESC
B3s6	Monoculture	MK560296	MN892310	*F. oxysporum*	FOSC
B3r1	Monoculture	MK560267	MN892303	FSSC	FSSC
B4s1	Monoculture	MK560305	MN892314	*F. meridionale*	FGSC
B4s3	Monoculture	MK560314	MN892333	FIESC	FIESC
B4s4	Monoculture	MK560315	MN892339	FIESC	FIESC
B4r1	Monoculture	MK560268	MN892283	FSSC	FSSC
B4r2	Monoculture	MK560269	MN892286	FSSC	FSSC
B5s1	Monoculture	MK560270	MN892302	FSSC	FSSC
B5s2	Monoculture	MK560297	MN892309	*F. oxysporum*	FOSC
B5s4	Monoculture	MN892352	MN892332	FSSC	FSSC
B6s1	Monoculture	MK560272	MN892301	FSSC	FSSC
B6s2	Monoculture	MK560316	MN892331	FIESC	FIESC
B6s3	Monoculture	MK560332	MN892347	*F. asiaticum*	FGSC
B7s2	Monoculture	MK560317	MN892340	FIESC	FIESC
B7s4	Monoculture	MK560318	MN892330	FIESC	FIESC
B7r2	Monoculture	MK560273	MN892300	FSSC	FSSC
B8s1	Monoculture	MK560333	MN892346	*F. asiaticum*	FGSC
B8s2	Monoculture	MK560274	MN892299	FSSC	FSSC
B8s4	Monoculture	MN892354	MN892308	FSSC	FSSC
B8r1	Monoculture	MK560276	MN892298	FSSC	FSSC
B9s1	Monoculture	MK560334	MN892345	*F. asiaticum*	FGSC
B9s2	Monoculture	MK560277	MN892282	FSSC	FSSC
B9s4	Monoculture	MN892353	MN892329	FIESC	FIESC

(continued)

Table 7.1 (continued)

Strain code	Planting patterns	GenBank accession number		Species name	Species complex
		EF-1α	*RPB2*		
B9s5	Monoculture	MK560330	MN892350	*F. commune*	*Fusarium nisikadoi* species complex, FNSC
B9r1	Monoculture	MK560278	MN892287	FSSC	FSSC
B9r2	Monoculture	MK560279	MN892297	FSSC	FSSC

Note: Species identification was determined by comparing the gene sequences corresponding to the *translation elongation factor EF-1α* and the *RNA polymerase II large subunit RPB2* gene sequences of each *Fusarium* isolate with those in the *Fusarium* MLST (http://www.wi.knaw.nl/Fusarium/Biolomics.aspx) and *FUSARIUM*-ID (http://isolate.fusariumdb.org/guide.php) databases

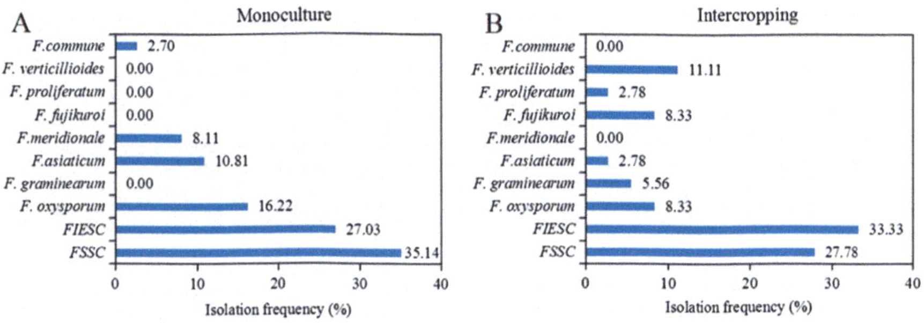

Fig. 7.3 Isolation Frequency of candidative pathogenic *Fusarium* species causing soybean root rot under intercropping and monoculture systems

frequencies of the highly pathogenic *F. oxysporum* and the *F. verticillioides* species complex were significantly lower in intercropping than in monoculture fields (Fig. 7.3).

The values on the bar chart represent the frequency of isolation of different *Fusarium* species causing soybean root rot under monoculture (A) and intercropping (B) systems. FSSC refers to the *Fusarium solani* species complex; FIESC refers to the *Fusarium incarnatum-equiseti* species complex. Meanwhile, *F. oxysporum, Fusarium asiaticum, F. meridionale, F. graminearum, F. fujikuroi, F. verticillioides*, and *F. proliferatum* are different species of *Fusarium*. Statistical analysis was performed using Fisher's exact test.

7.1.1.2.3 Correlation Analysis between Root Rot Disease Incidence and Pathogen Virulence under Strip Intercropping

Using the sorghum grain inoculation method, the pathogenicity of the candidative *Fusarium* isolates causing soybean root rot were tested. The results showed that all representative strains of pathogenic *Fusarium* species effectively infected soybean

Fig. 7.4 Symptoms of candidative pathogenic *Fusarium* infection in the roots of soybean cultivar NanDou12 inoculated with root rot disease. The respresetative isolates of *Fusarium* from strip intercropping included A8r2 of FSSC, A9s3 of FIESC, A4s2 of *F. oxysporum*, A3s1 of *F. asiaticum*, A2s4 of *F. graminearum*, A9s5 of *F. proliferatum*, A6s1 of *F. fujikuroi*, A9s2 of *F. verticillioides*. The respresentative isolates of *Fusarium* from monoculture included B3r1 and B4r2 of FSSC, B7s2 of FIESC, B3s1 of *F. oxysporum*, B8s1 of *F. asiaticum*, B2s2 of *F. meridionale*, and B9s5 of *F. commune*. *T respresentative isolates of Fusarium from strip intercropping*

roots to varying degrees at 15 days after inoculation, causing poor plant development, browning and rotting of the main roots, and poor or no development of the lateral roots (Fig. 7.4). Different *Fusarium* species and different representative isolates of the same *Fusarium* species showed varying degrees of pathogenicity, but there was no significant difference in the pathogenicity of the same *Fusarium*

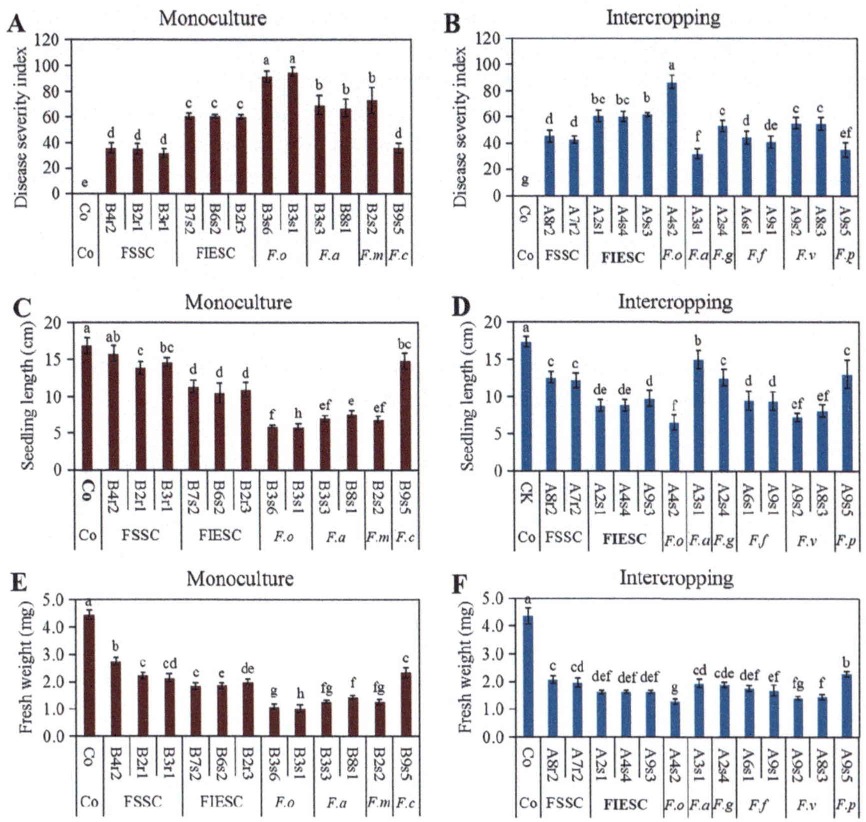

Fig. 7.5 Soybean growth parameters and root rot disease index after inoculation with different *Fusarium* species

species between monoculture and intercropping. After inoculation, plant height and fresh weight of soybean seedlings were significantly reduced. Highly pathogenic *F. oxysporum* and members of the *F. fujikuroi* species complex were the most pathogenic species to soybeans both in monoculture and intercropping, with the isolate *F. meridionale* B2s2 as well as isolates B3s3 and B8s1 of *F. asiaticum* ranking after *F. oxysporum*. However, the soybean disease index was lower after inoculation of *F. graminearum* (A2s4) and *F. asiaticum* (A3s1) in intercropping. In addition, the pathogenicity of FIESC isolates did not differ significantly between the two cropping systems; however, the pathogenicity of FSSC was lower in intercropping. There was a difference in pathogenicity among the three *Fusarium* pathogens causing soybean root rot disease in strip intercropping, namely *F. fujikuroi*, *F. verticillioides*, and *F. proliferatum*. Additionally, the pathogenicity of the specific pathogen *F. commune* in soybean monoculture was lower (Fig. 7.5) than that observed in strip intercropping. Therefore, the high frequency of highly pathogenic pathogens isolated in soybean monoculture may be one of the primary reasons for the more severe disease symptoms of soybean root rot observed in monoculture relative to those observed in strip intercropping.

In the figure, Corepresents the control group without *Fusarium* inoculation; FSSC refers to the *Fusarium solani* species complex; FIESC refers to the *Fusarium incarnatum-equiseti* species complex; *F. o* stands for *Fusarium oxysporum*; *F. a* for *Fusarium asiaticum*; *F. m* for *Fusarium meridionale*; *F. c* for *Fusarium commune*; *F. g* for *Fusarium graminearum*; *F. f* for *Fusarium fujikuroi*; *F. v* for *Fusarium verticillioides*; *F. p* for *Fusarium proliferatum*. Different lowercase letters in the figure represent significant differences ($P < 0.05$).

7.1.1.3 Effects of Strip Intercropping on Microbial Community Structure and Diversity in the Rhizosphere Soil

7.1.1.3.1 Effects of Strip Intercropping on the Population Structure of Rhizosphere *Trichoderma* and *Fusarium*

Using a soil dilution culture method, we analyzed the population of *Fusarium* and *Trichoderma* present in the rhizosphere soil of healthy and diseased soybean plants in monoculture and intercropping fields. The results showed that the abundance of *Fusarium* in intercropped contaminated soil was significantly higher than that in uncontaminated soil. However, the abundance of *Trichoderma* in the intercropped rhizosphere was higher than that in the rhizosphere of monoculture (Fig. 7.6). Moreover, the abundance of *Trichoderma* in intercropped uncontaminated soil was higher than that in the contaminated soil. Further, there was no significant difference in the abundance of *Fusarium* between monoculture plots in uncontaminated and contaminated soils. Therefore, intercropping changed the abundance of *Fusarium* and *Trichoderma* in the soybean rhizosphere soil, and the incidence of root rot disease tended to induce the accumulation of *Fusarium* in the rhizosphere, thus resulting in a decrease in the abundance of *Trichoderma*. Compared with monoculture cropping, the accumulation of highly abundant *Trichoderma* may be the primary reason for the decrease incidence of soybean root rot under intercropping.

Using the dilution plate method, pathogenic *Fusarium* and potential biocontrol *Trichoderma* were isolated and identified from the rhizosphere soil of healthy and diseased soybean plants under intercropping and monoculture. Sequencing and phylogenetic analysis revealed significant differences for the *Fusarium* population in soybean rhizosphere soil under the two cropping systems. In the intercropped rhizosphere soil, *Fusarium* species included *F. solani* species complex (FSSC), *F. equiseti* species complex (FIESC), and *F. oxysporum*, whereas in the monoculture rhizosphere soil, FSSC, *F. oxysporum,* and *F. commune* were identified. Among all rhizosphere soils, FSSC showed the highest isolation frequency. However, *F. oxysporum* and *F. acuminatum* were isolated more frequently in contaminated soil. In turn, *T. harzianum*, *T. virens*, and *T. afroharzianum* were isolated from both cropping systems. However, the composition and frequency of *Trichoderma* differed depending on the rhizosphere soil affected by root rot disease. Thus,

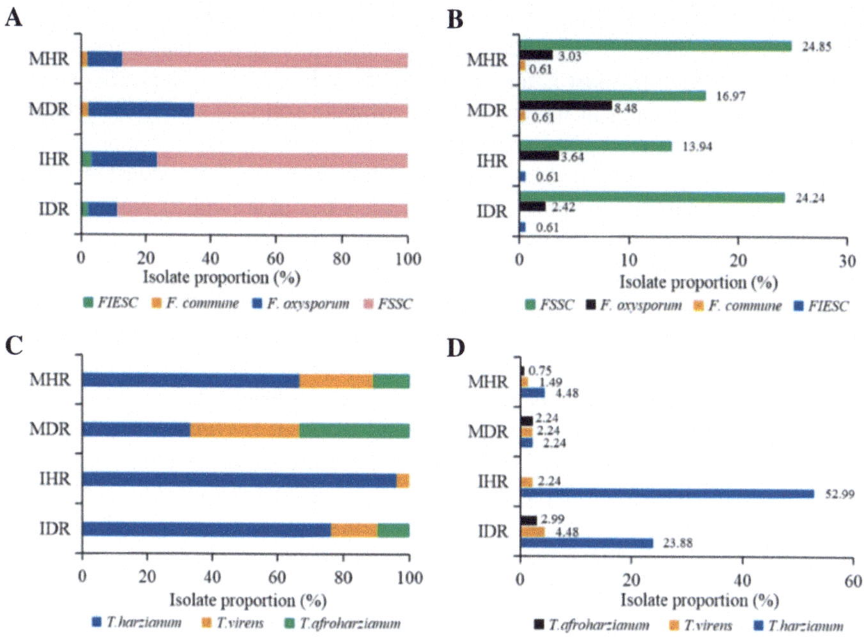

Fig. 7.6 Composition and isolation frequency of *Fusarium* and *Trichoderma* species in the rhizosphere soil of healthy and diseased soybean plants of root rot under strip intercropping and monoculture. (**A**) Composition proportion of *Fusarium* species; (**B**) Isolation frequency of *Fusarium* species; (**C**) Composition proportion of *Trichoderma* species; (**D**) Isolation frequency of *Trichoderma* species. IDR refers to rhizosphere soil of diseased soybean plants of root rot under strip intercropping; IHR refers to rhizosphere soil of healthy soybean plants under strip intercropping; MDR refers to rhizosphere soil of diseased soybean plants of root rot under monoculture cropping; MHR refers to rhizosphere soil of healthy soybean plants under monoculture cropping. Statistical analysis between treatments was performed using Fisher's exact test

T. hazianum was the dominant species, with an isolation frequency of 95.9% in intercropped uncontaminated soil and 76.2% in contaminated soil, which was significantly higher than in monoculture. Intercropping reduced the frequency of *T. virens* in the rhizosphere soil, whereas soil from monoculture fields favored the accumulation of *T. afroharzianum*. Thus, intercropping changed the community composition of *Trichoderma* in the soybean rhizosphere soil, significantly increasing the frequency of *T. harzianum*. However, root rot disease inhibited its accumulation.

Pathogenic *Fusarium* in the rhizosphere is an important source of infection for soil-borne soybean root rot disease. The pathogenicity of *Fusarium* isolates from soybean rhizosphere soil under strip intercropping and monoculture was tested on soybean cultivar "Nandou 12." We found that four *Fusarium* species, namely the *Fusarium equiseti* species complex (FIESC), the *Fusarium solani* species complex (FSSC), *F. oxysporum*, and *F. commune*, were all able to cause root rot disease in

Nandou 12, with poor seedling development and browning, and rotting of the main and lateral roots; however, the pathogenicity varied among different isolates, with *F. oxysporum* being the most virulent to soybean seedlings (7A). The pathogenicity of *F. oxysporum* isolates in monoculture rhizosphere soil was significantly higher than that in strip intercropping, severely affecting growth and development and causing root rot. The FSSC had the highest isolation frequency but showed moderate pathogenicity to soybean seedlings, with a disease index of around 50. No significant difference was observed in the pathogenicity among different isolates under the two cropping systems (Fig. 7.7b).

Trichoderma is a common soil-borne biocontrol fungus. Three *Trichoderma* isolates isolated from soybean rhizosphere soil were tested for their inhibitory effects on *F. oxysporum* causing soybean root rot using the plate confrontation culture. The results showed that *T. harzianum*, *T. virens*, and *T. afroharzianum* significantly inhibited the hyphal growht of *F. oxysporum*, and 38.1% of *Trichoderma* isolates had inhibition rates of 60–70% (Fig. 7.8A). Among them, the inhibitory rates of *T. virens* in monoculture and intercropping rhizosphere soil differed significantly. The average inhibitory rate of *T. virens* in intercropping rhizosphere soil was 72.0%, which was significantly higher than that observed in monoculture (60.30%) (Fig. 7.8B). Microscopic observation revealed that the three *Trichoderma* isolates interacted with *F. oxysporum* through spatial competition, adsorption, and entanglement of fungal hyphae or by directly invading the hyphae and breaking them down, leading to cell lysis. Thus, we concluded that the accumulation of *T. virens* and *T. afroharzianum* in the soybean rhizosphere soil was affected by the cropping pattern and was correlated with a lower incidence of root rot disease under intercropping.

7.1.2 Effects of Strip Intercropping on the Incidence Pattern of Soybean Mosaic Virus Disease and the Molecular Evolution of the Virus

Soybean mosaic virus (SMV) is one of the most important viruses affecting soybean production globally, as it causes variegated seed coat, reduced yield and quality, and has a high transmissibility rate. SMV infection typically reduces soybean yields by 8–35%. In severe cases, it can result in up to 94% yield loss. Resistance to SMV is a key requirement for the approval of commercial soybean varieties. We focused on soybean viral diseases in Sichuan Province, China, and collected samples from major cultivation areas to conduct virus-type identification and population genetic analysis, and to study the response of soybeans to SMV infection under shading. Our research provided a sound theoretical basis for the prevention and control of the SMV disease in intercropping soybeans.

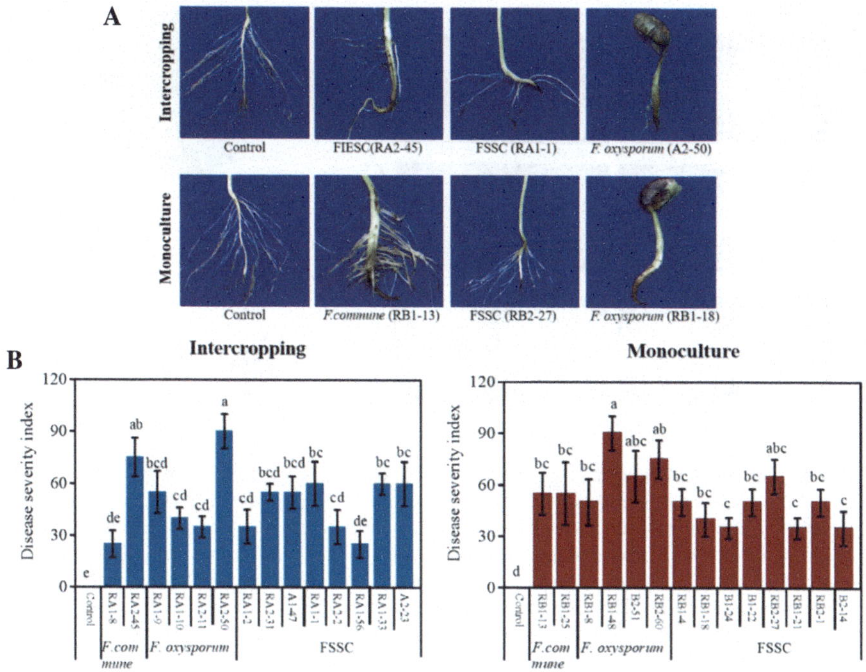

Fig. 7.7 Detection of the Pathogenicity of *Fusarium* Species in Rhizosphere Soil of Soybean Plants Under Strip Intercropping and Monoculture. (**A**) Disease symptoms in soybean seedlings after inoculation with representative isolates of *Fusarium* spp. including *Fusarium incarnatum-equiseti* species complex (FIESC) RA2-45, the *Fusarium solani* complex (FSSC) RA-1 and RB2-27, *F. oxysporum* A2-50 and RB18, and *F. commune* RB1-13, when the non-inoculated seedlings were used as the control; (**B**), Disease index of soybean root rot after inoculation. Different lowercase letters indicate significant differences among treatments ($P > 0.05$)

7.1.2.1 Effects of Shading on Soybean Resistance to Virus Disease in Strip Intercropping

7.1.2.1.1 Phenotype of SMV-Infected Soybean Plants

After infection with SMV, typical mosaic symptoms appear on soybean leaves. Further, changes in the light environment and virus infection have a significant impact on soybean growth and development. Thus, while under normal light conditions, soybean plants have an average height of 23.6 cm and a stem thickness of 3.24 mm, under shading, the average height increases to 58 cm and stem thickness decreases to 2.23 cm. Both height and stem thickness decrease due to SMV infection. Specifically, under normal light conditions, SMV-infected plants have an average height of 22 cm and stem thickness of 3.06 mm, indicating a considerable reduction caused by the infection, particularly of stem diameter. In contrast, shading caused SMV-infected soybean plants to increase average height to 37.16 cm, whereas average stem thickness decreased to 2.03 cm. Analysis using RT-PCR

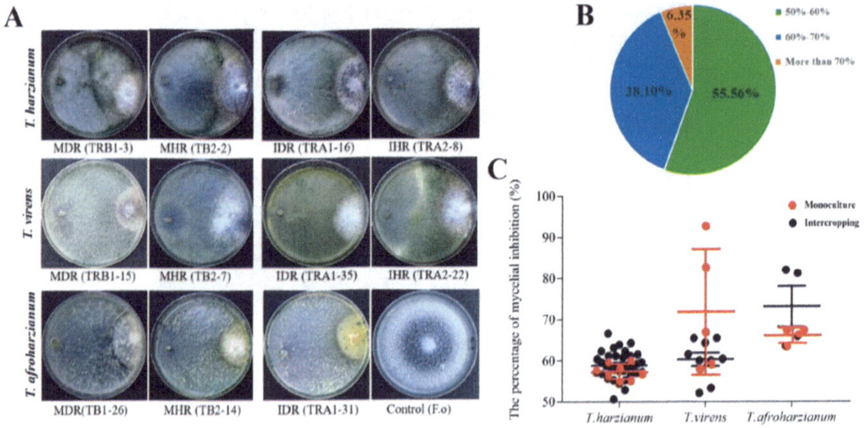

Fig. 7.8 Growth inhibition of *Fusarium oxysporum* causing soybean root rot by *Trichoderma* sp. (**A**) Plate confrontation culture of *Trichoderma* sp. and *F. oxysporum* in rhizosphere soil; IDR, rhizosphere soil of root rot-infected plants in strip intercropping; IHR, rhizosphere soil of healthy soybean plants in strip intercropping; MDR, rhizosphere soil of root rot-infected plants in monoculture; MHR, rhizosphere soil of healthy soybean plants in monoculture; (**B**) Statistical analysis of the inhibition rates of *Trichoderma* sp. on the hyphal growth of *F. oxysporum* in three ranges: 50–60%, 60–70%, and greater than 70%; (**C**) Analysis of the inhibition rates of *Trichoderma* sp. isolated from soybean rhizosphere soil in strip intercropping and monoculture on hyphal growth of *F. oxysporum*. Red represents monoculture, and black represents strip intercropping

confirmed the infection with SMV. Soybean main stem elongation is observed under shading, whereas resistance genes were downregulated, causing an increase in the pathway of long-distance virus transport. Further, under double stress by shading and SMV infection, the defense system of intercropped soybeans reduced virus accumulation by approximately 50% after 10 days of infection. Proper downregulation of resistance enhanced the relief of inflammation, thereby increasing tolerance to the virus (Fig. 7.9).

7.1.2.1.2 Downregulation of Soybean Defense Gene Expression

Under normal light conditions, soybeans infected with SMV showed 3548 differentially expressed genes (DEGs) compared with the control, 2228 of which were upregulated and 1320 downregulated. After shading treatment, a total of 4319 genes were affected by virus infection, with 2167 upregulated and 2152 downregulated (Fig. 7.10a). Further overlapping analysis of DEGs revealed 380 upregulated (Nu-Lu, Fig. 7.13b) and 66 downregulated (Nd-Ld, Fig. 7.10c) genes under both light conditions. Furthermore, 225 genes were downregulated under normal light conditions but upregulated under low light conditions (Nd-Lu, Fig. 7.10b). In contrast, 490 genes were upregulated under normal light conditions but downregulated under low light conditions (Nu-Ld, Fig. 7.10c).

SMV-induced DEGs showed similar functional classifications under normal light (NC vs. NS) and shading conditions (LC vs. LS). However, the number of

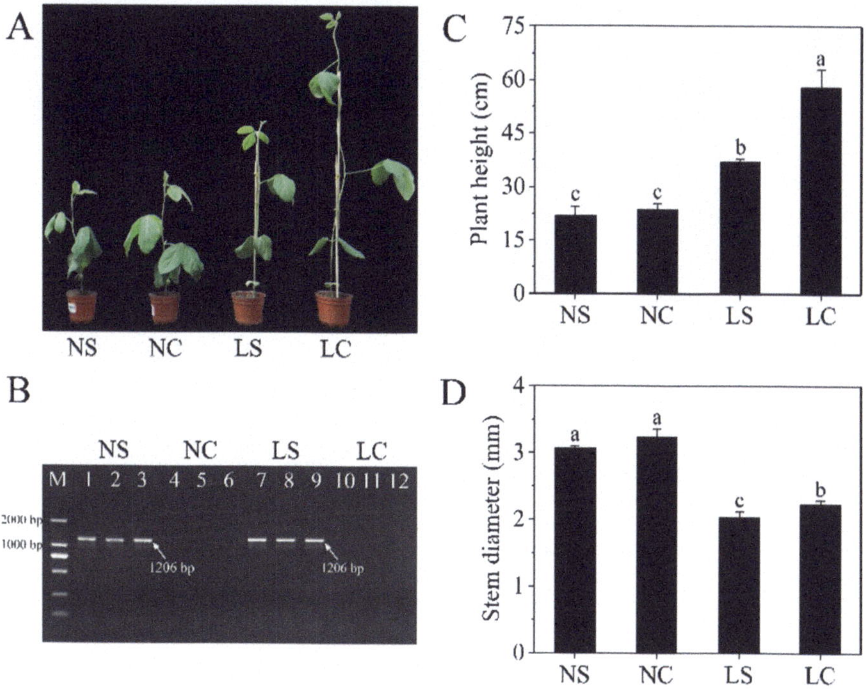

Fig. 7.9 Soybean phenotype after treatment and detection of SMV (reprinted with the permission of viruses-basel). (**a**) Growth of soybean seedlings under different treatments for 10 days; (**b**) RT-PCR detection of soybean infection with SMV; M—2000 bp DNA molecular weight marker; 1–3: Virus-infected leaf sample under normal light (NS); 4–6: Uninfected leaf sample under normal light (NC); 7–9: Virus-infected leaf sample under shading (LS); 10–12: Uninfected leaf sample under shading (LC); (**c**) Plant height 10 days after SMV infection; (**d**) Stem thickness 10 days after SMV infection. *NC* no virus inoculation under normal light, *NS* virus inoculation under normal light, *LC* no virus inoculation under shading, *LS* virus inoculation under shading. Data are means ± standard deviation. Different lowercase letters indicate significant differences at $P = 0.05$

downregulated genes among DEGs significantly increased under shading conditions (Fig. 7.11). Additionally, upregulated genes under normal light conditions were mainly induced by SMV infection, whereas under shading conditions, the number of downregulated genes significantly increased, far exceeding the number of upregulated genes. This finding indicated that light plays a positive regulatory role in defense response-related gene expression.

The top 15 metabolic pathways with the highest enrichment of DEGs are shown in Fig. 7.12. Under normal light conditions, SMV infection significantly induced the phenylalanine biosynthesis pathway. This was followed by plant-pathogen interaction, linoleic acid metabolism, and phenylalanine metabolism. Conversely, processes such as endoplasmic reticulum protein processing, sugar metabolism, nitrogen metabolism, and vitamin metabolism were inhibited. In contrast, under shading conditions, SMV infection induced linoleic acid metabolism, sugar

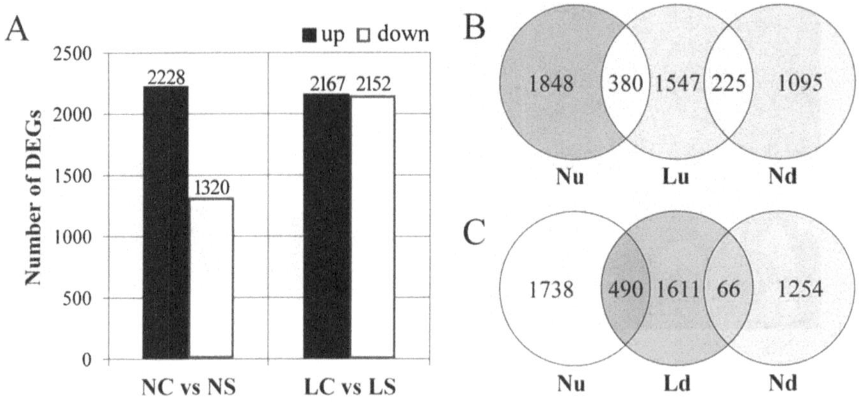

Fig. 7.10 Statistical analysis of differentially expressed genes (reprinted with the permission of viruses-basel). (**a**) Differentially expressed genes between SMV-infected and control soybean plants; (**b** and **c**) Overlapping analysis of DEGs; Nu and Nd refer to upregulated and downregulated genes between NC and NS, respectively; Lu and Ld refer to upregulated and downregulated genes between LC and LS, respectively. *NC* normal light without virus inoculation, *NS* normal light with virus inoculation, *LC* shading and no inoculation, *LS* shading and virus inoculation. DEGs between treated and control plants were determined based on the criteria of |log2Fold Change| > 1 and *p*-adj (adjusted *p*-value) < 0.05

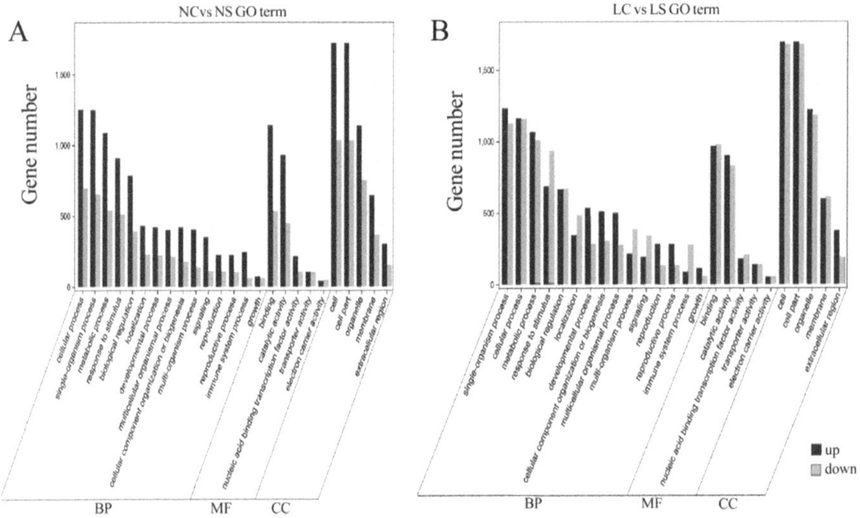

Fig. 7.11 GO enrichment analysis of differentially expressed genes. (reprinted with the permission of viruses-basel). (**a**) GO enrichment analysis of differentially expressed genes between NC and NS; (**b**) GO enrichment analysis of DEGs between LC and LS. MF, BP, and CC refer to Molecular Function, Biological Process, and Cellular Component, respectively. *NC* normal light without virus inoculation, *NS* normal light with virus inoculation, *LC* shading without virus inoculation, *LS* shading with virus inoculation

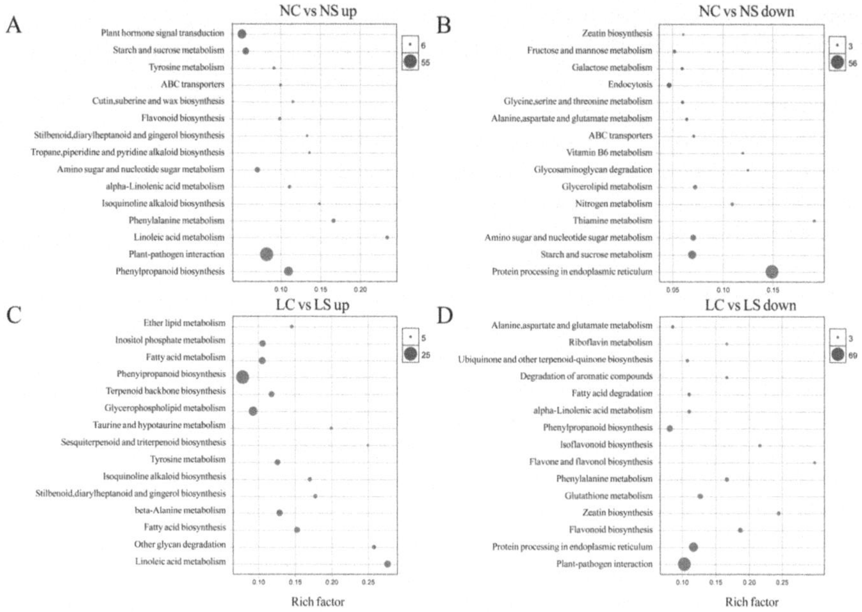

Fig. 7.12 KEGG pathway enrichment analysis of differentially expressed genes. (reprinted with the permission of viruses-basel). (**a**) upregulated and (**b**) downregulated genes between NC and NS; (**c**) upregulated and (**d**) downregulated genes between LC and LS. *NC* normal light but no virus inoculation, *NS* normal light and virus inoculation, *LC* shading but no virus inoculation, *LS* shading and virus inoculation

degradation, and fatty acid biosynthesis pathways in soybean. However, plant-pathogen interaction and some secondary metabolic pathways, such as those related to flavonoids, isoflavonoids, and zeatin, were suppressed. Furthermore, under normal light, SMV infection activated soybean immune-related pathways, whereas shading suppressed these defense pathways, indicating that light had a positive regulatory effect on plant defense.

7.1.2.1.3 Differential Gene Analysis of Plant-Pathogen Interaction Pathways

Together, plant immunity and pathogen invasion constitute the basis of the plant-pathogen interaction. The ability of plants to regulate their responses to stress rapidly and effectively is the basis for successfully resisting pathogen invasion. Therefore, here, we analyzed DEGs enriched during plant-pathogen interaction pathways. Our results showed that, under normal light conditions, 76 DEGs were enriched in this pathway, with 55 genes being upregulated and 21 genes downregulated (Table 7.2). In turn, 83 genes were differentially expressed under shading conditions, including 15 upregulated and 69 downregulated. Further, overlap analysis revealed that 24 DEGs were under both light and shading treatments, including WRKY transcription factors, MYB transcription factors, and calcium-binding proteins. Among them, the expression of most overlapping genes was upregulated under normal light

Table 7.2 Overlapping analysis of differentially expressed genes in the plant-pathogen interaction pathway under normal light and shading conditions

Gene number	Description	Log2FC (NS/NC)	Log2FC (LS/LC)
GLYMA_05G234600	MYB transcription factor MYB84	5.33	−3.24
GLYMA_18G056600	WRKY transcription factor 62	5.21	−3.66
GLYMA_09G038900	MYB transcription factor MYB13	5.06	1.74
GLYMA_06G187300	Protein EDS1L	4.01	−2.16
GLYMA_20G034200	Uncharacterized LOC100526868	3.94	−2.59
GLYMA_16G218300	Probable cyclic nucleotide-gated ion channel 20	3.83	−1.29
GLYMA_02G270700	Chitin elicitor receptor kinase 1-like	3.76	−1.16
GLYMA_19G214900	MYB transcription factor MYB111	3.20	2.40
GLYMA_09G210600	Disease resistance protein RPM1	3.10	−1.92
GLYMA_06G034700	Probable calcium-binding protein CML41	2.90	−2.53
GLYMA_20G209700	MYB/HD-like transcription factor	2.81	−3.06
GLYMA_10G180800	MYB29 protein	2.25	−3.85
GLYMA_18G208800	Probable WRKY transcription factor 33	2.23	−3.83
GLYMA_03G042700	Probable WRKY transcription factor 33	1.99	−2.70
GLYMA_14G222000	Calcium-dependent protein kinase 29	1.97	1.79
GLYMA_05G119500	BRASSINOSTEROID INSENSITIVE 1-associated receptor kinase	1.80	−1.45
GLYMA_10G230000	Protein SGT1 homolog B-like	1.73	−1.51
GLYMA_02G244600	MYB transcription factor MYB20	1.65	1.14
GLYMA_06G187400	Protein EDS1-like	1.55	−1.01
GLYMA_19G255300	Cyclic nucleotide-gated ion channel 1	1.35	−1.20
GLYMA_02G059600	Putative calcium-binding protein	1.13	−1.11
GLYMA_14G156300	Calcium-binding EF-hand family protein	−2.18	−1.76
GLYMA_16G178800	Heat shock protein 90-A2	−5.23	−1.33
GLYMA_09G131500	Heat shock protein 83	−7.64	−2.66

Reprinted with the permission of viruses-basel

NC normal light without virus inoculation, *NS* normal light with virus inoculation, *LC* shading without virus inoculation, *LS* shading with virus inoculation

treatment and downregulated under shading. This indicated that light may help activate plant-pathogen interaction pathways, thereby promoting plant resistance to pathogen invasion. However, under shading, plants cannot respond to pathogen invasion effectively, resulting in more severe disease development.

7.1.2.1.4 Analysis of Plant-Hormone Signal Transduction-Related Differential Gene Expression

Plant hormones play a very important role in plant growth and development, with hormones such as salicylic acid (SA), jasmonic acid (JA), and ethylene (ETH) playing a decisive role in plant immune responses. Specifically, under normal light conditions, SMV infection upregulated the expression of most genes in the soybean

SA, JA, and ETH signal transduction pathways. Among them, the upregulation of ethylene-responsive transcription factor ERF and pathogenesis-related protein PR1 was the highest. However, under shading, most DEGs in the SA, JA, and ETH signal transduction pathways were downregulated. In particular, ERF showed the largest decrease in expression, with a Log2FC value of −6.71, followed by PR1, with a Log2FC value of −5.91. In turn, the MYC and JAR genes in the JA pathway also showed a significantly downregulated expression, indicating that SMV infection inhibited the soybean defense hormone signal transduction network under shading (Table 7.3).

7.1.2.1.5 Real-Time Fluorescent Quantitative PCR Detection of Differential Gene Expression

To validate the expression patterns of genes observed in RNA sequencing data, 15 DEGs were selected for qRT-PCR, including PSK1, A-ARR, BAK1, AUX1, PIF4, PR1, PhyB, JAR1, HSP70, WRKY62, PhyA, DELLA, and ERF1. Overall, the expression patterns observed using qRT-PCR were similar to those observed in RNA sequencing. Namely, the expression of many defense genes in the SA and JA pathways was significantly downregulated under shading (Fig. 7.13).

Light effects on soybean resistance against SMV were elucidated by comparing the plant response to SMV infection under two light environments; further exploration was conducted by analyzing the effect of exogenous SA treatment on soybean resistance to SMV under shading. Due to the effects of pathogen infection, plant growth and defense are usually negatively correlated. Indeed, soybean growth and development are restricted by SMV infection and plant dwarfism is often observed under these circumstances. Further, shading during SMV infection worsened SMV-induced damage to soybean growth, with a 35.93% reduction in height and an 8.97% decrease in stem diameter, ultimately reducing yield. Meanwhile, under normal light, phenylalanine biosynthesis, plant-pathogen interaction, and signal transduction pathways were significantly induced, and protein synthesis and vitamin and sugar-nitrogen metabolism were inhibited, which was consistent with previous reports. Phenylalanine is a precursor of a series of secondary metabolites, and the upregulation of its synthesis pathway is beneficial to the formation of downstream defense products, playing an important role in plant stress responses. Similarly, plant-pathogen interaction pathways were also induced, indicating the activation of soybean defense responses under normal light treatment. In contrast, compared with normal light, under shading, genes related to soybean responses to stress, signal transduction, and immune system pathways were significantly downregulated. The plant-pathogen interaction pathway was significantly inhibited under shading, followed by secondary product and vitamin metabolism. Additionally, some important regulatory factors, such as WRKY/MYB transcription factors, CaM, KCS, and EDS1 were induced under normal light but suppressed under shading. Specifically, KCS is involved in the synthesis of long-chain fatty acids, which has a direct impact on the formation of the epidermal tissues in plants. The epidermis covers the plant body and is the first line of defense against pathogens and other biotic and abiotic stress stimuli. EDS1 is a key regulatory gene in the SA defense pathway. When

Table 7.3 Differential expression of genes involved in plant defense hormone signal transduction pathways under full light and shading conditions

Gene number	Description	Log2FC (NS/NC)	Log2FC (LS/LC)	Pathway
GLYMA_10G186800	Ethylene-responsive transcription factor 1B	2.71	−1.21	Ethylene
GLYMA_04G147000	EIN3-binding F-box protein 1	2.70	–	
GLYMA_20G203700	Ethylene-responsive transcription factor 1B	2.59	−2.19	
GLYMA_02G006200	Ethylene-responsive transcription factor 1B	–	−6.71	
GLYMA_10G036700	Ethylene-responsive transcription factor 1B	–	−2.88	
GLYMA_10G007000	Ethylene-responsive transcription factor 1B	–	−2.68	
GLYMA_19G248900	Ethylene-responsive transcription factor 1B	–	−1.55	
GLYMA_18G018400	Putative ETHYLENE INSENSITIVE 3	1.67	2.24	
GLYMA_13G166200	EIN3-binding F-box protein 1	1.58	–	
GLYMA_20G202200	Ethylene receptor 2	1.23	–	
GLYMA_13G076800	ETHYLENE INSENSITIVE 3-like 1 protein	1.15	–	
GLYMA_10G188500	Ethylene receptor	1.07	–	
GLYMA_16G020500	Transcription factor MYC2	2.49	–	Jasmonic acid
GLYMA_17G209000	Transcription factor MYC2	–	−1.52	
GLYMA_13G112000	Jasmonate ZIM domain-containing protein	1.69	–	
GLYMA_16G026900	Jasmonic acid-amido synthetase JAR1	–	−1.17	
GLYMA_18G030200	Coronatine-insensitive protein 1	−1.31	–	
GLYMA_15G062300	Pathogenesis-related protein 1-like protein	2.7	–	Salicylic acid
GLYMA_15G062400	Pathogenesis-related protein 1	–	−5.91	
GLYMA_15G062700	Pathogenesis-related protein 1	–	−3.48	
GLYMA_15G062500	Pathogenesis-related protein 1	–	−1.97	
GLYMA_09G020800	NPR1–1 protein	–	−1.05	
GLYMA_14G031300	Regulatory protein NPR3	2.31	–	
GLYMA_02G283300	Regulatory protein NPR3	1.04	–	
GLYMA_03G128600	Regulatory protein NPR5	–	1.99	
GLYMA_05G182500	Transcription factor TGA1-like	−1.06	–	
GLYMA_18G020900	Transcription factor TGA4-like	–	−1.1	
GLYMA_14G167000	Transcription factor TGA7	–	1.44	
GLYMA_13G085100	Transcription factor bZIP83	–	1.08	

Reprinted with the permission of viruses-basel

Note: *NC* normal light without virus inoculation, *NS* normal light with virus inoculation, *LC* shading without virus inoculation, *LS* shading with virus inoculation

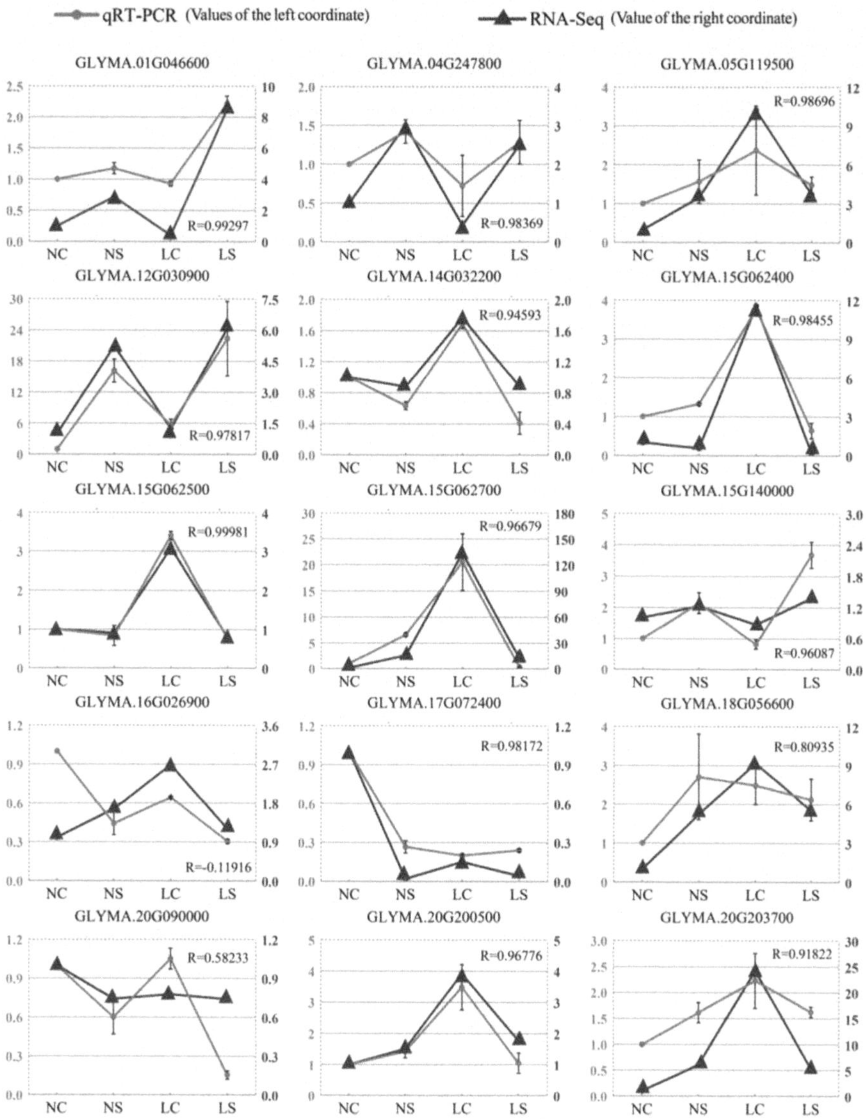

Fig. 7.13 Real-time fluorescent quantitative PCR and RNA sequencing comparative analysis of 15 DEGs under full light and shading treatments (reprinted with the permission of viruses-basel). The left vertical axis represents the gene expression level detected by qRT-PCR, and the right vertical axis represents the gene expression level detected by RNA sequencing; R is the correlation coefficient between qRT-PCR and RNA sequencing. *NC* normal light without virus inoculation, *NS* normal light with virus inoculation, *LC* shading without virus inoculation, *LS* shading with virus inoculation

plants were infected with SMV under shading, the expression of this series of regulatory genes was downregulated, indicating that the plant defense network could not be activated normally under such conditions.

Overall, both the SA and JA signal transduction pathways were suppressed in soybeans infected with SMV under shading conditions, thereby affecting plant resistance to SMV. This decrease in resistance was beneficial for plants with weaker growth status during the symbiotic period between maize and soybeans. Such relatively weak resistance can slow down the inflammatory response, reduce virus replication, alleviate flower-leaf symptoms, and reduce the severity of the disease.

7.1.2.2 Effects of Strip Intercropping on the Evolution of SMV

7.1.2.2.1 Study of SMV and Analysis of Population Genetic Diversity in Major Soybean Planting Areas in Southwest China

The soybean planting area in Sichuan Province is currently the fourth largest in China. In turn, the SMV population in China shows high genetic diversity. To better study and compare the effects of monoculture and strip intercropping on the genetic diversity of SMV and conduct an evolutionary analysis of SMV, we collected 251 soybean samples suspected of virus infection from 12 counties in nine cities in Sichuan, including Chengdu and Ya'an, and isolated and identified the soybean mosaic virus strains present in all these samples, among which, 77 from five counties tested positive for SMV, with the highest detection rate in Yucheng, Ya'an. Additionally, 39 samples from eight counties tested positive for Bean common mosaic virus (BCMV), with the highest detection rate in Zhongjiang, Deyang. Lastly, 24 samples from five counties tested positive for Cucumber mosaic virus (CMV). Results from virus identification tests are summarized in Table 7.4. Additionally, 16 samples were infected with multiple viruses, including ten samples of SMV and CMV coinfection, five samples of BCMV and CMV coinfection, and one sample of SMV and BCMV coinfection. These are summarized in Table 7.5. The main types of viruses causing soybean diseases in Sichuan were preliminarily identified as SMV and BCMV.

Based on the construction of neighbor-joining phylogenetic trees using the P1 gene, we determined that the SMV-RI and SMV-NI populations are located on different evolutionary branches. Additionally, the SMV-NI populations from the northeast, southeast, and southwest regions are also located on different evolutionary branches, indicating that the P1 gene contains more genetic evolutionary information, which can effectively distinguish among the sources of SMV populations (Fig. 7.14). For the CP gene, except for isolate GXQZ001, all other SMV-RI isolates clustered together in the same branch. However, there is no apparent evolutionary correlation between the SMV-NI population and the geographic origin of the virus. Phylogenetic analysis further showed that, regardless of the P1 or CP gene, the interstrain recombination-type SMV population (SMV-RI) was located on an independent branch. As for the non-recombination-type SMV population (SMV-NI), the clustering of its P1 gene was related to its geographic origin, whereas the CP gene showed no geographic correlation. In turn, SMV from North America and Asia can

Table 7.4 Virus identification

Order number	Collection site	Sample size	SMV		BCMV		CMV	
			Amount	Relevance ratio	Amount	Relevance ratio	Amount	Relevance ratio
1	Jialing	4	0	–	0	–	0	–
2	Xichong	8	0	–	0	–	0	–
3	Suining	19	0	–	9	47.4%	3	15.8%
4	Deyang	16	2	12.5%	10	62.5%	5	31.2%
5	Wenjiang	22	13	59.1%	0	–	0	–
6	Chongzhou	8	1	12.5%	5	62.5%	0	–
7	Ya'an	75	58	77.3%	1	1.3%	8	10.7%
8	Ren shou	44	3	6.82%	1	2.3%	0	–
9	Rich shun	19	0	–	5	26.3%	0	–
10	Rongxian	13	0	–	2	15.4%	7	53.8%
11	Yibin	17	0	–	6	35.3%	1	5.9%
12	Luzhou	6	0	–	0	–	0	–
–	Total	251	77	30.7%	39	15.5%	24	9.6%

Table 7.5 Viral coinfections

Order number	Sample name	Virus species	Collection site
1	SN3	BCMV+CMV	Suining
2	SN8	BCMV+CMV	
3	ZJ3	BCMV+CMV	Deyang
4	ZJ5	BCMV+CMV	
5	ZJ11	SMV + CMV	
6	ZJ12	SMV + CMV	
7	ZJ13	BCMV+CMV	
8	YA53	SMV + CMV	Ya'an
9	YA73	SMV + CMV	
10	YA78	SMV + CMV	
11	YA80	SMV + BCMV	
12	YA82	SMV + CMV	
13	YA90	SMV + CMV	
14	YA91	SMV + CMV	
15	YA92	SMV + CMV	
16	YA93	SMV + CMV	

also be classified according to the P1 gene, which is the most variable gene in all Potyvirus genomes. We speculate that, except for some northeast isolates clustering with the southwest branch, which may be due to soybean seed trade or germplasm diversity, such variation may help the virus adapt to new environments and hosts. Although the samples from the southwest region show the largest number of different strains, their genetic diversity is not the highest, which may be related to the promotion of strip intercropping soybeans and the scattered planting habits of soybeans in some hilly areas. Large-scale monoculture can increase the recombination rate (see 1.2), promote genetic variation of SMV, and be more conducive to the generation of new strains and the evolution of viruses. It is worth noting that the interstrain recombination population SMV-RI, which currently only occurs in soybean production areas of southern China, also shows high genetic diversity.

7.1.2.2.2 Effect of Cropping System on the Formation of Recombinant SMV

Recombination plays an important role in the evolution of RNA viruses. Many studies have reported intraspecies recombination in SMV, and the recombination hotspots are mainly concentrated in CI, NIa, and HC-Pro genes. Intercropping with maize reduced the interaction and infection of multiple viruses in soybean plants, suppressing the interspecies recombination rate by more than 90%, thus slowing down the process of viral genomic variation and evolution (Fig. 7.15). We detected intraspecies recombination events in SMV, as well as two interspecies recombination types of SMV formed by SMV and BCMV (named SMV-RI to distinguish them from intraspecies recombination) from southwest China. Notably, recombination analysis revealed that WMV and SMV-RI share the same recombination parents (BCMV and SMV) and almost identical recombination breakpoints (750–762 nt).

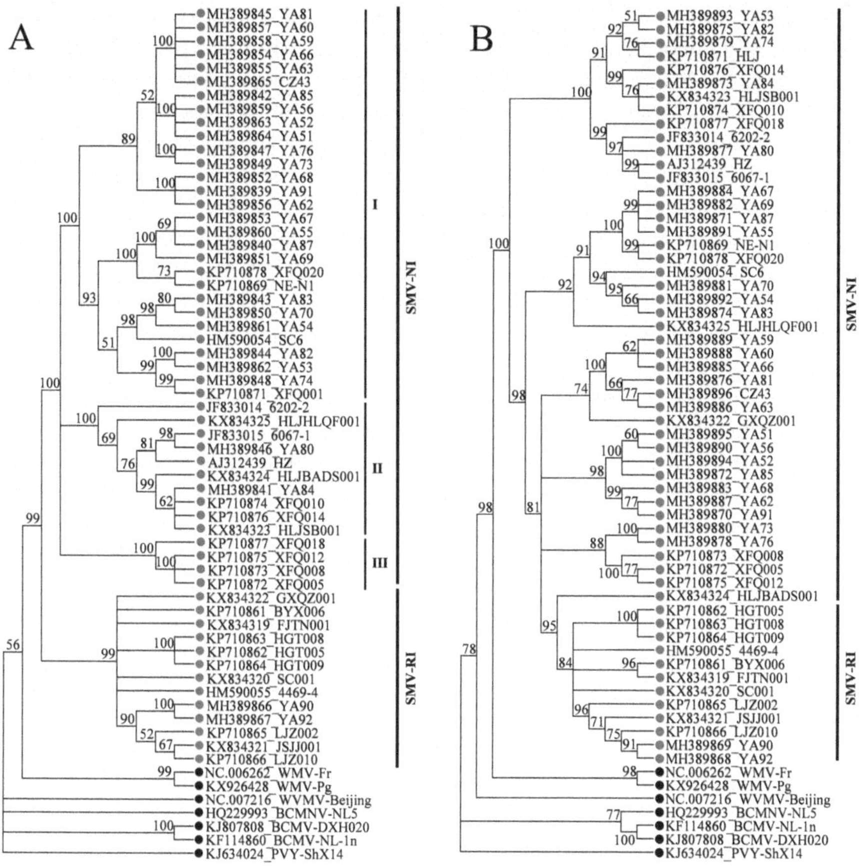

Fig. 7.14 (**a**) Systematic evolutionary analysis of the SMV P1 gene and (**b**) the coat protein CP gene. (Reprinted with the permission of viruses-basel)

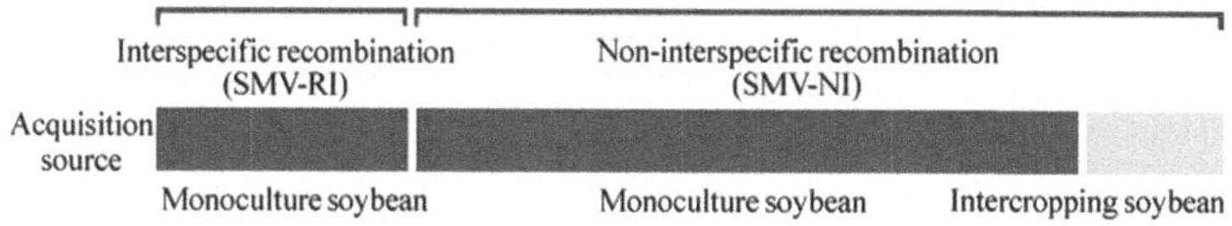

Fig. 7.15 Comparison of the number of interspecies recombinant and nonrecombinant viruses isolated from strip intercropping and monoculture soybeans

Although the nonrecombinant sequences (CP gene) of SMV-RI showed high similarity with SMV-NI, there was significant genetic differentiation between the two populations. Compared with SMV-NI, SMV-RI also has a wider host range. According to our identification, the inter-seeding recombination body formed by BCMV and SMV can also systemically infect tobacco. Through inoculation experiments, we found that among the 21 SC1-SC21 strains of SMV, only recombinant strain SC7 infected tobacco. The recombinant isolate SMV-P can systematically infect plants of the Araceae family, and yeast two-hybrid experiments revealed that the N-terminus of the SMV-P P1 gene can interact with the Fe/S protein of its host plants, Araceae, soybeans, and water chestnut, but not with its nonhost plant Arabidopsis, suggesting that the P1 gene may play an important role in virus-host adaptation. However, the AG repeats were highly repetitive before and after the recombination breakpoints in SMV-RI (747–769 nt), and these repeat sequences may contribute to recombination. Multiple recombination breakpoints were also detected in the SMV-RI population, including 750, 754, 757, and 762 nt, suggesting that the recombination region may have occurred more than once in the ancestor recombination and, indeed, may have undergone multiple recombination events, thus forming the SMV-RI population.

All interspecies recombinant viruses were collected from monoculture soybean fields, in which virus types increased in monoculture fields. The color bar shows the relative quantity of virus samples. Red indicates the number of samples collected from monoculture fields, and yellow indicates the number of samples collected from strip intercropping soybeans.

Soybean fields and their surroundings include other legume species, such as peanuts, kidney beans, peas, and green beans, as well as species from other plant families, such as pumpkins and peppers. The complex host environment is directly related to the pathogen population structure, creating conditions for inter-virus recombination. In addition to SMV, a large number of BCMV and CMV have also been detected in the collected soybean samples. Soybeans infected with SMV-RI showed mild symptoms but higher virus accumulation, indicating that SMV-RI was less pathogenic but more adaptable, and an expansion of the host range. In summary, large-scale monoculture tends to promote genetic variation of virus strains and facilitate virus evolution, whereas an adequate intercropping planting system will slow down virus evolution.

7.1.2.3 Effects of Strip Intercropping on Vector-Borne Diseases

In nature, the transmission of viruses is extremely complex, and many factors determine the field prevalence of viruses, creating favorable conditions for disease outbreaks. Specifically, dozens of aphid species can carry CMV and SMV for nonpersistent transmission, and the virus can be successfully transmitted in as little as 96 seconds. Research has shown that aphid infestations in soybean fields are often found near vegetable fields. For example, the overwintering hosts of CMV are mainly field weeds, perennial trees, and overwintering vegetables from which it is spread by winged aphids and becomes the initial source of infection on soybeans the following year, whereas SMV primarily overwinters on infected seeds or diseased plants in the field.

Soybean aphids prefer to feed on soybeans rather than maize. Due to the higher position of the maize canopy, the spread of aphids carrying viruses is limited, and the habitats of natural enemies increase. The beneficial ratio of ladybugs to aphids is greater than 1:162, which is beneficial for controlling aphid-borne viruses (Fig. 7.16, Table 7.6).

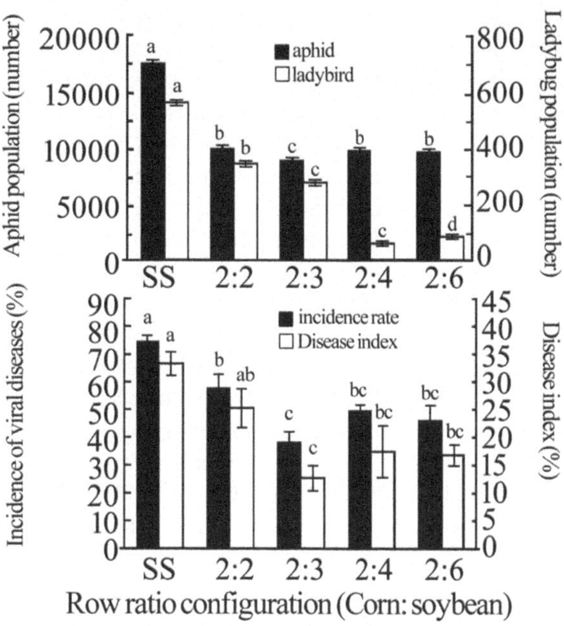

Fig. 7.16 Effect of different field configurations on the suppression of strip intercropping SMV disease transmitted by soybean aphids. (Reproduced with permission of Journal of Sichuan Agricultural University)

Table 7.6 Ladybug headcount and benefit-harm ratio

Treatment	Numbers of ladybug specimens	Number of individuals/ 100 plants	Ratio of natural enemy to pest specimens
SS	560.09 ± 4.14a	5.60 ± 0.04a	1:31.24 ± 0.18a
2:2	345.10 ± 2.55b	3.45 ± 0.03b	1:28.67 ± 0.29a
2:3	277.01 ± 2.65c	2.77 ± 0.03c	1:31.68 ± 0.29a
2:4	60.24 ± 2.04e	0.60 ± 0.02e	1:162.68 ± 5.40c
2:6	86.96 ± 1.51d	0.87 ± 0.02d	1:110.00 ± 1.66b

Reproduced with permission of Journal of Sichuan Agricultural University

Thus, under strip intercropping, the pest control effect is best at a row ratio of 2:3 or 2:4, which also affects the aphid population. The number of aphids is lowest when the row ratio is 2:3. As the prevalence of diseases is positively correlated with the occurrence of aphids, the disease index of CMV is lowest when the row ratio is 2:3. Additionally, the shape index and yield of soybeans are closest to net cultivation when the row ratio is 2:3 or 2:4.

7.1.3 Main Diseases and their Patterns of Incidence in Strip Intercropping Maize

7.1.3.1 Incidence of Major Maize Diseases in Strip Intercropping

Maize-soybean and maize-sweet potato strip intercropping are the two main intercropping modes in Sichuan. This study found that the most important maize diseases under these two cropping patterns were leaf spot (large and small), stripe rust, stalk rot, ear rot, and rust. Among them, the incidence of leaf spot, stripe rust, and ear rot was relatively high, and there was no significant difference in the incidence rate of related maize diseases between the two patterns. The incidence of stalk rot in maize under strip intercropping with soybeans or sweet potatoes was studied in four investigation sites: Yucheng District in Ya'an City, Renshou County in Meishan City, Langzhong in Nanchong City, and Rong County in Zigong City. The incidence of stalk rot in Renshou County and Rong County was the highest at 16.03% and 11.52%, respectively, whereas other investigation sites showed moderate incidence rates (Table 7.7).

7.1.3.2 Isolation and Identification of *Fusarium* Species Causing Stalk Rot in Maize Strip Intercropping Fields

Stalk rot is one major disease affecting maize yield in maize-producing regions of Sichuan and often occurs in strip intercropping. *Fusarium* is an important pathogen causing maize stalk rot. From 2016 to 2017, stalk rot samples were collected from maize strip intercropping in Renshou, Chongzhou, Ya'an, and Rong counties in Sichuan. Through tissue isolation and purification, morphological characterization, and sequence analysis based on *rDNA-ITS* and *EF-1α*, the *Fusarium* species causing stalk rot was identified. Pathogenicity test was also performed to clarify the *Fusarium* population causing stalk rot in maize strip intercropping.

7.1.3.3 Isolation and Identification of Pathogenic *Fusarium* Causing Maize Stalk Rot

Seventy-one isolates of *Fusarium* were obtained from 103 samples of maize stalk rot. The ribosomal-DNA internal transcribed spacer (*rDNA ITS*) and the translation elongation factor *EF-1α* gene fragments of each isolate were amplified using PCR. The sequences were analyzed, and a multigene phylogenetic tree was constructed using the MEGA 6.0 software. The results showed that the *Fusarium* isolates related to stalk rot were divided into seven groups. The first group (Y9, Y12, Z1, and C10) showed a sequence similarity of up to 100% to the reference strain sequence of *F. graminearum*. The other groups clustered with *F. equiseti*, *F. verticillioides*, *F. oxysporum*, *F. solani*, *F. fujikuroi*, and *F. proliferatum*, with sequence similarities of over 54%. The isolation frequency of the seven *Fusarium* species was compared. *F. proliferatum* showed the highest isolation frequency at 35.21%, followed by *F. solani* and *F. verticillioides* at 16.9% and 15.49%, respectively. Additionally, *F. graminearum*, *F. oxysporum*, *F. fujikuroi*, and *F. equiseti* showed the lowest isolation frequencies. The diversity and isolation frequency of different *Fusarium* species varied with the regions, but *F. proliferatum* was the dominant species across regions. In conclusion, under strip intercropping, the dominant *Fusarium* species causing stalk rot in maize were *F. proliferatum*, *F. verticillioides*, *F. solani*, and *F. graminearum* (Table 7.8).

Table 7.7 Incidence of major maize diseases in strip intercropping during 2016–2017

Cropping pattern	Stem rot		Banded sclerotial blight		Leaf spot		Rust disease		Spike rot		Other diseases	
	Disease incidence (%)	Disease index	Disease incidence (%)	Disease index	Disease incidence (%)	Disease index	Disease incidence (%)	Disease index	Disease incidence (%)	Disease index	Disease incidence (%)	Disease index
Corn and soybean ribbon compound planting	17.5	2.78	46.5	11.06	100	19.67	2.05	3.57	55.97	5.01	1.03	1.02
Corn sweet potato ribbon-shaped compound planting	18.5	2.03	40.3	20.11	99.75	10.23	3.51	3.00	58.40	3.49	2.03	1.45

Table 7.8 Isolation frequency of *Fusarium* species causing maize stalk rot under strip intercropping in Sichuan Province

Fusarium species	Average isolation frequency (%)	Isolation frequency (%)			
		Ya'an	Ren shou	Chongzhou	Rong
F. proliferatum	35.21 a*	39.14 a*	33.33 a*	38.45 a*	29.41 a*
F. solani	16.9 b	17.39 b	16.67 c	15.39 b	17.65 c
F. verticillioides	15.49 b	13.04 c	22.22 b	15.39 b	11.76 d
F. graminearum	11.27 c	13.04 c	0 e	7.69 c	23.53 b
F. oxysporum	8.45 d	4.35 e	16.67 c	15.39 b	0 f
F. fujikuroi	7.04 d	8.69 d	11.11 d	0 d	5.89 e
F. equiseti	5.63 e	4.35e		7.69 c	11.76 d

Means within columns followed by different lowercase letters indicate significant differences at $P < 0.05$

Fig. 7.17 Maize disease symptoms caused by *Fusarium* spp. Visible symptoms at 3 days after inoculating maize seedlings. **A**, *Fusarium verticillioides*; **B,** *Fusarium proliferatum*; **C**, *Fusarium graminearum*; **D**, *Fusarium equiseti*; **E**, *Fusarium solani*; **F**, *Fusarium avenaceum*; **G**, *Fusarium oxysporum*; **Con**, Un-inoculated control

7.1.3.3.1 Determination of the Pathogenicity of Candidative *Fusarium* Spp. Causing Maize Stem Rot

Pathogenicity test was performed using the seedling inoculation method. For three days after inoculation, there were significant differences in the pathogenicity of different *Fusarium* isolates, among which *F. graminearum* showed the strongest pathogenicity, with brown, sunken lesions that markedly expanded (Fig. 7.17). *F. proliferatum* and *F. verticillioides* caused water-soaked lesions and browning with rotting symptoms on the stem, and a large quantity of mycelia was produced. Meanwhile, *F. equiseti* and *F. oxysporum* caused partial water-soaked and discolored lesions on the maize stem, with little obvious lesion expansion and some mycelial production. *F. fujikuroi* and *F. solani* also caused stem discoloration, but the symptoms were not obvious. All *Fusarium* isolates obtained were pathogenic to maize seedlings, with *F. graminearum* being the most pathogenic (95.0% incidence rate) followed by *F. proliferatum* and *F. verticillioides*, whereas *F. oxysporum* and *F. equiseti* were less pathogenic. *F. fujikuroi* and *F. solani* showed the lowest pathogenicity, with incidence rate of less than 70% and disease index of lower than 1.0 (Fig. 7.19).

7.1.3.4 Infection Relationship Between *Fusarium* Causing Root Rot in Soybean and *Fusarium* Causing Stem Rot in Maize in Strip Intercropping

Fusarium is a pathogenic fungus causing various diseases in maize and soybean under strip intercropping, including stem rot and ear rot in maize, and seedling blight, root rot, and pod rot in soybean. Exploring the interaction between soybean root rot and maize stem rot under strip intercropping has important implications for the comprehensive prevention and control of *Fusarium*-related diseases in strip intercropping.

7.1.3.4.1 Pathogenicity of *Fusarium* Root Rot on Maize

The isolation and identification of pathogenic *Fusarium* species in Sichuan showed that several species, including *F. solani, F. oxysporum, F. equiseti, F. avenaceum, F. graminearum, F. commune*, and *F. proliferatum*, can cause soybean root rot, with *F. solani* and *F. oxysporum* being the dominant pathogens. Using the sorghum grain inoculation method, we found that, except for *F. solani*, the other six *Fusarium* species infected maize seedlings and reduced maize growth with different degrees of pathogenicity (Fig. 7.18). For example, *F. graminearum* severely inhibited root growth, significantly reduced plant height and fresh root weight as compared with the control, displaying the strongest pathogenicity. *F. commune* and *F. proliferatum* showed lower pathogenicity on maize seedlings, with disease indexes of 67.5 and 75.0, respectively, although plant growth was still significantly hindered as compared with control plants. Meanwhile, *F. oxysporum, F. equiseti,* and *F. avenaceum* showed lower disease indexes in maize seedlings, and the growth of both the root and the aboveground plant tissues was not significantly different from that of the controls. *F. solani* did not affect plant growth after inoculation (Fig. 7.19). Moreover, using a spore suspension injection method, the strongly pathogenic *F. graminearum* causing soybean root rot was inoculated intomaize plants in the fields, and caused brown necrosis on maize stems, with the disease symptoms spreading from the inoculation site towards up and down sides. In summary, these results showed that *Fusarium* species causing soybean root rot effectively invaded maize to result in stem rot.

7.1.3.4.2 Pathogenicity Test of *Fusarium graminearum* Causing Maize Stem Rot on Soybean

To investigate the pathogenicity of *Fusarium* species associated with maize stem rot on soybean, two isolates of pathogenic *F. graminearum*, R5-4 and R1-4, were selected and inoculated onto soybean using both the stem puncture and the sorghum grain inoculation methods. Results from stem puncture inoculation showed that

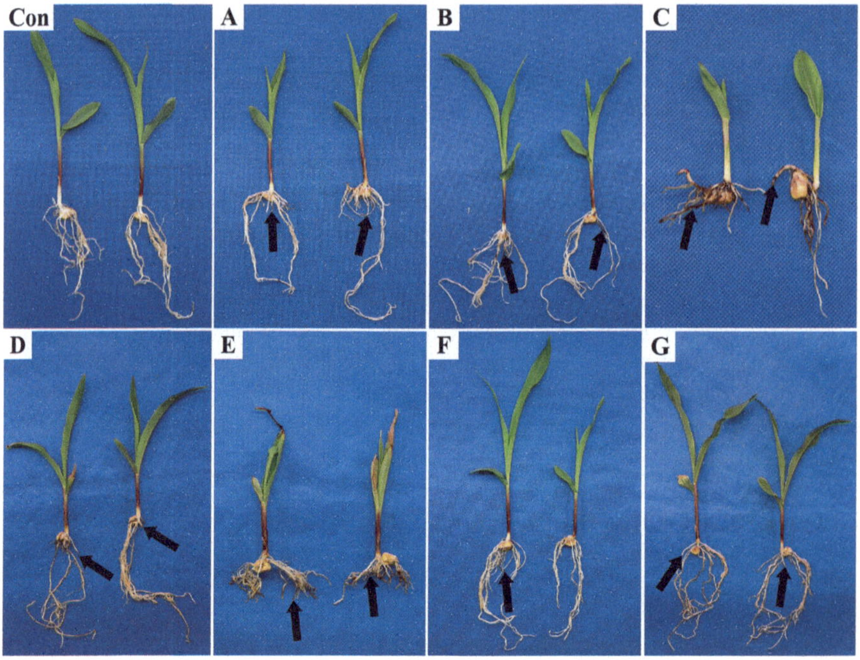

Fig. 7.18 Symptoms in maize seedlings infected with *Fusarium* causing soybean root rot. Maize symptoms were observed at 10 days after inoculation through infected sorghum grain. **Con,** the uninoculated control; **A**, *Fusarium moniliforme*; **B**, *Fusarium verticillioides*; **C**, *Fusarium graminearum*; **D**, *Fusarium commune*; **E**, *Fusarium proliferatum*; **F**, *Fusarium subglutinans*; **G**, *Fusarium avenaceum*

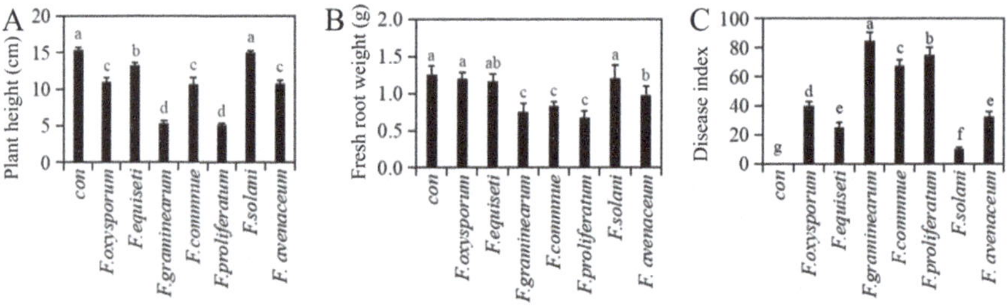

Fig. 7.19 Growth and disease development of maize seedlings inoculated with *Fusarium*-infected sorghum grain in the pot experiments. The height (**A**), root fresh (**B**), and disease index (**C**) were measured in maize seedlings at 10 d after inoculation with infected sorghum grains. **Con** refers to the control group without inoculation. Different lowercase letters indicate significant differences among *Fusarium* species at $P < 0.05$ level

maize stem occured discoloration and rot symptoms at the inoculation site of the two *F. graminearum* isolates. During three days after inoculating soybean seedlings, both isolates of *F. graminearum* caused browning and necrosis in the soybean stem, with obvious expansion of the infected stem section (Fig. 7.20). In turn, through sorghum grain inoculation *F. graminearum* was able to infect both maize and soybean seedlings during 10 days after inoculation, which led to the reduction of germination rate and inhibition of plant height and fresh root weight. Furthermore, compared with controls, *F. graminearum* caused soybean root rot, and R1-4 showed stronger pathogenicity than R5-4, but both isolates severely suppressed the main root and lateral root rotting and elongation of the main root (Fig. 7.20).

A, Symptoms in maize and soybean plants three days after stem puncture inoculation; B, symptoms in maize and soybean 10 days after inoculation with sorghum grain. Panels A and B show the severity of the disease caused by *Fusarium graminearum* strains R1-4 and R5-4, respectively. Con represents the uninoculated control group.

7.2 Main Pattern of Insect Pest Incidence

Research on soybean pest incidence under strip intercropping with maize is relatively scarce, especially regarding the systematic study of pest population distribution in different field configurations of soybean and maize. Our study revealed that under maize-soybean strip intercropping, the most common pests attacking soybean plants were *Spodoptera litura*, *Ergania doriae yunnanus*, *Paraluperodes suturalis nigrobilineatus*, *Nezara viridula*, Locusts, Aphids, Snails, and Nail snails. Other pests included *Coreidae*, *Stollia montivagus* Distant, *Conocephalidae*, *Agriolimax agrestis* Linnaeus, *Thripidae*, *Delphacidae*, *Geometridae* and *Ascotis selenaria*

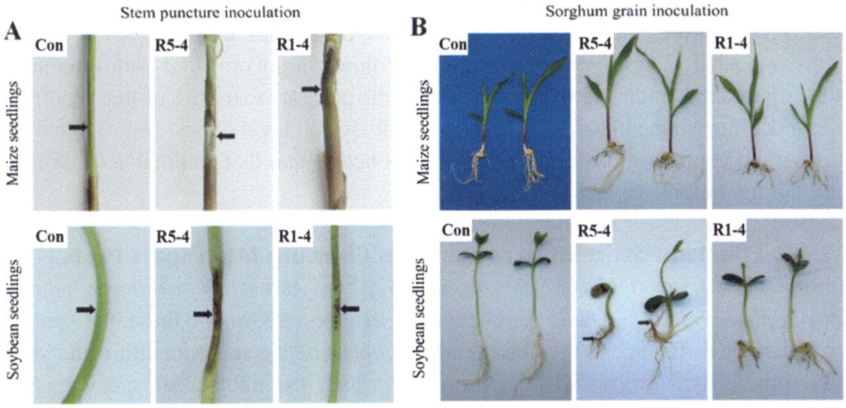

Fig. 7.20 Symptoms observed in soybean after inoculation with *Fusarium graminearum*

Schiffermüller et Denis. The main predators of these insect pests are carnivorous ladybugs, followed by *Chrysopidae*, *Labidura riparia* Pallas, *Araneida,* and Syrphidae. Strip intercropping increased the heterogeneity of interspecific relationships, changed the microclimate in the field, increased biodiversity, and enhanced ecosystem stability. Further, it significantly reduced the incidence of severe pest outbreaks. Strip intercropping reduced the degree of pest damage. Pest incidence and damage rates were significantly lower than those of monoculture soybean. Additionally, there were significant differences in pest rates among maize-soybean row ratios. Our systematic study revealed the incidence pattern of the main pests in maize-soybean strip intercropping, providing a solid theoretical basis for the formulation of comprehensive pest control technology and ultimately providing technical support for the large-scale promotion of strip intercropping.

7.2.1 Effects of Strip Intercropping on the Distribution of Low-Level Insect Pests

The main pests with low-level distribution are *Spodoptera litura* larvae, *Aphis glycines*, *Helicoverpa armigera*, and *Agriolimax agrestis* Linnaeus. The low-level pest distribution was significantly inhibited by the barrier and allelopathic avoidance effects.

7.2.1.1 Effects of Strip Intercropping on the Distribution of Major Insect Pests

Strip intercropping significantly reduced the number of *Spodoptera litura* larvae, *H. armigera*, *A. agrestis* Linnaeus, and *A. glycines*. The highest inhibition rates were 7%, 23.1%, 16.5%, 17.9%, and 50.2%, respectively, relative to the corresponding number of specimens of each of these pests in monoculture soybean. The dilution effect and spatial barrier effect due to the taller maize plants in the strip intercropping system reduced the damage caused by major soybean pests, especially by *S. litura*, *A. glycines*, and *H. armigera* before and after flowering. Compared with monoculture, strip intercropping significantly reduced the pest rate to 47.6% of that recorded in monoculture soybean fields. Furthermore, the overall level of pest control attained at row ratio configurations of 2:3 and 2:4 was better than that attained at other row ratio configurations tested (Table 7.10).

7.2.1.2 Effects of Strip Intercropping on Chewing Mouthpart Pests

Among the surveyed insect pest species, *S. litura* larvae, *H. armigera* adults, *L. oryzophilus*, and *A. glycines* are chewing insect species. Among these, the species for which the difference in the number of specimens between monoculture and strip intercropping was largest was *S. litura*, for which the number of specimens in monoculture was 7.37 times that observed in strip intercropping. Furthermore, among the different row ratios tested, the lowest number of *S. litura* specimens was recorded at a row ratio of 2:3. The number of *S. litura* specimens in monoculture was 14 times that recorded at this row ratio. In contrast, the number of *H. armigera*

Table 7.10 Overall effect of monoculture and strip intercropping on chewing insect pest populations

Treatment		Total number of plants	Plants with insects	Percentage of plants with insects(%)	Number of damaged plants	Percentage of damaged plants(%)
Monoculture soybean		10,000	6574.30 ± 37.43a	65.74 ± 0.37a	7682.87 ± 47.45a	76.83 ± 0.47a
Row ratio of maize and soybean	2:2	10,000	3497.33 ± 7.13d	34.90 ± 0.07d	3839.48 ± 11.74d	38.39 ± 0.12d
	2:3	10,000	3989.10 ± 14.03c	39.89 ± 0.14c	4071.76 ± 30.86c	40.72 ± 0.31c
	2:4	10,000	4216.70 ± 10.42b	42.17 ± 0.10b	4389.39 ± 4.29b	43.89 ± 0.04b
	2:6	10,000	3130.40 ± 29.50e	31.30 ± 0.30e	3496.87 ± 9.88e	0.35 ± 0.10e

specimens and that of specimens per 100 plants did not significantly differ between the row ratios of 2:4 and 2:6. The number of *H. armigera* specimens under strip intercropping with a row ratio of 2:2 was significantly higher than that under strip intercropping at row ratio values of 2:4 or 2:6. With respect to row ratio, the number of *L. oryzophilus* specimens in monoculture fields ranked as follows: >2:3 > 2:6 > 2:2 > 2:4. Furthermore, the number of *L. oryzophilus* specimens per 100 soybean plants at row ratios of 2:2 and 2:4 was less than one, which was lower than that of the control target. Similarly, the number of *A. glycines* per 100 soybean plants at a row ratio of 2:4 was less than one, and the damage caused by the species at this row ratio was minimal (Table 7.11).

7.2.1.3 Effects of Strip Intercropping on Piercing-Sucking Insect Pests

Green leafhoppers, brown planthoppers, and aphids are piercing-sucking insects belonging to the Order Hemiptera. Except for row ratio 2:2, the number of green leafhopper and brown planthopper specimens was less than one at row ratios of 2:3, 2:4, and 2:6 in maize-soybean intercropping. Although there was a significant difference in the number of brown planthopper individuals between row ratios of 2:4 and 2:6 on the one hand and 2:2 and 2:3 on the other, the number of brown planthopper specimens under different row ratios was smaller than the control target, except when the row ratio was 2:2. In turn, the number of aphid individuals under maize-soybean intercropping was significantly larger than that under strip intercropping, with more than 100 per 100 plants under monoculture, and no significant difference among different row ratios, except for 2:3 (Table 7.12).

7.2.1.4 Effects of Strip Intercropping on the Distribution of Major Soft-Bodied Insect Pest Species

The number of snail individuals in monoculture was 6.06 times that in maize-soybean intercropping fields configured at a row ratio of 2:2; 2.97 times that at a row ratio of 2:3; 4.33 times that at a row ratio of 2:4, and 4.54 times that at a row ratio

Table 7.11 Effect of net cropping and strip intercropping on the populations of chewing insect pests

Treatment	*S. litura* larvae		*H. armigera* adults	
	Insect specimen count	Number of specimens per 100 plants	Insect specimen count	Number of specimens per 100 plants
SS	40444.94 ± 241.75a	404.45 ± 2.42a	2168.67 ± 26.60a	21.69 ± 0.27a
2:2	3510.20 ± 65.73c	35.10 ± 0.66c	1333.33 ± 27.64b	13.33 ± 0.28b
2:3	2825.48 ± 99.52d	28.25 ± 1.00d	720.22 ± 3.05c	7.20 ± 0.03c
2:4	5481.93 ± 62.10b	54.82 ± 0.62b	500.56 ± 6.57d	5.01 ± 0.07d
2:6	3347.83 ± 3.65c	33.48 ± 0.04c	521.74 ± 4.40d	5.22 ± 0.04d
Treatment	*L. oryzophilus*		*A. glycines*	
	Insect specimens count	Number of specimens per 100 plants	Insect specimens count	Number of specimens per 100 plants
SS	347.83 ± 4.38a	3.48 ± 0.04a	415.69 ± 3.34a	4.16 ± 0.03a
2:2	89.45 ± 2.60d	0.89 ± 0.03d	206.63 ± 2.58b	2.07 ± 0.03b
2:3	249.31 ± 5.54b	2.49 ± 0.06b	198.00 ± 1.53c	1.98 ± 0.01c
2:4	60.24 ± 1.38e	0.60 ± 0.01e	86.96 ± 2.05e	0.87 ± 0.02e
2:6	144.61 ± 2.22c	1.45 ± 0.02c	121.00 ± 2.00d	1.21 ± 0.02d

SS net soybean yield was analyzed for significant differences using Duncan's test. Data are means ± standard error. Different lowercase letters within rows indicate significant differences at the 5% level ($P < 0.05$)

of 2:6. Further, the difference among row ratio configurations was significant. The number of nail snail individuals was highest in monoculture and lowest at a row ratio of 2:4 for maize-soybean intercropping (Table 7.13).

7.2.1.5 Effects of Strip Intercropping on the Population of Aphids

The number of ladybug specimens was highest in monoculture, and the difference between maize-soybean intercropping row ratio configurations was significant. Furthermore, the number of ladybug individuals in different maize-soybean intercropping row ratio configurations was in the order of 2:2 > 2:3 > 2:6 > 2:4, and that per 100 plants at a row ratio of 2:4 was less than one. Lastly, the benefit-harm ratio of ladybugs and aphids at different row ratio configurations was greater than 1:200 (Fig. 7.21, Table 7.14).

Table 7.12 Effects of net crop versus banded set row ratio on the population size of piercing-sucking insect pest species

Treatment	Green stink bug		Edged stink bug		Aphids	
	Insect specimen count	Number of specimens per 100 plants	Insect specimen count	Number of specimens per 100 plants	Insect specimen count	Number of specimens per 100 plants
SS	509.80 ± 4.17a	5.10 ± 0.04a	149.02 ± 2.10a	1.49 ± 0.02a	17498.64 ± 27.70a	174.99 ± 0.28a
2:2	141.38 ± 1.45b	1.41 ± 0.01b	38.06 ± 2.31b	0.38 ± 0.02b	9894.12 ± 28.12b	98.94 ± 0.29b
2:3	83.10 ± 0.59c	0.83 ± 0.01c	34.37 ± 0.69b	0.34 ± 0.01b	8776.00 ± 4.36c	87.76 ± 0.04c
2:4	60.24 ± 1.75d	0.60 ± 0.02d	28.43 ± 2.12c	0.28 ± 0.02c	9778.00 ± 8.15b	97.78 ± 0.08b
2:6	86.96 ± 0.54c	0.87 ± 0.01c	32.27 ± 1.19bc	0.32 ± 0.01bc	9565.22 ± 286.46b	95.65 ± 2.86b

Reproduced with permission of Journal of Sichuan Agricultural University
SS net soybean yield was analyzed for significant differences using Duncan's test. Data are means ± standard error. Different lowercase letters within rows indicate significant differences at the 5% level ($P < 0.05$)

Table 7.13 Effects of net crop versus banded set row ratio on the population size of mollusks

Treatment	Snail		Oncomelania	
	Insect specimen count	Number of specimens per 100 plants	Insect specimen count	Number of specimens per 100 plants
SS	1565.22 ± 7.15a	15.65 ± 0.07a	1043.48 ± 12.97a	10.43 ± 0.13a
2:2	258.50 ± 3.75e	2.59 ± 0.04e	557.82 ± 1.70b	5.58 ± 0.02b
2:3	526.32 ± 4.24b	5.26 ± 0.04b	277.01 ± 1.15c	2.77 ± 0.01c
2:4	361.45 ± 2.94c	3.61 ± 0.03c	180.72 ± 3.69d	1.81 ± 0.04d
2:6	344.83 ± 3.44d	3.45 ± 0.03d	200.11 ± 4.62d	2.00 ± 0.05d

Reproduced with permission of Journal of Sichuan Agricultural University
Note: *SS* net soybean yield was analyzed for significant differences using Duncan's test. Data are means ± standard error. Different lowercase letters within rows indicate significant differences at the 5% level ($P < 0.05$)

According to the above results, the number of individuals of the two-striped leaf beetle, the edge bug, the green bug, aphids, and locusts with high mobility was smallest at a row ratio of 2:4 or 2:3 for maize-soybean intercropping. Meanwhile, the number of individuals of oblique borer larvae, high-long elephant beetles, snails, and nail snails with weaker mobility was smallest at a row ratio of 2:2 and 2:3 for maize-soybean intercropping, except that the number of snail individuals and oblique borer larvae was smallest at a row ratio of 2:2 and 2:3, respectively. The other two harmful insects were the fewest at a row ratio of 2:4 for maize-soybean intercropping.

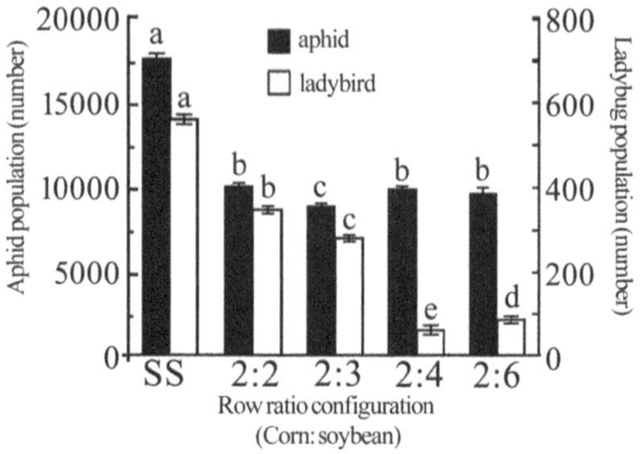

Fig. 7.21 Number of ladybeetle and aphid specimens (reproduced with permission of Journal of Sichuan Agricultural University). *SS* net soybean yield was analyzed for significant differences using Duncan's test. Data are means ± standard error. Different lowercase letters indicate significant differences at the 5% level ($P < 0.05$)

Simultaneously, mixed planting increased the incidence of parasitic and predatory natural enemies of these insect pests. Furthermore, our research showed that the benefit-harm ratio of ladybugs and aphids was lowest at a row ratio of 2:4 for maize-soybean intercropping, whereas it was highest at a row ratio of 2:2 but close to that of monoculture at a row ratio configuration 2:3, in which case, the number of aphids was the lowest and the control effect on aphids was best. The control effects on aphids reached 99.75% when the benefit-harm ratio was 1:700. The benefit-harm ratios in all planting systems were consistently greater than 1:700. Therefore, exotic ladybugs effectively control soybean aphid numbers. Taking all factors into consideration, a row ratio of 2:3 or 2:4 for maize-soybean intercropping showed better pest control effects, and soybean plants maintained a better growth and development status. Among mixed planting techniques, these two row ratios can be promoted as the preferential field configurations.

7.2.2 Effects of Strip Intercropping on Migration of High-Altitude Migratory Insect Pests

Monochromatic insecticidal lamps can target the phototaxis of insect pests to specific wavelength bands and are specifically effective in pest control. In particular, LED light sources have less impact on insect ecology and have great application potential. We used 13-wavelength LED insecticidal lamps to lure and kill high-altitude migratory pests throughout the entire growth period of maize and soybean plants in an intercropping system. We analyzed the species and population dynamics of high-altitude migratory pests under maize-soybean intercropping to explore the influence of strip intercropping on high-altitude migratory insect pests.

Table 7.14 Main high-altitude migratory insect pest in the middle of the field is the strip-shaped sleeve (data on different insect trap lamp kills)

	Number of trapped specimens per day													
Insect species	401 nm	397 nm	405 nm	407 nm	393 nm	411 nm	395 nm	375 nm	350 nm	385 nm	378 nm	389 nm	403 nm	Compound UV
Grapholita molesta	2.95	2.58	2.55	1.16	3	1.79	2.34	2.89	2.87	3.91	2.04	2.46	2.21	2.68
Spodoptera litura	2.16	3.34	3.04	2.25	3.67	1.67	6.41	2.64	3.07	2.48	2.83	3.84	1.2	2.88
Metabolus flavescens Brenske	1.98	4.53	3.63	2.14	2.3	2.53	2.19	1.52	2.85	1.58	2.84	2.41	5.19	3.52
Holotrichia parallela Motschulsky	3.95	3.28	7.88	14.8	3.67	3.11	8.6	1.63	1.67	1	1.4	1.19	2.57	2.42
Heteroptera	1.5	2.39	4.18	2.05	1.78	2.41	1.85	1.17	1.29	1.54	1.09	1.52	3.62	3.65
Carabid	2.85	3.3	4.58	3.44	3.02	3.32	3.99	1.84	3.33	2.36	2.18	2.68	3.6	3.44
Coccinellidae (predator)	1.44	1.97	2.42	1	1.59	1.58	2.47	1.41	1.2	1.56	1.41	1.96	1.36	3.02
Neutral insects	7.93	6.16	6.23	5.07	5.71	5.58	5.88	2.92	2.34	3.02	2.81	3.93	10.61	30.44

7.2.3 Effects of Strip Intercropping on Migration of High-Altitude Migratory Insect Pests

During the full reproductive period of the banded sleeve (140 days if rainy days are excluded), the main insect pests killed by different wavelength LED insecticidal lamps included the peach fruit moth *Dichocrocis punctiferalis* Guenée, the oriental leafworm *S. litura* Fabricius, the small yellow gill turtle *Metabolus flavescens* Brenske, the dark gill turtle *Holotrichia parallela*, and various bugs; additionally, predatory insects killed included ladybugs, *Coccinellidae*, and ground beetles, *Carabidae*; neutral insects killed included earwigs *Labidura riparia*, cicadas *Cicadidae*, water scorpions *Belostomatidae*, dung beetles *Geotrupidae*, diving beetles *Dytiscidae*, burying beetles *Silphidae*, and water turtles *Hydrophilidae* (Table 7.14). Combined with the results of field surveys from previous years, our data indicated that oriental leafworms, peach fruit moths, golden turtles, and bugs are important insect pests in the banded sleeve fields in this area.

7.2.3.1 Effect of Strip Intercropping on the Temporal Dynamics of High-Altitude Migratory Insect Pest Populations

This experiment showed that the main Lepidoptera pests that were attracted by different wavelength LED insecticidal lamps in the banded sleeve fields are peach fruit moths and oriental leafworms. Among them, the single-wavelength insecticidal lamps of 385, 393, 401, and 375 nm have better killing effects on peach fruit moths, which are 3.91, 3, 2.95, and 2.89 heads/day, respectively (Fig. 7.22). The daily number of killings of each insecticidal lamp on peach fruit moths was summarized to reflect the occurrence dynamics of peach fruit moths, which showed no evident patterns. Further, there is a strong generational overlap of peach fruit moths in this area. Only one concentrated peak of moth occurrence appeared in mid-August

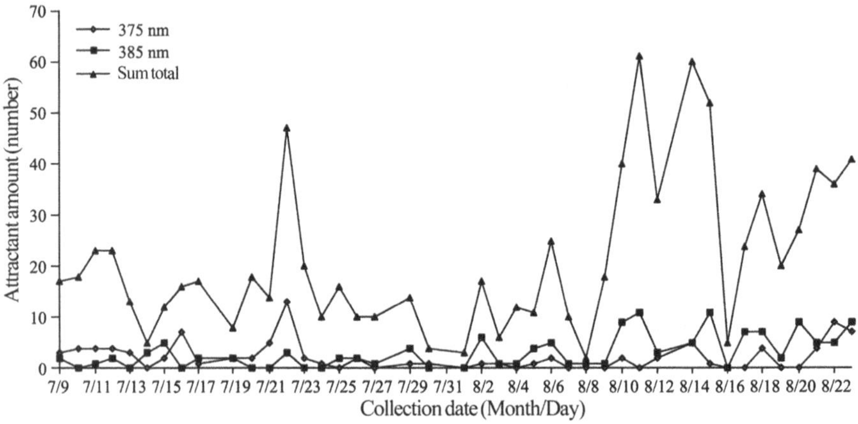

Fig. 7.22 Temporal dynamics of the oriental fruit moth population

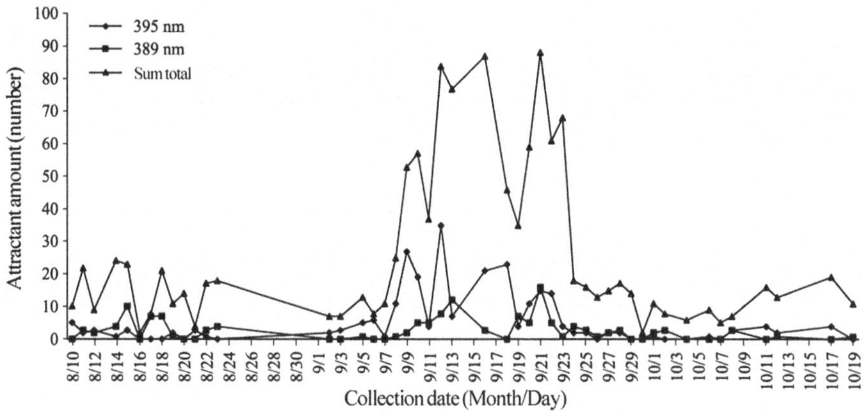

Fig. 7.23 Temporal dynamics of the *Spodoptera litura* population

during the full reproductive period. Two single-wavelength insecticidal lamps with good killing effects on peach fruit moths and minimal impact on natural enemies were selected: 385 and 375 nm. The killing dynamics of the two single-wavelength insecticidal lamps on peach fruit moths were plotted. The results showed that the killing dynamics of the two single-wavelength insecticidal lamps were similar to the trend of occurrence dynamics of peach fruit moths.

Analysis of the effect of different insecticidal lamps on the attraction of *Spodoptera litura* shows that 395 and 389 nm wavelength insecticidal lamps had the best effects on the attraction of *Spodoptera litura* (Table 7.14). Using the same method as for *Grapholita molesta*, the occurrence dynamics of *Spodoptera litura* were obtained, and the attraction dynamics of *Spodoptera litura* by two wavelength insecticidal lamps were analyzed. The peak of moth occurrence appeared in mid-August and mid-September, and the attraction dynamics of *Spodoptera litura* by two wavelength insecticidal lamps were similar to the trend of occurrence dynamics of *Spodoptera litura* (Fig. 7.23).

7.2.3.2 Effects of Strip Intercropping on the Temporal Dynamics of Coleoptera Pest Populations

Different wavelength LED insecticidal lamps mainly attracted *small yellow scaly beetles* and *dark scaly beetles* in the banded trap field (Table 7.14). Among them, the single-wavelength insecticidal lamps of 403 and 397 nm showed the greatest killing effect on small yellow scaly beetles at 5.19 and 4.53 individuals/day, respectively. Similarly, the single-wavelength insecticidal lamp with a wavelength of 407 nm had the greatest killing effect on dark scaly beetles, followed by 395 nm, at 14.8 and 8.6 individuals/day, respectively. The daily number of kills of small yellow scaly beetles and dark scaly beetles was summarized according to each insecticidal lamp, and the dynamic attraction of small yellow scaly beetles and dark scaly beetles was obtained. The peak of small yellow scaly beetle emergence under intercropping was in mid-to-

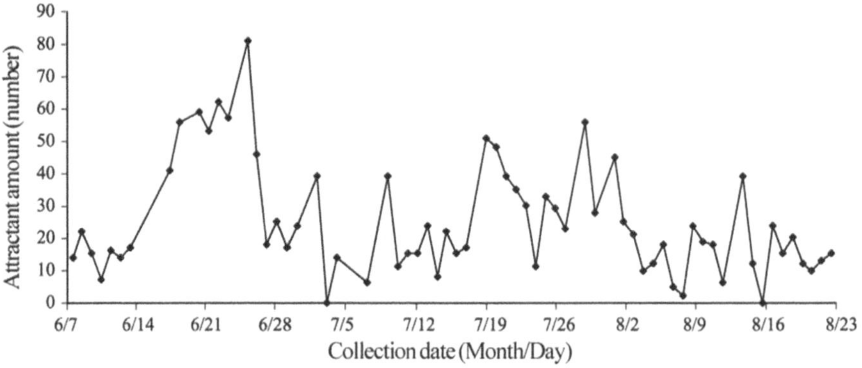

Fig. 7.24 Temporal dynamics of the yellow pond-turtle population

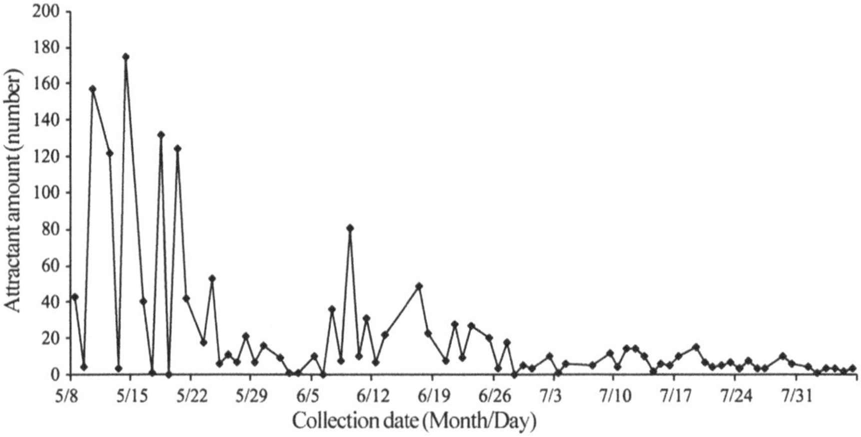

Fig. 7.25 Temporal dynamics of the darkling beetle population

late June, and there were two small peaks in late July and early August (Fig. 7.24). The peak of dark scaly beetle emergence was in mid-May, followed by a continuous decrease in the number of adults attracted (Fig. 7.25).

7.2.3.3 Effects of Strip Intercropping on the Temporal Dynamics of the Pentatomidae Insect-Pest Population

This experiment showed that different wavelength LED insecticidal lamps mainly attracted *hemipteran insect pests* in the banded trap field. The single-wavelength insecticidal lamps of 405 and 403 nm, as well as composite-wavelength insecticidal lamps, showed the best killing effect on hemipteran pests, with 4.18, 3.62, and 3.66 individuals/day, respectively (Table 7.14). The number of hemipteran specimens

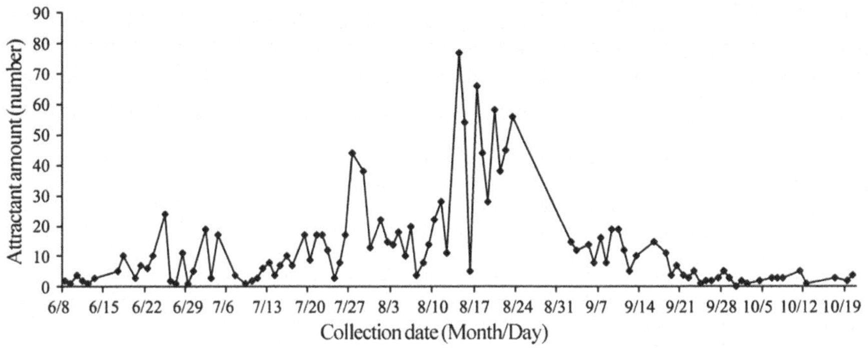

Fig. 7.26 Temporal dynamics of the Pentatomidae insect pest population

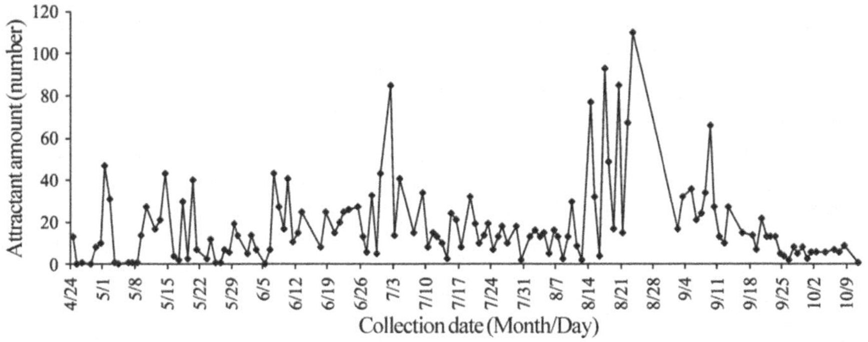

Fig. 7.27 Temporal dynamics of the *Carabidae* population

killed daily was summarized according to each insecticidal lamp, and the dynamic attraction of hemipteran pests throughout the whole growth period of banded intercropping was obtained to reflect the dynamics of hemipteran pest incidence. The activity time span of hemipteran pests under intercropping was wide, with late July to late August as the peak period of harm (Fig. 7.26).

7.2.3.4 Temporal Dynamics of Predator and Neutral Insect Populations in Strip Intercropping

The main natural enemies of the observed banded intercropping field-insect pests were ***Carabidae*** and ***Coccinellidae***. The dynamic analysis of the attraction of *Carabidae* throughout the whole growth period is shown in Fig. 7.27. Single-wavelength insecticidal lamps effectively attracted *Carabidae* from late April to October, and the largest number of individuals were attracted in mid-to-late August, indicating that this period was the peak period of *Carabidae* activity (Table 7.14). In addition, we attracted neutral insects that are unrelated to maize or soybeans, including mainly ***Thysanoptera*** and seven other insect categories. The effect of composite-wavelength insecticidal lamps on neutral insects was much greater than that of single-wavelength insecticidal lamps, with 30.44 individuals/day killed, and the single-wavelength insecticidal lamp with the largest number of neutral insects attracted (403 nm single-wavelength insecticidal lamp) caught one-third of what the composite-wavelength insecticidal lamp caught.

Lepidoptera insect pests usually cause damage only during the larval stage, although they have multiple generations in a year. Insecticidal lamps can attract adult Lepidoptera insect pests and reduce oviposition. By attracting adult Lepidoptera pests with insecticidal lamps, the field moth dynamics can be obtained, and then the peak and late period of moth emergence can be calculated. Combined with the preoviposition and oviposition periods, the hatching peak period (peak and late periods) of the next generation of larvae can be predicted to guide pest control. Thus, from the field moth dynamics of the peach fruit borer, it can be seen that there is a phenomenon of overlapping generations in the peach fruit borer in the banded intercropping field. Only one peak of moth emergence appeared in mid-August, when maize had already entered the maturity stage, and when the next generation of larvae hatched, maize had already been harvested. Therefore, insecticidal lamps cannot be used to predict the time when peach fruit borers will cause damage in banded intercropping fields in this area. Field studies have revealed that ***S. litura*** can cause severe damage over multiple generations in banded intercropping soybean fields, greatly affecting soybean yields. An analysis of the dynamics of *S. litura* moth emergence showed that *S. litura* had two concentrated peaks of moth emergence and was deemed suitable for short-term prediction. Insect pests within *Gelechiidae* usually occur once a year and are not suitable for short-term prediction. Therefore, only a summary of the attraction dynamics is made to reflect their approximate field dynamics. In turn, the peak period of emergence for ***Ostrinia furnacalis*** occurred in mid-May, followed by a gradual decrease in attraction. Meanwhile, ***Ostrinia scapulalis*** showed a more concentrated peak of emergence in late June but still showed multiple small peaks thereafter. The similarity between the dynamics of the two *Ostrinia* species and previous research is related to the waiting period after emergence for suitable soil conditions. It is possible that ***O. scapulalis*** has a smaller body size and is more affected by rainfall during its waiting period than ***O. furnacalis,*** which has a larger body size and is less affected by rainfall, resulting in a more concentrated attraction peak for ***O. furnacalis,*** whereas ***O. scapulalis*** showed multiple attraction peaks. Thus, our study of the attraction dynamics of Pentatomidae pests found that the peak period of Pentatomidae pest damage can cover both soybean flowering and pod-setting periods, thereby causing soybean flower and pod drop, consistently with field results.

7.2.4 Effects of Strip Intercropping on Co-occurrence and Facultative Feeding of Insect Pests During Maize-Soybean Symbiosis

The number of insect pest species with both symbiotic and facultative feeding periods increased, but their transmission and spread were inhibited. The main pest species during the symbiotic period were Lepidoptera grass moths (mainly peach fruit borers), Coleoptera scarab beetles, Hemiptera cicadas, and bugs. For example, green bugs, edge bugs, *S. litura* larvae, ***Oxya chinensis*** adults, ***Melanotus cribricollis***, locusts, and mollusks (e.g., nail snails). Among these pests, those showing a facultative feeding capability included maize snails, peach fruit borers, and aphids, among others. Maize and soybean aphids showed strong host selectivity and occasionally cross-feeding. Further, Maize showed a strong barrier effect on soybean aphids, which greatly hindered the transmission and spread of soybean aphids carrying viruses, with an inhibition rate of 59.3% (Fig. 7.28).

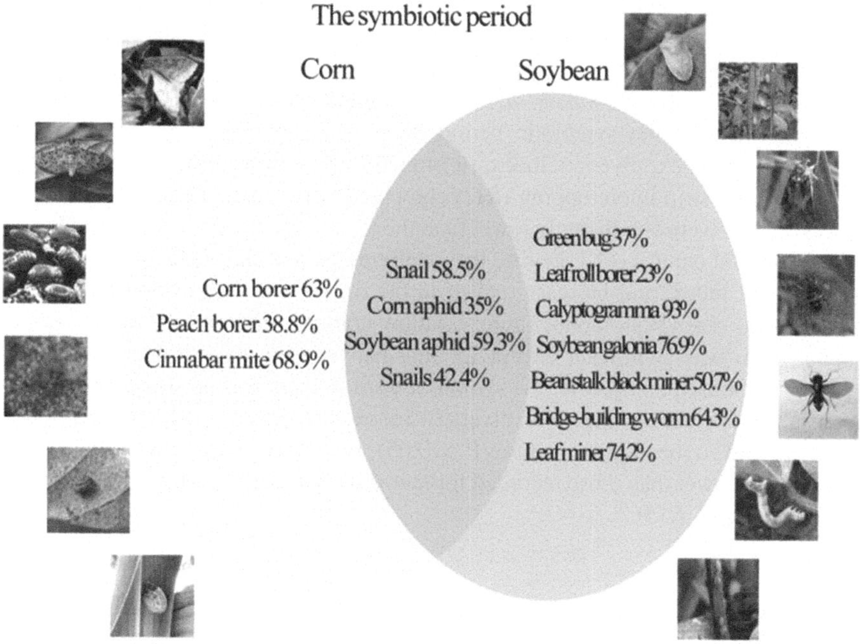

Fig. 7.28 Suppression rate of major insect pest species during the symbiotic period in maize-soybean strip intercropping systems compared with that in monoculture

7.3 Patterns of Incidence of Major Weeds in Maize-Soybean Intercropping Systems

It is well known that weeds can severely restrict the development of successful soybean-maize strip intercropping. Hence, clarifying the patterns of weed occurrence in maize-soybean strip intercropping systems and conducting research on weed control technology for such systems is of paramount importance. By mastering the occurrence patterns of major weeds and screening stem-leaf herbicides that can be used for soybean-maize strip intercropping, it is possible to achieve weed control in soybean-maize strip intercropping systems and eliminate weed-induced damage to crops and crop yields. The issue has great practical significance for the sustainable development of soybean-maize strip intercropping.

7.3.1 Effects of Strip Intercropping on Weed Abundance

The most common types of weeds observed in the maize-soybean strip intercropping system were three grass weeds, namely, crabgrass, dogtail grass, and goosegrass, and 11 broadleaf weeds including *Amaranthus spinosus*, bittercress, *Eclipta prostrata*, *Eupatorium adenophorum*, *E. odoratum*, *Polygonum barbatum*, *Solanum nigrum*, *S. rostratum*, *Sonchus arvensis*, *Xanthium strumarium*, and *Xanthium sibiricum*). Before the symbiotic period, there was no significant difference in weed richness index, diversity index, or biomass between strip intercropping (RI1) and traditional strip intercropping (RI2) compared with maize monoculture (MM); however, they were significantly lower than those in soybean monoculture (SM) and the fallow land control (FL). During the symbiotic period, the weed richness index, the diversity index, and biomass ranked in the order: strip intercropping < maize monoculture < soybean monoculture < fallow control; furthermore, the difference among them was significant except for that between the two strip intercropping treatments. Subsequently, after the symbiotic period, there was no significant difference in weed richness index or diversity index between the strip intercropping treatment and soybean monoculture ($P > 0.05$), but the corresponding values were significantly lower than those recorded for maize monoculture and the fallow control treatment (Table 7.15).

7.3.2 Effects of Strip Intercropping on Weed Biomass during the Maize-Soybean Symbiotic Period in Intercropping Systems

From the reproductive period, weed biomass in soybean monoculture and fallow land was significantly higher than that of other planting systems throughout the whole growth period, followed by maize monoculture. During the symbiotic and soybean pod period, weed biomass under each planting system increased significantly (Fig. 7.29).

Table 7.15 Weed abundance and diversity indices in farmland under different planting systems

Index	Treatment	Maize growth period		Maize and soybean symbiotic period	Soybean growth period	
		Small bell mouth period	Silking period		Pod-filling stage	Maturity stage
Abundance index	RI1	10.00 ± 1.000 a	8.67 ± 0.577 b	8.00 ± 1.000 d	8.67 ± 0.577 b	7.67 ± 0.577 b
	RI2	9.67 ± 0.577 a	8.67 ± 0.577 b	7.67 ± 0.577 d	8.33 ± 0.577 b	8.00 ± 0.000 b
	MM	9.67 ± 0.577 a	8.33 ± 0.577 b	10.00 ± 0.000 c	12.67 ± 0.577a	10.00 ± 1.000 a
	SM	11.00 ± 1.000 a	12.33 ± 0.577 a	12.33 ± 0.577 b	9.00 ± 2.646 b	6.33 ± 0.577 c
	FL	11.00 ± 0.000 a	13.00 ± 1.000 a	15.00 ± 1.000 a	13.00 ± 1.000 a	10.33 ± 0.577 a
Diversity index	RI1	1.99 ± 0.050 a	1.82 ± 0.057 b	1.67 ± 0.114 d	1.81 ± 0.050 b	1.71 ± 0.090 b
	RI2	1.96 ± 0.043 a	1.76 ± 0.200 b	1.65 ± 0.043 d	1.81 ± 0.024 b	1.89 ± 0.061 b
	MM	2.03 ± 0.096 a	1.83 ± 0.182 b	1.92 ± 0.033 c	2.20 ± 0.077 a	1.95 ± 0.011 a
	SM	2.01 ± 0.045 a	2.06 ± 0.011 a	2.15 ± 0.088 b	1.81 ± 0.138 b	1.65 ± 0.115 b
	FL	2.05 ± 0.031 a	2.18 ± 0.048 a	2.30 ± 0.032 a	2.23 ± 0.038 a	2.06 ± 0.074 a

RI1 strip intercropping, *RI2* traditional strip intercropping, *MM* monoculture maize, *SM* monoculture soybean, *FL* fallow land

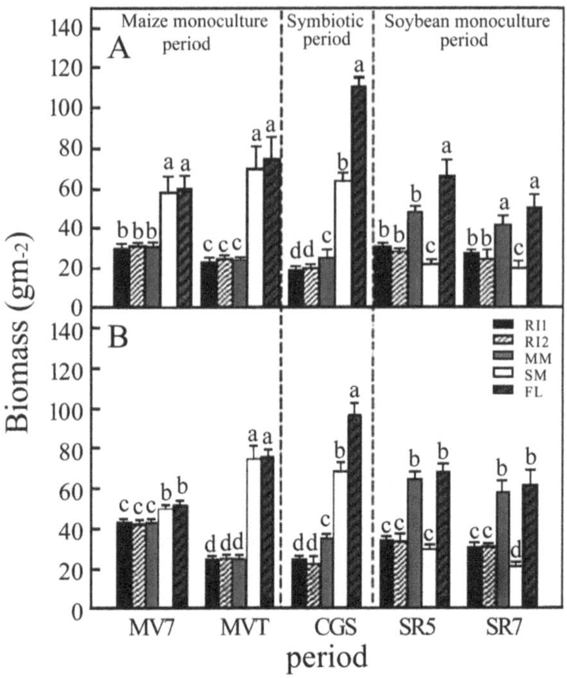

Fig. 7.29 Weed biomass at different growth stages and under different planting systems. *RI1* strip intercropping, *RI2* traditional intercropping, *MM* monoculture maize; SM: monoculture soybean, *FL* fallow land. *MV7* maize trumpet stage, *MVT* maize silking stage, *CGS* symbiotic period, *SR5* soybean pod-setting stage, *SR7* soybean maturity stage. (**a**: Data from 2012; **b**: Data from 2013)

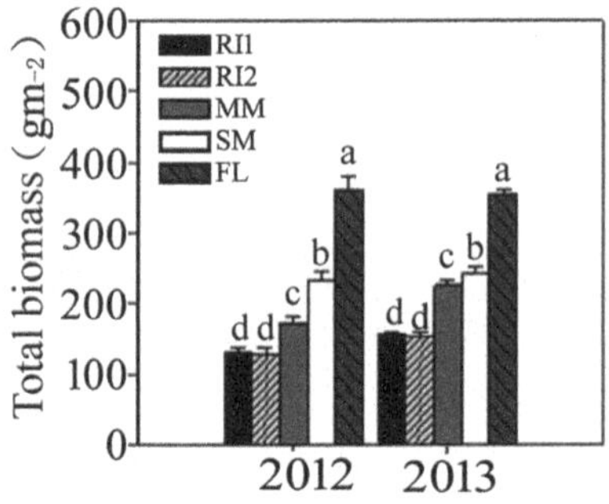

Fig. 7.30 Total weed biomass under different planting patterns. RI1: Strip intercropping; RI2: Traditional intercropping; MM: Monoculture maize; SM: Monoculture soybean; FL: Fallow land. MV7: Maize trumpet stage; MVT: Maize silking stage; CGS: Symbiotic period; SR5: Soybean pod-setting stage; SR7: Soybean maturity stage. (**a**: Data from 2012; **b**: Data from 2013)

Overall, compared with monoculture, weed population diversity and biomass in the strip intercropping systems tested (2:2 MS and 1:1 MS) were lower than those of maize or soybean monoculture during the symbiotic period. Total weed biomass during the whole growth period was reduced by 29%, 41%, and 61% compared with maize monoculture, soybean monoculture, and fallow land, respectively. The change was consistent between 2012 and 2013 (Fig. 7.30).

Core Technologies to Fulfill the Sustainable Productive Potential of Maize–Soybean Strip Intercropping Systems

8

Xiaochun Wang, Xuyang Zhao, Fu Jin, Yanwei Ma, Junlin Zou, Weizhao Bu, Liang Cui, Jinxi Zeng, Yongfu Ren, Li Li, Fengzhi Lu, Ting Xu, Yuankai Chen, Dunping Liao, Qun Zhang, Feng Yang, Ying Lou, Yun Hu, and Zhengyi Zhang

Abstract

Establishing appropriate field layouts is crucial for boosting production and efficiency in intercropping systems, as it can help address the various contradictions that may arise within and between crops. Field configuration encompasses the combination, spatial distribution, and interdependence of co-existing crop populations in the agroecosystem, including vertical and horizontal structures, such as various combinations of plant population density, row ratio (number of rows), row spacing, width, interval, belt width, and orientation. In traditional intercropping, numerous challenges exist, such as unsuitable crop combinations, improper variety selection, unsuitable spacing, strip width, or row ratio configuration. These issues lead to considerable differences in high-density intercropping and monoculture, resulting in low resource utilization, low population yield, poor mechanization, low production efficiency, and the inability to rotate crops. Modern intercropping techniques inherit and innovate upon traditional methods to achieve "high yield, mechanization, and sustainability." Based on the "high position priority, high-low synergy" theory for efficient use of solar energy, strip intercropping offers a viable solution. Innovations in technologies of maize–soybean strip intercropping, such as "selection, expansion, and contraction," have been developed to select and match varieties, expand inter-row light interception, and contract plant spacing to maintain privacy. The application of core technology has enabled the successful maintenance of maize yields and the addition of one-season soybean crop, while ensuring adequate mechanization and the integration of intercropping into the rotation.

X. Wang (✉) · X. Zhao · F. Jin · Y. Ma · J. Zou · W. Bu · L. Cui · J. Zeng · Y. Ren · L. Li · F. Lu · T. Xu · Y. Chen · D. Liao · Q. Zhang · F. Yang · Y. Lou · Y. Hu · Z. Zhang
College of Agronomy, Sichuan Agricultural University, Chengdu, China
e-mail: xchwang@sicau.edu.cn; zhao_xuyang@stu.sicau.edu.cn; Zengjinxi@stu.sicau.edu.cn; Liaodunping@stu.sicau.edu.cn; f.yang@sicau.edu.cn; Zhangzhengyi@stu.sicau.edu.cn

W. Yang et al. (eds.), *Theories and Application in Maize-Soybean Strip Intercropping System*, https://doi.org/10.1007/978-981-97-8723-4_8

Keywords

Relay-intercropping · Field configuration · Varieties selection · Density

8.1 Selection of Varieties

To establish an optimal scientifically based spatial layout and a rational population structure within a cropping system, a primary concern is the selection of appropriate shade-tolerant soybean varieties suitable for strip intercropping planting and selectively planting compact or semi-compact maize varieties that can reduce shading levels on soybean plants, as the degree of shading directly depends on the maize plant type. The end objective is to facilitate efficient airflow within the system to maximize soybean yield and economic benefits. However, given the large differences in growth characteristics among soybean varieties, their response to maize shading is unpredictable and inconsistent. Therefore, selecting appropriate shade-tolerant soybean varieties specifically suited for strip intercropping with maize is crucial to prevent plant lodging and the sprawling of soybean stems and vines. Indeed, the selection of appropriate shade-tolerant soybean varieties provides a strong foundation for obtaining high yields from maize–soybean intercropping systems.

8.1.1 Appropriate System for Evaluating Maize and Soybean Varieties Suitable for Strip Intercropping

In a strip intercropping system, maize plants are positioned in a wide–narrow pattern to maximize the edge-row advantage. In contrast, soybean crops are low-positioned and subject to shading stress caused by the taller maize plants. This can lead to excessive stem elongation and lodging, significantly impacting yield, mechanized harvest, and overall quality. In regions with lower light intensity conditions, compact maize varieties may need more illumination, while flat maize varieties can shade soybean plants excessively and reduce their yield. Therefore, selecting suitable maize and soybean varieties is essential to achieve the benefits of this planting system.

8.1.1.1 Indicators for Screening Suitable Maize Varieties for Strip Intercropping

Based on dry matter yield, at a level of 7500 kg/ha as a threshold, the tested maize varieties were divided into three yield categories, namely those that were high-yielding both in strip intercropping and monocropping, those that were low-yielding in strip intercropping but high-yielding in monocropping, and those that were low-yielding both in strip intercropping and monocropping. After 2 years of testing, more than 20 varieties were included in the first category. An analysis of the morphological and physiological indicators of a series of varieties established an effective morphological and physiological evaluation system for high-efficiency strip intercropping of maize varieties.

Leaf angle of the second leaf above the ear displayed the greatest coefficient of variation among the morphological parameters assessed, followed by leaf area and ear height, all exceeding a variability of 10% (Table 8.1). These findings suggest that significant variation exists for leaf angle of the second leaf above the ear, for leaf area, and ear height of maize varieties under strip intercropping. Furthermore, high-yielding maize varieties under strip intercropping demonstrated a significant quadratic relationship between the leaf angle of the second leaf above the ear ($R^2 = 0.5071$*), plant height ($R^2 = 0.7604$*), ear height ($R^2 = 0.6509$*), and yield (Table 8.2). Additionally, multivariate quadratic regression analysis ($Y = 3021.240–4.319X1^2 + 0.4601X1X2 + 0.148X2X3$, $R^2 = 0.9948$**) revealed that when the yield of high-yielding maize varieties in strip intercropping reached its peak, the corresponding values for leaf angle of the second leaf above the ear, plant height, and ear height were 19.26°, 270.58 cm, and 114.99 cm, respectively (Ma et al. 2019a). Hence, plant height, ear height, and leaf angle of the second leaf above the ear are key morphological indicators for selecting high-yielding monocropping maize varieties under strip intercropping. In contrast, leaf area may be a crucial limiting factor for improving yield in low-yielding maize varieties under strip intercropping.

Table 8.3 lists the key physiological parameters analyzed in this study. At the tasseling stage, the coefficient of variation of chlorophyll a, b, carotenoids, and total chlorophyll content exceeded 10%, indicating significant differences among maize varieties under strip intercropping. Similarly, leaf upper and lower epidermal thickness and leaf thickness showed variations above 10%.

Further, we found that chlorophyll a, total chlorophyll content, and leaf thickness at the grain-filling stage displayed a significant ($R^2 > 0.7$) quadratic relationship with yield under strip intercropping (Table 8.4). Specifically, the quadratic regression model revealed that high-yielding maize varieties had higher chlorophyll-a content, total chlorophyll content, and greater leaf thickness than low-yielding ones. Among the different physiological indicators studied, chlorophyll-a, total chlorophyll contents, and leaf thickness appeared to be crucial factors limiting maize yield under strip intercropping (Ma et al. 2019a, b). Consequently, they could serve as primary indicators for screening high-yielding maize varieties under this cropping system.

8.1.1.2 Evaluation Criteria for Soybean Varieties Suitable for Strip Intercropping

8.1.1.2.1 Evaluation Criteria for Shade Tolerance of Relay-Intercropping Soybean During Seedling Stage

In this study, we compared the growth and yield characteristics of 82 summer soybean varieties from Sichuan and Chongqing and one variety (Guixia No. 3)

Table 8.1 Basic parameter estimates of morphological characteristics in maize varieties categorized based on yield performance under strip intercropping and monoculture

Indictors	Leaf area (cm^2)	Leaf angle (°)		Stem diameter (mm)			Internode length (cm)				Plant height (cm)	Ear height (cm)
		Ear leaf first above	Ear leaf second above	3rd	4th	5th	3~5	3 leaves of ear	Above leaves of ear	Node of above ear		
MAX	7712	35	36	23.5	23.9	22.9	54	46	112	6	311	124.6
MIN	4423	17	14	17.5	17.1	16.2	28	33	60	3	225	63.0
CV (%)	14.41	19.49	21.54	7.22	8.97	8.98	11.28	8.48	11.20	10.97	7.80	13.68
Mean	6169	27	24	20.4	19.8	18.8	42	38	83	5	266	104.7

Note: *MAX* maximum value, *MIN* minimum value, *SD* standard deviation, *CV* coefficient of variation, *Mean* average value

Table 8.2 Correlation between morphological characteristics and yield of high-yielding corn varieties under strip intercropping

Production level	Indicators		Regression equation	R^2	*P*
High-yielding strip intercropping with monocropping of high yields	Leaf area		$y = 0.0005x^2–6.4056x + 28{,}740$	0.1256	0.4736
	Leaf angle	Ear leaf first above	$y = 12.163x^2–668.34x + 16{,}944$	0.5924	0.1298
		Ear leaf second above	$y = 3.5058x^2–238.97x + 11{,}631$	0.5071	0.0249
	Stem diameter	3rd	$y = -154.93x^2 + 6433.2x—58{,}559$	0.432	0.3527
		4th	$y = -53.394x^2 + 1952.2x—9583.5$	0.343	0.1049
		5th	$y = -167.27x^2 + 6149.9x—48{,}349$	0.332	0.167
	Internode length	3–5	$y = 6.6233x^2–584.67x + 20{,}677$	0.3256	0.2755
		3 leaves of ear	$y = 73.977x^2–5473.6x + 109{,}139$	0.1343	0.4598
		Above leaves of ear	$y = 0.5198x^2–95.184x + 12{,}322$	0.0783	0.4818
		Node of above ear	$y = 603.21x^2–5517.5x + 20{,}559$	0.0909	0.4718
	Plant height		$y = -1.8935x^2 + 1018x—12{,}466$	0.7604	0.0321
	Ear height		$y = -2.0124x^2 + 458.56x—17{,}663$	0.6509	0.0298

from Guangxi under monocropping and relay-intercropping agroenvironments during the seedling and maturity stages. Additionally, we analyzed the correlation between the two data sets to identify key traits influencing soybean yield in relay-intercropping scenarios. Using this information, we calculated the sensitivity index for each quality trait in response to shade during the soybean seedling stage, and we developed a method for evaluating soybean shade tolerance in relay-intercropping systems.

The data summarized in Table 8.5 shows that relay intercropping led to a significant increase in the main stem height of soybean plants during the initial flowering period, as well as the average internode length and length of the lower axis by 125.5%, 164.7%, and 62.9%, respectively. However, internode and stem diameters were reduced by 15% and 43.9%, respectively. Additionally, stem bending resistance and dry weight were 56% and 38.2% of those in monoculture. Among the seven traits measured, the internode length response index (IR) was the highest during the soybean seedling period, followed by main stem length, indicating that these two traits were the most sensitive to shading caused by relay intercropping. On

Table 8.3 Estimates of the basic parameters of physiological characteristics of different maize varieties under strip intercropping

Indicators	Total carbon (%)	Total nitrogen (%)	C/N	Chl-a (mg/m^2)	Chl-b (mg/m^2)	Carotenoids (mg/m^2)	a/b	Total Chl (mg/m^2)	Upper epidermis thickness (μm)	Lower epidermis thickness (μm)
MAX	46.95	4.26	15.41	503.64	130.71	93.39	4.71	727.73	27.90	22.96
MIN	42.36	2.76	10.49	287.51	66.31	51.37	3.60	405.18	16.80	14.71
CV (%)	2.60	11.30	9.85	13.31	16.93	14.54	6.26	13.71	11.78	10.94
Mean	44.05	3.50	12.74	384.45	92.56	70.75	4.17	547.75	23.43	18.82

Note: *MAX* maximum value, *MIN* minimum value, *SD* standard deviation, *cv* coefficient of variation, *Mean* average value

Table 8.4 Relationship between leaf physiological characteristics in different maize varieties varying in yield level under strip intercropping

Production level	Indicators	Regression equation	R^2	P
High-yielding in strip intercropping and monocropping	C/N ratio	$y = -170.36x^2 + 4365.8x-19{,}848$	0.1854	0.5168
	Chl-a	$y = -0.046x^2 + 42.254x-1239.7$	0.8766	0.0045
	a/b	$y = -288.2x^2 + 2103.6x + 4254.5$	0.0185	0.7226
	Total Chl	$y = -0.0231x^2 + 30.282x-1351.1$	0.8978	0.0139
	Leaf thickness	$y = 0.4395x^2-103.37x + 13{,}408$	0.784	0.0007

the contrary, the number of internodes was the least sensitive and exhibited the smallest response index (IR).

In monocropping and relay intercropping, the coefficient of variation of seven traits exceeded 13%. Stem weight displayed the most significant variability, reaching 52.67% and 53.97% under monocropping and relay-intercropping agroenvironments, respectively. Following closely was stem bending resistance, which varied by 36.79% and 45.88% under the same conditions, respectively. Conversely, the variation in juvenile internode length and number was comparatively minor. Further, two-way ANOVA revealed that the differences observed in all seven traits between varieties and planting systems were highly significant, with a significant or highly significant effect from the interaction between the two factors.

Based on the findings summarized in Table 8.6, soybean plants grown under relay intercropping showed significant increases in plant height, branch grain weight, single-plant grain weight, and branch grain weight at maturity, compared to monoculture. Notably, the number of branches, average branch length, and main stem grain weight significantly decreased. These results suggest that strip relay intercropping initially affected branch formation. However, the edge advantages associated with relay intercropping are fully utilized to stimulate branch yield, which is the foundation for soybean yield in this type of intercropping system.

Regarding the variation among materials in monocropping, branch length showed the most significant coefficient of variation, followed by 100-grain weight and plant height. In contrast, the smallest coefficient of variation was observed for single-plant grain weight and branch grain weight ratio. On the contrary, under relay-intercropping conditions, the highest coefficient of variation was observed for main stem grain weight, followed by branch grain weight and plant height, whereas the smallest coefficient of variation was recorded for the number of branches and branch grain weight ratio. In terms of yield-related traits, the value for the IR of branching grain weight was the highest, followed by those of plant height and individual grain weight. The impact of relay intercropping on 100-grain weight, however, was minimal. The variation coefficient for IR among varieties was above 50%, with plant height having the highest coefficient, which reached 83.46% (Zou et al. 2015). This indicates that, during soybean growth, the differences among varieties were progressively widened when experiencing pre-shading stress and later growth under two different light environments.

Table 8.5 Different soybean varieties' stem morphology and biomass in the seedling stage under monocropping and relay-intercropping conditions

Treatments	Parameters	Main stem length (cm)	Internode number	Hypocotyl length (cm)	Internode length (cm)	Stem diameter (cm)	Bending strength (N)	Stem weight (g/plant)
Monoculture	Mean	56.77	15.79	3.34	3.57	0.66	310.86	8.42
	SD	18.73	2.46	0.61	1.00	0.10	114.35	4.44
	CV (%)	32.99	15.57	18.17	28.07	15.3	36.79	52.67
Relay intercropping	Mean	128.57	13.44	5.44	9.45	0.37	136.63	5.21
	SD	36.43	1.79	0.78	1.88	0.06	62.69	2.81
	CV (%)	28.33	13.34	14.39	19.89	16.04	45.88	53.97
IR	Mean	1.36	0.14	0.67	1.74	0.43	0.55	0.37
	SD	0.6	0.08	0.29	0.6	0.08	0.14	0.17
	CV (%)	44.46	56.82	43.51	34.69	19.06	25.62	45.01

Note: IR represents the response index of various soybean traits to shading caused by relay intercropping, calculated as $IR = |1 - Xr/Xs|$, where Xr and Xs are the measured values of each trait under strip relay intercropping and monoculture conditions, respectively

Table 8.6 Yield characteristics of different soybean varieties under monocropping and relay-intercropping conditions

Treatments	Parameters	Plant Height (cm)	Branch number	100-grain Weight (g)	Branch grain weight (g/plant)	Main stem grain weight (g/plant)	Single-plant grain weight (g/plant)	Grain weight proportion (%)
Monocropping	Mean	77.14	5.76	15.07	19.06	8.94	28.01	67.57
	SD	23.97	1.06	4.96	4.51	2.06	4.91	7.45
	CV (%)	31.08	18.42	32.93	23.64	23.02	17.52	11.02
Relay intercropping	Mean	121.80	5.14	15.26	33.65	7.84	41.49	80.38
	SD	40.37	0.96	4.69	11.25	4.02	11.15	9.85
	CV (%)	33.15	18.64	30.75	33.44	51.31	26.88	12.26
IR	Mean	0.65	0.18	0.07	0.81	0.39	0.51	0.22
	SD	0.54	0.11	0.05	0.58	0.29	0.36	0.18
	CV (%)	83.46	59.60	78.95	71.90	74.95	70.21	78.77

Note: Grain weight ratio refers to the percentage of branch grain weight relative to single-plant grain weight

Table 8.7 Correlation coefficient between soybean stem response index during the seedling stage to relay intercropping and yield response index

Plant height	Internode number	Hypocotyl Length	Internode length	Stem diameter	Bending strength	Stem weight
0.53**	0.54**	−0.13	0.52**	0.31*	0.55**	0.28

Note: "*" and "**" were significantly correlated at $p = 0.05$ and $p = 0.01$, respectively

We conducted a partial correlation analysis to examine the relationship between stem characterization indices of soybean seedlings under strip relay-intercropping shading stress and individual plant yield (Table 8.7). We found that plant height, internode number, length, stem width, and lodging resistance all showed a significant and highly positive correlation with yield. In contrast, no correlation was observed between hypocotyl length and stem weight. These findings suggest that these five seedling-stage indicators accurately reflect soybeans' response level to relay-intercropping shading (Zou et al. 2015).

8.1.1.2.2 Evaluation Criteria for Shade Tolerance of Intercropping Soybean

Ninety-three soybean varieties from different regions of the country were selected for evaluation in a two-year field strip-intercropping experiment. This experiment aimed to measure the phenotypic and yield characteristics of soybean genotypes under different intercropping conditions, identify significant trait correlations that could affect soybean yield, and perform a correlation analysis. The results in Table 8.8 show that the variation coefficients of nine traits were all greater than 10% in 2013, and all but those of stem diameter and reproductive growth period were greater than 10% in 2014. Plant height and grain number showed the largest variation, followed by pod number per plant, indicating that they were most affected by variety and intercropping environment. From the correlation analysis in Tables 8.9 and 8.10 over the 2 years, it is evident that plant height was not related to the reproductive growth period but was observed to be significantly or highly significantly correlated with all other traits. Yield traits, such as yield per plant and grain weight per plant, significantly correlated with plant height, stem diameter, branching number, and nutritional growth period. However, the reproductive growth period did not correlate with yield formation, indicating that the length of the vegetative growth period and morphological development during that period are crucial factors in yield formation (Bu 2015).

To examine the shade tolerance of soybean in intercropping, we conducted net and intercropping comparison experiments with a total of 124 soybean genotypes in 2021, sourced from the Huang-Huai-Hai and southwestern regions, and we measured relevant phenotypic and yield traits. Figure 8.1 shows the corresponding correlation-analysis results. X1 single-plant grain weight positively correlated with dry leaf weight, above-ground dry matter weight, and bending resistance. However, it was unrelated to plant height, lodging index, or field lodging condition. On the contrary, lodging index positively correlated with plant height, bending resistance,

Table 8.8 Analysis of phenotypic characteristics of intercropped soybean in 2013–2014

	Mean		Max		Min		SD		CV%		VR	
Indicators	2013	2014	2013	2014	2013	2014	2013	2014	2013	2014	2013	2014
Plant height (cm)	95.9	94.1	170.6	139.7	57.1	53.3	22.3	22.6	23.3	24.0	113.5	86.5
Stem diameter (mm)	5.7	6.2	7.8	7.5	3.9	4.9	0.7	0.5	13.0	7.4	3.9	2.6
Branch number	2.4	1.5	4.8	3.5	1.0	0.4	0.8	0.5	35.1	33.4	3.8	3.1
Pods per plant	37.2	27.4	80.4	54.8	19.2	14.6	10.0	7.0	26.8	25.4	61.2	40.1
Grains per plant (g)	62.7	57.1	126.1	103.6	21.8	36.7	20.3	12.2	32.4	21.4	104.3	66.9
Grain weight per plant (g)	11.5	11.0	24.5	16.8	3.6	6.1	4.2	2.1	36.5	18.7	20.9	10.7
100-grain weight (g)	18.9	19.1	32.4	30.3	10.6	13.2	3.4	2.8	18.0	14.7	21.8	17.1
Vegetative growth period (d)	34.0	31.0	53.3	44.0	23.0	22.0	7.1	6.4	20.9	20.6	30.3	22.0
Reproductive growth period (d)	66.8	71.7	88.0	85.0	53.3	59.0	7.9	5.4	11.9	7.5	34.7	26.0

Note: "*" and "**" were significantly correlated at $p = 0.05$ and $p = 0.01$, respectively

Table 8.9 Correlation analysis of the main phenotypic traits of intercropping soybean in 2013

Agronomic traits	Plant Height	Stem diameter	Branch number	Pods per plant	Grains per plant	Grain weight per plant	100-grain weight	Vegetative growth period	Reproductive growth period
Plant height	1.000								
Stem diameter	0.590**	1.000							
Branch number	0.265*	0.289*	1.000						
Pods per plant	0.409**	0.717**	0.288*	1.000					
Grains per plant	0.427**	0.766**	0.300**	0.835**	1.000				
Grain weight per plant	0.564**	0.859**	0.391**	0.765**	0.844**	1.000			
100-grain weight	0.338**	0.411**	0.132	0.070	−0.044	0.421**	1.000		
Vegetative growth period	0.581**	0.669**	0.618**	0.435**	0.511**	0.702**	0.507**	1.000	
Reproductive growth period	0.030	−0.083	−0.381**	0.063	−0.142	−0.153	0.055	−0.438**	1.000

Note: "*" and "**" were significantly correlated at $p = 0.05$ and $p = 0.01$, respectively

Table 8.10 Correlation analysis of the main phenotypic traits of intercropping soybean in 2014

Agronomic traits	Plant height	Stem diameter	Branch number	Pods per plant	Grains per plant	Grain weight per plant	100-grain weight	Vegetative growth period	Reproductive growth period
Plant height	1.000								
Stem diameter	0.353**	1.000							
Branch number	0.177	0.105	1.000						
Pods per plant	0.336**	0.301**	0.516*	1.000					
Grains per plant	0.305**	0.266**	0.356**	0.857**	1.000				
Grain weight per plant	0.257*	0.542**	0.326**	0.592**	0.714**	1.000			
100-grain weight	0.047	0.324**	0.043	−0.347**	−0.448**	0.163	1.000		
Vegetative growth period	0.568**	0.381**	0.283**	0.321**	0.180	0.299**	0.306**	1.000	
Reproductive growth period	0.175	0.110	−0.133	−0.022	0.122	0.093	−0.048	−0.249**	1.000

Note: "*" and "**" were significantly correlated at $p = 0.05$ and $p = 0.01$, respectively

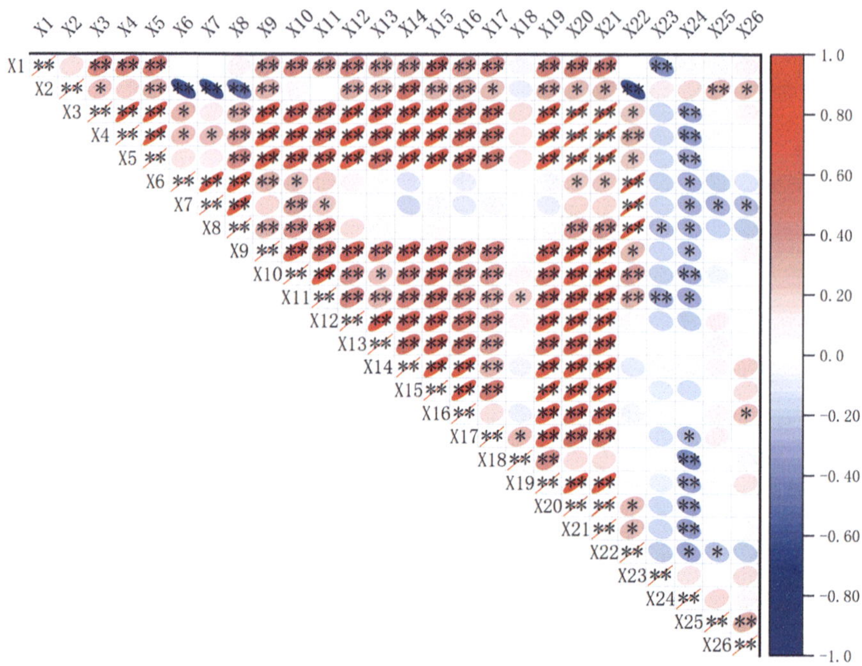

Fig. 8.1 Correlation analysis of the main phenotypic traits of different soybean varieties in intercropping systems. Note: X1 is single-plant grain weight, X2 is plant height, X3 is lodging resistance of the 1st internode, X4 is lodging resistance of the 3rd internode, X5 is lodging resistance of the 5th internode, X6 is lodging index 1, X7 is lodging index 2, X8 is lodging index 3, X9 is the stem diameter of the 1st internode, X10 is the stem diameter of the 3rd internode, X11 is the stem diameter of the 5th internode, X12 is the stem weight of the 0–1 internodes, X13 is the stem weight of the 2–4 internodes, X14 is the remaining stem weight, X15 is the leaf weight, X16 is the petiole weight, X17 is the branch weight, X18 is the pod weight, X19 is the aboveground dry weight, X20 is the lodging resistance sum of the 1st, 3rd, and 5th internodes, X21 is the square root of the lodging resistance sum of the 1st, 3rd, and 5th internodes, X22 is the lodging index sum of the 6th, 7th, and 8th internodes, X23 is the yield reduction per plant compared to net planting, X24 is the lodging resistance reduction compared to net planting, X25 is the lodging status at the grain-filling stage, and X26 is the lodging status at the harvest. "*" and "**" were significantly correlated at $p = 0.05$ and $p = 0.01$, respectively

and stem thickness, but correlated minimally with the dry weight of various organs, yield reduction, and field lodging condition. Further, bending resistance is significantly correlated with single-plant grain weight, plant height, lodging index, stem thickness, and dry weight of various organs. However, it was unrelated to pod dry weight, yield reduction, and field lodging condition. Furthermore, yield reduction was significantly and negatively correlated with single-plant grain weight, lodging index, and stem thickness but had no correlation with any other indicators. These findings contribute to a better understanding of the shade tolerance of soybeans in intercropping and provide valuable insights for future research in this area.

Based on correlation analysis results, we undertook further exploration into the key traits that determine lodging resistance and yield. In this regard, path analysis was performed where X1 was the single grain weight, X21 was the bending resistance of the first, third, and fifth internodes, X22 was the lodging resistance index of the sixth, seventh, and eighth internodes, and X23 was the decrease in yield per plant in comparison to the net yield (Table 8.11). The analysis showed that dry leaf weight had the most significant effect on the single-grain weight values. In the case of X21, aboveground dry weight, plant height, and lodging resistance index had the largest direct effect. On the contrary, resistance to lodging had the highest direct impact on X22, followed closely by X21, plant height, aboveground dry weight, and stem diameter of the first internode. In turn, single-grain weight had the most direct impact on X23. Therefore, when evaluating whether intercropped soybeans are shade-tolerant, careful assessment of plant height, stem diameter, aboveground dry weight, and bending resistance is important, in addition to a direct examination of yield.

8.1.2 Suitable Matching of Intercropping Maize and Soybean Varieties in Strip Intercropping

8.1.2.1 Choose Compact, Dense, and Lodging-Resistant Maize Varieties

8.1.2.1.1 Differences in Plant Shape Among Maize Varieties

Based on the findings of two consecutive years of field experiments, we studied the differences in yield and light energy utilization among various maize varieties (Fig. 8.2) in maize–soybean strip intercropping systems. Our results indicate that the compact maize variety, Denghai 605, performed exceptionally well on the soybean canopy during the coexistence period compared to the semi-compact (Chuandan 418) and loose (Yayu 13) maize varieties in strip intercropping. Specifically, Denghai 605 showed an average increase of 3.6% and 8% in light transmittance, leading to a 54.4% and 90.8% increase in active photosynthetic radiation in the soybean canopy, relative to Chuandan 418 and Yayu 13, respectively (Fig. 8.3, Table 8.12). Consequently, conducive light environments led to soybean yield increases of 631.2 and 901.75 kg/ha for Denghai 605, i.e., 56.7% and 107.0%, increased relative to Chuandan 418 and Yayu 13, respectively (Cui et al. 2014). Moreover, the compact maize variety had significantly higher land equivalent ratios and light energy utilization rates than the semi-compact and the loose ones (Table 8.13).

Table 8.11 Pathway analysis of the main shade tolerance- and lodging resistance-related traits of intercropped soybeans

Indicators	Regression equation	Path analysis
Single grain weight	X1 = 9.152 + 0.711X15–4.725X23	Px15y = 0.515, Px23y = −0.327
Bending resistance of the first, third, and fifth internodes	X21 = -0.607 + 0.242X19 + 0.073X2 + 7.624X22 + 0.905X12–0.436X10–0.188X26 + 0.142X14	Px19y = 0.594, Px2y = 0.512, Px22y = 0.669, Px12y = 0.100, Px10y = −0.098, Px26y = −0.060, Px14y = 0.079
Lodging resistance index of the sixth, seventh, and eighth internodes	X22 = -1.936–0.013X2 + 0.140X9–0.039X19 + 0.355X21–0.007X20	Px2y = −0.749, Px9y = 0.260, Px19y = −0.737, Px21y = 2.467, Px20y = −1.626
Decrease in yield per plant	X23 = 0.539–0.026X1	Px1y = −0.373

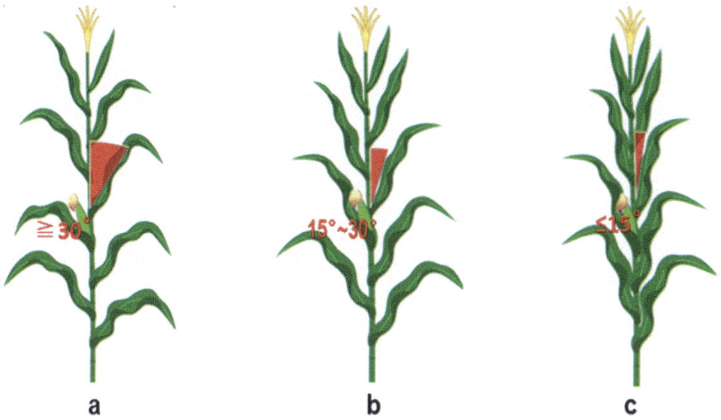

Fig. 8.2 Illustrations of different corn cultivars with various plant types. Note: a, b, and c represent loose, semi-compact, and compact maize cultivars, respectively

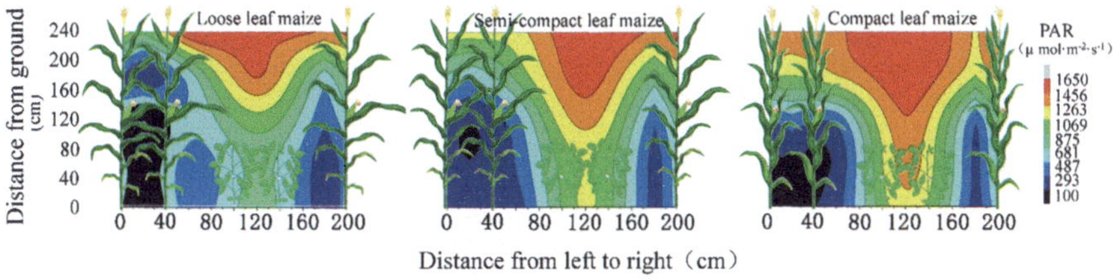

Fig. 8.3 Comparison of light interception in soybeans under strip intercropping with contrasting maize plant types

8.1.2.1.2 Selection of Maize Varieties with Resistance to High-Density Planting and Lodging by Using Strip Intercropping

Over the course of 2 years, 40 maize varieties widely cultivated in southwest China were collected and tested in field experiments conducted in monoculture and relay-intercropping modes. Based on a threshold yield of 7500 kg/ha, the tested maize varieties were categorized into three yield groups: high-yielding in monoculture only, high-yielding in both monoculture and relay intercropping, or low-yielding in both monoculture and relay intercropping. The results showed that the maize varieties that showed high yields in both monoculture and relay intercropping were Rongyu 1210, Chuandan 189, Miandan 581, Chuandan 418, and Chengdan 30 (Ma et al. 2019a, b). On the contrary, the varieties with high yields in monoculture but low yields in relay intercropping included Zhongdan 901, Chuandan 428, Zhongwangyu 18, Miandan 118, and Denghai 605 (Fig. 8.4).

Table 8.12 Population distribution characteristics of three maize varieties with contrasting plant type

		Tasseling		Grain-filling		Maturity	
Year	Maize varieties	PAR ($mol \cdot m^{-2} \cdot s^{-1}$)	Transmittance (%)	PAR ($mol \cdot m^{-2} \cdot s^{-1}$)	Transmittance (%)	PAR ($mol \cdot m^{-2} \cdot s^{-1}$)	Transmittance (%)
2011	Compact	764.19b	60b	534.59b	53b	435.77b	49b
	Semi-compact	552.32c	55c	354.25c	48c	222.52c	47c
	Loose	442.13d	51d	297.83d	44d	172.17d	43d
2012	Compact	741.71b	58b	518.93b	51b	423.2b	48b
	Semi-compact	536.05c	54c	332.78c	47c	216.02c	46c
	Loose	429.17d	49d	283.27d	43d	167.09d	41d

Note:PAR: photosynthetically active radiation. Lowercase letters indicate a significant difference at the $p < 0.05$ under all treatments

Table 8.13 Differences in crop yield, land equivalent ratio, and light energy utilization among maize varieties with contrasting plant types

Year	Maize varieties	Yield			LER	LUE%	
		Maize (kg/ha)	Soybean (kg/ha)	Total (kg/ha)		Maize	Soybean
2011	Compact	7586.4c	1812.3a	9398.7a	1.94a	3.09c	1.67a
	Semi-compact	7718.9b	1104.7b	8823.6b	1.57b	3.15b	1.54b
	Loose	7914.5a	831.3c	8745.8c	1.43c	3.20a	1.49c
2012	Compact	7658.3c	1676.2a	9334.5a	1.88a	3.13c	1.69a
	Semi-compact	7755.3b	1121.4b	8876.7b	1.59b	3.17b	1.56b
	Loose	7929.7a	853.7c	8783.4c	1.45c	3.23a	1.41c

Note: *LER* land equivalent ratio, *LUE* light use efficiency. Lowercase letters indicate a significant difference at the $p < 0.05$ under all treatments

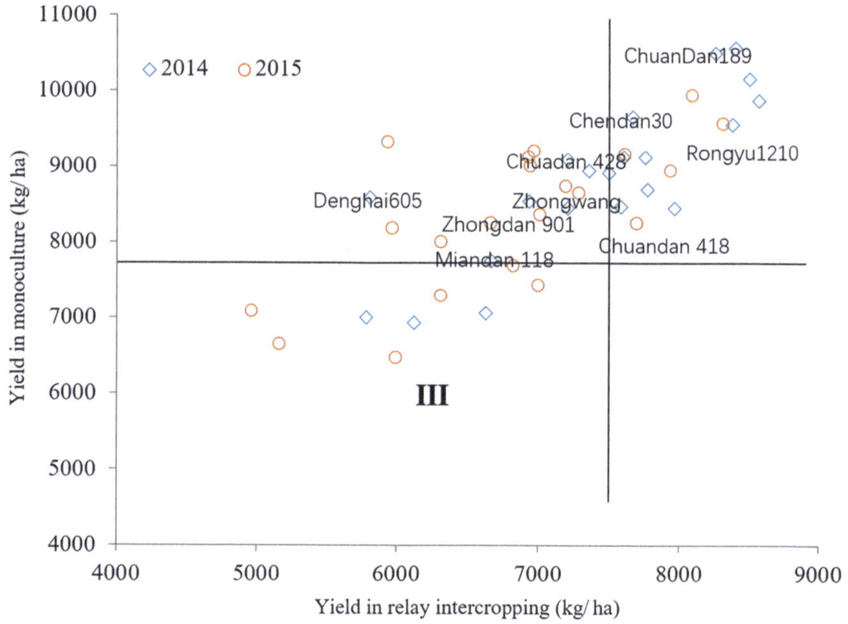

Fig. 8.4 Yield classification of different maize varieties under monoculture and relay intercropping. Note: I, II, III, and IV indicate, high-yielding in monoculture and relay intercropping, high-yielding in monoculture but low-yielding in relay intercropping, low-yielding in monoculture and relay intercropping, and low-yielding in monoculture and high-yielding in relay-intercropping, respectively

8.1.2.1.3 Suitable Strip Intercropping of Maize Varieties with Characteristics of Tolerance to High Plant Population Density and Lodging Resistance

Difference in Yield

In this study, we used two distinct types of maize varieties, namely Rongyu 1210, a high-yielding variety in both monoculture and relay intercropping, and Zhongwangyu 18, a high-yielding monoculture variety, which were previously chosen during the screening stage. Yield differences between these two varieties were subsequently analyzed and compared. Specifically, we observed that the average maize yield for high-yielding monoculture varieties and relay intercropping was 8343.98 kg/ha, i.e., 1.56% lower than that of monoculture maize. On the contrary, the average yield of high-yield monoculture and low-yielding varieties in maize relay intercropping was 6826.31 kg/ha, i.e., 20.27% lower than that of monoculture maize (Table 8.14). Clearly, different maize varieties seem to possess varying adaptability levels to environmental changes. Additionally, we identified that maize varieties that are ideally suited for relay intercropping exhibited a greater ability to adapt to the unique light environment of mixed planting (Ma et al. 2019a, b).

Difference in Leaf Area

Considering specifically the third leaf, Rongyu 1210 showed a leaf area that was 7.81% and 7.07% smaller than those of Zhongwangyu 18 at tasseling and grain-filling, respectively (Table 8.15). However, the leaf area of the lower plant body was 6.93% higher during tasseling and 12.63% higher during grain-filling than that of Zhongwangyu 18. In both varieties, leaf area decreased with growth, but the decline in Rongyu 1210 was 4.00% and 9.70% lower at tasseling and grain-filling, respectively, whereas leaf area reduction in Zhongwangyu 18 was 12.00% and 16.85% higher relative to the same growth stages. These findings suggest that Rongyu 1210 is more appropriate for relay intercropping, as it retains its leaves longer than Zhongwangyu 18 variety. Moreover, the leaf area of both varieties in relay intercropping showed an upward trend compared to monoculture. Further, Rongyu 1210 registered increases of 4.77% and 3.85% during tasseling and grain-filling, respectively, while Zhongwangyu 18 registered increases of 2.42% and 0.86%,

Table 8.14 Differences in yield among maize varieties contrasting for yield under monoculture and relay intercropping

Treatment		2014	2015	2016	2017	Mean
Rongyu 1210	Relay intercropping	8375.04	9550.05	7880.06	7570.75	8343.98
	Monoculture	8306.29	9562.55	7987.84	8049.60	8476.57
Zhongwangyu 18	Relay intercropping	6937.53	6943.80	6701.32	6722.60	6826.31
	Monoculture	8531.29	9000.00	8206.08	8516.97	8563.59

Table 8.15 Leaf area of different maize varieties under monoculture and relay intercropping

Treatment		Leaf area of a single plant (cm^2)		Area of the 3d leaf above the ear (cm^2)		Leaf area of the lower plant body (cm^2)	
		Tasseling	Grain-filling	Tasseling	Grain-filling	Tasseling	Grain-filling
Rongyu 1210	Relay intercropping	6938.17a	6671.00a	1843.23a	1877.43a	3377.51a	3078.97a
	Monoculture	6607.46b	6414.14a	1850.80a	1928.56a	3366.67a	3049.03a
Wangzhongyu 18	Relay intercropping	7105.03a	6344.06a	1987.10b	2010.12b	3143.41a	2690.18a
	Monoculture	6933.18 a	6289.27a	2126.57a	2148.69a	3025.01a	2295.53b

Note: Lowercase letters indicate a significant difference at the $p < 0.05$ under all treatments

respectively. Conversely, the trend followed by the third leaf area was the opposite. Rongyu 1210 experienced no significant difference relative to monoculture, whereas Zhongwangyu 18 had a significant fall of 7.02% and 6.89%, during tasseling and grain-filling, respectively, implying that intercropping seemingly affects Zhongwangyu 18 more significantly than Rongyu 1210 (Zeng 2018a, b).

Difference in Photosynthesis Parameters

When the same two maize varieties were grown at wider row spacing in relay intercropping, they registered significantly higher photosynthetic rates than in monoculture or in narrower row-spacing relay intercropping (Table 8.16). These results suggested that variety Rongyu 1210 should be highly recommended for relay intercropping, as it showed greater improvement in photosynthetic rate in wider row spacing compared to monoculture and experienced a lower light-inhibition effect in narrow row spacing when compared to the Zhongwangyu 18 variety (Zeng 2018a, b). Furthermore, the daily variation experiment in photosynthetic rate (Fig. 8.5) revealed that the photosynthetic rate of Rongyu 1210 was higher than that of Zhongwangyu 18 at 4:00 pm, and that the decline in photosynthesis activity proceeded at a slower rate. This finding indicates that the two varieties differ in rate of weak-light utilization (i.e., at 8–10 am, and 4–6 pm), with Rongyu 1210 showcasing a higher utilization rate than Zhongwangyu 18 (Ren et al. 2019a, b).

Distribution of Dry Matter

In this study, we compared the narrow-row and wide-row ^{13}C labeling in the leaves of the maize varieties, Zhongwanyu 18 and Rongwanyu 1210, which are characterized by higher and lower yield reduction rates in relay intercropping, respectively. Specifically, we calculated the proportion of ^{13}C distributed in each organ (Table 8.17) and found that, when narrow-row labeled leaves were used, ^{13}C partitioning decreased toward the ear but increased toward the stem and labeled leaves. Furthermore, the proportion of ^{13}C-labeled dry matter allocated to the ear was significantly higher for both narrow-row and wide-row labeled leaves in maize varieties suitable for relay intercropping than in monoculture varieties (Li et al. 2020a, b, c, d). This finding suggests that high-yielding relay intercropped maize varieties may ensure that a higher proportion of photosynthate is distributed to the developing grains, thus leading to higher yields.

8.1.2.2 Selection of Soybean Varieties Resistant to Shade and Lodging

8.1.2.2.1 Differences Among Soybean Varieties Differing for Shade Resistance

In a two-year field experiment, two soybean varieties significantly different in shade tolerance were compared for performance in a relay-intercropping system. The results showed significant differences in lodging rates and yield traits between the two varieties in the relay intercropping system. Specifically, both varieties exhibited lodging to varying degrees from the seedling stage (v2). However, the highly shade-tolerant variety (Nandou 12) had a significantly lower rate of increase in lodging as growth proceeded, compared to the weakly shade-tolerant variety, Nan032-4. The

Table 8.16 Changes in photosynthesis parameters of different maize varieties under monoculture and relay intercropping

Treatment		Pn (μmol m^{-2} s^{-1})		g_s (mol H_2O m^{-2} s^{-1})		Ci (μmol mol^{-1})		Tr (mmol H_2O m^{-2} s^{-1})	
		Tasseling	Grain-filling	Tasseling	Grain-filling	Tasseling	Grain-filling	Tasseling	Grain-filling
Rongyu 1210	Narrow row of RI	28.65c	26.45c	0.17c	0.16c	102.53a	99.86a	2.85c	3.58c
	Wide row of RI	34.86a	33.27a	0.24a	0.23a	67.90b	68.54b	5.42a	5.88a
	Monoculture	30.79b	29.44b	0.22b	0.20b	98.46a	101.97a	4.19b	4.72b
Wang Zhong 18	Narrow row of RI	24.06c	23.52c	0.14c	0.16c	117.97a	130.94a	2.63c	3.95c
	Wide row of RI	32.39a	32.83a	0.23a	0.25a	79.90b	79.94c	4.24a	5.91a
	Monoculture	29.59b	29.28b	0.19b	0.19b	117.08a	109.08b	3.04b	4.87b
F	A	83.24**	6.137	147.19**	8.89	2.94	35.37*	126.93**	11.95
	B	185.14**	109.50**	147.10**	26.80**	11.32**	21.22**	156.21**	115.46**
	A*B	6.19**	2.423	6.80**	0.705	0.2	1.367	5.56**	3.81*

Note: *Pn* net photosynthetic rate, g_s stomatal conductance, *Ci* intercellular CO_2 concentration, *Tr* transpiration rate, *RI* relay intercropping. Lowercase letters indicate a significant difference at the $p < 0.05$ under all treatments

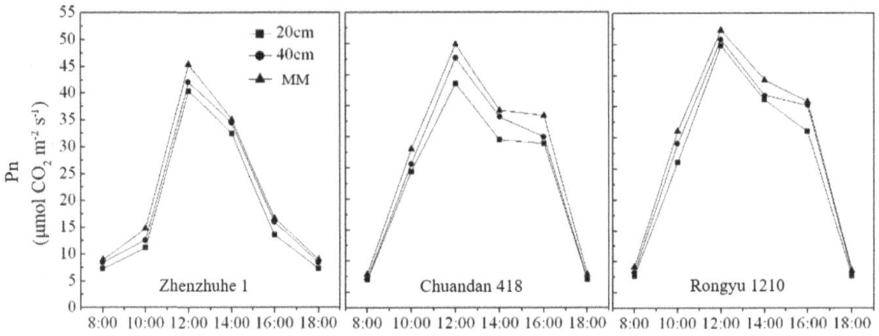

Fig. 8.5 Differences in the daily course of photosynthetic rate in different maize varieties in monoculture and relay intercropping. Note: *Pn* photosynthetic rate, *MM* monoculture maize

Table 8.17 Distribution percentage of ^{13}C to various plant organs (%)

Treatment	Upper leaves in wide rows	Lower leaves in wide rows	Upper leaves in narrow rows	Lower leaves in narrow rows	Stem	Ear	Ear Leaf
ZWY 18 marks narrow row	0.54a	2.58a	0.26a	0.43c	12.14a	83.75c	0.26a
ZWY 18 marks wide row	0.17c	0.18c	0.15b	1.16b	3.44b	94.80b	0.06c
RY1210 marks narrow row	0.12c	1.61b	0.11c	0.11d	3.16b	94.65b	0.21a
RY1210 marks wide row	0.22b	0.16c	0.26a	1.80a	1.85c	95.51a	0.15b

Note: Lowercase letters indicate a significant difference at the $p < 0.05$ under all treatments

average lodging rate over the entire growth period of the shade-tolerant variety was 30% lower than that of the shade-susceptible variety, which showed a lodging rate of over 90% (Fig. 8.6, Table 8.18). Severe lodging led to a significant decrease in the number of grains per plant and the 100-grain weight, with the weakly shade-tolerant variety showing a reduction of up to 83 grains and 46.89% in the number of grains and yield, respectively, in relay intercropping, compared to monoculture (Zou et al. 2015). In contrast, the highly shade-tolerant variety reached 80% of the yield per plant recorded under monoculture (Table 8.19).

Fig. 8.6 Field lodging of soybean varieties differing in shade tolerance under relay intercropping. A: Variety with weak shade tolerance (Nan 032-4). B: Variety with strong shade tolerance (Nandou 12)

8.1.2.2.2 Selecting Shade-Tolerant and Lodging-Resistant Soybean Varieties for Strip Intercropping

Relay Intercropping

In this study, the lodging resistance index was determined based on various factors, such as main stem length and thickness, aboveground biomass, and stem bending resistance 35 days after sowing. These factors were used to evaluate the ability of soybean plants to resist lodging in a relay-intercropping system with maize relative to soybean monoculture. The results obtained highlighted the fact that there was a significant difference in the lodging resistance index among soybean varieties, ranging between 0.1089 and 0.7874 (Table 8.20). Moreover, based on the lodging resistance index, the test materials were divided into three categories: high (≥0.60), intermediate (0.30–0.60), and low (≤0.30) resistance, with four genotypes falling

Table 8.18 Seedling lodging rate (%) for soybean varieties differing for shade tolerance

Year	Planting pattern	Varieties	Growth stage				
			V2	V3	V4	V5	V6(R1)
2013	Monoculture	Nan 032-4	0	0	0	0	0
		Nandou 12	0	0	0	0	0
	Relay intercropping	Nan 032-4	22.19	65.26**	74.09**	83.64**	94.17**
		Nandou 12	22.37	33.80	32.53	44.14	48.73
2014	Monoculture	Nan 032-4	0	0	0	0	0
		Nandou12	0	0	0	0	0
	Relay intercropping	Nan 032-4	22.75	59.77**	72.98**	87.72**	93.67**
		Nandou 12	22.16	34.71	37.72	38.93	51.11

Note: **Represents a significant difference in lodging rate at 1% level between Nan 032-4 and Nandou 12, under relay intercropping

into the high resistance group, including NanDou12, Local August Yellow, Gongxuan 1, and small yellow soybean. Furthermore, 16 genotypes were in the intermediate lodging resistance group, and 10 others were categorized as low-lodging resistant.

Intercropping

We conducted a clustering analysis on a total of 124 soybean genotypes to assess shade tolerance and lodging resistance. The intersection between indicators of yield reduction and lodging strength was computed to identify materials showing low yield reduction combined with high lodging resistance (i.e., in all, 13 genotypes), low yield reduction, and low lodging resistance (i.e., five genotypes), high yield reduction combined with high lodging resistance (11 genotypes), and high yield reduction combined with low lodging resistance (17 genotypes). The selection of suitable materials for intercropping was based on the above criteria (Zou et al. 2015). Finally, only those genetic materials that met the standard of a single-grain weight equal to or greater than 13 g (i.e., high yield) were promoted for intercropping (Table 8.21).

8.1.2.2.3 Characteristics of Shade-Tolerant/Lodging-Resistant Soybean Varieties Suitable for Strip Intercropping

Differences in Stem Morphology

Based on data from the 2013 and 2014 experiments summarized in Fig. 8.7, relay intercropping of maize and soybeans significantly increased main stem length in soybean seedlings during the early growth phase, while stem thickness and above-ground biomass were significantly reduced. Compared to monoculture, relay intercropping resulted in the following main stem length increases at the end of

Table 8.19 Different yield-related traits in soybean varieties differing for shade tolerance

Planting pattern	Soybean variety	Effective branch number	Main stem pods	Number of branch pods	Grain number of single plant	100-grain Weight (g)	Yield per plant (g/plant)
Monoculture	Nan 032–4	5.96a	19.53a	78.81a	170.8a	10.23b	16.12a
	Nandou 12	5.11a	22.97a	48.31b	117.65b	17.12a	20.12a
Relay intercropping	Nan 032–4	5.13a	10.97b	32.36b	87.40b	9.52b	8.56b
	Nandou 12	4.87a	17.54a	38.70a	95.51a	16.61a	16.33a

Note: Lowercase letters indicate a significant difference at the $p < 0.05$ under all treatments

Table 8.20 Shade tolerance and lodging resistance indices and their components in different soybean varieties

Soybean variety	Main stem length (cm)	Stem diameter (mm)	Bending strength (N)	Shoot dry weight (g)	Lodging resistance index	Lodging rate (%)	Lodging level
Nandou 12	53.91	2.73	28.32	1.82	0.7874	41.78	3
Bayuehuang	56.36	2.71	31.47	1.98	0.7659	45.94	3
Gongxuan-1	52.62	2.79	28.55	1.98	0.7637	38.46	3
Xiaohuangdou	60.8	2.83	30.52	2.15	0.6605	48.94	3
Jianyang Jiuyuehuang-2	47.33	2.35	19.85	1.71	0.5762	55.41	4
Shengfodadou	55.41	2.59	24.63	2.01	0.5728	53.41	4
Jianyang Lvpidou	58.74	2.74	22.16	1.98	0.5219	59.88	4
Jiuyuehuang	67.43	2.38	23.6	1.62	0.5157	58.65	4
Qianshan Wudou	70.14	2.46	23.75	1.69	0.4925	63.24	4
Yaan Heidou	63.41	2.46	23.58	1.88	0.4866	66.54	4
Pingwudadou-2	48.33	1.78	16.13	1.26	0.4741	64.46	4
Nan 256-1	72.33	2.48	20.93	1.56	0.4607	67.41	4
Nandou 021-1	69.98	2.32	21.68	1.57	0.4589	67.87	4
Texuan 13	68.84	2.52	18.17	1.47	0.452	65.69	4
Gongqiudou 04-2	71.99	2.3	22.38	1.74	0.4107	71.38	4
Yongshengheidadou	71.43	1.92	16.31	1.25	0.3506	72.41	4
Qindou-1	71.22	2.49	24.2	2.44	0.3467	69.98	4
Dahuangzhu	86.13	2.26	17.88	1.4	0.3351	62.55	4
Guixia 3	84.5	2.26	16.22	1.35	0.3211	74.41	4
Dayuandou	85.53	2.18	15.98	1.27	0.3204	72.39	4
Gongdou 2	58.99	2.63	18.06	2.81	0.2865	76.25	5
Mengqing 6	82.2	2.27	15.47	1.51	0.2825	71.74	4
Guang15	88.56	2.25	15.63	1.41	0.2818	75.44	5
Caidou	83.03	2.26	15.5	1.51	0.2794	77.44	5

Nandou 032-4	80.15	1.93	15.13	1.41	0.2576	79.14	5
Chongyuebaimao	85.44	1.91	18.17	1.91	0.2126	76.41	5
Dazhou 1	71.33	1.57	10.53	1.14	0.2033	80.14	5
Jiande Baimaodou 1	101.35	2.53	21.35	2.77	0.1924	84.03	5
Zihuadou	80.99	1.71	12.65	1.48	0.1804	78.81	5
Xiaobaimao	108.74	1.68	14.66	2.08	0.1089	86.23	5

Table 8.21 Recommended shade-tolerant and lodging-resistant intercropping soybean genotypes

Genotypic category	Soybean varieties
Low yield reduction combined with high lodging resistance	Jidou 12, Suike 23, Suike 67, Suike 43, Handou 19, Longhuang 3, Shengdou 10, Handou 15, Suike 66, Suike 998
Low yield reduction combined with low lodging resistance	Qihuang 34, Suike 75, Chengdou 18
High yield reduction combined with high lodging resistance	Pudou 820, Zhonghuang 302, Jindou 188, Zhonghuang 42, Zhonghuang 328
High yield reduction combined with low lodging resistance	Zhonghuang 74, Zhonghuang 39, Dedou 99–6

the symbiotic period: 88.16% for weakly shade-tolerant, 138.57% for moderately shade-tolerant, and 49.01% for strongly shade-tolerant soybean varieties. Additionally, stem thickness decreased by 55.79%, 53.14%, and 51.21%, and aboveground biomass decreased by 81.93%, 75.35%, and 74.72%, respectively (Zou et al. 2015). These findings suggest that the variation range of weakly shade-tolerant soybean varieties is larger, and therefore, they may be more susceptible to lodging in the field.

Differences in Physical Characteristics of the Stem

In the relay-intercropping system, the stem bending strength of each soybean variety began to differ at 34 days. Specifically, the strongly shade-tolerant variety had a remarkably higher bending strength than the weakly or moderately shade-tolerant ones at 48 days. At the end of the two experimental years, we found that the stem anti-bending strength of weakly, moderately, and strongly shade-tolerant soybean varieties under relay intercropping was reduced by 72.39%, 63.90%, and 58.03%, respectively, compared to monoculture (Table 8.22). This finding indicates that relay intercropping had a significant impact on the stem bending strength of soybeans, while strongly shade-tolerant lodging-resistant varieties were relatively less affected.

Relay intercropping significantly reduced the soybean lodging-resistance index value, and different soybean varieties showed varying performances in terms of lodging resistance index (Table 8.23). Furthermore, strongly shade-tolerant varieties showed better lodging-resistance index compared to moderately and weakly shade-tolerant ones, implying that they possess better adaptability and regulation capabilities in shaded environments. Among the various soybean varieties tested, Nandou12, a strongly shade-tolerant lodging-resistant variety, was less affected by shading, showing the smallest decrease in stem bending strength and a high lodging-resistance index. As a result, the variety had the lowest field lodging rate, making it an excellent choice for relay-intercropping soybean production (Zou et al. 2015). Therefore, to increase yield and income in relay-intercropping soybean production, it is crucial to select varieties with strong shade tolerance and comprehensive stem traits.

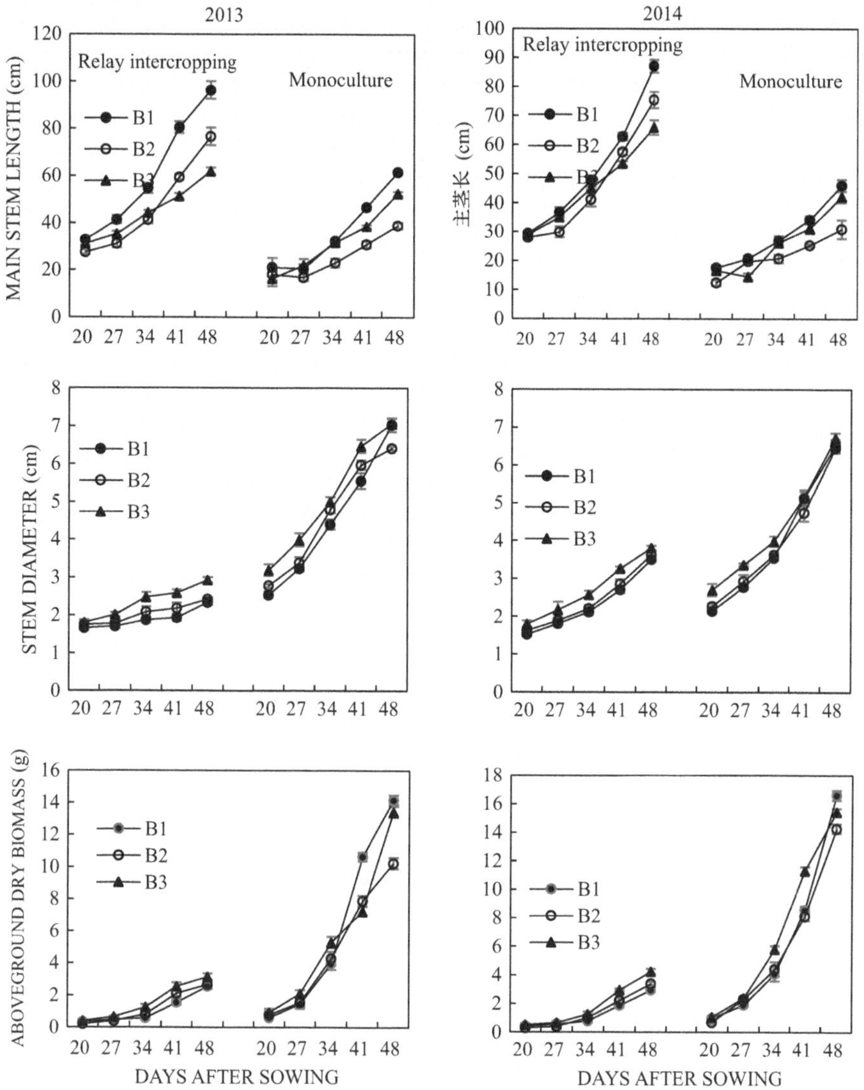

Fig. 8.7 Differences in morphological characteristics between monoculture and relay-intercropping soybean crops. Note: B1 (Nandou 032-4, weakly shade-tolerant and lodging-resistant); B2 (Jiuyuehuang, moderately shade-tolerant and lodging-resistant); B3 (Nandou 12, strongly shade-tolerant and lodging-resistant)

Table 8.22 Changes in bending strength of soybean stems

		2013					2014				
		Days after sowing (d)					Days after sowing (d)				
Planting pattern	Variety	20	27	34	41	48	20	27	34	41	48
Relay Intercropping	Nan 032-4	3.43c	7.85c	10.59d	18.82e	28.14e	3.33c	9.42d	15.35d	27.79d	35.60e
	Jiuyuehuang	3.74c	8.50c	13.88d	26.80d	38.16d	3.80c	9.58d	17.39cd	30.78cd	45.57d
	Nandou 12	3.90c	10.45c	27.96c	48.91c	63.12c	5.06c	12.08d	21.15c	34.48c	56.57c
Monoculture	Nan 032-4	13.13b	36.20b	49.81b	66.85b	93.33b	12.06b	32.96c	66.37b	95.48b	141.95b
	Jiuyuehuang	17.47a	36.37b	53.20b	67.64b	100.31b	13.57b	39.04b	69.71b	94.87b	133.40b
	Nandou 12	18.53a	45.33a	60.57a	81.04a	128.26a	18.73a	52.33a	83.13a	117.12a	162.91a

Note: Lowercase letters indicate a significant difference at the $p < 0.05$ under all treatments

Table 8.23 Changes in soybean lodging resistance index

		2013					2014				
		Days after sowing (d)					Days after sowing (d)				
Planting pattern	Variety	20	27	34	41	48	20	27	34	41	48
Relay Intercropping	Nan 032-4	0.651e	0.706d	0.538e	0.304e	0.270e	0.511e	0.905e	0.755e	0.673e	0.493f
	Jiuyuehuang	0.833d	1.040c	0.858d	0.470e	0.461d	0.737d	1.091d	0.869de	0.700e	0.657e
	Nandou 12	0.685e	1.028c	1.284c	0.998d	0.964c	0.701d	1.148d	1.004c	0.812d	0.783d
Monoculture	Nan 032-4	3.566c	4.843b	2.113b	1.355c	0.915c	2.071c	2.469c	2.173b	1.582c	1.207c
	Jiuyuehuang	5.156a	5.721a	3.190a	2.013a	1.975a	3.082a	4.175a	2.814a	2.202a	1.957a
	Nandou 12	3.881b	4.917b	2.196b	1.620b	1.559b	2.896b	3.850b	2.196b	1.719b	1.688b

Note: Lowercase letters indicate a significant difference at the $p < 0.05$ under all treatments

8.2 Expand Spacing and Increase Light

To optimize the use of aboveground light resources and encourage root development in maize–soybean strip intercropping, it is important to identify the appropriate maize–soybean strip width and spacing. Widening the maize rows and increasing soybean spacing reduced shading and boosted aboveground photosynthesis, promoting effective transfer of carbohydrates for underground growth, thus simultaneously improving nitrogen fixation capacity and facilitating intercropping between root systems, which resulted in better nutrient uptake efficiency in maize and enhanced aboveground growth. The strip width and spacing affected both the light environment and the system population density; hence, selecting the optimal soybean–maize relay-intercropping spacing is essential. Machinery movements must also be considered when deciding the size of strip width and spacing.

8.2.1 Production Unit Width

The maize–soybean strip intercropping configuration involves planting two rows of maize and two to six rows of soybean, alternating with each other. This configuration forms a unit in the production system. The entire field is made up of multiple production units, each with a width that is equal to the maize row width, the soybean row width, and the sum of the two spacings. The field configuration and parameters such as row number, row spacing, strip width, plant spacing form the core components of the system for achieving high yields, mechanization, and sustainability in maize–soybean relay intercropping.

The production unit width determines the overall field population structure and other parameters. Thus, having the right width is crucial. Under different ecological conditions, the appropriate width of the production unit for different intercropping configurations should be selected such as to enhance the increase in overall crop yield. A narrow production unit width will lead to intense mutual interference between maize and soybean plants, ultimately affecting soybean yields. In contrast, a wide production unit width will result in reduced edge rows, reduced maize advantages, population density reduction, and intercropping loss. The appropriate unit width is determined based on the characteristics of the maize and soybean varieties used, the specific climatic conditions, usage, symbiotic period length, and the width of farming machinery.

For areas with compact and short maize plant types, strong soybean-shade tolerance, unlimited light intensity, a short symbiotic period, and suitable farming machinery, a production unit width of 2 m is appropriate. On the contrary, in areas with poor light intensity, loose maize plant types, weaker soybean shade tolerance, or wider farming machinery, the maximum unit width should be 2.8–3.0 m, but it should not exceed 3.0 m. By having the appropriate production unit width in maize–soybean relay intercropping, higher yields, mechanization, and sustainability are guaranteed.

8.2.1.1 Relay Intercropping

The soybean–maize relay-intercropping system features a gradually increasing production unit width, from 1.6 to 2.2 m. As a result, the light distribution within the canopy, crop yield, land equivalent ratio, and nutrient uptake-competition ratio between the maize and soybean crops shows significant differences. The middle part of the maize canopy experiences gradually decreasing light transmittance as the production unit width increases, while the soybean canopy experiences an increase in light transmittance. The increase in production unit width leads to a decrease in maize yield and a gradual increase in soybean yield (Tables 8.24 and 8.25), although the crop yield follows a downward trend. Based on several years of experimental results, a production unit width of approximately 2 m is adequate to achieve high yields in the relay-intercropping system, regardless of the crop variety (Lu 2014; Ma et al. 2019a, b). Moreover, the nutrient accumulation advantage of this system is greatest at this particular width (Table 8.26).

8.2.1.2 Intercropping

In the Huang-Huai-Hai summer soybean-summer maize intercropping region, a long co-cultivation period of maize and soybean was studied using a relay-intercropping system. Monoculture units of various widths were established including 2.2 m with three soybean rows, 2.5 m with 4 soybean rows, and 2.8 m with five soybean rows. In each monoculture unit, two maize rows were spaced 40 cm apart, while the distance between maize and soybean was 60 cm and the soybean row spacing was 30 cm. The study was conducted over two consecutive years (Table 8.27). The results revealed that the width of the monoculture unit had a significant impact on the yield and value of both maize and soybean, as well as the overall crop. Increasing the width of the monoculture unit had an adverse effect on maize yield, with a significant drop of 4.7% and 6.7% observed for the 2.8 m width as compared to the 2.2 m width, respectively. In contrast, soybean yield exhibited an opposite trend, with an average increase of 24.8% and 26.8% observed from 2.2 to 2.8 m width, respectively. Based on the overall crop yield and value, the 2.- m width was found to be the most effective width for this region in Huang-Huai-Hai.

Further analysis revealed that maize grain number decreased significantly as the production unit width increased. Thus, a production unit width of 2.8 m reduced the maize grain number by 4.1% and 4.7% in the 2 years, respectively, compared to the 2.2-m unit width (Table 8.28). Additionally, the effective soybean plant number was significantly impacted by the production unit width, with no significant differences observed in soybean 100-grain weight. Furthermore, the soybean grain number per plant significantly increased by 21.8% and 21.1%, respectively, for the 2 years under the 2.8-m width compared to that of 2.2 m (Table 8.29).

In the spring soybean-spring maize intercropping area in northwest China, a series of different production-unit field configurations were tested on a large scale in field experiments. The two-year results (Table 8.30) indicated that increasing the production unit width led to a decrease in maize yield. Specifically, maize yield decreased significantly by an average of 7.1% and 6.4% in each of the 2 years, respectively, for a production unit width of 2.8 m, compared to a unit width of 2.2 m.

Table 8.24 Effects of production unit width on maize and soybean yield and land equivalent ratio under monoculture and relay intercropping

	2012				2013			
Treatment	Yield (10^3 kg/ha)				Yield (10^3 kg/ha)			
Production Unit width (m)	Maize	Soybean	System output	LER	Maize	Soybean	System output	LER
1.6	9.13a	0.48e	9.61a	1.32e	6.33a	0.71f	7.04b	1.56d
1.7	8.67ab	0.69d	9.37ab	1.38cde	6.11ab	0.85e	6.96b	1.63d
1.8	8.41abc	0.74d	9.15bc	1.37de	5.80bc	1.06d	6.87bc	1.73c
1.9	8.26bc	1.12bc	9.39ab	1.55b	5.72c	1.41b	7.13ab	1.97b
2.0	8.33bc	1.29b	9.62a	1.65a	5.87	1.52a	7.39a	2.08a
2.1	7.77bc	1.03c	8.80cd	1.45c	5.31d	1.28c	6.59c	1.81c
2.2	7.42cd	1.06c	8.48d	1.42cd	5.22d	0.92e	6.13d	1.53d
Monoculture maize	8.45abc		8.45d	1.00f	6.07ab		6.07d	1.00e
Monoculture soybean		1.96a	1.96e	1.00f		1.38bc	1.38e	1.00e

Note: *LER* land equivalent ratio, the maize variety is Cuandan 418, and the soybean variety is Gongxuan 1. Lowercase letters indicate a significant difference at the $p < 0.05$ under all treatments

Table 8.25 Effects of production unit width on crop yield (kg/ha)

Treatment	2017			2018		
	Maize	Soybean	System	Maize	Soybean	System
2.0 m	9382.67a	2149.80c	11,532.46a	8620.86b	1422.21c	10,043.06a
2.4 m	7633.07b	2563.62b	10,196.69b	8024.90b	1453.59c	9478.49a
2.8 m	6217.12c	2735.49ab	8952.61c	6770.96c	1733.79b	8504.75b
Monoculture maize	9568.77a	–	9568.77bc	9856.96a	–	9856.96a
Monoculture soybean	–	2881.65a	2881.65d	–	2186.95a	2186.95c

Note: Maize variety is Rongyu 1210 and the soybean variety is Nandou 12. Lowercase letters indicate a significant difference at the $p < 0.05$ under all treatments

Table 8.26 Effects of production unit width on the accumulation of nitrogen (N), phosphorus (P), and potassium (K) on crop yield (kg/ha)

Treatment	N		P		K	
Strip width	Maize	Soybean	Maize	Soybean	Maize	Soybean
1.6	231.7b	43.9g	45.1ab	5.9f	157.7ab	25.4f
1.7	230.8b	60.5f	44.6b	8.3e	154.1abc	34.7e
1.8	226.2bc	66.9e	44.1bc	8.8e	154.1abc	35.6e
1.9	223.9bc	92.6d	43.8bc	12.4d	151.7abc	50.2d
2.0	223.2bc	113.3b	42.1bc	16.0b	148.9bc	60.6b
2.1	219.8bc	106.3c	42.0bc	14.8c	148.2bc	55.3c
2.2	217.7c	102.4c	41.5c	14.2c	146.4c	53.7c
Monoculture maize	251.1a	–	47.8a	–	162.1a	–
Monoculture soybean	–	171.5a	–	22.3a	–	83.2a

Note: Lowercase letters indicate a significant difference at the $p < 0.05$ under all treatments

In contrast, increasing the production unit width significantly improved soybean yield, with an average increase of 29.7% and 23.6% each of the 2 years, respectively, for the production unit width of 2.8 m compared to that of 2.2 m.

Furthermore, expanding the production unit width resulted in a reduction in maize grain number, with a production unit width of 2.8 m having significantly smaller maize grain numbers compared to grain numbers obtained when the production unit width of 2.2 m. Although the 1000-grain weight of maize showed a decreasing trend with increasing population density, it was not significant (Table 8.31). In contrast, significant differences in the effective plant number of soybeans were observed due to the production unit width, while the 100-grain weight was not significantly affected. Furthermore, single soybean plant grain number was shown to be significantly affected by production unit width. In the production unit width of 2.8 m, single-plant grain number increased by 26.8% and 21.1% in the two experimental years, respectively, compared to a width of 2.2 m (Table 8.32).

8.2.2 Configuration of Row Ratio

In terms of land equivalent ratio, the soybean and maize row numbers can be compared in terms of actual rows. For example, in relay intercropping with three soybean rows and two maize rows, the row ratio is 3:2. According to the theory of "higher main crop, high-low synergy" in relay intercropping, a system with two rows of maize marginally advantageous in every row demands efficient allocation of row ratio for the lower soybean crop to utilize light energy and achieve higher group yield. To solve this problem, field experiments were conducted with varying row ratios of soybeans for two consecutive years from 2017 to 2018. As seen in Fig. 8.8, the light energy use efficiency of each relay-intercropping treatment group was significantly higher than that of monoculture maize or monoculture soybean

Table 8.27 Effects of production unit width on maize and soybean yield, and overall system output

Production unit width	2021				2022			
	Maize (kg/ha)	Soybean (kg/ha)	System (kg/ha)	Output (yuan/ha)	Maize (kg/ha)	Soybean (kg/ha)	System (kg/ha)	Output (yuan/ha)
2.2	7481.94a	1368.23c	8850.17b	29,158.83b	8091.77a	1110.47c	9202.25a	29,319.81b
2.5	7305.88b	1656.27b	8962.14a	30,394.06a	7963.78a	1319.42b	9283.20a	30,215.12a
2.8	7131.22c	1707.65a	8838.87b	30,213.34a	7553.56b	1408.00a	8961.56b	29,597.97b

Note: The price of maize and soybean is 2.4 yuan/kg and 6 yuan/kg, respectively. Lowercase letters indicate a significant difference at the $p < 0.05$ under all treatments

Table 8.28 Effects of production unit width on maize yield components

Production unit width (m)	2021 Effective ear number	Grain number	1000-grain weight (g)	2022 Effective ear number	Grain number	1000-grain weight (g)
2.2	65,655a	421.17a	273.12a	66,005a	386.14a	319.92a
2.5	66,034a	412.37b	270.48b	65,588ab	381.58a	320.66a
2.8	65,805a	403.93c	270.31b	648,149b	367.95b	320.46a

Note: Lowercase letters indicate a significant difference at the $p < 0.05$ under all treatments

Table 8.29 Effects of production unit width on soybean yield components

Production unit width (m)	2021 Effective plant (plant/ha)	Seed number per plant	100-seed weight (g)	2022 Effective plant (plant/ha)	Seed number per plant	100-seed weight (g)
2.2	126,732b	39.53c	27.36a	111,717c	39.87c	24.94a
2.5	131,503a	46.57b	27.07a	114,612b	45.48b	25.32a
2.8	130,497a	48.16a	27.20a	117,198a	48.27a	24.76a

Note: Lowercase letters indicate a significant difference at the $p < 0.05$ under all treatments

(Xu et al. 2014). With an increase in bandwidth, the light energy use efficiency of the maize and soybean groups in the relay-intercropping system gradually declined, with the maximum values appearing in T1. In 2017, the light energy use efficiency in T1 was significantly higher (12.61%, 14.68%, 32.98%, 31.58%, 40.45%, and 257.43%) than that of T2, T3, T4, T5, CKM, and CKS, respectively. The trend observed in 2018 was similar to that in 2017 (Chen et al. 2019). According to Table 8.33, the group yield and land equivalent ratio were the highest in the T1 treatment, indicating that the row ratio of maize and soybean in relay intercropping should be 2:2.

8.2.3 Configuration of Row and Plant Spacing

8.2.3.1 Row Spacing of Maize Strip

Row spacing is the distance between rows in the same crop strip. We designed a two-year field experiment using a 2-meter-wide relay-intercropping method and set six maize intercropping distances at 20, 30, 40, 50, 60, and 80 cm (Table 8.34). Results showed that, as intercropping distance increased, maize yield reduction decreased. Specifically, the reduction was ranged from 20.61% to 17.97% for the 20–30 cm range, 11.59% and 9.81% for the 40–50 cm range, and 6.27% and 1.40% for the 60–100 cm range, respectively. In contrast, soybean yield decreased as the intercropping distance increased, ranging from 25.30% to 28.67% for the 20–40 cm range, and 37.04% to 59.80% for the 50–80 cm range. The trends of population yield, land equivalent ratio, and actual yield loss values showed an initial increase, followed by a decrease (Liao 2015). Considering the insignificant differences in

Table 8.30 Effects of production unit width on maize, soybean yield, and crop output

Production unit width	2021 Maize (kg/ha)	2021 Soybean (kg/ha)	2021 System (kg/ha)	2021 Output (yuan/ha)	2022 Maize (kg/ha)	2022 Soybean (kg/ha)	2022 System (kg/ha)	2022 Output (yuan/ha)
2.2	15,335.57a	1380.38c	16,715.95a	51,221.88ab	14,428.12a	1589.04c	16,017.16a	49,932.95b
2.5	14,969.98b	1676.61b	16,646.59a	51,975.59a	14,270.8a	1829.75b	16,100.55a	50,936.74a
2.8	14,246.00c	1789.89a	16,035.89b	50,628.16b	13,500.62b	1964.47a	15,465.09b	49,588.56b

Note: The price of maize and soybean is 2.4 yuan/kg and 6 yuan/kg, respectively. Lowercase letters indicate a significant difference at the $p < 0.05$ under all treatments

Table 8.31 Effects of production unit width on maize yield components

Production unit width (m)	2021 Effective ear number	Grain number	1000-grain weight (g)	2022 Effective ear number	Grain number	1000-grain weight (g)
2.2	70,788a	519.2891	420.35a	69,731.82	524.604	398.07a
2.5	70,813a	505.8543	420.50a	70,257.33	513.436	399.49a
2.8	70,571a	484.3872	420.17a	69,802.98	491.1657	396.82a

Note: Lowercase letters indicate a significant difference at the $p < 0.05$ under all treatments

Table 8.32 Effects of production unit width on soybean yield components

Production unit width (m)	2021 Effective plant number (plant/ha)	Seed number per plant	100-seed weight (g)	2022 Effective plant population (plant/ha)	Seed number per plant	100-seed weight (g)
2.2	123,122c	50.82c	22.07a	124,684c	48.63c	26.20a
2.5	124,148b	62.59b	21.59a	127,397b	56.00b	25.65a
2.8	126,934a	64.42a	21.87a	129,190a	58.91a	25.81a

Note: Lowercase letters indicate a significant difference at the $p < 0.05$ under all treatments

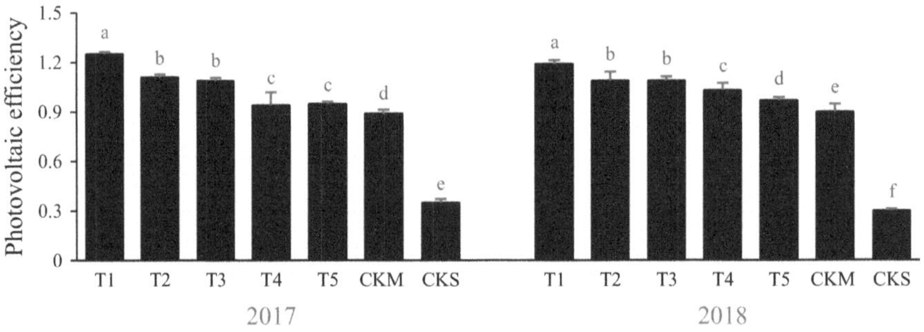

Fig. 8.8 Effects of different row ratios on light energy utilization of maize and soybean populations. Note: The spacing between maize and soybean is 40 cm. T1: Production unit width 2 m, row ratio 2:2; T2: Production unit width 2.4 m, row ratio 2:3; T3: Production unit width 2.4 m, row ratio 2:4; T4: Production unit width 2.8 m, row ratio 2:3; T5: Production unit width 2.8 m, row ratio 2:4; CKM: Maize monoculture (even spacing cultivation, row spacing of 70 cm); CKS: Soybean monoculture (even spacing cultivation, row spacing of 50 cm). Lowercase letters indicate a significant difference at the $p < 0.05$ under all treatments

various indicators within the 40–60 cm range, we found that the best option was to choose maize intercropping with a distance of 40 cm, which provided more space for soybeans.

Maize spacing is a crucial factor that affects the growth environment of the crop population. As per regression analysis, the light transmittance of the upper, middle, and lower parts of the canopy on the marginal side exhibited a significant positive

Table 8.33 Effects of row ratio on yield of maize–soybean strip intercropping system

Year	Treatment	Yield (10^3 kg/ha)			Contribution rate (%)		LER
		Maize	Soybean	System	Maize	Soybean	
2017	T1	9.38a	2.15d	11.53a	81.36	18.64	1.73a
	T2	7.63b	2.56c	10.20b	74.86	25.14	1.69ab
	T3	7.21b	2.54c	9.74c	73.96	26.04	1.64bc
	T4	6.22c	2.74ab	8.95d	69.44	30.56	1.61bc
	T5	6.03c	2.65bc	8.68d	69.50	30.50	1.55c
	CKM	9.54a	–	9.54c	100.00	–	1.00d
	CKS	–	2.88a	2.88e	–	100.00	1.00d
2018	T1	8.78b	1.42c	10.20a	86.06	13.94	1.54a
	T2	8.02c	1.45c	9.47c	84.65	15.35	1.48ab
	T3	7.92c	1.39c	9.32c	85.03	14.97	1.44bc
	T4	6.77d	1.73b	8.50d	79.61	20.39	1.48ab
	T5	5.96e	1.70b	7.66e	77.76	22.24	1.38c
	CKM	9.86a	–	9.86b	100.0	–	1.00d
	CKS	–	2.19a	2.19f	–	100.00	1.00d

Note: The spacing between maize and soybean is 40 cm. T1: Production unit width 2 m, row ratio 2: 2; T2: Production unit width 2.4 m, row ratio 2:3; T3: Production unit width 2.4 m, row ratio 2:4; T4: Production unit width 2.8 m, row ratio 2:3; T5: Production unit width 2.8 m, row ratio 2: 4; CKM: Maize monoculture (even spacing cultivation, row spacing of 70 cm); CKS: Soybean monoculture (even spacing cultivation, row spacing of 50 cm). Lowercase letters indicate a significant difference at the $p < 0.05$ under all treatments

correlation with maize spacing. The order of light transmittance determined by the spacing on the marginal side was found to rank: lower > middle > upper. In contrast, light transmittance of the upper, middle, and lower parts of the canopy on the inner side of the maize row was negatively correlated with spacing, with the order of light transmittance affected by the spacing ranking: middle > lower > upper (Zhang 2014). At a spacing of 40 cm, the upper part of the maize canopy experienced maximum light transmittance (Fig. 8.9). Additionally, the air temperature and CO_2 concentration on both the marginal and inner sides of the maize row showed a decreasing trend followed by an increasing trend with increasing spacing (Tables 8.35 and 8.36).

8.2.3.2 Spacing Between Maize and Soybean Bands

The spacing between maize and soybean strips is defined as the spacing between alternating maize and soybean rows in each production unit in the maize–soybean intercropping model. After conducting a two-year experiment with different maize–soybean spacings while maintaining maize row spacing at 0.4 m, the results shown in Table 8.37 indicate that treatment with a maize–soybean spacing of 0.6 m had the highest land equivalent ratio (LER). As the spacing decreased, LER markedly decreased. These findings suggest that a maize belt with a reasonable row spacing of 0.4 m and an expanded maize–soybean spacing of 0.6 m was beneficial to yield increase in relay intercropping (Yang et al. 2015a, b).

Table 8.34 Maize crop yield and yield components under different configurations of maize row spacings

Year	Treatment	Yield (kg/hm^2)			LER			AYL		
		Maize	Soybean	System	LER*m*	LER*s*	LERt	AYLm	AYLs	AYLt
2012	20:180	5041.26d	1479.27b	6520.53c	0.78d	0.74a	1.52a	0.74	0.35	1.09
	30:170	5185.40d	1449.87b	6635.27c	0.80cd	0.73a	1.53a	0.69	0.39	1.08
	40:160	5721.28c	1401.17b	7122.45a	0.89b	0.70a	1.59a	0.77	0.41	1.18
	50:150	5749.40c	1203.93c	6953.33b	0.89b	0.60b	1.49a	0.7	0.27	0.97
	60:140	6109.20b	1081.31c	7190.51a	0.95a	0.54b	1.49a	0.72	0.21	0.93
	80:120	6308.61ab	816.41d	7125.02a	0.98a	0.41c	1.39b	0.63	0.02	0.65
	100:100	6386.49ab	612.86e	6999.35b	0.99a	0.31d	1.30b	0.98	−0.38	0.59
	MM	6456.11a	–	6456.11d	–	–	–	–	–	–
	MS	–	1992.51a	1992.51e						
2013	20:180	5095.71f	1398.58b	6494.29d	0.81e	0.76a	1.56a	0.79	0.37	1.16
	30:170	5289.11ef	1329.31bc	6618.42c	0.84de	0.72ab	1.56a	0.76	0.36	1.13
	40:160	5568.48de	1346.50b	6914.98a	0.88cd	0.73a	1.61a	0.76	0.45	1.21
	50:150	5767.28cd	1221.44cd	6988.72a	0.91c	0.66bc	1.58a	0.74	0.38	1.12
	60:140	5858.65bcd	1134.84d	6993.49a	0.93bc	0.62c	1.54a	0.69	0.36	1.04
	80:120	6052.89abc	732.27e	6785.16b	0.96ab	0.39d	1.35b	0.6	−0.02	0.58
	100:100	6204.09ab	496.46f	6700.55b	0.98a	0.27e	1.25b	0.97	−0.47	0.5
	MM	6313.07a		6313.07d	–	–	–	–	–	–
	MS		1859.81a	1859.81e						

Note: LERm, LERs, and LERt represent the partial land equivalent ratio of maize, the partial land equivalent ratio of soybean, and the total land equivalent ratio of the system, respectively. AYLm, AYLs, and AYLt represent the actual losses of maize, soybean, and system yield, *MM* monoculture maize, *MS* monoculture soybean. Lowercase letters indicate a significant difference at the $p < 0.05$ under all treatments

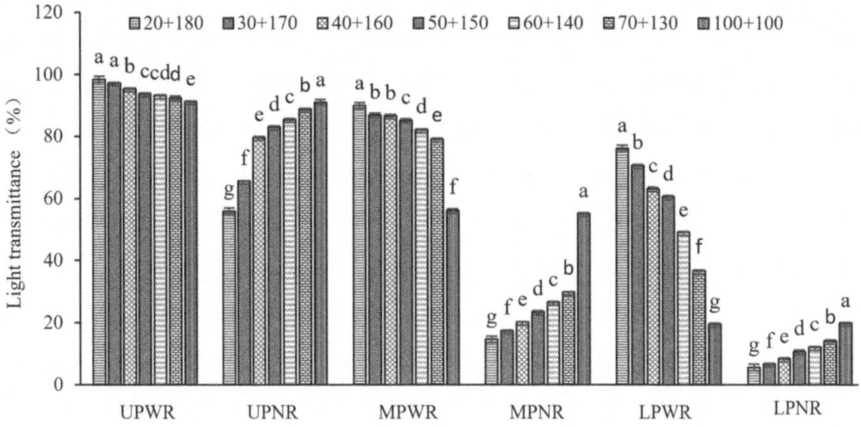

Fig. 8.9 Impact of maize row spacing on light transmittance through the maize canopy. Note: *UPWR* upper part of wide row, *UPNR* upper part of narrow row, *MPWR* middle part of wide row, *MPNR* middle part of narrow row, *LPWR* lower part of wide row, *LPNR* lower part of narrow row. Lowercase letters indicate a significant difference at the $p < 0.05$ under all treatments

Table 8.35 Effect of maize row spacing on maize canopy-air temperature (°C)

	Marginal side			Inside		
Maize + Soybean	Upper	Middle	Lower	Upper	Middle	Lower
20 + 180	33.37Dd	32.77Bb	31.50Cd	30.43Dd	29.33Dd	29.23Dd
30 + 170	32.20Ee	30.83Cc	29.87De	28.87Ee	28.37Ee	28.27Ee
40 + 160	28.80Gg	27.93Dd	27.70Fg	27.40Ff	26.83Ff	27.03Ff
50 + 150	30.37Ff	28.23Dd	28.47Ef	27.73Ff	26.63Ff	26.80Ff
60 + 140	35.20Bb	33.23Bb	32.00Cc	31.53Cc	30.50Cc	30.53Cc
70 + 130	36.33Aa	35.17Aa	34.73Aa	34.50Aa	32.23Bb	32.27Bb
100 + 100	34.07Cc	33.07Bb	33.37Bb	33.83Bb	33.10Aa	33.30Aa

Note: Lowercase and capital letters indicate a significant difference at the $p < 0.05$ and $p < 0.01$ under all treatments, respectively

Table 8.36 Effects of maize row spacing on air CO_2 concentration (ppm) in the maize canopy

	Marginal side			Inside		
Maize + Soybean	Upper	Middle	Lower	Upper	Middle	Lower
20 + 180	333.7Dd	327.7Bb	315.0Cd	304.3Dd	293.3Dd	292.3Dd
30 + 170	322.0Ee	308.3Cc	298.7De	288.7Ee	283.7Ee	282.7Ee
40 + 160	288.0Gg	279.3Dd	277.0Fg	274.0Ff	268.3Ff	270.3Ff
50 + 150	303.7Ff	282.3Dd	284.7Ef	277.3Ff	266.3Ff	268.0Ff
60 + 140	352.0Bb	332.3Bb	320.0Cc	315.3Cc	305.0Cc	305.3Cc
70 + 130	363.3Aa	351.7Aa	347.3Aa	345.0Aa	322.3Bb	322.7Bb
100 + 100	340.7Cc	330.7Bb	333.7Bb	338.3Bb	331.0Aa	333.0Aa

Note: Lowercase and capital letters indicate a significant difference at the $p < 0.05$ and $p < 0.01$ under all treatments, respectively

Table 8.37 Effects of different spacings of maize and soybean on the population land equivalent ratio

	2012			2013		
	Yield (kg/ha)			Yield (kg/hm^2)		
Treatment	Mazie	Soybean	LER	Maize	Soybean	LER
A1	5044d	1472b	1.52	5099c	1396b	1.56
A2	5718c	1399b	1.59	5570ab	1348b	1.61
A3	6107b	1082c	1.49	5855ab	1136c	1.54
A4	6313ab	819d	1.38	6049ab	730d	1.35
MS	–	1993a	–	–	1854a	–
MM	6451a	–	–	6311a	–	–

Note: A1–4 represent maize–soybean spacing treatments, namely 70, 60, 50, and 40 cm, respectively. MS and MM stand for soybean and maize monoculture, respectively. Lowercase letters indicate a significant difference at the $p < 0.05$ under all treatments

Increasing spacing between maize and soybean plants leads to a corresponding decrease in shading of soybeans by the aboveground maize parts. Nonetheless, the interaction between maize and soybean root systems is also affected. Research indicates that decreasing spacing results in a significant decline in N, P, and K accumulation in soybean root systems in strip intercropping. Consistently, in our study, N, P, and K contents in soybean root systems under strip intercropping treatments were considerably lower compared to monoculture (Fig. 8.10). Further, N, P, and K accumulation in maize root systems decreased as growth proceeded, with N steadily and continuously decreasing, P initially increasing and then decreasing, and K decreasing steadily in the late stages of maize growth. Notably, a significant reduction in the accumulation of N, P, and K in maize root systems was observed when maize row spacing was further reduced. This decrease was most prominent in the 70 and 65 cm treatments, unlike soybean root systems, which showed the opposite trend (Lou 2014). Therefore, optimal row spacing is essential in coordinating intra- and inter-species root competition between the two crops to minimize mutual interference that limits nutrient accumulation. Both excessively wide and excessively narrow row spacing are unfavorable for balanced root growth and efficient nutrient uptake and accumulation.

8.3 Reduction of Plant Spacing to Ensure Plant Density

To achieve "equivalent yield of maize and monoculture while increasing the yield of one season soybean" in maize–soybean relay intercropping, it is crucial to reduce the plant spacing of maize and soybean, change from "replacement" to "addition" planting method, maintain a maize population density equivalent to that of local monoculture, and adjust soybean population density based on the length of the period of coexistence of the two crop species, while ensuring a minimum of 70% to a maximum of 100% of monoculture population density for each.

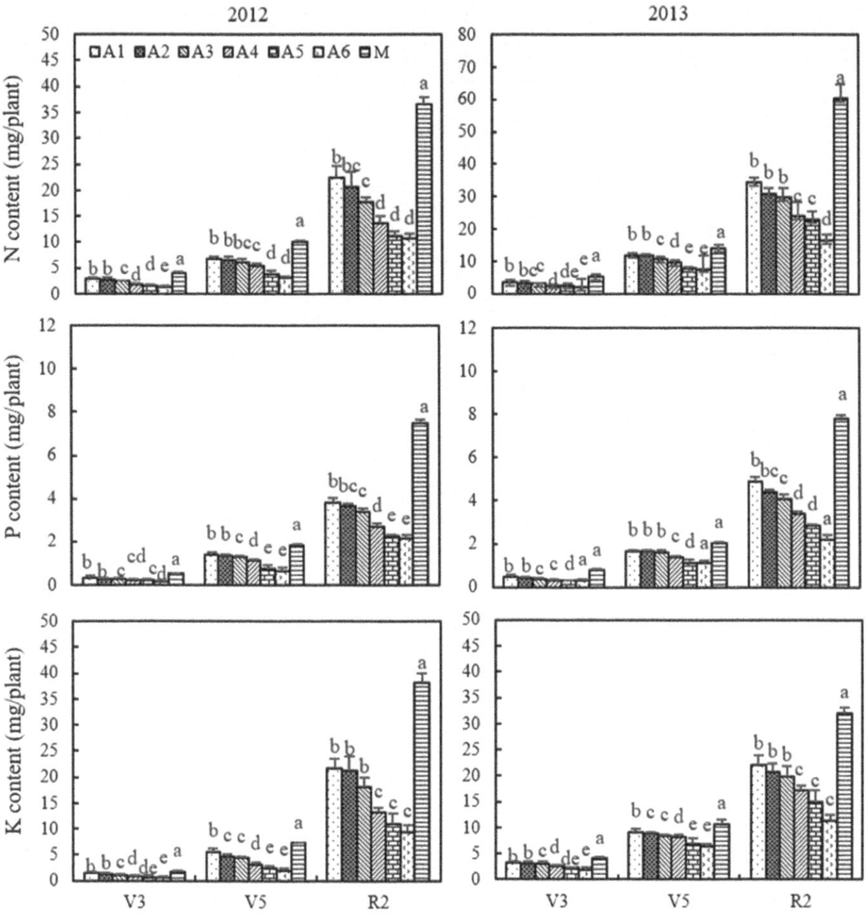

Fig. 8.10 Nutrient content in maize and soybean root systems at different growth stages. Note: Treatments with different maize–soybean spacing were denoted as A1-A6 to designate spacing at 0.7, 0.65, 0.6, 0.55, 0.5, and 0.4 m, respectively. M is the maize monoculture-control treatment at 0.7 m spacing. Lowercase letters indicate a significant difference at the $p < 0.05$ under all treatments

8.3.1 Maize Plant Spacing and Plant Population Density

Relay intercropping can result in a significant increase in maize yield when there is an increase in plant population density (Table 8.38). In Renshou, in South Sichuan, and Lezhi, in Central Sichuan, maize yields of 8.6 and 9.2 t/ha were obtained with maize population density increasing up to 60,000 plants/ha. Consistently, at Pingchang, in northeast Sichuan, maize crops produced 9.2 t/ha at a peak population density of 75,000 plants/ha (Li et al. 2019a, b, c). A field experiment conducted over two consecutive years from 2019 to 2020 compared the yield differences between monoculture and relay intercropping under high population density conditions (Table 8.39). Results revealed that yields of relay intercropping were comparable

Table 8.38 Effects of planting density on yield of relay-intercropping maize at various test sites

Density (Plants / ha)	Renshou				Lezhi				Pingchang			
	Grain number	1000-grain weight (g)	Effective ear number (10^3 ear / ha)	Yield (10^3 kg/ ha)	Grain number	1000-grain weight (g)	Effective ear number (10^3 ear / ha)	Yield (10^3 kg /ha)	grain number	1000-grain weight (g)	Effective ear number (10^3 ear / ha)	Yield (10^3 kg/ ha)
45,000	667.8a	252.9a	44.0c	7.4c	624.1a	278.1a	42.1c	7.3c	598.5a	266.9b	43.0c	6.8c
60,000	608.1b	241.9b	58.2b	8.6a	592.6b	277.5a	55.0b	9.2a	495.8b	271.5a	59.1b	7.9b
75,000	529.4c	240.7b	64.8a	8.3b	486.9c	266.8b	69.8a	8.9b	492.4b	262.6b	71.0a	9.2a

Note: The width of the production unit was 2 m, the population density of maize was 45,000, 60,000, and 75,000 plants/ha; plant spacing was 22.2, 16.7, and 13.3 cm, respectively. Lowercase letters indicate a significant difference at the $p < 0.05$ under all treatments

Table 8.39 Yield differences between monoculture and relay-intercropping maize at high population density

		Effective ear (103 ear/ha)		Grain number per plant		1000-grain weight (g)		Yield (103 kg/ha)	
Year	Density (Plant/ha)	RI	M	RI	M	RI	M	RI	M
2019	60,000	56.91a	55.71a	538.74a	555.39a	289.46b	297.07a	8.81a	9.08a
	75,000	68.44a	69.14a	478.47a	480.61a	284.71a	276.17a	9.09a	9.21a
2020	60,000	57.11a	57.08a	377b	517a	265a	267a	6.18b	7.49a
	75,000	69.28a	61.81b	363b	471a	261a	256a	7.14a	7.75a

Note: *RI* relay intercropping, *M* monoculture. Lowercase letters indicate a significant difference at the $p < 0.05$ under all treatments to those of monoculture maize when maize population density was maintained consistent with that in monoculture (60,000–75,000 plants/ha) (Hu 2021).

The results of intercropping in the Huang-Huai-Hai region (specifically, at the Linying test site in Henan Province) revealed that an increase in maize plant density resulted in a significant increase in maize yield (Table 8.40). The highest yield, i.e., 8.2–9.0 t/ha, was achieved at a population density of 82,500 plants/ha. Interestingly, higher planting density reduced the yield of monoculture maize. Conversely, soybean yield decreased significantly with increasing maize population density, but the highest crop yield and economic value were obtained at the higher population density level. A maize population density of 82,500 plants/ha garnered the highest yield value, leading to an average increase in revenue of 7050.8 yuan/ha more than monoculture. Similarly, in the northwest-Baotou test site in Inner Mongolia (Table 8.41), maize yield significantly increased with increasing population density. Thus, at 90,000 plants/ha, intercropping maize reached a high yield of 16.2–16.9 t/ha. The highest yield value of maize in the northwest region was also observed at a density of 90,000 plants/ha, which resulted in an average increase in revenue of 6718.8 yuan/ha more than monoculture.

8.3.2 Soybean Plant Spacing and Population Density

8.3.2.1 Soybean Row Number and Population Density

The results of population density and row-spacing configuration experiments conducted in Ya'an and Shehong (Table 8.42) revealed that soybean yield was primarily influenced by population density followed by row spacing configuration and the interaction between the two. Soybean yield showed an initial increase followed by a decrease with increasing population density under various row spacing configurations. Thus, high rainfall at the Ya'an experimental site allowed soybean yields to reach 2471.47 kg/ha on average in a relay-intercropping configuration combining a population density of 90,000 plants/ha with rows spaced at 29 cm × 3, while maize monoculture was used. In turn, yields of 2475.89 kg/ha were registered at the Shehong experimental site located in the central hills of Sichuan Province, in an intercropping configuration consisting in a combination of 13.5×10^4 plants/ha and rows spaced at 39 cm × 2 (Zhang 2008).

8.3.2.2 Population Density and Sowing Period

Our findings in two consecutive field experiments (Table 8.43) suggest that timely late sowing had an enhancing effect on the economic coefficient of the soybeans crop. However, excessively late sowing was seemingly detrimental to an increase in soybean yield. Additionally, the results indicated that crop yield peaked at a plant population density threshold beyond which, yield declined. Specifically, our data revealed that sowing soybeans on June 25th at a population density of 14,250 plants/ha will allow a higher crop yield to be achieved (Xu et al. 2014).

Table 8.40 Effects of planting density of maize on crop yield, land equivalent ratio, and crop value in Linying, Henan

Planting pattern	Density (plant/ha)	2021 Maize yield (kg/ha)	2021 Soybean yield (kg/ha)	2021 LER	2021 Output (yuan/ha)	2022 Maize yield (kg/ha)	2022 Soybean yield (kg/ha)	2022 LER	2022 Output (yuan/ha)
Intercropping	52,500	6038.6f	1661.5a	1.39	26,877.2c	6568.2f	1344.9a	1.24	26,460.2c
	67,500	7466.4d	1589.8b	1.40	30,444.4b	7982.6d	1301.4a	1.28	30,159.8b
	82,500	8414.0ab	1480.9c	1.42	32,444.7a	9058.3b	1191.6b	1.31	32,512.9a
Monoculture	52,500	6796.6e	–	–	19,030.5f	7765.9e	–	–	21,744.6e
	67,500	8078.7c	–	–	22,620.4e	8883.7c	–	–	24,874.3d
	82,500	8678.1a	–	–	24,298.6d	9484.8a	–	–	26,557.3c

Note: Lowercase letters indicate a significant difference at the $p < 0.05$ under all treatments

Table 8.41 Effects of maize planting density on crop yield, land equivalent ratio, and crop value in Baotou, Inner Mongolia

		2021				2022			
Planting pattern	Density (plant/ha)	Maize yield (kg/ha)	Soybean yield (kg/ha)	LER	Output (yuan/ha)	Maize yield (kg/ha)	Soybean yield (kg/ha)	LER	Output (yuan/ha)
Intercropping	60,000	12,563.7e	1715.1a	1.28	45,468.8cd	12,135.9d	1891.5a	1.29	45,329.2d
	75,000	15,097.9c	1614.3b	1.30	51,960.0b	13,859.3c	1808.3a	1.30	49,655.9b
	90,000	16,889.9b	1517.5c	1.31	56,396.8a	16,204.3a	1683.5b	1.30	55,473.2a
Monoculture	60,000	14,736.9d	–	–	41,263.4d	13,750.0c	–	–	38,500.0f
	75,000	16,929.7b	–	–	47,403.0c	15,213.3b	–	–	42,597.3e
	90,000	18,163.3a	–	–	50,857.3b	16,991.1a	–	–	47,575.0c

Note: Lowercase letters indicate a significant difference at the $p < 0.05$ under all treatments

Table 8.42 Effects of population density and row spacing on yield of relay-intercropping soybean (kg/ha)

Location		B1	B2	B3	AVG
Yaan	A1	2055.11bcABC	1933.40bcdBCD	1764.85cdeCDE	1917,79aAB
	A2	2093.04abcABC	2471.47aA	1853.84cdeBCD	2139.45aA
	A3	2258.55abAB	2105.15abcABC	1801.37cdeBCD	2055.02aA
	A4	1772.64cdeBCDE	1642.52deCDEF	1476.37efDEF	1630.51bB
	A5	1258.16fgEFG	1126.93fgFG	921.79gG	1102.29cC
	AVG	1887.50aA	1855.89aA	1563.64bB	
Shehong		B1	B2	B3	AVG
	A1	2249.90abABC	2198.9bcABC	1792.73eD	2080.51abA
	A2	1899.25deCD	2420.78abA	2278.25abAB	2199.43abA
	A3	2475.89aA	2374.55abA	1945.67cdeBCD	2265.37aA
	A4	2165.85bcdABC	1966.06cdeBCD	1760.60eD	1964.17bAB
	A5	1913.09deBCD	1705.69eD	1167.89fE	1595.55cB
	AVG	2140.80aA	2133.20aA	1789.03bB	

Note: The main factor is soybean population density with five levels: A1 (45,000 plants/ha), A2 (90,000 plants/ha), A3 (135,000 plants/ha), A4 (180,000 plants/ha), and A5 (225,000 plants/ha); row spacing was used as sub-factor with three levels: soybean row spacing × row number): B1 (39cm × 2), B2 (29cm × 3), and B3 (23cm × 4). Lowercase and capital letters indicate a significant difference at the $p < 0.05$ and $p < 0.01$ under all treatments, respectively

Table 8.43 Effects of plant population density and sowing date on intercropping maize yield

		Economic coefficient					Yield (kg/ha)				
Year	Treatment	B1	B2	B3	B4	Mean	B1	B2	B3	B4	Mean
2012	A1	0.39a	0.35c	0.36b	0.35c	0.36b	1583.41ab	1683.42a	1733.42a	1516.74b	1629.25a
	A2	0.38b	0.38b	0.38b	0.39a	0.38a	1583.42b	1633.42ab	1866.76a	1500.08b	1645.92a
	A3	0.32c	0.36b	0.38a	0.38a	0.36b	1200.06a	1283.40a	1300.07a	1166.73a	1237.57b
	Mean	0.36b	0.36b	0.37a	0.37a		1455.63b	1533.41ab	1633.42a	1394.52b	
2013	A1	0.43a	0.41c	0.42b	0.41c	0.42b	1530.67b	1657.60a	1665.07a	1526.93b	1595.07a
	A2	0.45c	0.45c	0.46b	0.51a	0.47a	1500.80c	1638.93b	1762.13a	1512.00c	1603.47a
	A3	0.40c	0.40c	0.43b	0.47a	0.43b	1356.00bc	1430.40ab	1468.00a	1326.13c	1395.13b
	Mean	0.43c	0.42d	0.44b	0.46a		1462.49c	1575.64b	1631.73a	1455.02c	

Note: Lowercase letters indicate a significant difference at the $p < 0.05$ under all treatments

9 Supporting Technologies to Ensure High Productivity and Sustainability in Maize–Soybean Strip Intercropping

Taiwen Yong, Weiguo Liu, Jing Shang, Xiaoli Chang, Jizhi Yang, and Wenyu Yang

Abstract

The main topic of this chapter is the fertilization techniques for relay strip intercropping of maize and soybean in China. We discuss the use of reduced application fertilization to increase crop yields while improving fertilizer utilization efficiency and reducing environmental pollution. We also investigated the impact of integrated nitrogen application amount reduction on maize and soybean yield, nitrogen absorption and systematic nitrogen uptake in kernels, and root nodulation and aboveground growth. Through field experiments, we established correlations between different indicators and identified optimal fertilizer-application distances and nitrogen fertilizer amounts suitable for maize–soybean intercropping. The findings suggest that proper reduction in nitrogen fertilizer and appropriate fertilizer-application distance can increase the yield of maize and soybeans, as well as improve nitrogen absorption and utilization. These findings provide a theoretical and practical basis for efficient fertilization technology in maize–soybean relay strip-intercropping systems.

Keywords

Reduced fertilizer application · Nitrogen · Phosphorus · Potassium · Co-application

T. Yong (✉) · W. Liu · J. Shang · X. Chang · J. Yang · W. Yang
College of Agronomy, Sichuan Agricultural University, Chengdu, People's Republic of China
e-mail: xl_changkit@sicau.edu.cn; mssiyangwy@sicau.edu.cn

W. Yang et al. (eds.), *Theories and Application in Maize-Soybean Strip Intercropping System*, https://doi.org/10.1007/978-981-97-8723-4_9

9.1 Fertilization Techniques

9.1.1 Integrated Reduced-Application Fertilization Technology

Maize–soybean relay strip-intercropping has become the preferred model for stabilizing maize production and expanding soybean cultivation in China as it allows for increasing yield while reducing fertilizer use—a most prominent advantage. However, efficient fertilization techniques for relay strip intercropping are lacking, and nutrient use efficiency in this cropping system is low, which is not conducive to realizing the annual yield potential of the system. Reduced-application fertilization is a new type of fertilizer-management technology that ensures stable crop yields while improving fertilizer use efficiency and reducing environmental pollution. Given the characteristics of relay maize–soybean strip-intercropping, making full use of the nitrogen (N) fixation capacity of soybeans allows the reduction of N fertilizer input, while ensuring the amount of phosphorus (P) and potassium (K) fertilizers used, thereby reducing the amount of N used by soybeans and ensuring the amount of N required for healthy and vigorous maize development. Hence, integrated fertilization requires consideration of both maize and soybean in the fertilization process, which increases the workload and operation steps relative to monocropping.

9.1.1.1 Determination of the Proportion of N Fertilizer to Apply as Basal and Topdressing Fertilization in Maize

9.1.1.1.1 Effects of N Fertilizer-Application Management on Maize Yield

Nitrogen fertilizer application and basal/topdressing fertilization (these latter at jointing and tasseling stages) significantly influenced the yield of relay strip-intercropped maize; furthermore, a significant interactive effect between the two factors on yield was detected (Table 9.1). Specifically, data from experiments conducted in 2010 and 2011 showed that the 3:2:5 treatment resulted in an average yield increase of 3.15% and 8.54% relative to the 5:0:5 and 5:2:3 treatments, respectively; furthermore, the 5:2:3 treatment resulted in an average yield increase of 4.79% and 7.17%, respectively. Under the same basal/topdressing ratio, the effect of different N-application rates on maize yield followed a consistent pattern in both years, showing an initial increase followed by a decrease with increasing N rate. The application of 180 and 270 kg N hm^{-2} significantly outperformed other N treatments in yield. Particularly, in 2010, the combination of 270 kg N hm^{-2} with the 3:2:5 basal/topdressing ratio resulted in the highest yield, with an 8.64% increase compared to traditional N fertilizer management using 270 kg N hm^{-2} and the 5:0:5 basal/topdressing ratio. However, yield results differed in 2011, in which case, the combination of 180 kg N hm^{-2} with the 3:2:5 basal/topdressing ratio produced the highest yield, significantly differing from other treatments, with a 10.35% increase compared to the combination of 270 kg N hm^{-2} with the 5:0:5 basal/topdressing ratio (Wang et al. 2014).

Table 9.1 Effects of nitrogen fertilizer management on maize yield

Treatment	A1	A2	A3	A4	Mean value
2010					
B1	6632.04b	7336.52a	7097.30ab	7021.65a	7021.88b
B2	6805.39a	7520.69a	7710.67a	6965.26b	7250.50a
B3	6327.92c	6991.08b	7129.36b	6076.00c	6631.09c
Mean value	6588.45c	7282.76a	7312.44a	6687.64b	
F value	A = 57.12**	B = 64.10**	A × B = 6.74**		
2011					
B1	6983.40a	7438.56b	7071.86b	6538.76b	7008.15b
B2	7098.60a	7803.79a	7528.30a	6992.15a	7355.71a
B3	6598.61b	7087.43c	7171.99b	6485.97b	6836.00c
Mean value	6893.54b	7443.26a	7257.38ab	6672.29c	
F value	A = 47.98**	B = 62.25**	A × B = 2.94*		

Note: Data are means of three replicates. Means within columns followed by different lowercase letters are significantly different at the 0.05 level; * indicates a significant difference at $P < 0.05$. ** indicates a significant difference at $P < 0.01$. A1, A2, A3, and A4 represent 90, 180, 270, and 360 kg N hm^{-2}, respectively. B1, B2, and B3 represent the ratios of base fertilizer:vegetative-growth fertilizer:reproductive-growth fertilizer as 5:0:5, 3:2:5, and 5:2:3, respectively

9.1.1.1.2 Effects of Nitrogen Fertilizer Management on Nitrogen Utilization in Maize

The N harvest index of maize–soybean relay strip intercropping was significantly higher than that of maize–sweet potato relay strip intercropping (Table 9.2). In both cropping systems, a similar trend consisting of an initial increase followed by a decrease with increasing N-application rates was observed. Further, both systems reached higher NHI values under the N90 and N180 treatments, with no significant difference among N-application ratios in the maize–soybean relay strip-intercropping system, whose effectiveness ranked in the following order: 3:2:5 > 5:2:3 > 5:0:5. In contrast, in the maize–sweet potato relay strip-intercropping system, the order of effectiveness was 5:0:5 > 5:2:3 > 3:2:5. Further, the N partial factor productivity (PFP) also showed significant differences between the two systems, with that for maize–soybean relay strip intercropping being higher than that for the maize–sweet potato system. Moreover, under low N conditions, maize–soybean relay intercropping exhibited greater advantages. Specifically, as the N-application rate increased, the value for N PFP significantly decreased, but there was no significant difference among N-application ratios. In the maize–soybean relay strip-intercropping system, the value for N PFP ranked in the following order: 3:2:5 > 5:0:5 > 5:2:3, while in the maize–sweet potato relay strip-intercropping system, the ranking order was 5:0:5 > 5:2:3 > 3:2:5. Additionally, N use efficiency (NUE) significantly decreased with increasing application rate, and the influence of N-application ratios on NUE in maize differed significantly. In the maize–soybean relay strip-intercropping system, NUE peaked at 28.2% with the 3:2:5 application ratio. Furthermore, significant interactions were detected among N agronomic efficiency, N PFP, N-application basal/topdressing ratios, and NHI in the maize–sweet

Table 9.2 Effects of nitrogen fertilizer management on maize nitrogen use efficiency and harvest index

Basal/topdressing ratio	N ($kg\ hm^{-2}$)	NHI (%)		N PFP ($kg\ kg^{-1}$)		N fertilizer agronomic efficiency ($kg\ kg^{-1}$)		NUE (%)	
		Maize/soybean	Maize/sweet potato	Maize/soybean	Maize/sweet potato	Maize/soybean	Maize/sweet potato	Maize/soybean	Maize/sweet potato
5:0:5	90	58.0a	62.8a	73.69a	60.01a	11.27a	13.20a	18.3a	27.9a
	180	60.5a	56.8b	40.76b	36.08b	9.55b	12.68a	12.5b	25.8a
	270	57.4a	58.1b	26.29c	22.88c	5.48c	7.28b	18.8a	20.5b
	360	59.1a	52.7c	19.50d	14.62d	3.90d	2.92c	12.1b	15.9c
	Mean value*	58.7a	57.6a	40.06a	33.40a	7.55b	9.02a	15.4c	22.6a
3:2:5	90	60.0a	54.1ab	75.62a	57.87a	13.19a	11.06a	39.8a	20.6a
	180	61.6a	57.0a	41.78b	31.70b	10.57b	8.30b	24.2b	18.0b
	270	59.5a	55.9a	28.56c	21.35c	7.75c	5.75c	26.3b	16.5b
	360	57.5b	52.3b	19.35d	14.83d	3.74d	3.13d	22.4b	14.0c
	Mean value*	59.6a	54.8b	41.33a	31.44a	8.81a	7.06b	28.2a	17.3b
5:2:3	90	59.0a	59.9a	70.31a	61.10a	7.89a	14.30a	26.3a	31.3a
	180	60.9a	57.0b	38.84b	33.54b	7.63a	10.14b	22.9b	23.0b
	270	59.6a	52.7c	26.41c	21.11c	5.60b	5.51c	17.3c	18.3c
	360	55.8b	52.9c	16.88d	14.76d	1.27c	3.06d	14.0d	12.6d
	Mean value*	58.9a	55.6b	38.11a	32.63a	5.60c	8.25ab	20.1b	21.3a

Average fertilization amount	0	50.4b	48.1d	–	–	–	–	–	–
	90	59.0a	58.9a	73.21a	59.66a	10.78a	12.85a	28.1a	26.6a
	180	61.0a	56.9ab	40.46b	33.78b	9.25a	10.37b	19.9b	22.3b
	270	58.8a	55.6b	27.08c	21.78c	6.28b	6.18c	20.8b	18.4c
	360	57.5ab	52.6c	18.58d	14.74a	2.97c	3.04d	16.2c	14.2d
F value	A	4.521*	9.874*	14.365*	7.648*	8.345*	21.654*	18.362*	31.521*
	B	5.647*	7.638*	4.687	3.278	9.287*	9.321*	14.605*	5.347*
	A × B	1.321	3.456*	2.417	3.109	4.805*	2.654	3.214	7.634*

Note: different lower case letters indicate significant differences among N application at levels of $P<0.05$; * indicate that the difference is significant at levels of $P < 0.05$

potato relay strip-intercropping system. The NHI value for maize decreased significantly with increasing N-application rate in both the N after-migration and balanced basal application treatments, while NHI peaked in the N180 treatment using the 3:2:5 application ratio. These findings indicate that NUE improved in maize–soybean relay strip-intercropping, while it increased in maize–sweet potato relay strip-intercropping due to the lower initial soil-fertility levels (Wang et al. 2014).

9.1.1.2 Determination of Integrated N-Application Amount for Reduction of Fertilizer Application

Through a two-year field experiment, plant dry matter accumulation and translocation, grain-filling characteristics in corn and soybeans, N absorption and utilization, corn, soybean, and system yields, as well as the effects on soybean root growth and N-fixation ability in a maize–soybean relay strip-intercropping system, were all studied under three N-application levels, i.e., 36% N reduction (RN1), 18% N reduction (RN2), and conventional N application (CN), and four fertilizer-application distances from the maize row, i.e., 0 cm (D1), 15 cm (D2), 30 cm (D3), and 45 cm (D4). Correlations for each indicator were determined, and the optimal fertilizer-application distance and N fertilizer amount suitable for maize–soybean integrated N-reduction technology and the mixed population production level for efficient relay strip intercropping were identified. This study provided a sound theoretical and practical basis for an efficient fertilization technology in maize–soybean relay strip-intercropping systems.

9.1.1.2.1 Effects on Maize and Soybean Yield

Proper reduction in N fertilizer and appropriate fertilizer-application distance can increase maize and soybean yields (Table 9.3). The highest yields for both maize and soybean were observed under RN2 treatment, with average increases of 4.95% and 7.74% compared to CN treatment over the two-year experimental period. Fertilizer-application distances that were excessively close or far from the maize rows were both unfavorable for increasing the yield of either crop. Under RN2 treatment, maize yield was the highest in the D2 treatment, with a 14.93% average increase compared to the D1 treatment over the 2 years. In turn, soybean yield was the highest in either the D2 or D3 treatment. Specifically, yields in D3 and D2 were the highest in 2012 and in 2013, respectively, with increases of 13.77% and 18.50%, respectively, compared to that recorded in D4. Further analysis of the total annual yield of maize and soybean showed that the highest overall yield was recorded in RN2, with an average increase of 5.47% compared to CN. Additionally, among the fertilizer-application distances tested, the 15–30 cm range from the maize rows (D2 and D3) registered higher yields, with D2 showing a 13.37% average increase compared to D1, and an 11.61% increase compared to D4 over the 2 years (Chen et al. 2016).

Table 9.3 Effects of nitrogen-application rate and fertilization distance on maize and soybean (Renshou) yield

		Maize yield (kg hm^{-2})				Soybean yield (kg hm^{-2})				System yield (kg hm^{-2})			
Year	Treatment	RN1	RN2	CN	Mean	RN1	RN2	CN	Mean	RN1	RN2	CN	Mean
2012	D1	7136.7a	7545.9b	7269.2a	7317.3bc	1277.0b	1489.7b	1303.2b	1356.6b	8413.8a	9035.6bc	8572.4b	8673.9b
	D2	7436.5a	8881.2a	7841.3a	8053.0a	1294.5b	1665.4ab	1785.0a	1581.6a	8731.0a	10546.5a	9626.3a	9634.6a
	D3	7649.0a	7789.3b	7537.4a	7658.6ab	1483.6ab	1887.0a	1713.6a	1694.7a	9132.6a	9676.4ab	9251.1ab	9353.3a
	D4	7003.8a	7046.6b	7233.9a	7094.8c	1590.7a	1658.6ab	1701.1a	1650.1a	8594.5a	8705.2c	8935.0ab	8744.9b
	Mean	7306.5b	7815.7a	7470.4ab		1411.4b	1675.2a	1625.7a		8718.0b	9490.9a	9096.2ab	
2013	D1	7474.4ab	7953.9b	7707.5b	7711.9b	1559.8b	1670.5d	1629.6b	1620.0c	9034.2c	9624.4b	9337.1b	9331.9b
	D2	8096.9a	8932.4a	8492.0a	8507.1a	1824.3a	2664.9a	2327.0a	2272.0a	9921.2ab	11597.3a	10819.1a	10779.2a
	D3	8008.6ab	8824.5a	8084.9ab	8306.0a	2037.5a	2364.2b	2226.8a	2209.5a	10046.0a	11188.7a	10311.7a	10515.5a
	D4	7264.7b	7936.0b	7683.3b	7628.0b	1930.1a	2110.8c	1711.3b	1917.4b	9194.8bc	10046.8b	9394.6b	9545.4b
	Mean	7711.2b	8411.7a	7991.9b		1837.9c	2202.6a	1973.7b		9549.1c	10614.3a	9965.6b	

Reprinted from Chen et al. (2016)

RN1: 36% N reduction, RN2: 18% N reduction, CN: conventional N input, D1: 0 cm, D2: 15 cm, D3: 30 cm, D4: 45 cm; different lower case letters indicate significant differences among four fertilizer-application distances at levels of $P<0.05$.

9.1.1.2.2 Effects on N Absorption and Systematic N Uptake by Maize and Soybean Kernels

Appropriate N fertilizer-application rates and fertilizer-application distances are beneficial for increasing N uptake by maize and soybean grains and total N accumulation of the aboveground plants in the maize–soybean relay strip-intercropping system (Table 9.4). Grain N uptake by maize and soybean plants, as well as that by the overall strip-intercropping system, was highest under RN2, with average increases of 4.65%, 6.91%, and 5.10%, respectively, compared to CN over the two-year period. Further, N application excessively close or far from maize rows was unfavorable for N uptake by maize or soybean plants, with the optimal distance being 15–30 cm from the maize rows. Specifically, maize grain N-uptake and total aboveground plant N uptake in maize–soybean relay strip intercropping reached a maximum under D2, with average increases of 8.24% and 16.59%, respectively, compared to conventional planting under D1, over the 2 years of study. In turn, soybean grain N uptake peaked under D3 in 2012 and under D2 in 2013, with increases of 2.64% and 4.95%, respectively, compared to D4. The optimal fertilizer-application distance for N uptake varied with crop and N-application rate. Thus, for maize, grain N uptake was generally the highest under D2 for different N rates, with a 14.92% increase under RN2, relative to the same N level in D1, over the 2 years. Meanwhile, for soybeans and the maize–soybean relay strip-intercropping system, grain N uptake was higher in the RN1/D3 combination treatment, while in RN2 and CN levels, the highest N uptake was registered in D2. On average, soybean grain and overall aboveground plant N uptake in the maize–soybean relay strip-intercropping system in RN2 were 12.04% and 18.35% higher, respectively, under D2 than under D4 over the two-year duration period of the study (Chen et al. 2016).

9.1.1.2.3 Effect of Reduced Integrated Nitrogen Application on Root Nodulation and Aboveground Growth

Effects on Sucrose Accumulation in the Roots

From flowering to grain filling, sucrose content in soybean nodules of the relay strip-intercropping system was higher than that in monocultured soybeans, with the highest content observed during the pod-filling stage, followed by that observed in the flowering stage, and the lowest during grain filling. Moderate N application was beneficial for sucrose accumulation in soybean nodules during the flowering stage but excess N fertilization caused a reduction in sucrose content in the nodules during grain filling (Fig. 9.1). However, differences between planting patterns were detected, which were associated with planting-pattern effects on aboveground growth and carbon fixation differences.

Effects on the Number of Nodules and N Fixation Capacity

The number of soybean nodules per plant showed a continuous increasing trend as plant growth proceeded (except for monocropping soybeans in 2019), with the highest mean number (233 nodules plant^{-1}) observed during grain filling (Fig. 9.2). Except for the fifth-leaf stage in 2020, the number of nodules in

Table 9.4 Effect of nitrogen-application rate and fertilization distance on maize and soybean nitrogen uptake

Year	Treatment	Total aboveground N-uptake in the maize-soybean relay strip-intercropping system (kg hm^{-2})				Grain N uptake (kg hm^{-2})							
						Maize				Soybean			
		RN1	RN2	CN	Mean	RN1	RN2	CN	Mean	RN1	RN2	CN	Mean
2012	D1	264.05a	292.88b	269.43c	275.45b	95.73a	99.98b	97.10a	97.60bc	98.19b	115.90b	100.77b	104.95b
	D2	268.01a	324.92a	315.84a	302.92a	97.88a	117.44a	103.71a	106.34a	99.91b	129.66ab	138.39a	122.65a
	D3	281.43a	320.57a	303.77ab	301.92a	99.28a	102.56b	98.67a	100.17ab	114.87ab	147.34a	133.29a	131.83a
	D4	274.69a	274.34b	278.45bc	275.83b	88.81a	91.45b	94.66a	91.64c	123.39a	129.5ab	132.44a	128.44a
	Mean	272.04b	303.18a	291.87a		95.43b	102.86a	98.54ab		109.09b	130.6a	126.22a	
2013	D1	295.55b	308.59d	311.57b	305.24d	111.52a	113.84b	113.64ab	113.00bc	112.60b	116.80c	118.00b	115.80c
	D2	339.87a	408.26a	374.14a	374.09a	115.61a	128.28a	120.96a	121.62a	138.10a	185.50a	165.80a	163.13a
	D3	339.91a	383.44b	362.43a	361.93b	112.55a	126.04a	114.63ab	117.74ab	141.60a	165.90b	158.80a	155.43a
	D4	322.40a	345.13c	313.49b	327.01c	107.3a	111.96b	108.58b	109.28c	136.90a	151.80b	121.10b	136.60b
	Mean	324.43c	361.36a	340.41b		111.75b	120.03a	114.45b		132.30c	155.00a	140.93b	

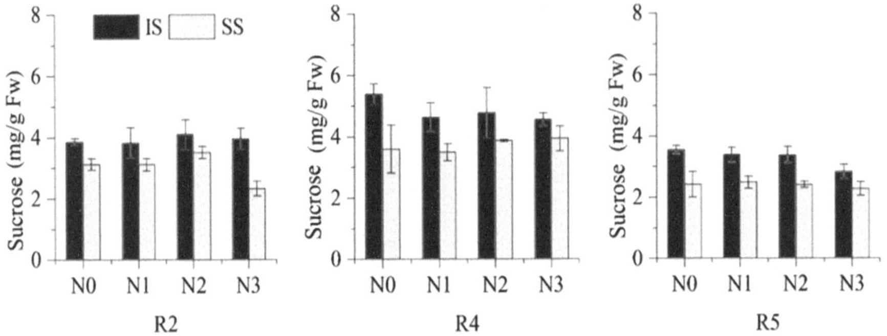

Fig. 9.1 Effects of nitrogen-application rates and planting patterns on soybean nodule sucrose content. N0: no N application, N1: reduced N application, N2: conventional N application, N3: increased N application, IS: relay strip-intercropping soybean, SS: soybean monoculture

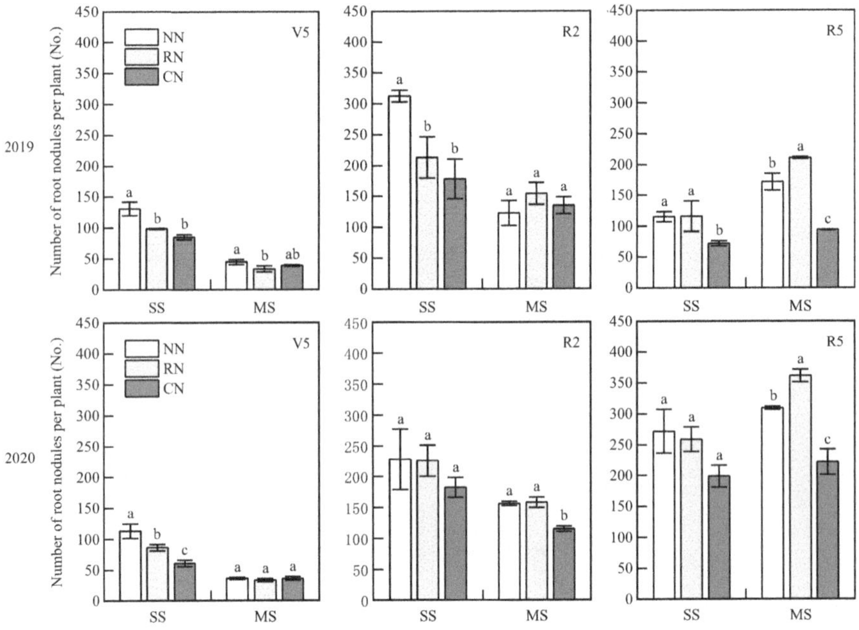

Fig. 9.2 Effects of nitrogen-application level and planting system on the number of root nodules in soybean plants. MS: relay strip-intercropping soybean, SS: monocropping soybean. NN: no N application, RN: reduced N application, CN: conventional N application. (Reprinted from Peng et al. 2022)

intercropped soybeans was significantly lower than that in monocropping soybeans during flowering and earlier stages. However, during grain filling—on average—the number of nodules in relay strip-intercropping soybeans was significantly 39.9% higher than that in monocropping soybeans. The influence of the N-application rate on the number of nodules was significant; specifically, under the monocropping system, it showed a decreasing trend with an increasing N-application rate. However, for relay strip-intercropping soybeans, the number showed an increasing trend followed by a decreasing trend after flowering, as the N rate increased. The differences were significant during grain filling, with N-reduction treatment favoring significantly higher numbers of nodules than those recorded when no N was applied or under conventional N application. In 2019, the number of nodules per plant under N-reduction treatment was 22.7% and 125.2% higher than under no N and conventional N application, respectively. Similarly, in 2020, the number of nodules under N-reduction treatment was 16.8% and 62.9% higher than that under no N and conventional N application, respectively (Peng et al. 2022).

The results shown in Table 9.5 indicate that adjusting fertilizer-application distance appropriately can significantly increase the number of soybean root nodules, nodule dry weight, and root dry weight, compared to conventional planting with fertilization in holes (A5). The number of nodules, nodule dry weight, and root dry weight during V5, R2, and R5 growth stages all initially increased and then decreased as distance from the soybean rows increased. Furthermore, the values of these parameters were higher across treatments than in CK plots with no fertilizer applied. Additionally, the highest values were observed when the distance from the maize rows was set at 15 cm. This finding indicates that appropriate adjustment of the fertilizer-application distance from the soybean rows effectively caused the number of soybean nodules, nodule dry weight, and root dry weight to increase (Yong et al. 2013).

Nitrogenase activity of soybean nodules increased and then decreased during different growth stages. The highest activity rate was observed during the R2 stage, while during R5, nitrogenase activity was close to zero in all treatments, and the differences among treatments were not significant (Fig. 9.3). At stages V5 and R2, there were significant differences in nitrogenase activity of the integrated nitrogen-application treatment compared to conventional planting with N fertilization in holes and the unfertilized controls. Specifically, during V5, the highest nitrogenase activity was observed in A4-treated plants, which was closest to the soybean rows. In A4, nitrogenase activity was 39.1% higher than that recorded in A5 and 122% higher than that recorded in CK. Furthermore, at R2, the highest nitrogenase activity was observed under A3, which was at the midpoint between maize and soybean rows. In this case, nitrogenase activity was 80.63% higher than that recorded in A5, and 130% higher than that observed in CK plants. Further, the analysis of soybean nodule N-fixation potential revealed that the highest potential was observed during the R2 stage (Fig. 9.4). In contrast, in terms of N-application methods, the integrated nitrogen-application treatment at a distance of 15–30 cm from the maize rows was significantly more efficient than conventional planting with nitrogen fertilization in holes or the unfertilized treatment. Specifically, during the V5 stage, the highest N-fixation potential was observed in the A2 treatment, while during the R2 and R5 stages, the highest N-fixation potential was observed in the A3 treatment, in which case it was 29.26% and 10.2% higher compared to that recorded in the A5 treatment, and 30.92% and 28.57% higher than that of the CK treatment, respectively. These findings indicate that an optimized integrated N application based on fertilization distance was beneficial in promoting N-fixation potential in soybean root nodules.

Table 9.5 Effect of the distance of fertilizer application on soybean nodule dry weight

Treatment	Number of nodules (A/plant)			Nodule dry matter (g/plant)			Root dry matter (g/plant)		
	V5	R2	R5	V5	R2	R5	V5	R2	R5
A1	7bc	48c	86c	0.045b	0.086b	0.240c	3.26c	9.07cd	10.45c
A2	14a	109a	119a	0.082a	0.226a	0.554a	4.64a	13.63a	17.92a
A3	15a	89b	123a	0.072a	0.247a	0.519a	3.64b	12.47b	15.77ab
A4	10b	56c	105b	0.047b	0.103b	0.502ab	3.26c	9.97c	11.61bc
A5	10b	55c	107b	0.059ab	0.171ab	0.462b	3.06c	9.17cd	9.80c
CK	6c	45c	77d	0.040b	0.081b	0.222c	2.67d	8.56d	9.36c

Reprinted from Yong et al. (2013)

A1–A4 represent the distance between maize and soybean-integrated fertilization from the maize row, 0 cm (A1), 15 cm (A2), respectively, 30 cm (A3) and 45 cm (A4); soybean N-fertilizer hole application (A5); and no N applied control (CK). Different lower case letters indicate significant differences among five treatments at levels of $P<0.05$.

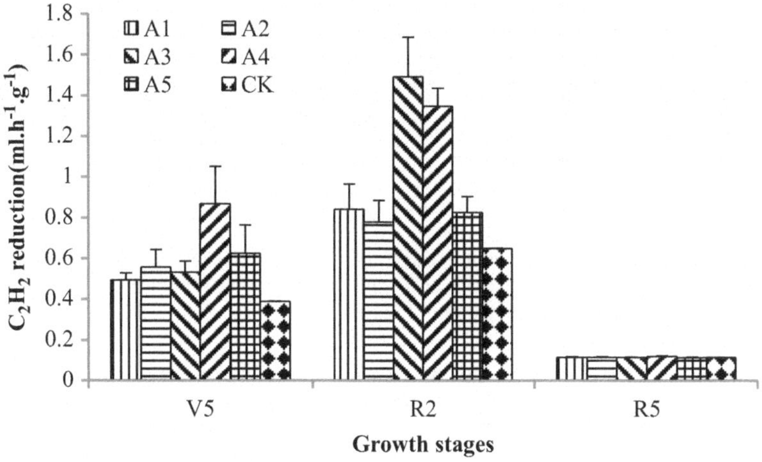

Fig. 9.3 Effect of fertilization distance on nitrogenase activity in soybean root nodules. A1–A4 represent the distance between maize and soybean integrated fertilization from the maize row, 0 cm (A1), 15 cm (A2), respectively; 30 cm (A3) and 45 cm (A4); soybean N fertilizer-hole application (A5); and no N applied control (CK). (Reprinted from Yong et al. 2013)

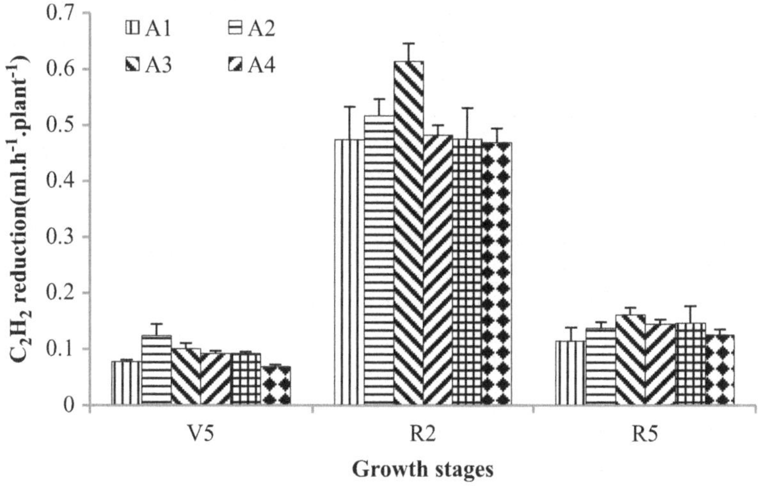

Fig. 9.4 Effect of distance of fertilization on nitrogen-fixation potential of soybean root nodules. A1–A4 represent the distance between maize and soybean integrated fertilization from the maize row, 0 cm (A1), 15 cm (A2), respectively, 30 cm (A3) and 45 cm (A4); soybean N fertilizer-hole application (A5); and no N applied control plants (CK). (Reprinted from Yong et al. 2013)

Effects on Photosynthetic Capacity and Biomass Accumulation in the Aboveground Plant Body

The amplitude of fluctuation of SPAD values in crop leaves was smaller under relay strip intercropping than that observed under monocropping, although it showed a

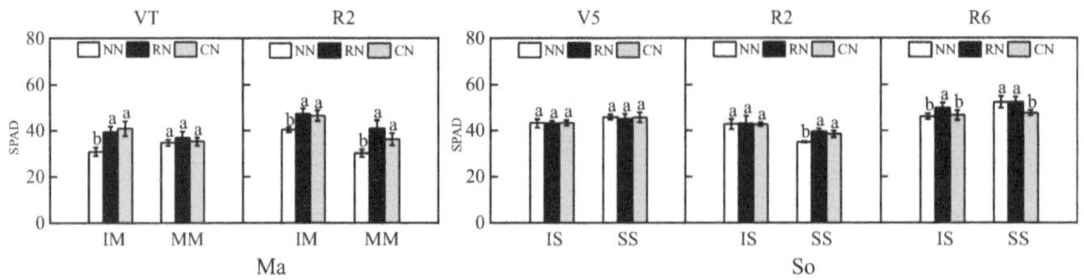

Fig. 9.5 Leaf SPAD values at different growth stages in maize and soybean. MM: monocropping maize, IM: relay strip-intercropping maize. NN: no N application, RN: reduced N application, CN: conventional N application. (Reprinted from Li et al. 2022)

stable upward trend (Fig. 9.5). Further, during the maize grain-filling stage, SPAD values in leaves in the relay strip-intercropping system were 34.52% higher than those recorded for leaves in monoculture maize. In turn, during the soybean flowering and grain-filling stages, SPAD values measured in relay strip-intercropping soybean leaves were 10.39% and 29.48% higher, respectively, than those of monocropping soybean. Additionally, N application increased leaf SPAD values, with the highest values observed under reduced N application. During grain filling, SPAD values of relay strip-intercropping maize leaves under reduced N application were 17.46% higher than those of relay strip-intercropping maize leaves without N application. Meanwhile, SPAD values of monocropping maize leaves under reduced N application were 35.02% higher than those without N application. Consistently, during soybean grain filling, SPAD values of relay strip-intercropping soybean leaves under reduced N application were 7.71% and 6.67% higher than those of relay strip-intercropping soybean without N application and relay strip-intercropping soybean with conventional N application, respectively.

During VT-R2, the grain net yield of relay strip-intercropping maize was 1.16 times that of monocropping maize. Although relay strip-intercropping maize starts accumulating dry matter in the grains later than monoculture maize, its growth rate accelerates during the grain-filling stage. Additionally, the accumulation of dry matter in relay strip-intercropping maize grains varied with N-application level, showing an initial increase followed by a decrease. In contrast, monoculture maize showed a gradual continuous increase with increasing N application. Among all treatments, relay strip-intercropping maize under reduced N application (RN) exhibited the highest grain dry matter accumulation during the R2 stage (Fig. 9.6). Further, dry matter accumulation in relay strip-intercropping maize grain was 18.52% and 8.7% higher under no N application (NN) and reduced N application, respectively, than that recorded in monoculture maize grain. In turn, the accumulation of dry matter in soybean grains decreased with increasing N-application rate. Thus, relay strip-intercropping soybeans under conventional N application (CN) showed 23.15% lower dry matter accumulation than that corresponding to relay strip-intercropping soybeans under reduced N application. Moreover, monocropping soybeans under reduced N application showed 97.62% lower dry matter accumulation than that shown by unfertilized monocropping

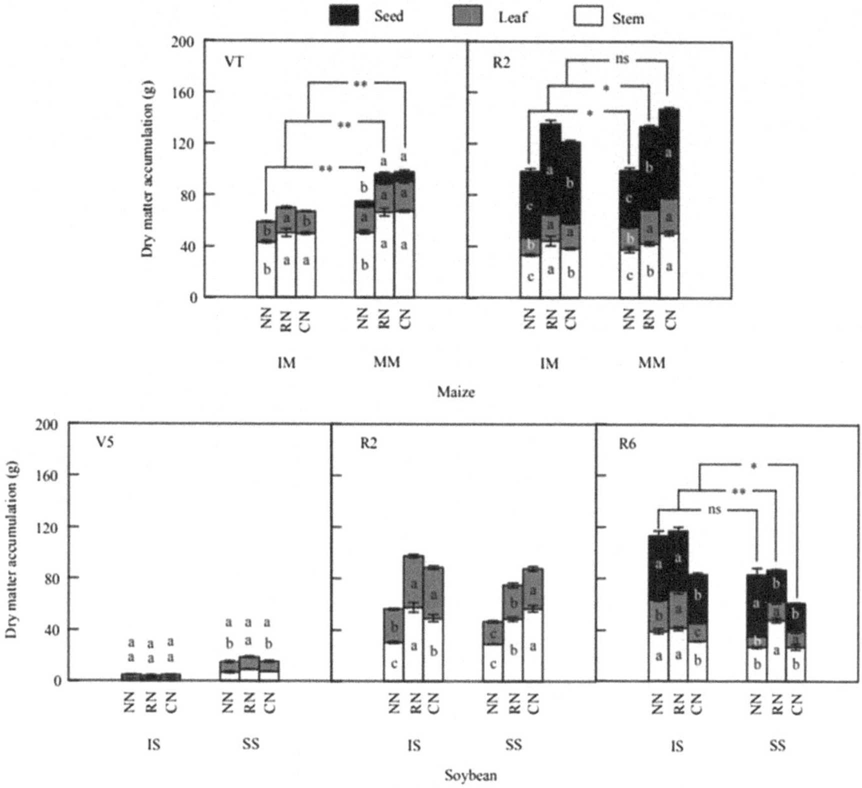

Fig. 9.6 Effects of nitrogen fertilization on aerial dry matter accumulation in maize and soybean in two planting systems. IS: relay strip-intercropping soybean; SS: monocropping soybean; NN: no N applied; RN: reduced N application; CN: conventional N application. (Reprinted from Li et al. 2022)

soybeans. Lastly, relay strip-intercropping soybeans showed 89.66% and 74.15% higher grain dry matter accumulation under reduced N and conventional N applications, respectively, than monocropping soybeans (Li et al. 2022).

9.1.1.3 Effects of Reduced Integrated Nitrogen Fertilization

9.1.1.3.1 Economic Coefficient of Maize and Soybeans

Reducing N fertilizer application and implementing appropriate fertilization distances effectively improved the economic coefficient (HI) of maize, soybeans, and the maize–soybean relay strip-intercropping system (Table 9.6). The HI value for maize varied between 0.49 and 0.55 among treatments, among which RN2D2 showed the highest values, namely, 0.55 and 0.53 in 2012 and 2013, respectively. Further, the HI recorded for RN2 was significantly higher than that for RN1 or CN. Additionally, D2 showed a significantly higher value for HI compared to D1 or

Table 9.6 Effects of nitrogen fertilizer amount and distance of application on the economic coefficients of maize, soybeans, and the maize–soybean strip-intercropping system in Renshou county, Sichuan province

		Maize				Soybean				Maize–soybean strip-intercropping system			
Year	Treatment	RN1	RN2	CN	Mean	RN1	RN2	CN	Mean	RN1	RN2	CN	Mean
2012	D1	0.50ab	0.51b	0.50a	0.50b	0.46b	0.49b	0.49b	0.48b	0.50ab	0.51c	0.49b	0.50c
	D2	0.51ab	0.55a	0.52a	0.53a	0.47b	0.53a	0.52a	0.51a	0.51a	0.55a	0.52a	0.53a
	D3	0.52a	0.53ab	0.52a	0.52a	0.47b	0.53a	0.51ab	0.50a	0.51a	0.53b	0.52a	0.52b
	D4	0.49b	0.50b	0.51a	0.50b	0.50a	0.50b	0.50b	0.50a	0.49b	0.50c	0.51ab	0.50c
	Mean	0.50b	0.52a	0.51ab		0.47b	0.51a	0.51a		0.50b	0.52a	0.51b	
2013	D1	0.50a	0.51b	0.50b	0.50b	0.52a	0.52a	0.49b	0.51bc	0.50a	0.51ab	0.50b	0.50b
	D2	0.50a	0.53a	0.52a	0.52a	0.54a	0.53a	0.52a	0.53a	0.51a	0.54a	0.52a	0.52a
	D3	0.49a	0.53a	0.51ab	0.51ab	0.53a	0.53a	0.51ab	0.52ab	0.50a	0.53ab	0.51ab	0.51ab
	D4	0.49a	0.51b	0.50b	0.50b	0.52a	0.51a	0.49b	0.51c	0.50a	0.51b	0.50ab	0.50b
	Mean	0.50c	0.52a	0.51b		0.53a	0.52a	0.50b		0.50b	0.52a	0.51b	

Reprinted from Chen et al. (2016)

RN1: 36% N reduction, RN2: 18% N reduction, CN: conventional N input, D1: 0 cm, D2: 15 cm, D3: 30 cm, and D4: 45 cm. Different lower case letters indicate significant differences among four treatments at levels of $P<0.05$.

D4 under RN2. Similarly, HI values for soybeans ranged from 0.47 to 0.54 among treatments, with RN2D2 having the highest value in 2012, while RN1D2 had the highest HI in 2013. Lastly, the economic coefficients of the maize–soybean relay strip-intercropping system varied between 0.49 and 0.55, with RN2D2 having the highest value in both 2012 and 2013, reaching 0.55 and 0.54, respectively. Thus, the total economic coefficient of the RN2 treatment was significantly higher than that of the RN1 or CN treatments. In turn, D2 showed a significantly higher total economic coefficient compared to D1 or D4 treatment.

9.1.1.3.2 Nitrogen Uptake and Use Efficiency in Intercropped Maize and Soybeans

Compared to conventional N-application treatments, reducing N application significantly improved N uptake and NUE in maize, soybeans, and the maize–soybean intercropping system (Table 9.7). Maize N uptake efficiency was 47.85% and 35.13% higher in RN1 and RN2, respectively, relative to CN treatment. Additionally, the N uptake efficiencies of soybean and the maize–soybean intercropping system were both highest in the RN2, under which treatment they were 55.37% and 43.62% higher, respectively, than in the CN treatment. Both excess or insufficient fertilizer-application distances had a negative impact on N uptake efficiency in both crops. The optimal distance was 15–30 cm from the maize row. In the RN2 treatment, maize and maize–soybean intercropping system N-uptake efficiencies were the highest in the D2 treatment, under which treatment it was 115.90% higher than under conventional planting (D1) treatment. Meanwhile, soybean N-uptake efficiency was the highest in the D3 treatment in 2012; specifically, 61.50% higher than in the D4 treatment. In turn, in 2013, the highest efficiency was observed in treatment D2, in which case it was 70.06% higher than in D4 (Chen et al. 2016).

9.1.1.3.3 Soil Nitrogen Contribution Rate in Maize and Soybeans

The effect of N-application rate on soil N contribution rate was significantly different for soybean and the maize–soybean intercropping system but not for maize (Table 9.8). Under all three N levels tested, soil N contribution rate in maize, soybean, and the maize–soybean intercropping system was the lowest for the RN2 treatment, which concomitantly was characterized by the highest plant N uptake, resulting in a decrease of 2.62%, 6.25%, and 4.52%, respectively, compared to the two-year average for the CN treatment. The trend of soil N contribution rate for each crop was opposite to that of plant N uptake, with the lowest values observed for treatments D2 and D3, regardless of the N fertilization level. Under the RN2 treatment, the soil nitrogen contribution rate for maize was the lowest in D2, resulting in a two-year average decrease of 11.32%, compared to D1.

The soil N contribution rate in soybean plants was the lowest for D3, with a two-year average decrease of 11.17%, compared to D4. Finally, for the maize–soybean intercropping system, the soil N contribution rate was the lowest for D2, resulting in a two-year average decrease of 15.60%, compared to D4. These findings indicate that reducing N-application rate and optimizing fertilization distance not

Table 9.7 Effect of N-application rate and fertilizer distance on N uptake efficiency in maize and soybeans in Renshou county, Sichuan province

Year	Treatment	Maize				Soybean				Maize–soybean strip-intercropping system			
		RN1	RN2	CN	Mean	RN1	RN2	CN	Mean	RN1	RN2	CN	Mean
2012	D1	33.32a	29.38ab	21.90a	28.20ab	1.42b	8.33b	1.84b	3.86c	34.75a	37.71bc	23.75b	32.07b
	D2	33.36a	35.78a	24.47a	31.20a	3.27b	13.79b	13.34a	10.14b	36.64a	49.58a	37.81a	41.34a
	D3	31.84ab	26.89b	21.74a	26.83b	11.18a	21.06a	12.41a	14.88a	43.03a	47.96ab	34.15ab	41.72a
	D4	26.20b	17.80c	14.87b	19.62c	13.62a	13.04b	11.60b	12.75ab	39.82a	30.84c	26.48b	32.38b
	Mean	31.18a	27.46b	20.74c		7.37b	14.05a	9.80b		38.56a	41.52a	30.55b	
2013	D1	13.94b	12.23b	10.61a	12.26b	4.35b	6.83d	5.88b	5.69c	18.28c	19.05d	16.49b	17.94c
	D2	20.09a	21.04a	13.77a	18.3a	19.3a	34.93a	21.68a	25.31a	39.39a	55.96a	35.45a	43.6a
	D3	18.84a	19.92a	12.36a	17.04a	20.58a	26.86b	19.55a	22.33a	39.41a	46.77b	31.9a	39.36a
	D4	13.98b	12.05b	9.85a	11.96b	17.09a	20.54c	7.23b	14.95b	31.07b	32.58c	17.07b	26.91b
	Mean	16.71a	16.31a	11.65b		15.33b	22.29a	13.59b		32.04b	38.59a	25.23c	

Reprinted from Chen et al. (2016)

RN1: 36% N reduction, RN2: 18% N reduction, CN: conventional N input, D1: 0 cm, D2: 15 cm, D3: 30 cm, and D4: 45 cm. Different lower case letters indicate significant differences among four treatments at levels of $P<0.05$.

Table 9.8 Effect of nitrogen level and fertilization distance on maize and soybean soil-nitrogen contribution rate

Year	Treatment	Maize				Soybean				Maize–soybean strip-intercropping system			
		RN1	RN2	CN	Mean	RN1	RN2	CN	Mean	RN1	RN2	CN	Mean
2012	D1	56.96a	54.32bc	56.16b	55.81b	97.10a	81.41a	94.17a	90.89a	72.33a	65.38ab	70.93a	69.55[a]
	D2	57.23a	48.98c	53.38b	53.19b	93.42a	72.89ab	69.13b	78.48b	71.41a	58.92c	60.50c	63.61b
	D3	58.18a	56.05b	56.3b	56.84b	81.96b	63.55b	71.22b	72.24c	68.29a	59.61bc	63.04bc	63.65b
	D4	62.76a	66.62a	65.50a	64.96a	77.62b	74.13a	72.07b	74.61bc	69.61a	69.93a	68.65ab	69.40a
	Mean	58.78a	56.49a	57.83a		87.52a	73.00b	76.65b		70.41a	63.46b	65.78b	
2013	D1	82.78a	81.15a	79.99ab	81.31a	92.80a	86.61a	85.96a	88.46a	87.03a	83.4a	82.65a	84.36a
	D2	76.98b	71.15b	75.78b	74.64b	74.35b	55.55c	62.14b	64.01c	75.74b	62.99c	68.79b	69.17c
	D3	77.89b	72.20b	77.49ab	75.86b	73.10b	62.63b	64.53b	66.75c	75.66b	67.20c	70.96b	71.27c
	D4	82.64a	81.05a	81.23a	81.64a	76.67b	67.92b	83.13a	75.91b	79.82b	74.51b	82.06a	78.8b
	Mean	80.07a	76.39b	78.62ab		79.23a	68.18c	73.94b		79.56a	72.03c	76.12b	

RN1: 36% N reduction, RN2: 18% N reduction, CN: conventional N input, D1: 0 cm, D2: 15 cm, D3: 30 cm, and D4: 45 cm. Different lower case letters indicate significant differences among four treatments at levels of $P<0.05$.

only improved plant N uptake but additionally, reduced crop soil-N uptake, thereby increasing NUE without depleting soil fertility.

9.1.1.4 Effects of Phosphorus Fertilizer Reduction

9.1.1.4.1 Sucrose Accumulation and Transport in Soybean Roots

During the 20 days following silking in 2017 and 2018, maize leaf and root sugar contents in a strip-intercropping system were significantly higher than those in monoculture (Fig. 9.7). Further, leaf sugar content was higher in the P-fertilized treatments compared to that in the P-unfertilized treatment, but the differences in root sugar content among the P-fertilizer and planting mode treatments were not significant. However, the root-to-leaf sugar ratio was significantly higher in the P-unfertilized, compared to the P-treated plants, and the ratio was significantly higher in the strip-intercropping system than in monoculture (Zhou 2021) (Fig. 9.7).

During the V5 stage, both soybean leaf and root sugar contents in the relay strip-intercropping system were significantly lower than those in monoculture. Leaf sugar content in the relay intercropping system did not significantly differ among P treatments, but it significantly increased under monoculture in the P20 treatment.

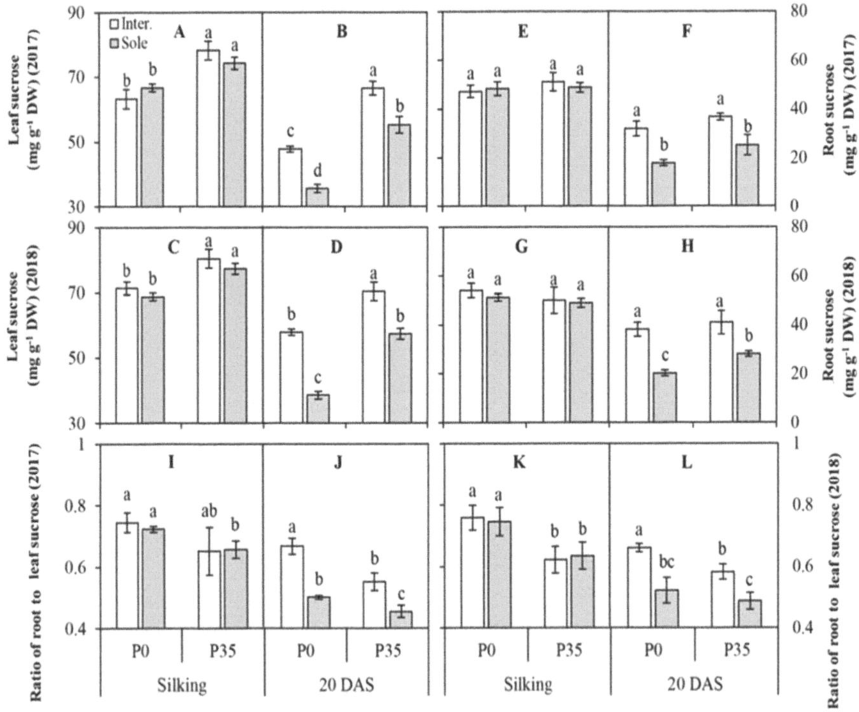

Fig. 9.7 Effects of phosphorus fertilization level and planting pattern on maize leaf- and root-sugar contents during silking and 20 days after silking. Inter.: relay intercropping; Sole: monocropping. P0: 0 kg P hm^{-2}; P35: 35 kg P hm^{-2}. (Reprinted from Zhou et al. 2021)

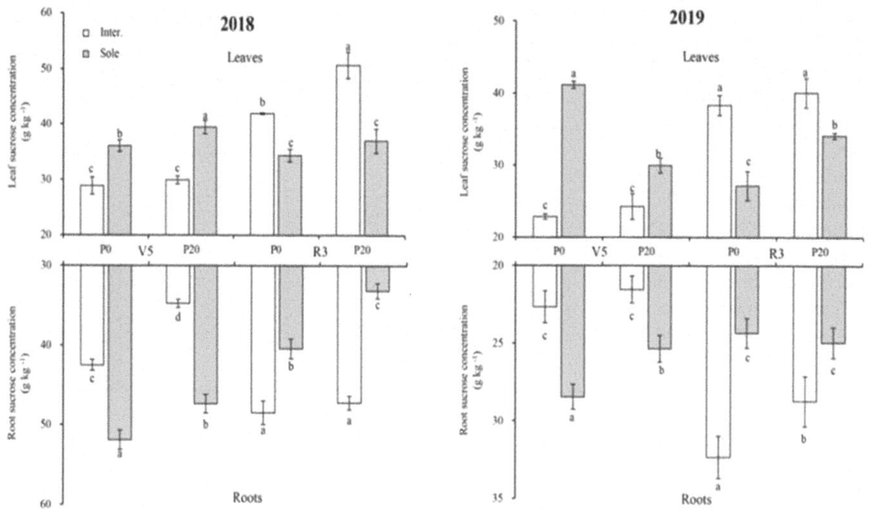

Fig. 9.8 Effects of phosphorus application level and planting pattern on soybean leaf and root sucrose contents. Inter.: relay intercropping; Sole: monocropping. P0: 0 kg P hm^{-2}, P35: 35 kg P hm^{-2}

Further, the root sugar content was significantly higher in the low P treatment compared to the P20 treatment, and both monocropping and relay intercropping followed the same pattern (Fig. 9.8). During the R3 stage, both soybean leaf and root sugar contents were much higher in the relay strip-intercropping system than in the monocropping system. Additionally, the root-to-leaf sugar ratio in the intercropping system was significantly higher than that in the monocropping system. From V5 to the R3 stage, soybean leaf-sugar content in monoculture remained relatively stable, with a significant decrease observed in the P-unfertilized treatment, in 2019. In contrast, the leaf sugar content of relay intercropping soybean plants significantly increased. In turn, the root sugar content decreased significantly in monoculture but increased significantly in the relay intercropping system. In both 2018 and 2019, the sugar content in various soybean tissues showed the same trend across all treatments.

9.1.1.4.2 Carbon Secretion from Maize Roots

Carbon secretion from maize roots varied between 0.22 and 0.49 μg C cm^{-1} 24 h^{-1} among treatments, showing significant variation among replicates. Specifically, P application significantly increased carbon secretion by 82.0% in the relay strip-intercropping system, compared to the P-unfertilized treatment, whereas in the monocropping system, a 70.0% increase was observed (Fig. 9.9). Carbon secretion in the relay strip-intercropping system was only higher than that in the monocropping system when P was applied at a rate of 35 kg P ha^{-1}, with no significant differences between the two systems in the no-P treatment.

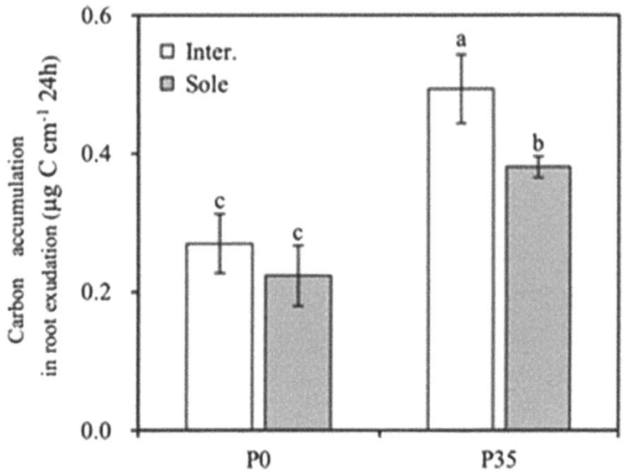

Fig. 9.9 The effect of phosphorus application rate and planting pattern on the 24-h carbon secretion of maize roots during silking. Inter.: relay intercropping, Sole: monocropping. P0: 0 P application, P35: 35 kg P hm^{-2} applied. Different lowercase letters indicate significant differences among treatments at a significance level of 0.05

9.1.1.4.3 Soybean Root Growth

Phosphorus had no significant effect on soybean root dry weight or root length, except at the R3 stage in 2019, when plants grown under the P0 S-L treatment showed significantly greater root length than those in the P100 S-L treatment (Fig. 9.10). At the V5 stage, plants in the L-S treatment showed significantly greater root dry weight and root length compared to those in the S-L treatment. Conversely, at the R3 stage, plants in the L-S treatment had significantly lower root dry weight and root length compared to those growing in the S-L treatment (Fig. 9.10a–d). Additionally, from V5 to the R3 stage, plants in the S-L treatment exhibited significantly faster and greater growth in root weight and length than those in the L-S treatment. Interestingly, at the R3 stage, the S-L treatment had a significantly shorter primary root length compared to the L-S treatment, and the no P treatment had an even shorter primary root length. On the other hand, lateral root length in plants growing in the S-L treatment was significantly higher than in the L-S treatment, but P-treated plants had shorter lateral roots than plants that did not receive any P application, showing a consistent trend in both light environments. Additionally, the proportion of lateral roots to total root weight at the R3 stage was significantly higher in the S-L treatment than in the L-S treatment, and the P-unfertilized treatment plants had a higher proportion of lateral roots, compared to the P-treated plants.

The aboveground light environment and P treatments significantly influenced the expression levels of genes associated with auxin synthesis and the response of the root system (Fig. 9.11). In a field experiment, the expression level of GmYUCCA14 in the intercropping system was significantly higher than that in the monocropping system at both V5 and R3 stages, particularly under low P treatment, but no significant differences were detected between monoculture treatments at the V5 stage under either P treatment. The expression pattern of GmTIR1C was similar to that of GmYUCCA14 in response to P and light treatments, but the advantage of the relay intercropping system over the monocropping system was significant only in the P0 treatment, with no significant difference between planting systems in the P

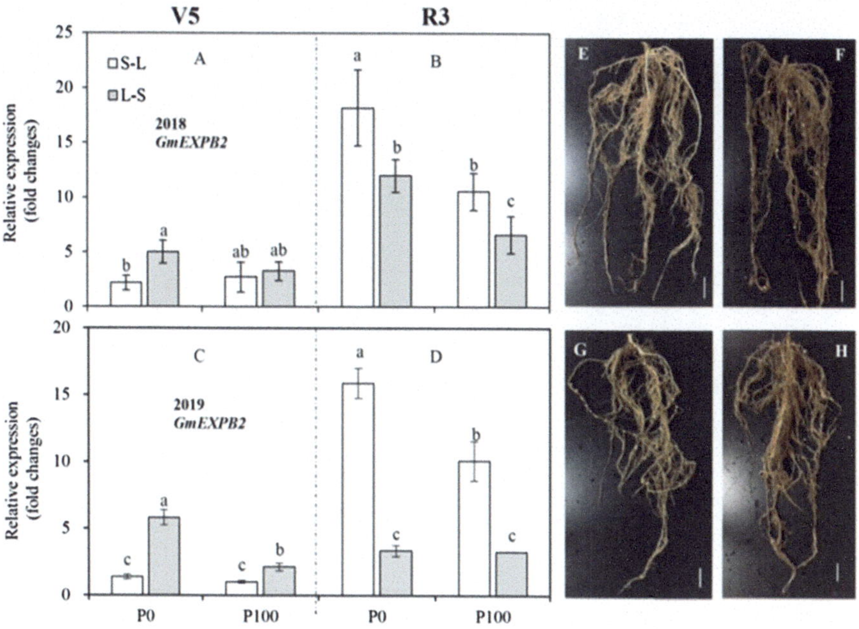

Fig. 9.10 Effects of phosphorus application and planting patterns on root growth. P0: No P application. P100: 100 kg P ha^{-1} applied. S-L: Early shading, where shading was imposed using a 50% light-transmitting shade net from emergence until the V5 stage of soybean, after which shading was removed. L-S: Late shading, where shading was initiated from the V6 stage of soybean using a 50% light-transmitting shade net and continued until the harvest period

treatments. At V5 and R3 stages, the expression level of GmARF05 was significantly higher in the relay intercropping system compared to the monocropping system (except for the P20 treatment at the V5 stage), although its expression was less influenced by P at the R3 stage, with no significant difference between planting systems in either P treatment; moreover, P-untreated plants in the monocropping system showed higher expression than P-treated plants.

9.1.2 Nitrogen–Phosphorus–Potassium (N-P-K) Fertilizer Co-application Technology

9.1.2.1 Nitrogen–Phosphorus (N-P) Fertilizer Co-application Technology

9.1.2.1.1 Effects of Nitrogen–Phosphorus (N-P) Co-application on Nitrogen and Phosphorus Accumulation in Maize

The results of the analysis of the effects of intercropping and N-P fertilizer application on N and P accumulation in maize (Table 9.9) indicated significant differences

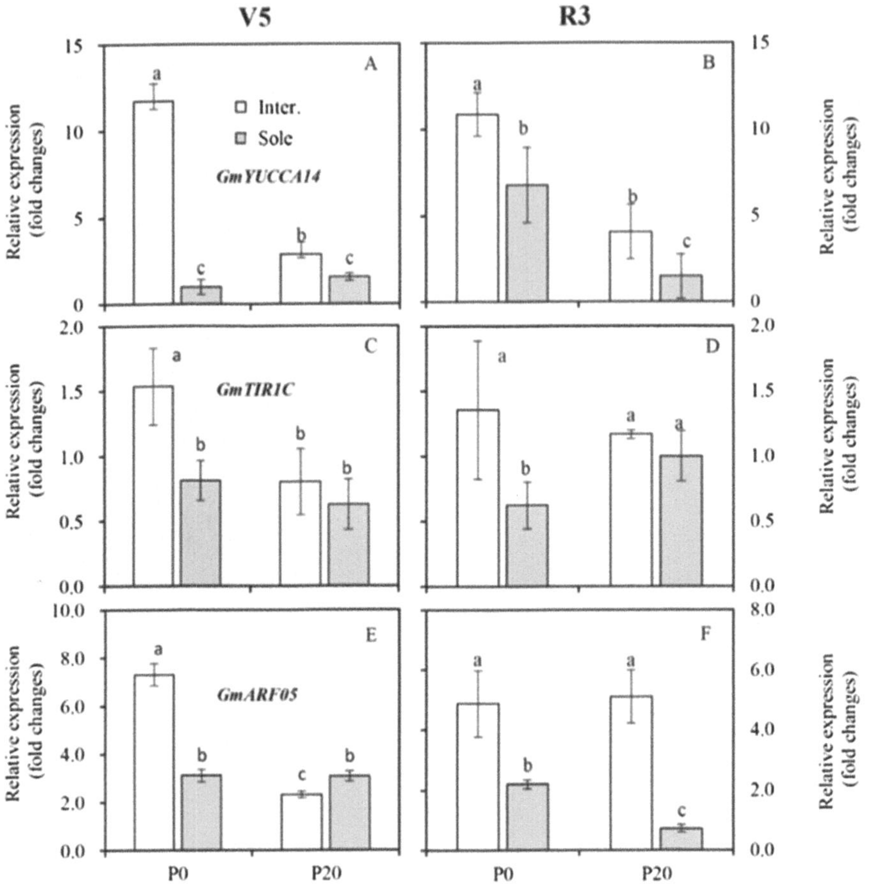

Fig. 9.11 Effects of phosphorus application and planting pattern on gene expression related to root auxin synthesis and response. P0: 0 kg P ha^{-1}; P20: 20 kg P ha^{-1}. (Reprinted from Zhou et al. 2021)

in N and P accumulation levels among different planting methods. The patterns of N and P accumulation in maize plants in root separation pot experiments and field plot trials were consistent. We observed that N and P accumulation in maize plants growing without root separation under conditions of relay intercropping was significantly higher than that of plants growing in treatments with complete root separation and in maize monoculture. Compared to the root separation approach, the strategy of not separating the roots and applying additional N fertilizer increased total N accumulation in maize plants by 28.31% over the level attained in the root separation experiment and by 38.22% in the field plot trial. Under relay intercropping conditions, P accumulation in maize plants was significantly higher than that observed in maize plants in monoculture. In the root separation experiment and the field plot trial, total P accumulation in maize plants was 27.56% (no phosphorus

Table 9.9 Effects of intercropping and nitrogen–phosphorus application on nitrogen and phosphorus accumulation in maize

Treatment			Relay strip intercropping maize (no separation)			Monocropping maize (complete separation)		
			N0	N2	Mean	N0	N2	Mean
N accumulation	Pot experiment	P0	2.43b	3.15a	2.79a	2.21b	2.56a	2.39b
		P2	2.55b	3.25a	2.90a	2.46b	2.94a	2.70a
	(g $plant^{-1}$)	Mean	2.49b	3.20a		2.34b	2.75a	
	Field experiment	P0	140.98b	191.17a	166.08b	133.25a	146.74a	140.00b
		P2	152.03b	213.84a	182.94a	146.01b	177.97a	161.99a
	(kg hm^{-2})	Mean	146.51b	202.51a		139.63b	162.36a	
P accumulation	Pot experiment	P0	0.54a	0.57a	0.5b	0.43a	0.48a	0.45b
		P2	0.67a	0.70a	0.69a	0.52b	0.61a	0.57a
	(g $plant^{-1}$)	Mean	0.61a	0.64a		0.48b	0.55a	
	Field experiment	P0	34.59b	39.50a	37.05b	31.06a	33.24a	32.15b
		P2	40.23b	44.50a	42.37a	34.17a	36.91a	35.54a
	(kg hm^{-2})	Mean	37.41b	42.00a		32.62b	35.08a	

Reprinted from Wen et al. (2015)

Note: Different lower case letters indicate significant differences among N application treatments at levels of $P<0.05$. N0: 0 kg N hm^{-2}, N2: 180 kg N hm^{-2}. P0: 0 P application (0 kg hm^{-2}), P2: 180 kg P hm^{-2}

application), 16.89% (70 kg ha^{-2} phosphorus application), 14.68% (no phosphorus application), and 19.73% higher (70 kg ha^{-2} phosphorus application). These findings indicated that maize–soybean relay intercropping enhanced N and P accumulation in maize plants compared to maize monoculture (Wen 2015).

9.1.2.1.2 Effects of Nitrogen–Phosphorus co-Application on N and P Accumulation in Soybeans

Nitrogen and P fertilizers have a significant impact on N and P accumulation in soybeans throughout the entire plant growth cycle. The influence of N-P on N accumulation in soybean stems becomes increasingly important as growth proceeds (Table 9.10). Leaf N accumulation before flowering is particularly affected by P fertilization, increasing initially and then decreasing as the P application rate increases. Leaf N accumulation peaked in the P3 treatment, reaching 32.76 kg/ha. Further, N content in mature pods was significantly affected by P. Specifically, high P application rates reduced the N content in pods while increasing the same in the grains. Additionally, P application rate had a highly significant effect on P accumulation in soybean grains, ranking as follows: P3 > P2 > P4 > P1. Grain P accumulation peaked at 14.18 kg/ha in the P3 treatment, which was 1.24 times higher than P accumulated when no P was supplied. The influence of N on P accumulation in the grains followed the order of N3 > N2 > N1 > N4, with N3 leading to a P accumulation of 14.03 kg/ha, 1.22 times higher than that observed in the absence of N fertilization.

9.1.2.2 Phosphorus (P-K) Fertilizer Co-Application Technology

9.1.2.2.1 Effects of Potassium Fertilizer Application During Seedling and Flowering Stages on Soybean Yield

Different potassium application rates and fertilization methods have a significant impact on soybean yield (Table 9.11). On comparing different K management strategies we found that the highest soybean yield was achieved in the B4 treatment, which involved equal split application of K fertilizer during the seedling and flowering stages. Conversely, the lowest yield was observed when all K was applied as a basal application. Under both low K (T1, K_2O 56.25 kg hm^{-2}) and high K (T3, 168.75 K_2O kg hm^{-2}) application rates, the highest soybean yield was obtained under the B4 treatment. In contrast, at an intermediate K rate (T2, K_2O 112.5 kg hm^{-2}), the highest yield was achieved in the B6 treatment, which involved splitting the total K dosage into basal application, seedling stage, and flowering stage fertilization in a 1:1:1 ratio (Ren 2019).

Soybean leaf Pn and g_s in the monoculture system were significantly higher (33.91% and 17.7%, respectively) than the corresponding values in the relay intercropping system (Table 9.12). However, relay-intercropping soybean plants showed higher Ci compared to monoculture soybean plants. Among K treatments, soybean plants in the K3 treatment showed the highest Pn, whereas those in K4 showed the lowest P_n. Leaf Pn in the K3 treatment was significantly higher than that

Table 9.10 Effects of nitrogen–phosphorus co-application on nitrogen and phosphorus uptake and accumulation in mature soybeans

		N				P				
Treatment			Stem	Leaf	Pod	Seed	Stem	Leaf	Pod	Seed
N1	P1	18.28a	14.44ab	26.14a	75.74c	1.86b	1.33bc	3.56a	9.53b	
	P2	18.95a	13.24b	9.10c	97.72a	2.19a	1.20c	1.38bc	13.42a	
	P3	14.12b	16.09a	13.42b	89.43b	1.67b	1.51ab	1.83b	13.45a	
	P4	18.51a	17.03a	8.15c	64.69d	1.86b	1.66a	0.96c	9.54b	
N2	P1	17.67a	16.11b	21.50a	76.70c	1.59b	1.46c	2.60b	10.88c	
	P2	16.99a	14.83b	23.91a	83.40b	1.61b	1.24d	3.04ab	12.21b	
	P3	19.51a	20.25a	23.66a	106.24a	2.06a	1.77b	3.42a	15.85a	
	P4	19.22a	23.73a	13.26b	110.43a	2.16a	2.30a	1.82c	17.03a	
N3	P1	18.37b	16.18a	20.97b	84.74b	1.79b	1.41b	2.85b	13.25b	
	P2	23.38a	17.62a	35.29a	87.35ab	2.29a	1.47ab	4.53a	14.00b	
	P3	17.20b	19.11a	15.23c	97.72a	1.92ab	1.69a	1.91c	16.01a	
	P4	16.85b	17.05a	20.15bc	93.10a	1.47cc	1.47ab	2.22bc	12.86b	
N4	P1	12.93ab	16.19a	16.99a	84.64a	1.58ab	1.35a	2.14a	12.07a	
	P2	15.63a	17.59a	17.18a	83.40a	1.77a	1.31a	2.27a	12.23a	
	P3	10.68b	14.5ab	11.90a	80.88a	1.27b	1.16ab	1.69a	11.42ab	
	P4	10.86b	11.83b	14.59a	66.37b	1.24b	0.95b	1.54a	9.59b	
N		16.4*	13.34*	13.87*	5.20*	7.64*	14.79*	13.75*	5.68*	
P		4.94*	1.82	11.18**	10.19**	4.10*	4.76*	12.54**	11.61**	
NXP		2.70*	3.76*	7.80**	12.30**	4.70*	5.09**	6.80**	8.25**	

Reprinted from Ren et al. (2019)

Note: Different lower case letters indicate significant differences among phosphorus application treatmens at levels of $P<0.05$. N1: 0 kg N hm^{-2}, N2: 120 kg N hm^{-2}, N3: 180 kg N hm^{-2}, N4: 240 kg N hm^{-2}; P1: 0 kg P hm^{-2}, P2: 35 kg P hm^{-2}, P3: 70 kg P hm^{-2}, P4: 105 kg P hm^{-2}

Table 9.11 Effects of potassium application rate and method on soybean yield

Treatment	T1	T2	T3	Subplot effect
B1	1626.63b	1402.75c	1542.25bc	1523.84c
B2	1327.11c	1653.85bc	1606.50bc	1529.84c
B3	1615.68b	1651.22bc	1788.39ab	1685.1abc
B4	1996.17a	1547.62bc	1968.84a	1837.55a
B5	1515.08bc	1767.60ab	1373.90cd	1552.19c
B6	1610.14b	2001.27a	1768.94ab	1793.45ab
B7	1684.03b	1943.16a	11969.22d	1607.86bc
	1624.96A	1709.64A	1468.18B	

Note: Uppercase and lowercase letters, respectively, indicate that the difference is significant at levels of $P < 0.01$, and difference is significant at levels of $P < 0.05$.
T1: 56.25 kg K_2O hm^{-2}, low K rate; T2: 112.5 kg K_2O hm^{-2}, intermediate K rate; T3: 168.75 K_2O kg hm^{-2}, high K rate. Treatments B4–7 include the ratio of application used in each case, namely, basal:seedling:flowering. B1: basal application; B2: topdressing at seedling stage (V5); B3: top-dressing at flowering stage (R2); B4, 0:1:1; B5, 1:0:1; B6: 1:1:1; B7, 1:1:0. Different lower case letters indicate significant differences among treatmens at levels of P<0.05

in either K0 or K4 treatment. Proper K fertilizer application enhanced leaf g_s and Pn in soybean plants (Table 9.12).

9.1.2.2.2 Effect of Phosphorus and Potassium Co-application on Soybean Lodging Resistance and Yield

Application of potassium fertilizer in combination with an adequate P dosage (0–17 kg· P_2O_5 ha^{-1}) increased soybean yield in relay strip-intercropping conditions, while excess P (25.5 kg· P_2O_5 ha^{-1}) reduced yield (Table 9.13). Compared to the control group (i.e., neither P nor K applied), P, K, and P-K combination treatments all increased pod number per plant, grain-filling rate, and grain yield of strip intercropped soybeans. Among these treatments, the combined P-K application showed the highest increase in all parameters. When·17 kg ha^{-1} P_2O_5 was applied in conjunction with 112.5 kg·ha^{-1} K_2O, the grain yield of strip intercropped soybeans peaked at 2832.04 kg·ha^{-1}, i.e., a 28.55% increase relative to the control plants, which were supplied with neither P nor K (Xiang et al. 2010).

Reasonable application of P and K fertilizers improved basal internode morphology of intercropped soybean plants and enhanced stem lodging resistance (Table 9.14). Specifically, at the same P level, basal internode length and main stem decreased with increasing K rate, while basal internode diameter, dry weight, stem cellulose and lignin contents, carbon-to-nitrogen ratio (C/N), and stem lodging resistance index (CLRI) showed the opposite trend. Adding K fertilizer at an adequate P level (0–17 kg P_2O_5 ha^{-1}) helped reduce the lodging rate in relay intercropping soybeans, but excess P (25.5 kg P_2O_5 ha^{-1}) had the opposite effect.

Table 9.12 Effects of potassium application rate and planting method on the characteristics of soybean photosynthetic activity

Treatment	Pn (μmol m^{-2} s^{-1})			g_s (mol m^{-2} s^{-1})			Ci (mol m^{-2} s^{-1})		
	Natural light	Shading by maize	Mean	Natural light	Shading by maize	Mean	Natural light	Shading by maize	Mean
K0	8.82	8.14	8.48bc	0.07	0.07	0.07c	195.77	219.51	207.64a
K1	12.09	8.78	10.43abc	0.14	0.08	0.11abc	174.9	237.93	206.41a
K2	14.8	8.61	11.71ab	0.22	0.08	0.15a	168.61	214.49	191.55ab
K3	13.94	9.73	11.84a	0.18	0.1	0.14ab	177.48	164.55	171.02b
K4	6.91	8.66	7.79c	0.07	0.1	0.08bc	165.92	199.42	182.67ab
K5	12.4	7.55	9.97abc	0.14	0.06	0.10abc	174.39	208.15	191.27ab
Mean	11.49a	8.58b		0.14a	0.08b		176.18b	207.34a	

Reprinted from Xiang (2013)

K0: 0 g K_2O/pot, K1: 0.92 g K_2O/pot, K2: 1.84 g K_2O/pot, K3: 2.76 g K_2O/pot, K4: 3.68 g K_2O/pot, K5: 4.6 g K_2O/pot. Different lower case letters indicate significant differences among treatmens at levels of $P<0.05$

Table 9.13 Effect of phosphorus and potassium co-application on soybean yield

Treatment	Number of pods per plant	Number of grains per plant	100-grain weight	Grain Plumpness
P0K0	71.02G	1.56	19.6	93.56I
P0K1	74.45E	1.59	19.41	94.28G
P0K2	77.00D	1.61	19.3	94.76DEF
P0K3	81.52B	1.61S	19.27	95.20BC
P1K0	72.07FG	1.56	19.77	93.79HI
P1K1	76.22D	1.59	19.42	94.47FG
P1K2	78.76C	1.63	19.3	94.99CD
P1K3	82.81A	1.64	19.21	95.40AB
P2K0	74.02E	1.6	19.89	93.90H
P2K1	76.59D	1.62	19.76	94.70DEF
P2K2	78.86C	1.75	19.45	94.99CD
P2K3	83.19A	1.75	19.22	95.70A
P3K0	71.18G	1.63	19.68	93.54I
P3K1	72.50F	1.73	19.37	94.63EF
P3K2	74.70E	1.79	19.18	64.84DE
P3K3	79.05C	1.8	19.12	95.66A

Reprinted from Xiang et al. (2010)
Note: Uppercase letters indicate that the difference is significant at levels of $P < 0.05$
P0, 0 P; P1, 8.5 kg P2O5 ha^{-1}; P2, 17 kg P_2O_5 ha^{-1}; P3, 25.5 kg P_2O_5 ha^{-1}; K0, 0 K; K1, 37.5 kg K_2O ha^{-1}; K2, 75 kg K_2O ha^{-1}; K3, K_2O 112.5 kg K_2O ha^{-1}

Table 9.14 Effect of phosphorus and potassium application rates on lodging resistance of soybean plants

Treatment	Cellulose	Lignin	C/N	CLRI
P0K0	14.67 ± 0.18h	18.71 ± 0.29gh	2.12 ± 0.04ij	116.32fgh
P0K1	15.58 ± 0.19fg	19.62 ± 0.28e	2.65 ± 0.20h	121.45f
P0K2	16.82 ± 0.15c	20.61 ± 0.22c	3.22 ± 0.17e	131.17de
P0K3	17.66 ± 0.10b	21.89 ± 0.42b	3.62 ± 0.11cd	134.91de
P1K0	14.86 ± 0.37h	19.01 ± 0.19f	3.62 ± 0.11cd	119.88fg
P1K1	14.86 ± 0.37h	20.49 ± 0.42cd	3.14 ± 0.14ef	130.53e
P1K2	17.02 ± 0.14c	22.21 ± 0.27b	3.47 ± 0.06d	136.16d
P1K3	17.02 ± 0.14c	23.13 ± 0.48a	3.98 ± 0.28b	150.32b
P2K0	17.02 ± 0.14c	19.26 ± 0.2ef	2.80 ± 0.13gh	121.37f
P2K1	17.02 ± 0.14c	20.72 ± 0.27c	2.96 ± 0.15fg	136.02d
P2K2	17.83 ± 0.53b	21.78 ± 0.17b	3.54 ± 0.12d	142.94c
P2K3	18.29 ± 0.09a	23.34 ± 0.09a	4.60 ± 0.13a	159.52a
P3K0	14.82 ± 0.22h	18.53 ± 0.28h	1.96 ± 0.13j	111.80h
P3K1	15.68 ± 0.03f	19.05 ± 0.13fg	2.35 ± 0.11i	132.67de
P3K2	15.90 ± 0.08ef	19.43 ± 0.13ef	3.22 ± 0.08e	133.22d
P3K3	16.87 ± 0.10c	20.08 ± 0.06d	3.83 ± 0.11bc	132.61de

Reprinted from Xiang et al. (2010)
Note: different lower case letters indicate significant differences among treatmens at levels of $P<0.05$. P0, 0 P; P1, 8.5 kg P2O5 ha^{-1}; P2, 17 kg P_2O_5 ha^{-1}; P3, 25.5 kg P_2O_5 ha^{-1}; K0, 0 K; K1, 37.5 kg K_2O ha^{-1}; K2, 75 kg K_2O ha^{-1}; K3, K_2O 112.5 kg K_2O ha^{-1}

9.2 Chemical Control Technology

9.2.1 Chemical-Based Soybean Seedling Enhancement Technology

When soybeans are strip intercropped with corn, they are shaded by the taller maize plants at different growth stages depending on planting dates. Such shading of soybean plants from the corn canopy causes excess internode elongation, increased plant height, and, in severe cases, the main stem adopts a vine-like growth habit, as it becomes thinner, showing reduced lignin and cellulose contents, which reduces its strength and makes it highly susceptible to lodging. Soybean plants that experience lodging during the seedling stage are more susceptible to pest and disease infections, whereby seedling mortality increases substantially, leading to a severe shortage of basic seedlings. Later on, mechanical harvesting of lodged soybeans becomes difficult, resulting in extremely high yield losses.

9.2.1.1 Uniconazole Dry Seed Dressing

9.2.1.1.1 Effects on the Vertical Growth of Strip Intercropping Soybean Above and Below the Ground

The following four uniconazole seed-dressing treatments were tested for their effect on the growth of intercropped soybean plants: 0 (B0, control), 2 (B1), 4 (B2), and 8 mg/kg (B3). After 21–42 days of sowing, the thiamethoxam seed treatment significantly inhibited the plant height and root length (Fig. 9.12, Table 9.15), and the inhibitory effect increased with increasing treatment concentration. Further, such treatment significantly increased the stem diameter and root volume (Fig. 9.13, Table 9.15). Specifically, after 42 days of sowing, plant heights in treatments B1, B2, and B3 were 32.2%, 36.8%, and 42.8% lower than that of the control B0, respectively. Similarly, the root length in treatments B2 and B3 was 23.1% and

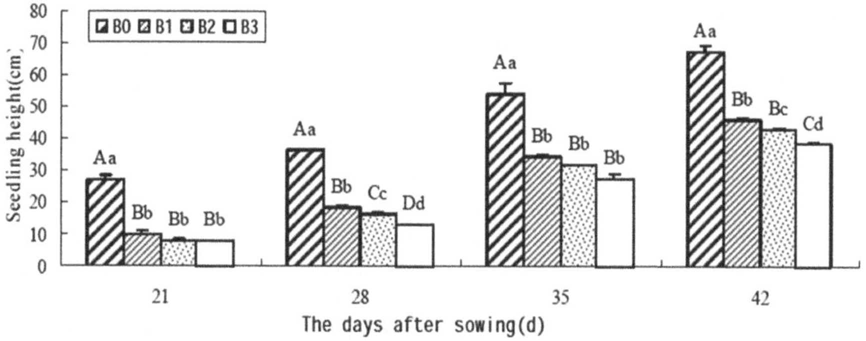

Fig. 9.12 Effect of dry seed mixed difenoconazole treatment on the growth of soybean seedlings in the banded sleeve. Note: B0, B1, B2, and B3 represent 0, 2, 4, and 8 mg/kg, of difenoconazole, respectively

Table 9.15 Effect of dry seed mixed uniconazole treatment on root length and root volume of soybean seedlings in the banded sleeve

Uniconazole treatment of dry seedeed mixing	Root length				Root volume			
	21d	28d	35d	42d	21d	28d	35d	42d
B0	29.63Aa	30.17Aa	37.63Aa	40.80Aa	1.4Cc	1.5Cc	3Cc	4.5Dd
B1	21.92ABb	23.50ABb	27.23Bb	38.93Aa	1.8ABab	2.5Bb	4.6Bb	6.8Bb
B2	19.57ABb	23.03ABb	24.33BCc	35.00Bb	2Aa	3.2Aa	5.6Aa	7.9Aa
B3	18.05Bb	20.03Bb	23.00Cc	31.37Cc	1.6BCbc	2.2Bb	4.5Bb	5.5Cc

Note: Capital and lowercase letters following means within columns represent significant differences at 1% and 5% levels. B0, B1, B2, and B3 are 0, 2, 4, and 8 mg/kg, uniconazole, respectively. 21–42d indicate days after sowing

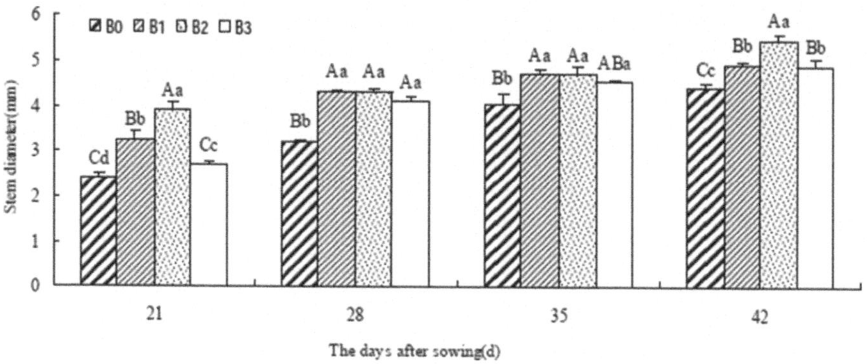

Fig. 9.13 Effect of dry seed mixed uniconazole treatment on stem diameter of soybean seedlings in the banded sleeve

14.2% lower than that of the control, respectively; stem diameters in B2, B1, and B3 treatments were 23.6%, 11.8%, and 10.6% higher than that of the control, respectively. In turn, root volumes in B2, B1, and B3 were 75.6%, 51.1%, and 22.2% higher than those recorded for controls, respectively. These findings indicated that an appropriate concentration (4 mg/kg) of thiamethoxam applied as seed dressing effectively improved soybean growth in strip intercropping, and inhibited the vertical growth of the aboveground parts while promoting the expansion of root volume and the growth of vigorous seedlings.

9.2.1.1.2 Effects on the Dry Matter Accumulation of Strip Intercropping Soybean in Seedling Stage

The dry-mixed seed treatment with two concentrations of uniconazole, B1 (2 mg/kg) and B2 (4 mg/kg), significantly increased the root and aboveground dry weights of soybean seedlings in the banded sleeve (28–42 days after sowing), with the highest dry matter accumulation observed in plants treated with B2 (Table 9.16). Further, there was no significant difference in root and aboveground dry weights of plants treated with B1 at 21 days after sowing, while dry matter accumulation in plants treated with B3 (4 mg/kg) was significantly higher than that recorded for the control plants only at 42 days after sowing, indicating a greater inhibitory effect of high concentrations of difenoconazole on the growth of soybean seedlings.

9.2.1.1.3 Effects on Lodging of Strip Intercropping Soybean in the Field

The effects of six concentrations (0, B1, control, 2.0, B2, 4.0, B3, 6.0, B4, 8.0, B5, and 10.0 mg/kg, B6) of the uniconazole seed-dressing treatment on lodging resistance were investigated in two soybean cultivars, Nandou 8 (A1) and Gongxuan 1 (A2), under conditions of maize–soybean strip intercropping. The results shown in Fig. 9.14 indicate that the control treatments of the two soybean cultivars both experienced lodging at 30 and 37 days after sowing, with lodging scores reaching 4 and 2.8, respectively, by 51 days after sowing. The low concentrations of propyzamide in B2 and B3 resulted in increased lodging scores in the later stage,

Table 9.16 Effects of dry seed mixed difenoconazole treatment on root and aboveground dry weight in soybean seedlings in the banded sleeve

Uniconazole dry seed dressing	Days after sowing (d)							
	Root dry weight (g $plant^{-1}$)				Aboveground dry weight (g $plant^{-1}$)			
	21d	28d	35d	42d	21d	28d	35d	42d
B0	0.082Bc	0.112Cc	0.204Cc	0.233Dd	0.331Cc	0.464Dd	1.08Cc	1.42Dd
B1	0.080Bc	0.212Bb	0.262Bb	0.525Bb	0.307Cc	0.837Bb	1.30Bb	2.78Bb
B2	0.186Aa	0.347Aa	0.384Aa	0.767Aa	0.647Aa	1.268Aa	1.82Aa	3.88Aa
B3	0.113Bb	0.119Cc	0.207Cc	0.337Cc	0.451Bb	0.489Cc	1.06Cc	1.82Cc

Note: Uppercase and lowercase letters, respectively, indicate that the difference is significant at levels of $P < 0.01$, and difference is significant at levels of $P < 0.05$

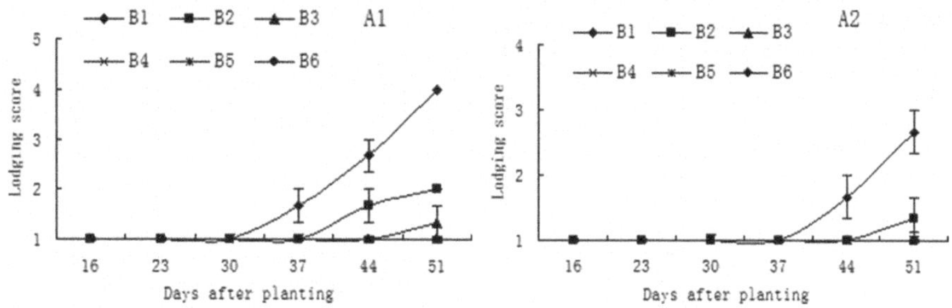

Fig. 9.14 Effects of different doses of enoxabenzene seed coating on lodging of soybean under maize–soybean strip intercropping. Note: B1–B6 are the following ethephon concentrations: 0, 2.0, 4.0, 6.0, 8.0, and 10.0 mg/kg. A1 and A2 represent soybean varieties Nandou 8 and Gongxuan 1, respectively

with lodging scores of 2 and 1.3, respectively, by 51 days after sowing, indicating that lodging occurred. Other concentrations maintained lodging scores at 1, exhibiting excellent lodging resistance.

9.2.1.1.4 Effects on the Yield of Intercropping Soybean

The dry-mixed seed treatment with uniconazole increased the yield of strip-intercropped soybeans (Table 9.17). The yield of Nandou No. 8 (A1) variety ranked as follows: B3 > B4 > B5 > B6 > B2 > B1, and analysis of variance showed that the differences in yield between B3, B4, and B5 relative to B1 were significant but those between B2 and B6 were not, with respect to B1. Treatments B3, B4, and B5 increased production by 28.50%, 21.55%, and 15.43%, respectively, compared to the control. In turn, the yield of variety Gongxuan No. 1 (A2) ranked as follows: B4 > B3 > B5 > B6 > B2 > B1. In this case, treatments B3, B4, B5, and B6 were significantly different compared to the control, B1; while the difference between B2 and the control was not significant. Treatments B3, B4, B5, and B6 increased production by 16.96%, 19.65%, 16.50%, and 13.76%, respectively, compared to the control.

Analysis of yield components showed that the effective pod numbers across mix-seed treatments were higher than those of the control. Particularly, the effective pod numbers of A1 in B3, B4, and B5 were significantly higher than those for the control group, whereas the other two dose treatments did not show any significant difference compared to the control. In turn, the effective pod numbers of A2 in B3, B4, and B5 were significantly higher than those for the control group, whereas the other treatments did not show any significant difference compared to the control. Although the uniconazole dry-mixed seed treatment also had an impact on the number of seeds per pod and 100-grain weight, the differences in these yield components between each treatment and the control were not significant. The effective pod number, number of seeds per pod, and 100-grain weight of A2 were all significantly higher than those of the A1 variety.

In summary, appropriate concentrations of uniconazole for dry-mixed seed treatment resulted in vigorous soybean seedlings in the strip-intercropping system;

Table 9.17 Effects of uniconazole dry seed dressing on yield and yield components of soybean in a striped intercropping field layout

Variety	Seed mixing dosage	Yield (kg/hm^2)	Number of plants (10,000 plants / hm^2)	Effective number of pods	Number of pods	100-grain weight (g)
A1	B1	117.21d	12.0	5.97d	1.40a	11.74ab
	B2	121.07cd	12.0	6.68cd	1.24a	12.16a
	B3	150.61a	12.0	8.07ab	1.30a	11.94a
	B4	142.47ab	12.0	8.54a	1.28a	11.06b
	B5	135.30abc	12.0	7.74abc	1.27a	11.44ab
	B6	128.50bcd	12.0	6.86bcd	1.33a	11.71ab
A2	B1	2345.29c	12.0	60.57c	1.55b	20.78ab
	B2	2520.21bc	12.0	64.06bc	1.58b	20.80ab
	B3	2742.99a	12.0	66.93ab	1.60b	21.32a
	B4	2806.22a	12.0	68.05a	1.72a	19.96b
	B5	2732.24a	12.0	67.88a	1.61b	20.88ab
	B6	2667.95ab	12.0	64.06bc	1.63ab	21.30ab

additionally, they reduced seedling lodging and increased soybean yield. Lastly, the optimal dosages for intermediate and late varieties were 4.0 and 6.0 mg/kg, respectively.

9.2.1.2 Uniconazole Foliar-Spray Treatment

9.2.1.2.1 Effect of Uniconazole on Soybean Plant Height, Stem Thickness, and Internode Length in a Strip-Intercropping Field Layout

Uniconazole concentrations included D0 (0 mg/kg, control), D1 (30 mg/kg), D2 (60 mg/kg), and D3 (90 mg/kg). Foliar spray treatment using these concentrations was applied to soybean seedlings at the fifth leaf stage (V5) in a strip-intercropping planting. The results showed that plant height, main stem height, and first internode length of uniconazole-treated plants were lower than those of the control, and decreased with increasing concentrations of the spray treatment (Table 9.18). Thus, the heights of soybean plants treated with D1, D2, and D3 were 21.8%, 30.4%, and 38.2% lower than those of the controls, respectively. In turn, the length of the first internode was 5.7%, 8.7%, and 10.1% shorter than that of the controls, respectively, while stem thickness for D1, D2, and D3 was 20.6%, 23.5%, and 15.7% greater than that of the controls, respectively.

9.2.1.2.2 Effects of uniconazole Foliar Spraying on Dry Matter Accumulation in Strip-Intercropped Soybean

As shown in Table 9.19, spraying uniconazole at the fifth leaf stage (V5) inhibited dry matter accumulation in the first pod period (R3) and the grain-filling period (R5) of strip intercropped soybeans at the higher concentrations tested, i.e., D4 and D5, whereas low concentrations significantly increased it. Further, application at the

Table 9.18 Effect of foliar application of uniconazole on the morphology of post-flowering soybean plants in a strip-intercropping planting layout

Spraying concentration of uniconazole	Height (cm)	First internode length (cm)	Stem diameter (cm)
D0	100.35Aa	7.85Aa	0.722Cc
D1	78.45Bb	7.40Bb	0.871Aba
D2	69.85Bb	7.17Bbc	0.892Aa
D3	62.07Cc	7.06Bc	0.835Bb

Note: Uppercase and lowercase letters, respectively, indicate that the difference is significant at levels of $P < 0.01$, and difference is significant at levels of $P < 0.05$. D is the concentration of uniconazole applied by foliar spraying, D0: 0 mg/kg, D1: 30 mg/kg, D2: 60 mg/kg; and D3: 90 mg/kg

R1 stage showed that in 2006, dry matter accumulation in plants treated with D3 and D4 was significantly higher than that recorded for the controls, but not in 2007, in which case, there was no significant difference. Spraying at a concentration of 25–100 mg/kg at the V5 stage enhanced dry matter accumulation at the R8 stage, while all treatments sprayed at the R1 stage caused higher dry matter accumulation relative to controls, reaching a significant level in the experiment of 2006. As for dry matter accumulation in plants sprayed after the R3 stage, these findings indicate that spraying 25–100 mg/kg of at the V5 stage was conducive to the accumulation of dry matter, while higher concentrations were needed for spraying at the R1 stage to be effective.

9.2.1.2.3 Effect on Soybean Yields in a Strip-Intercropping Planting Layout

Spraying uniconazole at the V5 stage with D1 (30 mg L^{-1}) resulted in the highest yield, followed by D2 (60 mg L^{-1}) and D3 (90 mg L^{-1}), all of which showed significantly higher yields than the control group (Table 9.20). Specifically, D1 increased yield by 35.5% compared to the control, while the yields of the high-concentration treatments D4 (120 mg L^{-1}) and D5 (150 mg L^{-1}) were lower than those of the control. Further, at the early flowering stage (R1), there was a slight difference in yield between treatments, except for that obtained in D5, which was significantly higher than that observed for the control group. Meanwhile, the effective pod number and 100-grain weight at each stage were consistent with the yield among the treatments. Thus, in D1 and D2 at the V5 stage, both were significantly higher than in the control, whereas in D4 and D5 they were lower than those in the control. As for the R1 stage, yield in D5 was significantly higher than that in the control. The order of performance among treatments with respect to the number of grains per pod was opposite to that of yield, which may be due to negative correlations observed among the various yield components (Yan 2010).

Table 9.19 Effect of foliar application of uniconazole on post-flowering dry matter accumulation of strip-intercropped soybean (g/plant)

Spraying period and concentration		R3		R5		R8		Accumulation of dry matter after R3 period	
		2006	2007	2006	2007	2006	2007	2006	2007
V5	D0	9.85Cc	38.18BCb	12.26Bb	52.87Dd	23.60De	90.25Dd	13.75Cc	52.07Dd
	D1	10.17Bb	40.53Aa	12.53Bb	63.64Aa	26.64Bc	107.85Aa	16.47Bb	67.32Aa
	D2	10.52Aa	38.91Bb	12.68Bb	59.50Bb	28.59Ab	103.27Bb	18.06Aa	64.36Bb
	D3	10.57Aa	38.74Bb	13.39Aa	54.20Cc	29.56Aa	99.45Cc	18.99Aa	60.71Cc
	D4	9.53Dd	36.88CDc	11.47Cc	53.06Dd	25.08Cd	90.22Dd	15.55Bb	53.35Dd
	D5		36.55Dc		49.55Ee		76.68Ee		40.13Ee
R1	D0	9.85Bb	38.18Aab	2.26Bb	52.87Aa	23.60Cc	90.25Aa	13.75Cc	52.07Aa
	D1	10.06ABb	37.74Ab	12.28Bb	52.46Ab	24.71BCb	90.87Aa	14.65BCbc	53.13Aa
	D2	10.18ABab	38.12Aab	12.34Bb	52.72Aab	25.46Bb	90.96Aa	15.28Bb	52.85Aa
	D3	10.41Aa	38.30Aab	12.66Aa	52.83Aab	25.78Bb	90.85Aa	15.37Bb	52.54Aa
	D4	10.48Aa	38.93Aa	12.71Aa	52.98Aa	27.34Aa	91.28Aa	16.86Aa	52.35Aa
	D5		39.00Aa		53.01Aa		91.87Aa		52.87Aa

Reprinted from Yan (2010)

Note: Uppercase and lowercase letters after means within the same column indicate the differences and stage at 1% and 5% levels of probability, respectively. 1C refers to the spraying stage, C1 for the V5 stage, and C2 for the R1 stage of growth; 2D refers to the concentration of uniconazole spray (mg L^{-1}), 2006 (D0: 0, D1: 25, D2: 50, D3: 75, and D4: 100 mg L^{-1}); 2007 (D0: 0, D1: 30, D2: 60, D3: 90, D4: 120, and D5: 150 mg L^{-1})

Table 9.20 Effects of foliar application of uniconnazole on soybean yield and yield components in a strip-intercropping planting layout

Spraying period	Spraying concentration	Effective number of pods	Pods	100 grain weight (g)	Yield (kg/hm^2)
V5	D0	60.0BCc	1.63Abb	23.88Cc	2335.5CDd
	D1	75.3Aa	1.50Cc	28.05Aa	3164.1Aa
	D2	65.3Bb	1.58BCb	26.69ABb	2759.3Bb
	D3	59.5BCc	1.63ABb	26.08Bb	2535.9BCc
	D4	53.9CDd	1.72Aa	23.86Cc	2211.8Dde
	D5	50.5Dd	1.72Aa	23.70Cc	2061.7De
R1	D0	60.0Bb	1.63Aa	23.85Cc	2332.3Bbc
	D1	60.5Abb	1.58Abb	23.89Cc	2283.6Bc
	D2	60.7Abb	1.56Bbc	24.42BCc	2312.4Bbc
	D3	60.8Abb	1.55Bbc	24.42BCc	2301.3Bbc
	D4	61.0Abb	1.54Bc	25.40ABb	2386.1Bb
	D5	62.2Aa	1.54Bc	26.37Aa	2530.6Aa

Note: Uppercase and lowercase letters after means within the same column are the same as in Table 4.1; 2D refers to the concentration of uniconazole sprayed, D0: 0, D1: 30, D2: 60, D3: 90, D4: 120, and D5: 150 mg L^{-1}

9.2.2 Soybean Growth Control by Growth Regulators and Lodging Prevention Technology

Under conditions of unrestricted water and fertilizer provision throughout the flowering and pod-setting phases, soybeans grown in strip intercropping exhibit exuberant growth, giving rise to robust, thriving vegetation characterized by significant ground coverage. Consequently, an abundance of microambient humidity and insufficient ventilation and light penetration may prevail, culminating in pronounced rates of floral and pod abscission as well as lodging, ultimately precipitating deleterious consequences on crop yield. Therefore, in this study, five different regulators were selected, and the chemical control of growth and lodging prevention technology for strip intercropped soybeans was studied by spraying twice during the V5 stage of soybean branching and once during the R1 stage of soybean flowering.

9.2.2.1 Effects on Plant Height

The five different chemicals tested significantly reduced plant height when sprayed both at branching stage V5 and at initial flowering R1 (Fig. 9.15). Particularly, treatment with 30% paclobutrazol + mepiquat chloride (75 mg/L + 150 mg/L) had the most marked effect, reducing the mean plant height by 25 cm compared to controls. Treatments with 30% uniconazole + mepiquat chloride (40 mg/L + 80 mg/L), 15% paclobutrazol + mepiquat chloride (75 mg/L + 150 mg/L), and 5% uniconazole + mepiquat chloride (40 mg/L + 80 mg/L) had similar effects on plant height, causing reductions of 19 cm, 19 cm, and 16 cm, respectively. During initial flowering stage R1, treatment with 30% paclobutrazol + mepiquat chloride (150 mg/L paclobutrazol), 30% uniconazole (80 mg/L), 15% paclobutrazol (150 mg/L), or

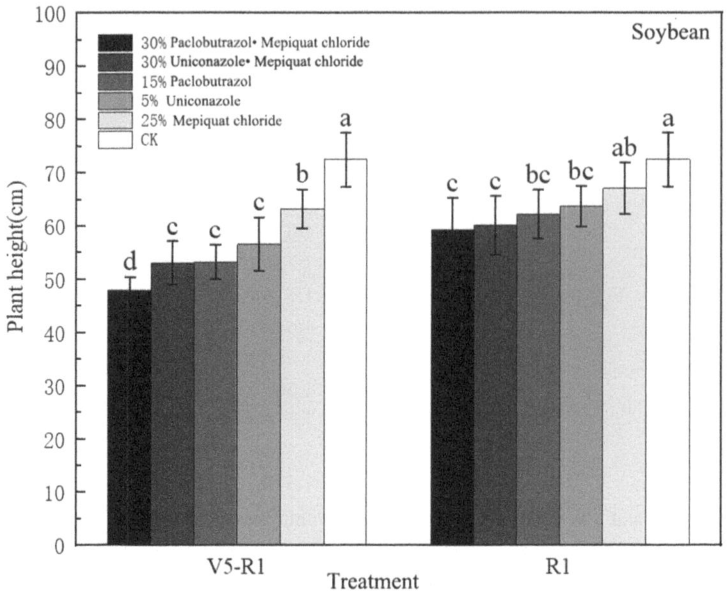

Fig. 9.15 Effect of growth regulators on soybean plant height in a strip-intercropping planting layout

5% uniconazole (80 mg/L), all significantly reduced soybean plant height, while treatment with 25% mepiquat chloride (400 mg/L) had only a slight effect.

9.2.2.2 Effects on Stem Diameter

When 30% paclobutrazol + mepiquat chloride (75 mg/L, 150 mg/L) and 30% uniconazole + mepiquat chloride (40 mg/L, 80 mg/L) were sprayed at the branching stage (V5) and initial flowering stage (R1), they both increased stem diameter highly significantly by 0.78 and 0.75 mm, respectively, compared to the controls. Similarly, spraying with the same treatment at the initial flowering stage (R1) also increased stem diameter highly significantly by 0.91 on average, over the controls. Meanwhile, the effects of the other four treatments tested were not significant (Fig. 9.16).

9.2.2.3 Effects on Lodging Resistance

Five chemicals applied to soybean plants at the branching stage V5 and early flowering stage R1 significantly improved soybean maturity and shortened average internode length. Specifically, spraying with 30% paclobutrazol + mepiquat chloride (75 mg/L, 150 mg/L) improved lodging resistance and shortened average internode length to the greatest extents, such that average internode length was shortened by 0.8 cm compared with controls. Meanwhile, spraying with 25% mepiquat chloride (200 mg/L, 400 mg/L) had a slightly milder effect (Fig. 9.17).

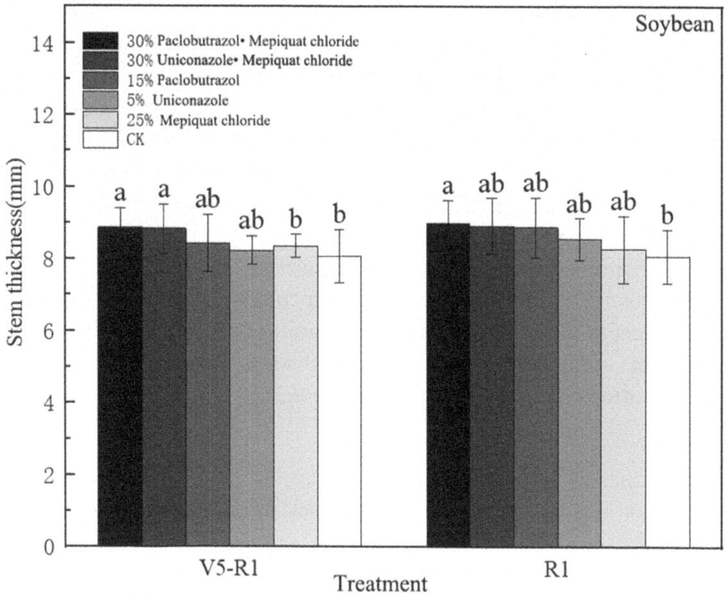

Fig. 9.16 Effect of growth regulator-spray treatment on soybean stem diameter in a strip-intercropping planting layout

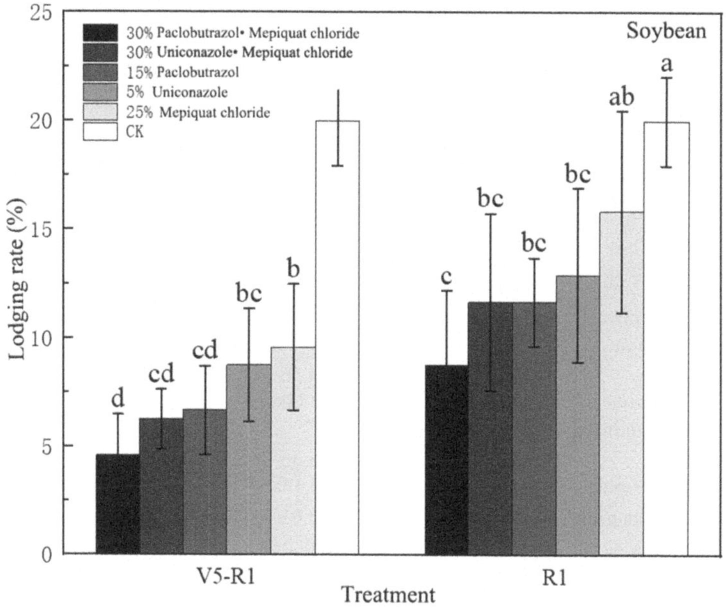

Fig. 9.17 Effect of growth regulators on soybean lodging resistance in a strip-intercropping planting layout

9.2.2.4 Effects on Yield

Simultaneously spraying the five types of chemicals at branching stage V5 and initial flowering stage R1 increased the number of grains per plant and 100-grain weight. Except for 25% mepiquat chloride (200 and 400 mg/L), the other chemicals increased 100-grain weight. Specifically, 30% paclobutrazol + mepiquat chloride (75 and 150 mg/L) showed the best effect. All five chemicals increased the yield, and the theoretical yield was the highest for 15% paclobutrazole (75 mg/L) and 5% uniconazole (40 and 80 mg/L). In turn, when the five types of chemicals were simultaneously sprayed at soybean initial flowering stage R1 on both corn and soybeans, treatment with 30% paclobutrazol + mepiquat chloride (75 and 150 mg/L) had the best effect in increasing the number of grains per plant and 100-grain weight in soybean plants, with the highest theoretical yield, followed by 30% paclobutrazol + mepiquat chloride (75 and 150 mg/L) (Table 9.21).

Table 9.21 Effects of growth regulators on soybean yield in a strip-intercropping planting layout

Date	Drug	Effective number of plants per mu	Number of grains per plant	100-grain weight (g)	Theoretical yield (kg/mu)
V5-R1	30% Paclobutrazol Meperidium	4622.55 ± 123.89a	109.45 ± 8.0077a	28.8 ± 0.82a	123.87 ± 4.87a
	30% Uniconazole · Mepyronium	4463.83 ± 59.51a	110.68 ± 10.78a	28.39 ± 0.73ab	119.07 ± 9.66a
	15% Paclobutrazole;	4444.00 ± 149.78a	117.21 ± 4.41a	28.57 ± 0.77ab	126.43 ± 4.76a
	5% Uniconazole	4563.03 ± 149.78a	113.63 ± 18.25a	28.3 ± 1.64ab	126.05 ± 30.16a
	25% Mepiquat chloride	4444.00 ± 181.83a	113.85 ± 9.06a	26.72 ± 1.51b	114.74 ± 9.49a
	CK	4503.51 ± 123.89a	102.15 ± 4.3a	26.98 ± 0.61ab	105.40 ± 1.93a
R1	30% Paclobutrazol · Meperidium	4801.10 ± 68.72a	104.66 ± 9.2a	26.67 ± 0.95ab	113.66 ± 4.99ab
	30% Uniconazole · Mepyronium	4642.39 ± 314.93a	113.56 ± 13.06a	28.55 ± 2.49a	127.62 ± 16.69a
	15% Paclobutrazole	4523.35 ± 157.46a	105.36 ± 3.15a	26.34 ± 1.04b	106.78 ± 7.74b
	5% Uniconazole	4523.35 ± 472.4a	103.61 ± 19.58a	27.32 ± 0.61ab	107.53 ± 10.65b
	25% Mepiquat chloride	4543.19 ± 363.66a	105.65 ± 1.98a	26.84 ± 0.47ab	109.48 ± 8.68b
	CK	4503.51 ± 123.89a	102.15 ± 4.3a	26.98 ± 0.61ab	105.40 ± 1.93b

In summary, in maize–soybean strip intercropping, spraying with 30% paclobutrazol + mepiquat chloride (75 and 150 mg/L) at branching stage V5 and early flowering stage R1 reduced plant height and average internode length, increased stem thickness and branching number, and effectively controlled excessive growth, thereby preventing lodging, and increasing yield.

9.2.3 Growth-Regulation Technology for Maize

As the taller crop in the maize–soybean strip-intercropping system, maize not only influences its own tolerance to high population density and its lodging resistance, but additionally, it changes the light environment, which in turn affects soybean growth. In order to investigate the effect of uniconazole on maize plant type and yield in the strip-intercropping system, using ChuanDan 418 as the plant material, uniconazole was sprayed on maize leaves at the 8-leaf stage (A1), 10-leaf stage (A2), 12-leaf stage (A3), and 14-leaf stage (A4); distilled water was sprayed as control (CK) treatment at each stage.

9.2.3.1 Effect of uniconazole on Maize Yield and Its Economic Benefits in the Strip-Intercropping System

Except for the number of effective ears, the effect of the time of application of uniconazole on maize leaves was significant ($P < 0.05$) on all indicators (Table 9.22). Particularly, 1000-grain weight and number of grains per ear of maize first increased and then decreased as the application was performed at the latter stages of growth. The maximum values recorded were reached with the application at A2, in which case parameters were significantly higher than those in A1 and CK, increasing by 14.04% and 7.25%, respectively. In turn, the number of grains per ear reached the maximum value with a spray application at A4, which caused the number to become 1.46% higher than that in CK plants. Lastly, the yield was significantly reduced with early application of spray (A1), and there was no significant difference among the other treatments. The yield reached the highest value when uniconazole spray was applied at A2, in which case it was 5.91% higher than the yield attained by the CK treatment groups.

The time of application of uniconazole on maize also affected soybean yield and its components. Specifically, soybean yield showed a linear downward trend as the application was performed at the latter stages of growth. Thus, the number of grains per plant, effective plant number, and 100-grain weight all peaked when uniconazole was sprayed on maize leaves at A1, in which case the three values recorded for the three parameters were significantly higher than those recorded in the CK treatment; specifically, the number of grains per plant, effective plant number, and 100-grain weights were 11.36%, 25.27%, and 9.60% higher than those in control plants, respectively. Furthermore, soybean yield differed significantly among treatments; thus, for example, yield in the A1 treatment was 809.72 kg/ha higher than CK yields, implying a 38.68% increase.

Table 9.22 Effect of stage of application of uniconazole on maize and soybean yield, and on overall intercropping-system yield

Index		Growth stage of application				
		A1	A2	A3	A4	CK
Maize	Effective panicle number (10,000 plants · hm^{-2})	5.11a	5.11a	5.07a	5.03a	5.02a
	Number of grains per spike	491.83c	539.69b	551.67ab	579.00a	570.55ab
	1000 grain weight (g)	252.79c	294.08a	282.63ab	277.10ab	272.77bc
	Yield ($kg \cdot hm^{-2}$)	5906.55b	7927.48a	7844.14a	7760.80a	7458.71a
Soybean	Number of grains per plant	110.64a	109.13a	104.21ab	101.39ab	98.07b
	Effective panicle number (10,000 plants · hm^{-2})	9.42a	8.89a	8.68ab	7.92b	7.04c
	1000 grain weight (g)	20.42a	20.12a	19.79ab	19.19bc	18.46c
	Yield ($kg \cdot hm^{-2}$)	2093.58a	1918.41b	1772.84b	1563.91c	1283.86d
Intercropping yield ($kg\ hm^{-2}$) System		8000.13c	9845.89a	9616.98a	9324.72ab	8742.57b

Note: Means within columns followed by different lowercase letters are significantly different ($P < 0.05$)

Overall yield in the strip-intercropping system was mainly affected by maize yield and showed a trend of first increasing and then decreasing as uniconazole was applied at the latter stages of growth. Thus, yield performance ranked in the order of A2 > A3 > A4 > CK > A1. Compared with A1 and CK, overall intercropping yield increased by 1845.76 and 1103.32 $kg \cdot hm^{-2}$ in each case, which represents an increase of 18.75% and 11.21%, respectively.

9.2.3.2 Effect of Uniconazole Application on Maize Plant Architecture

Effects of uniconazole application on maize plant height and stem diameter differed significantly ($P < 0.05$) among treatments (Fig. 9.18). Application before the 12th leaf stage significantly reduced maize plant height, especially in treatment A1, in which case mean plant height decreased by 17.28% compared with that observed in CK groups. In turn, the application of uniconazole significantly promoted stem diameter, with an increase of 8.21% in treatment A1 compared with the CK groups.

Maize ear height and upper ear height showed a decreasing trend compared with CK, as uniconazole application was performed at the latter growth stages. Furthermore, all treatments exhibited significant differences (Table 9.23). Particularly, treatment A1 showed the lowest values for ear and upper ear height, with significant

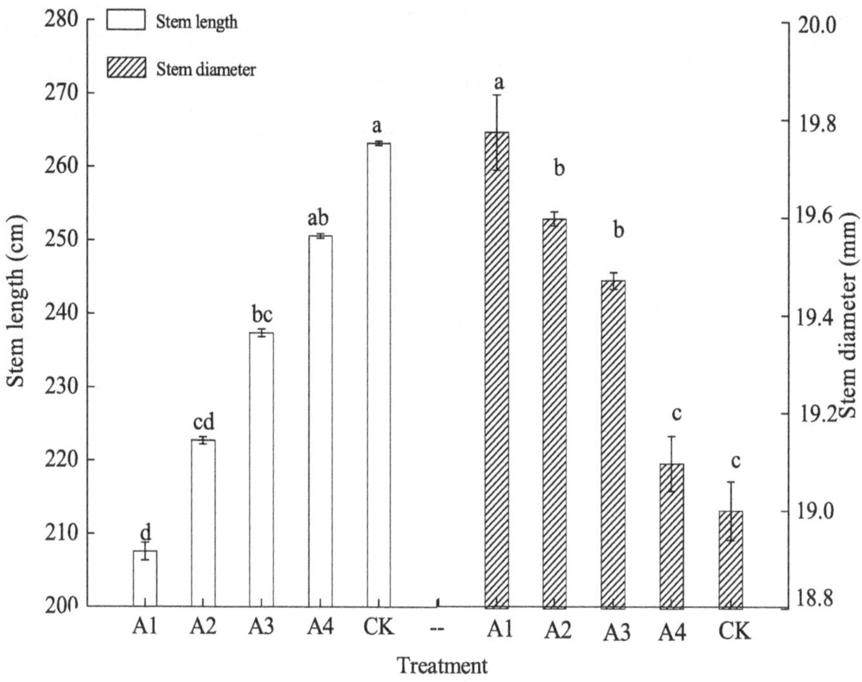

Fig. 9.18 Effect of uniconazole application timing on maize plant height and stem diameter

reductions of 22.85% and 19.81%, respectively. Ear leaf angle and upper ear leaf angle also showed a decreasing trend as application timing was advanced, with the lowest values observed in treatment A1. Compared with CK, they decreased by 18.12% and 27.19%, respectively. Ear leaf area and upper ear leaf area were the largest in treatment A4, with increases of 38.58% and 27.08%, respectively, compared with the values recorded in treatment A1. These results indicate that appropriate timing of uniconazole application leads to an adequate maize plant architecture for maximizing system output.

9.2.3.3 Effect of Uniconazole Application on Light Transmittance in the Soybean Canopy in Maize–Soybean Intercropping

Photosynthetically active radiation (PAR) intensity is an important indicator of the light environment and plays an important role in crop growth and development. As uniconazole application on maize was performed at the latter stages of growth, the light transmittance of the intercropped soybean canopy showed a significant decreasing trend (Fig. 9.19), reaching a maximum value of 49.97% in treatment A1, which was significantly higher (by 60.30%) than the minimum value in the CK treatment, i.e., 19.84%.

Table 9.23 Effect of uniconazole application timing on maize plant architecture

Handle	Ear height (cm)	Upper ear height (cm)	Angle between leaf and ear position (°)	Angle between upper leaf and leaf of spike (°)	Leaf area at ear position (cm^2)	Upper leaf area of spike (cm^2)
A1	89.27c	118.28b	24.44c	15.82c	618.13d	632.56c
A2	100.23bc	122.47b	24.89bc	17.87c	740.70c	759.60b
A3	113.30ab	124.08b	25.83abc	18.59bc	877.98b	800.71ab
A4	111.92ab	138.70a	29.20ab	21.38ab	1006.42a	867.45a
CK	115.71a	147.50a	29.75a	21.72a	926.37ab	851.92ab

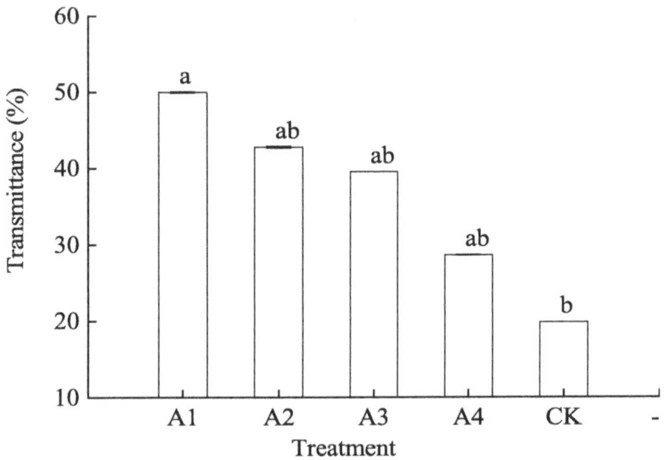

Fig. 9.19 Effect of uniconazole application timing on light transmittance at the top of the soybean plants

9.2.3.4 Effect of Uniconazole Application on Soybean Growth in Maize–Soybean Intercropping

Spraying corn with uniconazole had a significant effect on soybean plant height (Fig. 9.20). The CK treatment showed the highest values throughout all sampling time points, whereas the early spraying treatment (A1) showed the lowest ones. Conversely, the trend followed by the stem diameter was the opposite; specifically, spraying corn with uniconazole significantly increased soybean stem diameter, and the earlier the spraying, the more significant the effect. The average stem diameter in A1 was 21.94% higher than that of CK at all sampling time points. Meanwhile, leaf area showed a similar trend as stem diameter, with the corresponding value in A1 (August 11) being significantly higher than that of CK by 17.65%. These results indicate that uniconazole spraying "adjusted" maize plant structure, thereby affecting the transparency of the top layer of soybean leaves. Early maize spraying was more conducive to constructing high-yielding soybean plants.

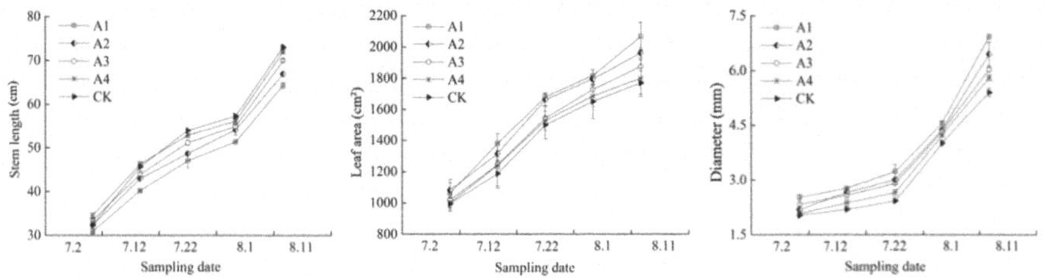

Fig. 9.20 Effect of uniconazole application at various maize growth stages on plant height, stem diameter, and leaf area of strip intercropped soybeans

9.2.3.5 Relationship Between Maize Plant Architecture and Soybean Canopy-Light Transmittance

A highly significant negative correlation ($P < 0.01$) between soybean canopy-light transmittance and maize ear height, height above the ear, leaf area of ear position, and leaf area of upper ear position, with correlation coefficients of −0.70, −0.79 and − 0.78, respectively, was noted (Table 9.24). Path analysis was conducted on the effect of each index on light transmittance to further investigate the effect of maize plant architecture on soybean canopy-light transmittance. The results showed that, except for the leaf angle of the upper ear, the comprehensive and direct effects of the other indicators were negative, indicating a significant negative effect on light transmittance. The direct effect of height above ear (−0.9195) on light transmittance was more significant than that of other indicators, indicating that the decrease in maize ear height improved the light conditions for soybean (Table 9.25).

9.2.3.6 Relationship Between Soybean Canopy-Light Transmittance and Plant Morphology

The effects of spraying maize with uniconazole on light transmittance and morphological characteristics of strip intercropped soybeans at the V1 stage (July 2nd), are summarized in Table 9.26. A negative correlation between light transmittance and plant height of intercropped soybeans was detected, with a correlation coefficient = − 0.97. In contrast, stem thickness and leaf area showed highly significant ($P < 0.01$) positive correlations with light transmittance, with correlation coefficients of 0.87 and 0.80, respectively. Further analysis of the effects of other morphological parameters was conducted using path analysis for plant height (X1), stem thickness (X2), and leaf area (X3) effects on light transmittance. The results showed that the path coefficient of plant height was negative, indicating that it had a significant negative effect on light transmittance; while stem thickness and leaf area both had positive path coefficients, indicating that selecting for high light transmittance can lead to larger leaf area (Table 9.27).

In summary, the application of uniconazole to maize leaves at the 10-leaf stage reduced maize plant height, and made the plant more compact, not only improving its own yield and enhancing its lodging resistance, but also, significantly increasing

Table 9.24 Correlation coefficients between maize plant architecture and light transmittance

Index	Transmittance	Ear height	Upper ear height	Angle between leaf and ear position	Angle between upper leaf and leaf of spike	Panicle leaf area	Upper leaf area of spike
Transmittance	1						
Ear height	−0.70**	1					
Upper ear height	−0.92**	0.59*	1				
Angle between leaf and ear position	−0.09	−0.32	0.23	1			
Angle between upper leaf and leaf of spike	0.13	−0.32	−0.09	0.06	1		
Panicle leaf area	−0.79**	0.62*	0.65**	−0.31	−0.17	1	
Upper leaf area of spike	−0.78**	0.81**	0.68**	−0.4	−0.18	0.83**	1

Note: * and ** represent significance at the 0.05 and 0.01 levels, respectively

Table 9.25 Path analysis for the relationship between maize plant architecture and light transmittance

Factor	Combined effect	Direct effect	Indirect effect					
			→X1	→X2	→X3	→X4	→X5	→X6
X1	−0.6962	−0.1939		−0.279	0.0606	0.0138	−0.2183	−0.0794
X2	−0.9195	−0.4705	−0.115		−0.0442	0.0039	−0.2277	−0.066
X3	−0.0924	−0.1897	0.0619	−0.1097		−0.0026	0.1082	0.0395
X4	0.126	−0.0435	0.0616	0.0421	−0.0112		0.0598	0.0172
X5	−0.7915	−0.3513	−0.1205	−0.3049	0.0584	0.0074		−0.0806
X6	−0.7793	−0.0976	−0.1577	−0.3182	0.0767	0.0077	−0.2902	

Note: X1: ear height; X2: height above ear; X3: leaf angle of ear position; X4: leaf angle of upper ear position; X5: leaf area of ear position; X6: leaf area of upper ear position

Table 9.26 Correlation coefficients between morphological characteristics and light transmittance of strip-intercropped soybeans

Index	Transmittance	Plant height	Stem diameter	Leaf area
Transmittance	1			
Plant height	−0.97	1		
Stem diameter	0.87**	−0.81	1	
Leaf area	0.80**	−0.69	0.96**	1

Note: ** represent significance at the 0.01 level

Table 9.27 Path analysis for morphological characteristics and light transmittance of strip-intercropped soybeans

			Indirect effect		
Index	Combined effect	Direct effect	→X1	→X2	→X3
X1	−0.98	−1.01		0.57	−0.54
X2	0.05	−0.70	0.82		0.75
X3	0.11	0.78	0.69	−0.67	

Note: X1 represents plant height, X2 represents stem diameter, and X3 represents leaf area

light transmittance of the top layer of the soybean canopy, thereby limiting soybean plant height, increasing stem diameter and leaf area, and ultimately, grain yield.

9.3 Green Prevention and Control Technology

Green prevention and control are important means for continuously controlling plant diseases and insect pests, thereby ensuring safe agricultural production, improving the quality and safety of agricultural products, and protecting the environment. The competition between corn and soybean farmland use in China is particularly prominent, with over 85% of market soybeans relying on imports. Hence, in response to the call from General Secretary Xi Jinping for "stabilizing corn and expanding soybeans," the research summarized herein clearly demonstrated that the combined planting of soybeans and corn is a key measure to achieve this goal. Further, in 2022, the Central Document No. 1 clearly stated that "the combined planting of soybeans and corn should be promoted in the Huang-Huai-Hai, Northwest, and Southwest regions," while the 14th Five-Year Plan for Agriculture and Rural Areas requires the promotion of combined planting of soybeans and corn in 100 million mu during the 14th Five-Year Plan period. In 2022, it will be promoted in 16 provinces, autonomous regions, and municipalities with a total area of 15 million mu, with Sichuan promoting it in an area of 3.1 million mu. However, in Sichuan Province, where this technology has been promoted for a long time, plant diseases and insect pests have been increasing year by year. Moreover, there is a lack of corn and soybean varieties suitable for this intercropping model, that are also resistant to diseases, pests, and herbicides. Neither the patterns of incidence of plant diseases and insect pests in different regions nor the main technical parameters for their prevention and control have been comprehensively elucidated, nor has the intensive technology system required been constructed. In particular, there is a lack of herbicides that can be used for both monocotyledonous and dicotyledonous crops, and simple, intelligent crop protection equipment. Therefore, we urgently need to screen and create new varieties that are resistant to diseases, pests, and herbicides, develop new drugs, and new green prevention and control technologies. Researching green prevention and control technology for maize–soybean intercropping systems, and integrating it into demonstration programs, plays an important role in fully tapping the potential for increased production of such systems to ensure large-scale promotion for increased yields.

9.3.1 Prevention and Control Strategies

A green prevention and control strategy has been developed to cope with pests and diseases in strip-intercropping systems. The goal is to promote agricultural safety production and reduce the use of chemical pesticides. The principle underlying the strategy is to use one tool for multiple purposes, one application for multiple treatments, and a combination of sealing and control. Specific measures include population monitoring, and ecological, biological, and physical control, scientific use of pesticides, and other environmentally friendly measures to combat pests and diseases. Key technologies for green pest control include predator–insect technology, biological pesticide technology, physical control technology, pheromone technology, sterilizing radiation technology, and transgenic insect-resistant technology, among others.

Based on the characteristics of pest incidence in corn-soybean strip-intercropping systems, the principles of biodiversity in the strip-intercropping system, crop variety layout, heterogeneous light environment, spatial isolation, dilution effect, natural enemy hypothesis, root exudation effect on rhizosphere microorganisms, and interplant competition are fully utilized. Following the plant protection policy of "prevention first and comprehensive control," the "one application for multiple treatments," "one tool for multiple purposes," and "sealing and control" strategies have been formulated, such that a comprehensive pest management system for corn-soybean strip intercropping has been elaborated, whose core content includes the following aspects:

"Prevention first." This refers to paying attention to reducing the initial source of pests before symbiosis. During symbiosis (i.e., corn-filling period and soybean branching period), multiple pests are treated with one chemical only. After corn harvest, soybean pest control is strengthened to control the total amount of overwintering harmful organisms.

"Comprehensive control." This refers to taking into account technology to reduce the source of initial invasion of the corn and soybean crops, and the joint use of integrated technology and disease warning technology, considering the responses of corn and soybean to multiple technologies for overall management.

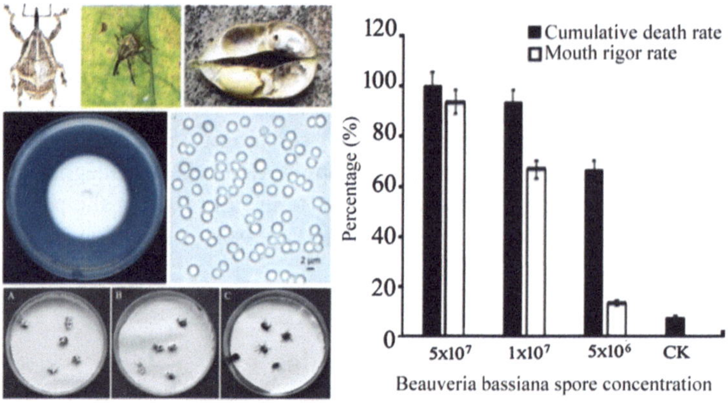

Fig. 9.21 Concentration-dependent control of soybean aphids by *Beauveria bassiana* isolate BEdy1

9.3.2 Prevention and Control Technology

9.3.2.1 Multiple Treatments with a Single Application

Prevention and control of common pests in corn and soybean plants using broad-spectrum biocontrol agents. Thus, for example, *Beauveria bassiana* is a natural fungal killer that can control the growth of many insect pests, including the high elephant beetle, a potentially devastating soybean pest (Fig. 9.21). *Beauveria bassiana* invades the body wall of insect pests and parasitizes them, causing insects to become stiff and die. The fungus can be used as a synergistic measure against multiple pests to achieve one-time application and control multiple pest species. The release of another natural agent, the seven-spot ladybug can also combat corn aphids and soybean aphids. With biological natural control at the core of this strategy, sustainable agricultural development can be achieved.

Further, prevention and control of common pests such as snails and nail snails can be attained by using broad-spectrum pesticides in the pre-symbiotic period in maize–soybean intercropping, including 4% high-chlorine methomyl emulsion + *Bacillus thuringiensis* or 20% triazophos emulsion or 20% chlorfenapyr or 1.8% avermectin emulsion or 10% flufenoxuron-avermectin and other highly efficient broad-spectrum insecticides and antibiotics for the prevention of corn earworms at the mouth of the corn horn, and in the field, and ditch sprinkling with 80% polyoxymethylene for timely prevention and control of dryland snails and nail snails.

9.3.2.2 One Chemical for Multiple Diseases and Pests During the Symbiotic Period, i.e., Corn-Filling Period and Soybean Branching Period

Bacterial leaf-spot disease should be the focus of prevention and control by continuous application of 72% agricultural streptomycin sulfate and 30% copper sulfate pentahydrate suspension. To prevent soybean high elephant beetle, aphids, bugs, peach borer, and other major insect pests, the same agent should be sprayed for a second time. One chemical for multiple diseases and insect pests during the soybean monoculture period after corn harvest.

The same agent should be sprayed a third time. An adjuvant (Jijian) should be added to the agent to improve its efficiency, reduce the pesticide dose, and achieve high rates of prevention and control with significant reduction of application and increase in efficiency.

9.3.2.3 Seed Coating Agents

Before sowing, a rapid inspection of seed-borne mixed pathogens on corn and soybeans was conducted. A 20% ene-pent-evil seed-dressing agent and an 8% ene-prop–aviomycin suspension seed-dressing agent were formulated with plant growth regulators, highly efficient, less toxic, and easily decomposable fungicides, and antibiotics as effective ingredients for soybean-seed treatment to prevent common diseases (e.g., root rot) and underground pests in corn and soybean roots. Indoor root-rot prevention reached over 81%, which was higher than that achieved by the market 5% pentoxazole suspension seed-dressing compound (Table 9.28); additionally, it effectively promoted seedling emergence and growth. The use of compound coating agents can simultaneously prevent multiple seedling diseases. The improvement of seedling resistance will greatly reduce the pressure of later-stage prevention and control, and significantly reduce by up to 2–3 the number of pesticide applications required.

9.3.2.4 One Insect, Multiple Traps

Using a collecting light source with similar bands to simultaneously lure and kill common insect pests in corn and soybean crops is a multi-purpose approach. Based on the patterns of incidence of pests and their phototaxis to different light wavelengths, five single-wavelength LED insecticidal lamps were selected to lure and kill major pests at different times and growth stages, such as to reduce the impact on natural enemies and neutral insects (Table 9.29) while achieving maximum prevention and control of pest insects. In addition, highly efficient and low-toxicity chemical pesticides were used to kill insect pests that could not be lured by the lamps. Multiple-color insect plates for the main pests during the soybean–corn co-growth were screened and mounted onto an adjustable device. A color-adjustable insect lure device (including multi-color insect lure plates and an insect lure-plate color adjustment device) was used to lure and kill pests that appeared at different stages during the co-growth period of corn and soybeans. Prevention and control were greater than 85%, thereby reducing the number of pesticide applications by up to four.

9.3.2.5 Binding and Sealing

Chemical weed control saves labor, shows high weeding efficiency, it is convenient to use, and renders high economic benefits. Indeed, it is an important symbol of agricultural modernization. However, there are certain difficulties in the selection of herbicides for composite planting systems where monocotyledonous and

Table 9.28 Fusarium root-rot preventive effect of a 20% seed-coating suspension of enilconazole·pentachloronitrobenzene·metalaxyl on potted soybean plants

Fusarium	Treatment	Dosage (g/kg)	Incidence (%)	Disease index (%)	Relative efficacy (%)
Fusarium graminearum	20% Enilconazole·Pentachloronitrobenzene·Metalaxyl FSC	0.2	31.07b*	20.50e*	67.35e*
		0.6	32.86b	14.75f	76.51d
		1.0	28.21b	15.25f	78.71cd
		2.0	20.00d	12.50fg	80.09c
	5%Tebuconazole FSC	4.0	24.00c	45.40c	80.80c
	Blank control	4.0	98.31a	62.78b	–
Fungi Imperficti	20% Enilconazole·Pentachloronitrobenzene·Metalaxyl FSC	0.2	30.95b	12.25 fg	70.50e
		0.6	31.95b	12.16fg	79.61c
		1.0	30.48b	9.04g	84.84b
		2.0	22.54c	11.73fg	83.33bc
	5%Tebuconazole FSC	4.0	24.60c	46.00c	82.30bc
	Blank control	4.0	100.00a	59.63b	–
Fusarium oxysporum	20% Enilconazole·Pentachloronitrobenzene·Metalaxyl FSC	0.2	28.25b	16.51f	80.06c
		0.6	26.19bc	14.05f	83.03bc
		1.0	23.06c	8.98g	89.15a
		2.0	11.20e	30.20d	90.50a
	5%Tebuconazole FSC	4.0	25.50bc	45.80c	83.80bc
	Blank control	4.0	100.00a	82.80a	–

Note: * and lowercase indicate that the difference is significant at levels of $P < 0.05$

Table 9.29 Trapping effect of LED insecticidal lamps of different wavelengths on different insect pests in maize–soybean strip-intercropping plots

Types of Insects	Number of insects captured per day at different bands													
	401 nm	397 nm	405 Nm	407 nm	393 nm	411 nm	395 Nm	375 nm	350 nm	385 nm	378 nm	389 nm	403 nm	compound
Grapholita molesta	2.95	2.58	2.55	1.16	3	1.79	2.34	2.89	2.87	3.91	2.04	2.46	2.21	2.68
Spodoptera litura	2.16	3.34	3.04	2.25	3.67	1.67	6.41	2.64	3.07	2.48	2.83	3.84	1.2	2.88
Metabolus flavescens Brenske	1.98	4.53	3.63	2.14	2.3	2.53	2.19	1.52	2.85	1.58	2.84	2.41	5.19	3.52
Holotrichia parallela Motschulsky	3.95	3.28	7.88	14.8	3.67	3.11	8.6	1.63	1.67	1	1.4	1.19	2.57	2.42
Heteroptera	1.5	2.39	4.18	2.05	1.78	2.41	1.85	1.17	1.29	1.54	1.09	1.52	3.62	3.65
Carabid	2.85	3.3	4.58	3.44	3.02	3.32	3.99	1.84	3.33	2.36	2.18	2.68	3.6	3.44
Coccinellidae (predator)	1.44	1.97	2.42	1	1.59	1.58	2.47	1.41	1.2	1.56	1.41	1.96	1.36	3.02
Neutral insects	7.93	6.16	6.23	5.07	5.71	5.58	5.88	2.92	2.34	3.02	2.81	3.93	10.61	30.44

Reprinted from Yan et al. (2018)

dicotyledonous crop species coexist, as differences in morphological structure and physiological characteristics between them lead to different herbicide metabolism and detoxification abilities. Approaches aimed at effectively achieving chemical weeding in maize–soybean strip intercropping, and screening corn–soybean compatible stem-leaf herbicides have a high practical significance for the sustainable development of corn–soybean strip intercropping.

In the twenty-first century, China entered the fast lane of application research on composite planting herbicides for monocotyledonous and dicotyledonous crops in strip-intercropping systems. Specifically, ethametsulfuron-methyl has been more commonly reported in application research on corn and soybeans. Additionally, fine isopropyl methanesulfonamide is a widely used herbicide that is relatively safe for many crops and shows a high level of control on numerous weed species. More recently, studies have shown that flufenacet has good control effects on Poaceae weeds and some broad-leaved weeds and has been shown to be safe for three soybean varieties, Zhonghuang 56, Ludou 4, and Zhonghuang 39. An effective dosage of flufenacet (less than 800 g/hm^{-2}) is safe for various maize varieties, including Zhengdan 958, Zhongdan 909, and Jinyangsun No. 7 maize varieties.

For common weeds in corn and soybean fields, broad-spectrum herbicides are used to treat multiple weeds all at once. Under intercropping conditions, five soil-treatment single agents and five stem-leaf treatment single agents were screened for corn and soybeans with only slight chemical damage from which plants readily recover. Among such soil agents, ethametsulfuron-methyl, quinclorac-ethametsulfuron-methyl, and pinoxaden-ethoxysulfuron show high levels of safety for corn and soybeans. The effect of fine isopropyl methanesulfonamide in terms of preventing stress and reduction in fresh weight is the highest, reaching 72% and 79%, respectively (Table 9.30). Meanwhile, stem-leaf treatment agents show the highest prevention effect with thifensulfuron-methyl, from whose damage plants recover readily (Table 9.31). Screening revealed that a mixture of 60% quinclorac-ethoxysulfuron-methyl with 48% glyphosate had a prevention effect greater than 75% on horseweed and *Amaranthus retroflexus*, with little chemical damage to corn and soybean plants. The treatment effectively prevented weeds during the coexistence period of corn with soybean plants in strip intercropping.

9.3.2.6 Pre-emergence Weed Control and Post-emergence Targeted Weed Control

Before sowing and after emergence, high-efficiency soil treatment agents such as 96% precision ethametsulfuron-methyl are used for close weed control to reduce the incidence of seedling weeds and thus alleviate prevention and control pressure. For post-emergence spraying, targeted spraying is performed according to the specific characteristics of monocotyledonous and dicotyledonous crop species. Thus, for corn, spraying 75% thifensulfuron-methyl at the 3–5 leaf stage is recommended, while for soybeans, spraying 25% flufenacet at the 5–6 compound leaf stage is recommended. High-level banded directional spraying is used to improve the accuracy of chemical application.

Table 9.30 Effects of soil treatments on weed control and phytotoxicity in strip intercropping

Treatment	7 days			14 days		
	Number of plants	Plant protection efficacy (%)	Number of plants	Plant protection efficacy (%)	Fresh weight (g)	Fresh weight protection efficacy (%)
Xylene pentol	2.00a	0b	6.00b	60.58ab	0.0183b	67.12ab
Astromepromethanomide	0.67a	72.22a	4.33b	71.96a	0.0095b	78.83a
Thyl Izl	1.00a	61.11a	7.00b	54.81b	0.0198b	60.33b
Zinidone ethanoxamine	1.33a	50.00a	6.00b	60.58ab	0.0229b	55.35b
Blank control	1.67a	–	15.00a	–	0.0568a	–
Treatment	Germination rate (%)		1-day drug damage index (%)		7-day drug damage index (%)	
	Corn	Soybeans	Corn	Soybeans	Corn	Soybeans
Xylene pentol	95	100	0	0	0	1
Astromepromethanomide	100	100	15	8	29	6
Thyl Izl	100	100	3	3	3	7
Zinidone ethanoxamine	100	100	4	4	7	2
Blank control	100	100	0	5	1	5

Table 9.31 Effects of stem and leaf herbicides on weed control and phytotoxicity in maize–soybean strip intercropping

Treatment	7 days		14 days		Fresh weight protection efficacy (%)
	Grass weeds	Broadleaf weeds	Grass weeds	Broadleaf weeds	
Bentazone	100.00a	62.89b	63.07bc	42.33b	83.82b
Thiophene sulphurone	**76.67b**	**73.50a**	**82.28ab**	**65.47a**	**96.10a**
Fluorane silvery	93.33a	70.32ab	84.75a	55.71ab	84.21b
Imidazole niacin	93.33a	68.65ab	57.19c	44.17b	91.22ab
Blank control	–	–	–	–	–
	7-day drug damage index (%)		14-day drug damage index (%)		
	Corn	Soybeans	Corn	Soybeans	
Bentazone	6	2	0	0	
Thiophene sulphurone	**16**	**30**	**12**	**12**	
Fluorane silvery	54	40	42	20	
Imidazole niacin	20	54	20	42	
Blank control	0	0	0	0	

Sowing and Fertilizing Machine in Maize–Soybean Strip Intercropping

10

Dandan Han, Lihua Zhang, and Wenyu Yang

Abstract

Aiming at the problems of fewer sowing machines, poor sowing single grain, uneven plant spacing, low operational efficiency, and intelligence level under the maize–soybean intercropping planting mode, this chapter designs and develops 2BF-5 maize–soybean intercropping precision seeder and 2BYDF-6 maize–soybean intercropping seeder, which can realize adjustable and controllable control of different plant spacing, sowing depth, suppression strength, and the needs of different fertilizer types and fertilizer amount of the intercropping mode for the widespread popularization and application of the maize–soybean intercropping mode. After the bench and field tests, all performance indexes meet the relevant standards and actual production requirements, providing support for the wide-scale popularization and application of the maize–soybean intercropping pattern.

Keywords

Maize · Soybean · Intercropping · Seeder · Sowing machine · Fertilizing machine

10.1 Agronomic Requirements

The maize–soybean strip intercropping mode consists of two rows of tightly spaced, densely planted maize belt, and 2~6 rows of soybean belt compound planting. A production unit consists of a maize belt, a soybean belt, and two maize–soybean

D. Han (✉) · L. Zhang · W. Yang
College of Mechanical and Electrical Engineering, Sichuan Agricultural University, Yaan, China

College of Agronomy, Sichuan Agricultural University, Chengdu, China
e-mail: mssiyangwy@sicau.edu.cn

W. Yang et al. (eds.), *Theories and Application in Maize-Soybean Strip Intercropping System*, https://doi.org/10.1007/978-981-97-8723-4_10

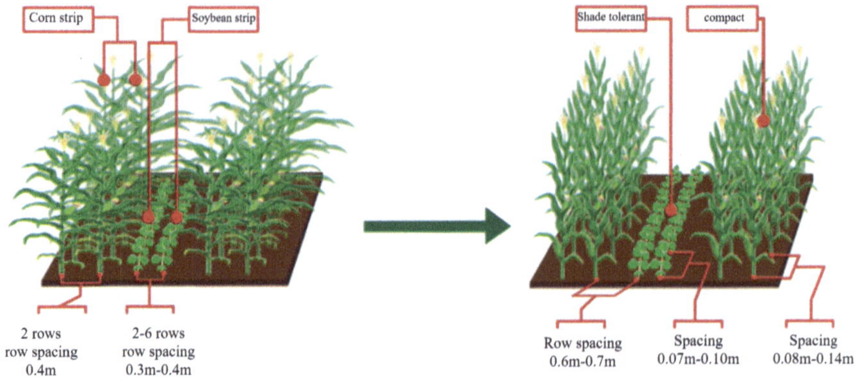

Fig. 10.1 Field configuration for maize–soybean ribbon composite planting

belts, with a production unit width of 2.0~3.0 m. The maize belt is composed of two rows with a spacing of 40 cm, while the soybean and maize–soybean row spacings are 30~40 and 60~70 cm, respectively. To achieve a sufficiently large maize yield, a mu of more than 100 kg of soybean harvest, the maize plant distance needs to be significantly reduced, for example, using a maize plant spacing of 8–14 cm. To ensure that the belt compound planting maize density and local net crop maize are equivalent, approximately 4000–5500 plants per acre and a soybean plant spacing of 7–12 cm are needed to ensure that a soybean planting density and local net crop soybean density of 70%–100%. Increasing the plant spacing to increase the lighting and reduce the plant confidentiality is key to taking full advantage of the side rows (Yang and Yang 2019a, b; Yang 2017; Lin et al. 2004) (Fig. 10.1).

To address the problems of nonsynergistic fertilization and low fertilizer utilization under maize–soybean strip intercropping conditions, the use of special fertilizers and reduced integration technology for banded composite planting can improve nitrogen and phosphorus utilization. In the maize–soybean strip intercropping mode, maize is applied with 16–24 kg of pure nitrogen per acre (depending on the local target yield and soil fertility), ensuring the same amount of single plant application as net crop maize. An equal amount of controlled-release compound fertilizer is applied to maize as that of a seed fertilizer between maize rows. For soybeans, 2–3 kg of nitrogen per acre and 10–25 kg of low nitrogen (14%) balanced compound fertilizer per acre are applied as a base fertilizer when preparing the land or as a seed fertilizer when sowing. Depending on the maize and soybean growth, a small amount of nitrogen, phosphorus, potassium, or microfertilizer can be applied as a supplement or for foliar follow-up.

Fig. 10.2 Schematic of fertilizer application sites for maize–soybean strip cropping

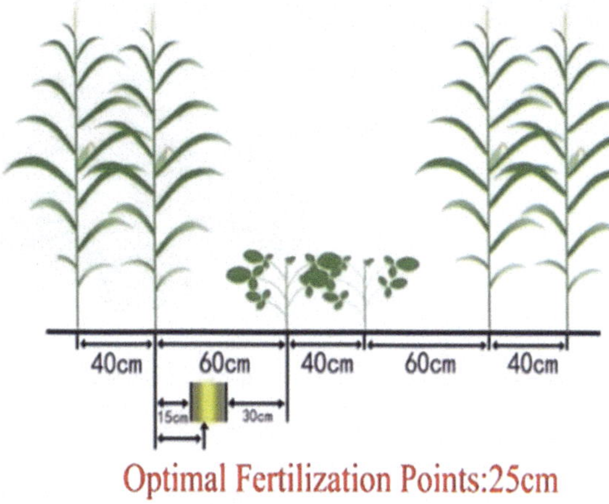

If maize and soybeans are sown simultaneously, a fertilizer is applied to the maize and soybeans at the time of sowing, and the maize fertilizer application site is located on the outside of the maize row between the two sowing strips, 25 cm near the maize row (Ding 2017; Du et al. 2017) (Fig. 10.2).

10.2 Maize and Soybean Banding Compound Planting, Seeding, and Fertilizing Machine

Based on the agronomic requirements of maize–soybean strip intercropping described in the previous section, combined with the geographical characteristics of the actual intercropping sowing operation and the different heterogeneous needs of the sown crops, 2BF-5 and 2BYDF-6 are used as examples to introduce the design ideas, machine performance, and part configuration of the maize–soybean strip intercropping sowing and fertilizer machine to meet the sowing needs of maize–soybean strip intercropping mode in different areas.

10.2.1 2BF-5-Type Maize–Soybean Intercropping Precision Seeder

10.2.1.1 Introduction to the Machine

The 2BF-5 maize–soybean intercropping seeder consists mainly of a frame, fertilizer application device, drive device, driving block, and seeding block, among other components, as shown in Fig. 10.3. The fertilizer and seeding drives are driven as a whole and powered by a group of drive units. The drive and seeding units were mounted on the rear beam of the frame, the fertilizer tank was mounted directly above the frame, and a fertilizer opener was mounted on the front beam of the frame (Ren et al. 2019a, b).

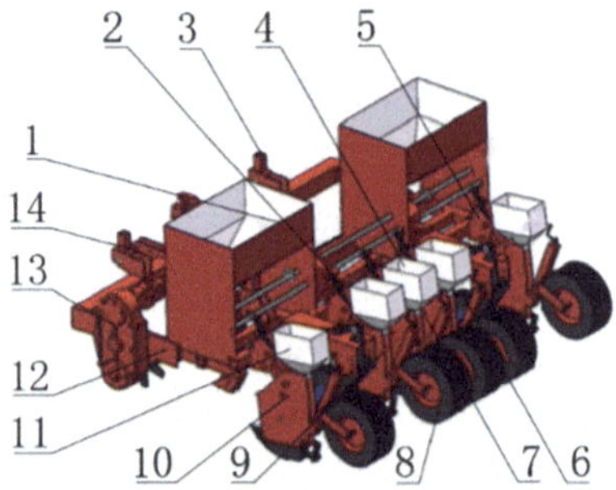

Fig. 10.3 2BF-5 maize–soybean strip intercropping seeder. 1. Fertilizer box. 2. Maize and soybean sub-regulation and sub-control system. 3. Maize hole spacing adjustment device. 4. Soybean seed box. 5. Adjustable monolithic shape. 6. Forward-tilting adaptive-driving ground wheels. 7. Finger clip seeder. 8. Drive wheel. 9. Soybean seeding unit. 10. Maize seeding unit. 11. Maize seed box. 12. Frame. 13. Strip-stubble-removal device. 14. Soybean hole-distance adjustment device

The maize–soybean intercropping precision seeder is a passive seed planter connected to a tractor using a three-point suspension. During operation, the drive wheels are mounted with a downward front tilt and a pre-loaded spring for elastic adjustment to ensure that the drive wheels are always on the ground. The driving units on both sides of the planter are equipped with overrunning clutch gears that are powered by faster-speed drive wheels and transmitted to the main driving shaft. Simultaneously, power is transmitted to the maize and soybean seeding units through the maize and soybean hole spacing adjustment devices, respectively, after changing the transmission ratio, which drives the embedded spoon disc-type rudder wheel hole sowers in each single-seeding unit for seeding through the sprockets. The maize and soybean seeding driving shafts are isolated from each other using bearings embedded in hexagonal tubes to achieve precise control of the respective maize and soybean hole distances (Ding. 2010; Qu et al. 2014).

10.2.1.2 Main Technical Parameters

The machine design parameters of the 2BF-5 maize–soybean strip intercropping seeder are listed in Table 10.1.

10.2.1.3 Main Component Configuration

10.2.1.3.1 Sowing Monomer Design

The seeding unit is primarily composed of a seeding opener, seed box, spoon-wheel seed-metering device, pressure density wheel, mulching wheel, and suppression wheel, among other components, as shown in Fig. 10.4.

A seeding unit with a spoon-wheel seed-metering device and seed box in the middle is used to reduce the working resistance of the planter and the adhesion between the soil and the seed opener, and an acute angle opener is used, which has a strong soil entry ability and low soil movement rate. After the seeds fall into the seed

Table 10.1 Technical parameters of the 2BF-5 type maize–soybean intercrop seeder

Parameter	Unit	Specifications
Structure	/	Shape-shifting seeding monolithic structure
Matching power	kW	>38
Dimensions (L × W × H)	mm	2400 × 1700 × 1250
Quality	kg	925
Number of maize/soybean rows	/	2/3
Bandwidth	mm	2200
Band spacing	mm	600
Maize sowing row spacing	mm	400
Maize sowing spacing	mm	78.5 (high planting density: 118,500 plants·hm^{-2}
		101.9 (Medium planting density: 91,500 plants·hm^{-2})
		121.6 (low planting density: 76,500 plants·hm^{-2})
Soybean sowing row spacing	mm	300
Soybean planting spacing	mm	62.8 (High planting density: 220,500 plants·hm^{-2})
		81.5 (Medium planting density: 169,500 plants·hm^{-2})
		97.3 (low planting density: 141,000 plants·hm^{-2})

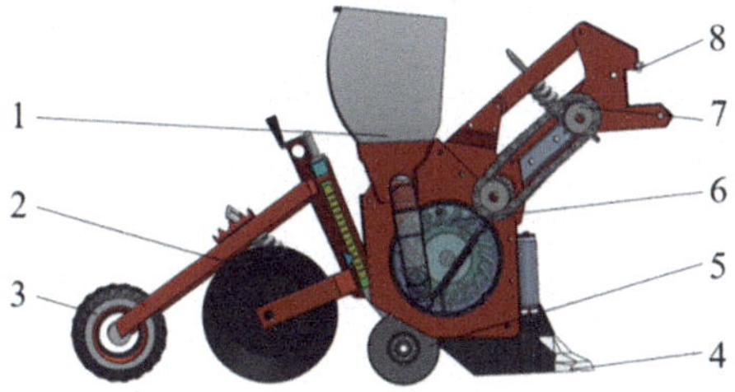

Fig. 10.4 Schematic of the maize sowing monomer structure. 1. Seed inlet 2. Mulching wheel 3. Suppression wheel 4. Seeding opener 5. Pressure density wheel 6. Spoon-wheel seed-metering device 7. Sprocket 8. Frame connection

furrow through a seed-guiding mechanism, to reduce the bouncing of the seeds in the seed furrow, a pressure density wheel is installed immediately afterward, which is flexible and can be in close contact with the ground. In this way, the trajectory of the seeds after being discharged from the seed-guiding mechanism overlaps with the pressure density wheel at the ground junction, thus limiting the bouncing of the seeds in the seed furrow and improving the uniformity of the seeding spacing. This mechanism is shown in Fig. 10.5.

10.2.1.3.2 Seed-Metering Device Design

The seed-metering device is the core component of the seeder, and its structure directly affects the working quality of the seeder. Spoon-wheel seed-metering devices are mechanical devices mainly used for the precise sowing of maize, soybean, and other crops. They have a compact structure, simple transmission, and good seeding quality and are among the most widely used precision seeders

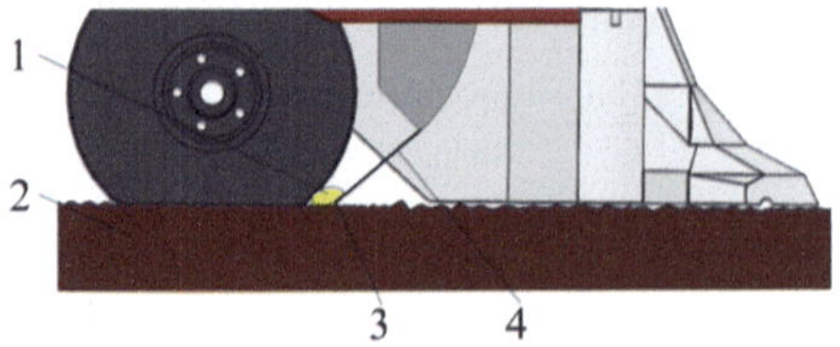

Fig. 10.5 Schematic of the pressed density wheel and seed fit. 1. Maize seed. 2. Soil. 3. Intersection between the compacting wheel and soil. 4. Maize discharge route

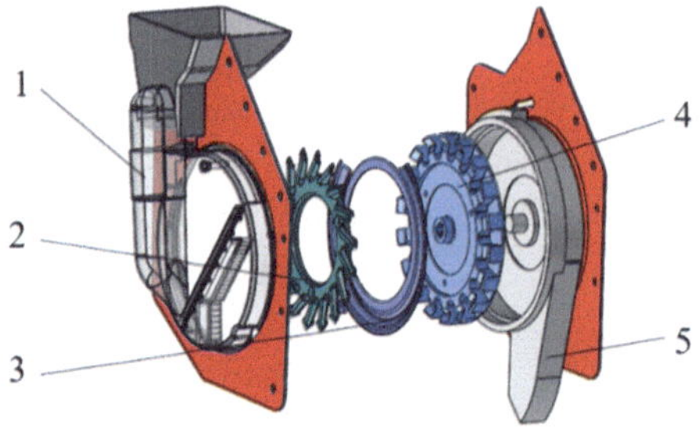

Fig. 10.6 Spoon-wheel seed discharger with a seed-guiding structure. 1. Housing. 2. Spoon-wheel seed-metering device. 3. Spacer. 4. Seed-guiding impeller. 5. Cavity shell with a seed-guiding mechanism

currently in use. To meet the demand for precision sowing of tightly spaced, densely planted maize and soybean under belt compound planting mode, the use of as many spoons as possible can improve the seeding frequency to meet the requirements of high-speed operation. In addition, to prevent the seed from bouncing after falling into the seed furrow from the seed guide opening during field operation, which leads to uneven seed distribution in the seed furrow and affects the seeding spacing uniformity, a seed-guiding mechanism is designed at the outlet of the scoop-wheel seeder to realize zero-speed seeding; the overall structure is shown in Fig. 10.6.

An ideal seed-guiding mechanism design should follow the principles of smooth operation and smooth seed passage, that is, the seed-guiding process does not affect the seed discharge sequence and interval and does not produce bouncing. According to the "zero-speed seed drop" principle, the seed-guiding mechanism is designed as a straight segment *AB* and a curved segment *BC*, as shown in Fig. 10.7. The force analysis for seed movement in the seed-guiding mechanism is shown in Fig. 10.7.

After theoretical analysis, the angle between the straight segment *AB* and the horizontal surface was determined to be 75°. To ensure that the seeds in the curved segment have a higher vertical velocity, the angle between the seeds and the circle centerline of curved segment *BC* and the intersection of the straight segment and the curved segment point *B* and the circle centerline is 35°. According to the seed-guiding mechanism design, the height above ground can be obtained from a circular arc. According to the height from the ground of the seed-guiding design, we can

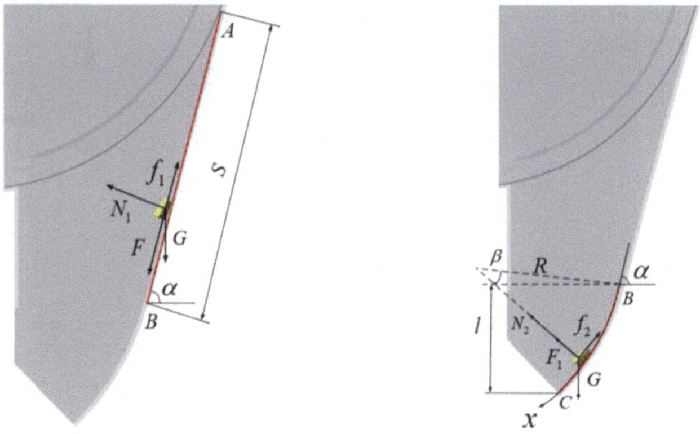

a. *AB* segment force analysis b. *BC* segment force analysis

Fig. 10.7 Force analysis for seed movement in the seed-guiding mechanism

obtain the vertical drop height of the arc section as $l \geq 6.5$ cm, so as to obtain the radius R of curved segment BC as $R > 12.8$ cm; we used $R = 13$ cm. Finally, according to the speed requirement of "zero-speed seed throwing," the length of straight segment AB is $S = 15$ cm.

10.2.1.3.3 Profiling Mechanism Design

The stability, uniformity, and accuracy of the sowing depth directly affect seed germination and emergence. Current research on sowing depth stability mainly focuses on the profiling mechanism, which is used to accurately place seeds in the seed furrow (Sheng. 1995). The planter adopts a rear-mounted rear-position parallel four-link copying mechanism, which fixes the seeding unit completely on the rear side of the four-bar mechanism and can effectively carry out ground-copying operations on high-sloped terrain, as shown in Fig. 10.8. The depth-limiting wheel of the rear-mounted profiling mechanism is located behind the furrow opener; that is, it does not over-compact the soil and can be used to suppress the soil after seeding and mulching (Wang 2020).

10.2.1.3.4 Power Transmission Mechanism

The drive system of the maize–soybean intercropping seeder is designed as a maize–soybean sub-regulation and sub-control system, which uses different variable-speed drive systems to independently drive the device. To realize the sub-regulation and sub-control of maize–soybean plant spacing and sowing depth, the power source is transferred in different directions, mainly including the main drive shaft, fertilizer drive shaft, maize–soybean grain-spacing adjustment device, maize–soybean seed-discharge drive shaft, single-drive sprocket, and bearings. Among them, the maize–soybean seeding and fertilizer power are provided by the seeder drive, the specific form of coupling is shown in Fig. 10.9. The maize-seeding drive shaft is a hexagonal solid shaft, while the soybean-seeding shaft is a hexagonal tube, with the former

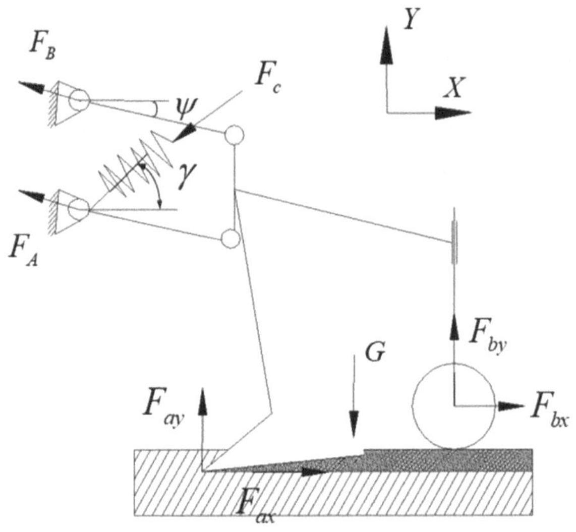

Fig. 10.8 Schematic of the posterior profiling structure

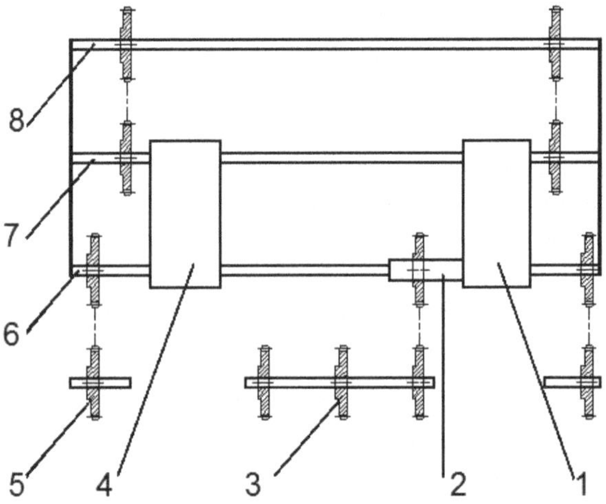

Fig. 10.9 Schematic of the transmission system.
1. Soybean grain-distance adjustment device
2. Soybean-seeding drive shaft 3. Single-drive soybean drive sprocket 4. Maize grain-distance adjustment device
5. Maize single-drive sprocket
6. Maize-seeding drive shaft
7. Main drive shaft
8. Fertilizer drive shaft

running through the latter and isolated from each other by bearings embedded in the hexagonal tube to ensure smooth transmission and achieve accurate control of the maize and soybean hole distances.

During operation, the main drive shaft is used to receive the power generated by the drive unit, part of which is transmitted to the respective maize- and soybean-seeding drive shafts and output to the seeding unit for seeding through the chain drive. The other part is transmitted to the fertilizer discharge device for precise discharge. Meanwhile, changing the fork of the grain-distance adjustment device can adjust the gear position to realize a change in the transmission ratio and precise control of the maize and soybean hole distances. The hole spacing of the intercrop seeder can be adjusted to 10–12–14 cm for maize and 8–10–12 cm for soybean, which improves the precision of single-grain sowing and ensures the sowing quality of maize and soybean under high-density small-spacing conditions.

10.2.2 2BYDF-6-Type Maize–Soybean Intercropping Precision Seeder

10.2.2.1 Introduction to the Machine

In view of the agronomic demand for the mechanized seeding of soybean–maize composite planting in the Yellow Huaihai Sea region, an intercrop precision seeder for composite planting in the belt was optimized. The machine can simultaneously sow two rows of maize and four rows of soybeans, and its structure is shown in Fig. 10.10.

The seeder is mainly composed of a frame, seeding unit, driving ground wheel, fertilizer discharge ground wheel, transmission system, and human–machine interaction display, among other components, and is capable of sowing and fertilizing maize and soybean simultaneously. When the planter operates, the maize- and soybean-seeding units are driven by different ground wheels, which are located between the seeding units and installed symmetrically at the rear of the frame. To avoid straw blockage, the soybean-seeding units are installed in staggered positions. The planter adjusts the plant distance between maize and soybean through the maize and soybean spacing adjustment device, adjusts the height of the depth-limiting wheel to control the depth of the double-disc opener, and adjusts the spring of the ballast wheel to control the sowing depth. When the planter operates, the fertilizer and seed openers open the fertilizer and seed furrows, respectively. The fan provides negative pressure to the seed rower, which makes the seed rower realize the process of sucking and seed rowing. Finally, the seed falls into the seed furrow through the seed-guiding tube and completes the mulching and suppression under the action of the V-type suppression wheel to complete the planting process.

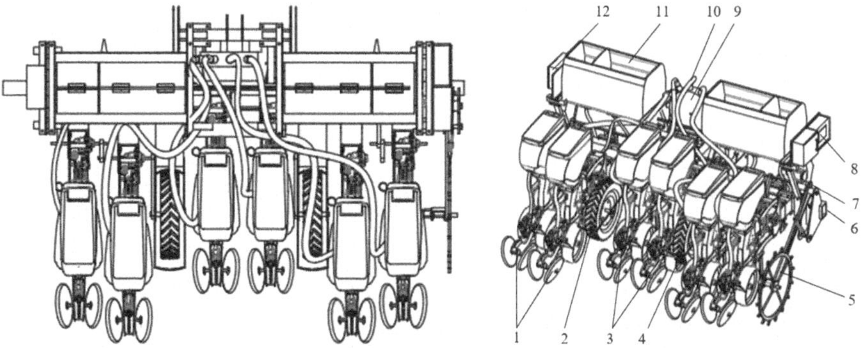

Fig. 10.10 2BYDF-6 Maize–Soybean Strip Intercropping Seeder. 1. Soybean-seeding unit. 2. Soybean unit driving ground wheel. 3. Maize-seeding unit. 4. Maize unit driving ground wheel. 5. Fertilizer discharge ground wheel. 6. Frame. 7. Transmission system. 8. Human–machine interaction display. 9. Fan. 10. Air duct. 11. Fertilizer box. 12. Main control box

10.2.2.2 Main Technical Parameters

The machine design parameters of the 2BYDF-6 maize–soybean strip intercropping seeder are listed in Table 10.2.

10.2.2.3 Main Component Configuration

10.2.2.3.1 Sowing Monomer Design

The sowing unit is an important component of the seeder. The designed sowing unit is primarily composed of a seed box, seed rower, mimic mechanism, pre-groover, mudguards, double-disc groover, seed-guiding tube, depth-limiting wheel, and suppression wheel. The sowing unit integrates the functions of grooving, seed rowing, depth limiting, mulching, and suppression, as shown in Fig. 10.11. The maize and soybean planting spacing can be changed by adjusting the transmission ratio of the gears in the maize and soybean spacing adjustment boxes. The sowing depths of maize and soybean are 3~5 and 2~3 cm, respectively, and by turning the depth limitation rocker, the sowing unit can achieve the sowing depth requirement according to ground conditions. Because the suppression force required by maize and soybean is different, different suppression force requirements can be achieved by changing the position of the suppression force adjustment lever (Gao et al. 2020; Li 2003).

10.2.2.3.2 Seed-Metering Device Design

Air-absorbing seeders have been widely used for the precision sowing of maize and soybean crops because of their high seeding accuracy, lack of seed damage, and high-speed operation capabilities (Liu et al. 2021; Ding et al. 2018; Shi et al. 2019;

Table 10.2 Technical parameters of the 2BYDF-6 maize–soybean strip intercropping seeder

Parameter	Unit	Specifications
Structure type	/	Suspension type
Matching power	kW	≥50
Machine size (L × W × H)	mm	1470 × 2500 × 1250
Operating speed	km/h	2~8
Work efficiency	hm^2/h	0.48~1.92
Seed discharger type	/	Air suction seed discharger
Fertilizer drainer type	/	Outer groove wheel-type fertilizer drainer
Number of maize/soybean rows	/	2/4
Working width	mm	2400
Maize—maize/soybean/soybean/soybean—maize seeding row spacing	mm	400/300/600
Maize/soybean planting spacing	mm	80~140/70~130
Maize/soybean seeding depth	mm	40~50/30~40
Fertilizing depth	mm	60~150

Fig. 10.11 Schematic of the seeder monomer structure. 1. Seed box. 2. Seed-metering device. 3. Mimic mechanism. 4. Suppression force adjustment device. 5. Pre-groove opener. 6. Mudflap. 7. Double-disc groove opener. 8. Seed-guiding tube. 9. Depth-limiting wheel. 10. Mud scraper. 11. Suppression assembly. 12. Suppression wheel. 13. Suppression wheel bracket. 14. Suppression force adjustment lever. 15. Tension spring. 16. Depth-limiting adjustment rocker. 17. Bracket. 18. Seeding monomer drive shaft

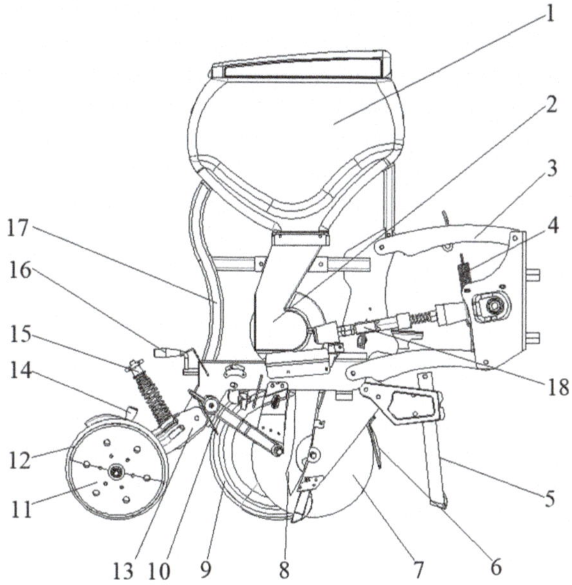

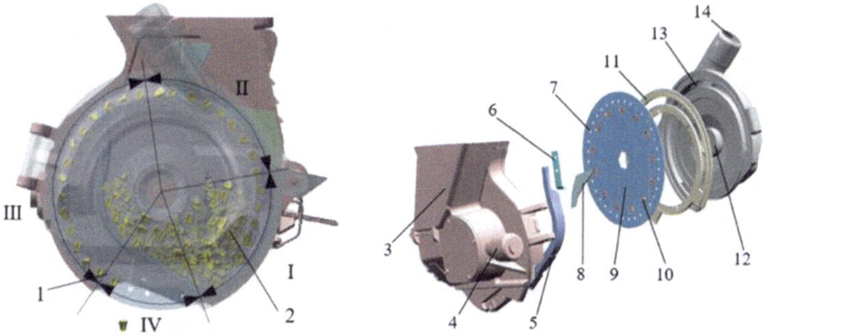

A. Assembly diagram of the seed releaser b. Enlarged view of the seed disperser

Fig. 10.12 Schematic and structure of the air-absorbing seed discharger. 1. Seed discharge port. 2. Seed inlet. 3. Front shell. 4. Drive shaft. 5. Seed stopper. 6. Internal seed-cleaning knife. 7. Type hole. 8. External seed-cleaning knife. 9. Seed discharge disc. 10. Seed-disturbing table 11. Sealing ring. 12. Rear shell. 13. Negative pressure chamber. 14. Air inlet. I. Seed-filling area. II. Seed-cleaning area. III. Seed-carrying area. IV. Seed-dropping zone

Li et al. 2019a, b, c). To meet the sowing demands of tightly spaced, densely planted maize and soybean, the maize and soybean intercropping machine adopts an air-absorbing seeder, which is mainly composed of a seeding disk, internal and external seed-cleaning knives, and front and rear shells, as shown in Fig. 10.12. When the seed releaser is operating, the seed-releasing disc rotates counterclockwise, and the seeds in the seed-filling area are adsorbed on the seed-releasing disc under the double action of mutual squeezing pressure and seed-disturbing tabs, which can play the role of disturbing the seeds and improve the seed-filling effect. By adjusting the extension length of the external seed-cleaning knife and the up and down positions of the internal seed-cleaning knife, the relative positions of the internal and external seed-cleaning knives and seed-releasing disc holes are adjusted, and the excess seeds on the holes are removed under the double action of the internal and external seed-cleaning knives. When the seeds reach the seed drop area, the seed baffle plate isolates the negative-pressure airflow, and the seeds fall into the seed trench through the seed-guiding tube under the action of gravity (Han et al. 2023; Yang et al. 2016).

10.2.2.3.3 Seed-Driving System Design

The drive system of the soybean–maize intercropping seeder can realize power-source transfer in different directions, and the maize and soybean drive wheels transfer power to their respective planting units through a chain drive, which is directly driven by the maize drive wheels through the drive shaft. The soybean-planting units are located symmetrically on both sides of the seeder, and the power of the soybean drive wheels on the left side is transferred to the soybean-planting units on the right side through the main drive shaft. The soybean-seeding unit is symmetrically located on both sides of the planter, and the power of the soybean drive wheel on the left side is transmitted to the soybean-seeding unit on the right side through the main drive shaft, where the two soybean-seeding units are staggered between the front and rear. The maize and soybean drive wheel fronts are equipped with a spacing-adjustment box, the sub-regulation and sub-control of the maize and soybean spacing is achieved by adjusting the transmission ratio, as shown in Fig. 10.13.

10.2.2.3.4 Fertilizer Discharge Drive System Design

According to the agronomic requirements of the soybean–maize intercropping pattern, the fertilizer application volumes for maize and soybean are 975–1500 and 300–600 kg/hm^2, respectively. To meet the different requirements of the intercropping planter regarding fertilizer application volume and fertilizer type for maize and soybean, the following design is adopted for the planter's fertilizer

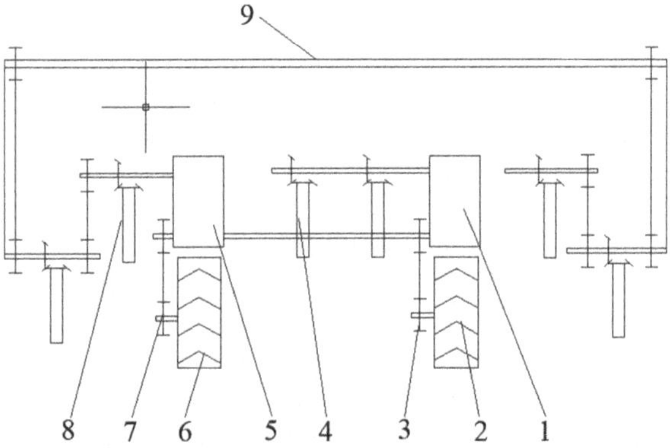

Fig. 10.13 Roadmap of the seeder drive system. 1. Maize spacing adjustment box. 2. Maize drive ground wheel. 3. Maize drive sprocket. 4. Maize-seeding unit drive shaft. 5. Soybean spacing adjustment box. 6. Soybean drive ground wheel. 7. Soybean drive sprocket. 8. Soybean-seeding unit drive shaft. 9. Main drive shaft

discharge system: (a) The fertilizer box is separated by a baffle plate, in which the space for soybean fertilizer application accounts for approximately. (b) Two fertilizer dischargers are used to simultaneously supply fertilizer to one maize sowing unit and one fertilizer discharger is used to simultaneously supply fertilizer to two rows of soybean sowing units, with the soybean fertilizer opener located between the two rows of soybean sowing units and the maize fertilizer opener located 10 cm to the side of the maize sowing unit. (c) The maize and soybean fertilizer-discharging systems have their own fertilizer-discharging shafts driven by different motors to achieve different rotation speeds and fertilizer discharge volumes.

The planter has an independent fertilizer discharge wheel, and the encoder rotates coaxially with the fertilizer discharge wheel to collect the rotation signal of the fertilizer discharge wheel as the planter advances and transmits the signal to the main control box. Parameters such as the fertilizer application amount per acre of maize and soybean are adjusted in the human–machine interaction display interface. The main control box analyzes the received speed signal and the fertilizer application amount per acre of maize and soybean, and then controls the rotation speed of different motors and drives the rotation of the fertilizer discharge shaft. The working process of the seeder fertilizer discharge system is shown in Fig. 10.14.

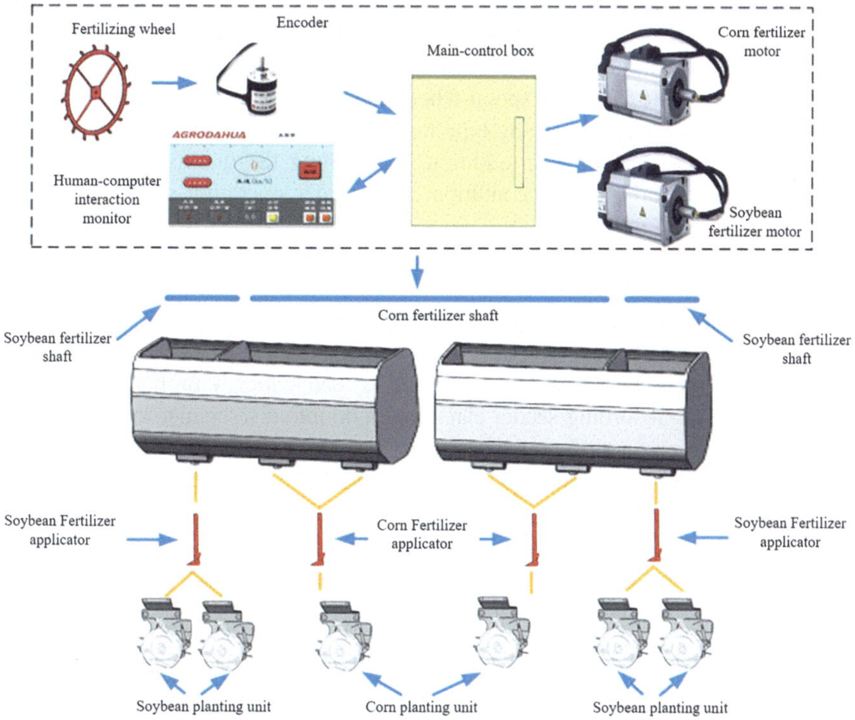

Fig. 10.14 Schematic of the fertilizer discharge drive system

10.3 Key Technologies

10.3.1 Transmission Technology Control and Regulation Division

To separately adjust the soybean and maize spacing, the drive of the intercropping seeder needs to be designed as a separate adjustment and control system, using different variable speed drive systems for independent driving, the specific drive mode of the seeding drive system can be referred to Figs. 10.9 and 10.13. Meanwhile, changing the fork of the grain-distance adjustment device can adjust the gear position to achieve a change in the transmission ratio and accurate control of the maize and soybean hole distances.

10.3.2 Small-Spacing Precision Sowing

To meet the demand for precision sowing under tightly spaced, densely plated maize and soybean belt compound planting conditions, selecting an appropriate seed rower is crucial. A mechanical seed rower can select the spoon wheel type, which has a simple structure and is convenient to use and maintain. A pneumatic seed rower can select the air suction type, which has good sowing uniformity, low seed-crushing rate, and high-speed operation capabilities. To satisfy the requirements of single-grain precision sowing under dense planting conditions, the number of seed spoons should be increased to 22 for the spoon-wheel seeder. The number of holes should be designed to be 36 for air-absorbing-type maize seeder sowing and 72 for the hole-type soybean seeder sowing. In addition, to improve the quality of the sowing operation under tightly spaced conditions, the seeder operating speed should be controlled within 5 km/h.

10.3.3 Sowing-Depth Suppression Control

A seeder equipped with a spoon-wheel seeder can adopt a profile sharp-angle opener, and an air-absorbing seeder can adopt a double-disc opener to adapt to the furrowing operation in different working environments. Because the maize and soybean sowing depths are approximately 3–5 and 2–3 cm, respectively, all seeders adopt an adjustable monomeric profiling device to ensure that the sowing depth of each monomeric row can be independently regulated. Owing to the different suppression strength needs of maize and soybean, seeders can adopt a combination of solid (maize) and inflatable (soybean) suppression wheels or large (maize) and small (soybean) V-shaped suppression wheels to improve crop emergence consistency (Ma. 2018).

10.3.4 Fertilizer Application Volume Control

As the planting density of intercrop sowing is large, to ensure that the amount of fertilizer applied to a single maize plant during intercrop sowing is the same as that of the net crop, that is, the amount of fertilizer used per acre is the same as that of the net crop, the amount of fertilizer applied per unit distance of the maize row during intercrop sowing must be at least twice that of the net crop. Therefore, the fertilizer discharge system of the seeder needs to be specially designed. First, the fertilizer box with a baffle separating the soybean fertilizer space needs to account for approximately one-third of the total fertilizer box volume. Second, to meet the different maize and soybean fertilizer volume needs, the maize sowing monomer must fertilize two maize rows simultaneously and the soybean sowing monomer must fertilize two soybean rows simultaneously. In addition, the soybean fertilizer opener is located between the two soybean sowing monomer rows, while the maize fertilizer opener is located 10 cm to the side of the maize seeding unit.

10.3.5 Straw Anti-blocking

During no-till sowing in stubble terrain, to avoid straw blockage, the sowing monomer can be installed in a front-to-back staggered position to ensure a larger gap between the two sowing monomer rows to facilitate the flow of straw when the seeder is working. A rotary tillage device can also be installed in front of the seeder to facilitate the breaking of the straw and improve its passage. In addition, to meet the sowing row spacing requirements, a super-narrow monomer can be used to ensure a larger gap between the two rows to prevent straw blockage.

10.4 Operational Effect and Outlook

10.4.1 Operational Effect

10.4.1.1 2BF-5-Type Maize–Soybean Intercropping Precision Seeder

The maize and soybean varieties used in the trial were Zhengdan 958 and Zhengdou 04024, respectively. The theoretical plant spacing was 10.19 cm and the theoretical sowing depths were 4 cm for maize and 6.28 and 3 cm for soybean. The fertilizer for maize was a special controlled-release fertilizer with an application rate of 360 kg/hm^2 applied at 15 cm near the maize row, whereas no fertilizer was applied to soybeans. The field management was the same as that used in the field. To verify the performance of the designed 2BF-5 maize–soybean intercropping precision seeder, and to study the effect of mechanical seeding on seedling quality under three different tillage methods: stubble no-till (SNT), stubble no-till (CSNT), and stubble rototill (CSRT), seeding tests were conducted in March 2017 in Zhongmou County, Zhengzhou City, China, with a 50-kW Dongfanghong tractor at a speed of 4 km/h, as shown in Fig. 10.15.

Fig. 10.15 Field operation of the 2BF-5 maize–soybean strip intercropping seeder

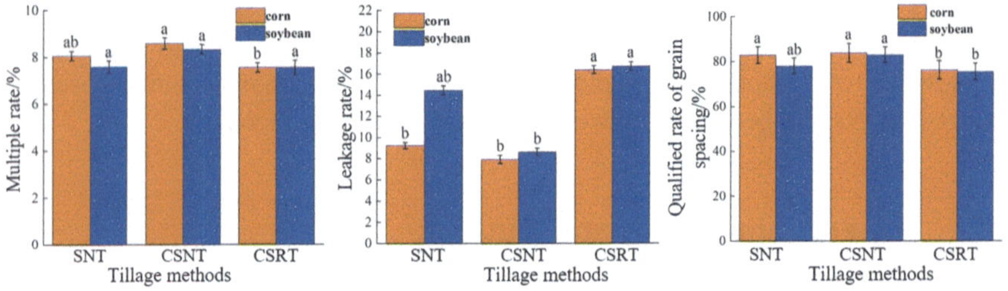

Fig. 10.16 Effect of tillage on the resowing index, omission index, and grain-spacing qualification rate

10.4.1.1.1 Sowing Quality Under Different Tillage Methods

The operational performance of the seeder can be reflected via comprehensive indexes, such as the reseeding index, omission index, and grain-spacing qualification rate. As can be seen from Fig. 10.16, the seeding spacing qualification rate of the 2BF-5 maize–soybean intercropping seeder under different spacings reached 83.57%, the reseeding index varied from 7.63% to 8.63%, the missed seeding index fluctuated between 7.8% and 9.17%. It is worth noting that all performance indexes met the relevant standards and actual production requirements. The reseeding index of maize and soybean sowing was the lowest under stubble rototill

(CSRT) treatment and highest under stubble no-till (CSNT) treatment, whereas the missed sowing index was the lowest under stubble no-till (CSNT) treatment and highest under stubble rototill (CSRT) treatment. Under stubble no-till (SNT) treatment, the residual straw and root stubble on the ground surface caused the seeding units to move up and down further, which negatively affected the seeding spacing consistency. Under stubble rototill (CSRT) treatment, the soil porosity was high and the seeds fell into the seed furrow and moved with the soil pore space under the action of gravity, which caused a decrease in spacing consistency. The test results showed that the seeding spacing consistency of the 2BF-5 maize–soybean intercropping precision seeder was the highest under stubble no-till (CSNT) treatment.

10.4.1.1.2 Average Grain Spacing and Seedling Emergence Under Different Tillage Methods

The average grain spacing of seeder sowing can, to some extent, reflect the slip rate of the driving ground wheel, which is a reflection of the seeder rower leakage. Analysis results of the average spacing of the seeder under different tillage methods showed that there was no significant difference in the average spacing and sowing volume between the stubble rototill, stubble no-till, and stubble no-till methods, which were closer to the set sowing spacing, indicating that the design of the machine drive with an additional drive wheel tilted forward and flexible against the ground could effectively solve the problems of missing seedlings and slippage. The average seeding distance under the stubble no-till (CSNT) treatment was the smallest and closest to the set seeding distance, whereas the stubble no-till (SNT) treatment had a more uneven straw distribution on the ground, resulting in an increase in slippage rate. Therefore, the best seeder operation was achieved under stubble no-till treatment.

As shown in Fig. 10.17, the seedling emergence rate of maize was not significantly different under the no-till and no-till stubble treatments (76.6% and 77.27%, respectively), whereas it was only 60.57% under the rototill stubble treatment, which

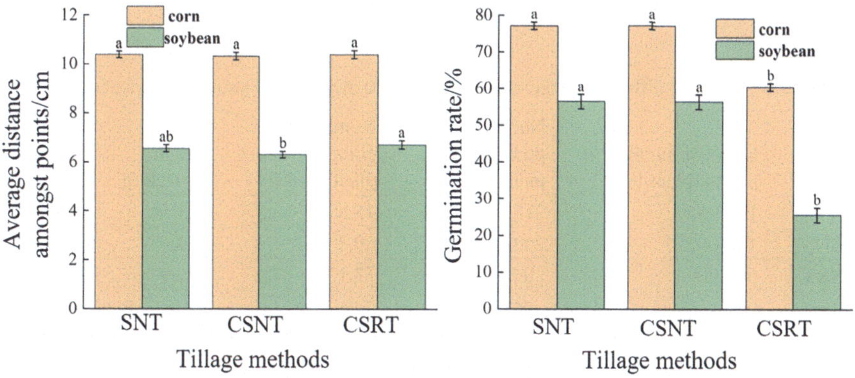

Fig. 10.17 Effect of tillage on the mean grain spacing and seedling emergence rate

was significantly lower than that of the other two treatments. The seedling emergence rate of soybean was not significantly different between the no-till and no-till stubble treatments (53.62% and 56.57%, respectively), both of which were significantly lower than that of maize. The low seedling emergence rate may be due to the fact that soybean seeds are dicotyledonous plants with poor soil breaking ability, and seed germination will consume more energy with increasing sowing depth while increasing seedling emergence resistance.

10.4.1.1.3 Maize and Soybean Yield Under Different Tillage Methods

From the results presented in Table 10.3, it can be seen that the effective number of maize plants was significantly lower under the stubble rototill treatment than under the no-till and no-stubble treatments, and the number of rows and ears did not show significant differences among the three tillage methods, while the 1000 grain weight index was the highest under no-till treatment. Maize yield was significantly lower under stubble rototill treatment than under the no-till and stubble-less treatments.

As shown in Table 10.4, the effective number of soybean plants under the different tillage methods differed significantly, with the highest and lowest numbers found under the no-tillage and rototill treatments, respectively. The number of pods per plant and the number of grains per plant were significantly higher under stubble-rotation tillage treatment than under the stubble-no-tillage and stubble-no-tillage treatments, whereas there was no significant difference in the 100-grain weight. In

Table 10.3 Effect of different tillage methods on maize yield and yield components

Tillage method	Number of effective plants/(plant-hm^{-2})	Number of spike rows/row	Number of row grains/grain	Number of grains/grain	1000-grain weight/g	Yield/(kg-hm^{-2})
SNT	66,375^{a}	14.7^{a}	33^{a}	483.3^{a}	296.83^{ab}	9513.0^{a}
CSNT	67,665^{a}	14.7^{a}	33^{a}	482.7^{a}	303.65^{a}	9912.0^{a}
CSRT	52,635^{b}	15.3^{a}	31.7^{a}	484.7^{a}	292.61^{b}	7480.5^{b}

Note: Data in the same column with different letters indicate differences between treatments up to $P < 0.05$

Table 10.4 Effect of different tillage methods on soybean yield and yield components

Tillage method	Effective number of plants/(plant-hm^{-2})	Number of pods/pod per plant	Number of grains/grain per plant	100-grain weight/g	Yield/(kg-hm^{-2})
SNT	109,935^{b}	21^{b}	45.8^{b}	21.5^{a}	1080.0^{a}
CSNT	118,065^{a}	20.6^{b}	50.2^{b}	21.9^{a}	1299.0^{a}
CSRT	50,790^{c}	29.3^{a}	77.1^{a}	20.9^{a}	823.5^{b}

Note: Data in the same column with different letters indicate differences between treatments up to $P < 0.05$

terms of yield per acre, soybean yield was the highest under no-till stubble treatment and was significantly higher than under the no-till stubble and no-till stubble treatments, which corresponded positively to the effective number of plants.

10.4.1.2 2BYDF-6 Maize–Soybean Intercropping Precision Seeder Operation

The maize and soybean varieties used in the experiment were Zhengdan 958 and Zhonghuang 39, respectively. The fertilizer for maize was a special fertilizer for strip intercropping maize, and the fertilizer for soybean was a special fertilizer for beans. The theoretical plant spacing for maize and soybean was set at 10 and 9 cm, respectively. The experiment was conducted in Linying County, Luohe City, Henan Province, China during May 10–16, 2021, with wheat as the previous crop and the stubble height not exceeding 15 cm. The machine used was a 112.54-kW Rize Fuvor RZ1804-s. When the maize seedlings had fully emerged and their height was approximately 50 cm (around July 12), seeding plots were randomly selected, the maize and soybean seedling spacing was continuously measured at 50-m intervals, and the results were recorded; each set of measurements was repeated five times to count the spacing pass rate, reseeding rate, omission rate, and seedling retention rate of maize and soybean after seedling emergence. The seedling retention rate was calculated as follows:

$$N = \frac{S}{W} \times 100\%, \tag{10.1}$$

where N is the seedling retention rate (%), S is the actual number of plants within 50 m, and W is the theoretical number of plants within 50 m. The seeder field trial and seedling emergence results are shown in Fig. 10.18.

During field operation, to ensure the sowing quality, the working pressure of the tractor supporting fan was 6 kPa or more, with the test results shown in Fig. 10.19. When sown in the field, the spacing qualification and reseeding rates of maize and soybean after seedling emergence decreased with an increase in the seeder working speed. The leakage rate increased gradually as the working speed increased. Statistically, the field sowing effect was lower than that of the bench test, which was due to

A: Maize and soybean field sowing effect B: Intercropping seeder field operation C: Maize and soybean seedling emergence effect

Fig. 10.18 Field operation and effect of the 2BYDF-6 maize–soybean strip intercropping seeder

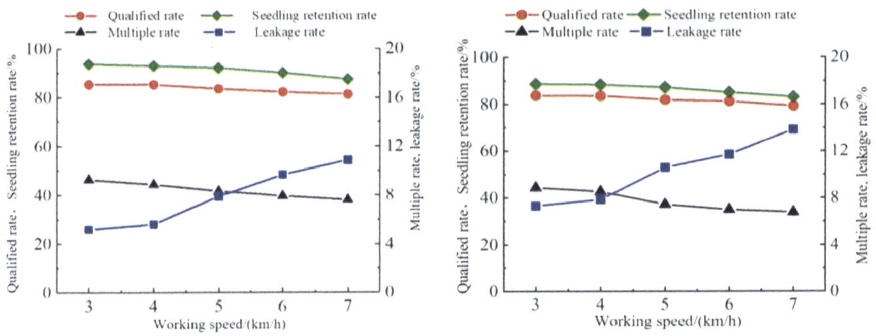

a. Maize field trial results

b. Soybean field trial results

Fig. 10.19 Field trial results for maize and soybean

the fact that straw and weeds tend to accumulate in front of the sowing unit during field operation, resulting in a poor furrowing effect of the furrow opener, and seeds could not fall into the seed furrow normally and evenly, affecting the seeding rate. In addition, the maize and soybean seed germination rate is a key factor affecting the plant spacing qualification rate after seeding.

When the working speed was 3–6 km/h, the spacing qualification rate of maize was 82.89~85.55%, the reseeding rate was 7.93~9.26%, the omission rate was 5.19~9.69%, and the seedling retention rate was 89.59~93.54%. The spacing qualification rate of soybean was 81.3~83.8%, the reseeding rate was 6.98~8.87%, and the seedling retention rate was 84.17~88.73%. All evaluation indexes conformed to the JB/T 10293-2013 "Technical conditions of single-grain (precision) seeder." In addition, considering the working rate and fan energy consumption of the seeder under field operation, the optimal working speed of this intercropping seeder was 5 km/h and the optimal working pressure was 7.5 kPa.

10.4.2 Outlook

Presently, the seeder mainly relies on the manual adjustment of the operating parameters, and the degree of intelligence is low. In the future, the application of intelligent technology in seeders should be strengthened, including the following. (1) Sowing and fertilizing quality detection and alarm technology, which can detect and alert missed sowing and fertilizing blockages in real time. (2) Automatic control technology, which changes the transmission of sowing and fertilization from the ground wheel drive to the motor drive to reduce ground wheel slippage and the associated vibration caused. (3) Variable sowing and fertilizer application technology, which can adjust the sowing density and fertilizer application in real time according to soil fertility and water content. (4) Adaptive control technology, which can adjust the furrowing depth and suppression strength in real time according to plot flatness and soil conditions to improve the seedling emergence consistency.

Plant Protection Equipment in Maize–Soybean Strip Intercropping System

11

He Li, Xuewen Liu, Changchang Yu, and Li Ding

Abstract

The pesticides must be sprayed separately in the plant protection operation of strip compound planting mode of soybean and maize, because the different types of pesticides used in soybean protection and corn protection, and improper spraying will cause serious damage (especially herbicides). On the other hand, there is a big difference between the height of soybeans and corn in the mid- to late-growing stage. It is difficult to carry out plant protection operation using the ordinary spraying machine. It is necessary to develop the plant protection equipment in a targeted manner. Combined with the actual geographical characteristics of the spraying operation and the needs of users, focusing on the agronomic requirements of strip compound planting mode of soybean and maize, a diesel engine-driven articulated steering unmanned plant protection vehicle (model 3WPZ-600), an electric-driven self-propelled unmanned plant protection vehicle (model 3WPZ-200), and a gasoline engine-driven hand-held plant protection vehicle (model 3WPZ-100) for strip compound planting mode of soybean and maize were developed. The design of the independent double spraying system facilitates the independent control of pesticides and spraying parameters required for soybeans and corn. The optimized selection of anti-drift directional nozzles and the installation of anti-drift partition boards and shields can achieve the physical isolation of pesticide droplets. The design of the high clearance chassis with an adjustable wheel track can adapt to different planting rows and soybean plant heights. The design of the adjustable spraying width structure can adapt to the different planting widths of the corn belt and the soybean belt. The overall height of the spray boom can be adjusted and the two ends of the spray boom can

H. Li (✉) · X. Liu · C. Yu · L. Ding
College of Mechanical & Electrical Engineering, Henan Agricultural University, Zhengzhou, China
e-mail: xwliu@henau.edu.cn; yuchang@henau.edu.cn; dingli@henau.edu.cn

W. Yang et al. (eds.), *Theories and Application in Maize-Soybean Strip Intercropping System*, https://doi.org/10.1007/978-981-97-8723-4_11

be partially deformed which can adapt to the difference between the height of soybeans and corn in the whole growing stage. The results of the prototype field test include the coefficient of variation of the single nozzle flow rate was less than 15%; the coefficient of variation of spray uniformity in the direction of the spray boom was less than 10% at operating heights of 50–60 cm and pressures greater than 3 bar; and the droplet drift increased with the increase of the traveling speed, the drift of pesticides sprayed in the soybean belt to the corn belt became severe when the forward speed reached 8 km/h, and the droplet deposition density reached 43.52 points per cm^2. It is recommended that in weed control, the operation speed should not exceed 7 km/h. Therefore, the boom sprayer designed for strip compound planting mode of soybean and maize could meet the plant protection operation requirements from the seedling weeding to the middle and late stage of pest control and also provide technical reference for the research and development of plant protection machinery in other compound planting modes.

Keywords

Intercropping · Plant protection equipment · Drone plant protection · Maize · Soybean

11.1 Agronomic Requirements

At present, the main zone-like maize and soybean compound planting pattern in China has a maize-to-soybean ratio of 2:3. The maize row spacing is 35–40 cm, with a plant spacing of 10–16 cm. The soybean row spacing is 35–40 cm, with a plant spacing of 7–8 cm. The spacing between maize and soybean is 65–70 cm. There are different pesticide types used on maize and soybean, different maize and soybean bandwidths and plant heights, and interference between liquid medicines, which lead to pesticide damage in the plant protection operation under the maize–soybean strip intercropping. Through the core technologies of double spray systems, anti-interference belt dividers, plant height adaptive spray rod mechanisms, wheel pitch adjustment mechanisms, and anti-drift sprinkler head optimization, the self-moving belt-spraying machine adapts to different requirements from weed control in the seedling stage to pest control in the mid and late stages and realizes the synchronous and efficient application of maize and soybean belts.

11.2 Design Ideas

11.2.1 3WPZ-1000 Strip Rod Sprayer

11.2.1.1 Overall Design

According to the agronomic requirements of the mechanized plant protection operation of maize and soybean belt compound planting in the Huang-Huai-Hai area, a self-propelled strip sprayer supporting belt compound planting was optimized. The machine's main self-propelled system is composed of a spray system, which mainly includes a four-wheel drive high-gap chassis, diesel engine, gearbox, and front- and rear-drive axles. The innovative design of the self-propelled system chassis has adjustable wheelbase and folding steering characteristics. The wheelbase is adjustable from 1.6 to 1.9 m, which can effectively adapt to different row ratio modes. The bending radius can be reduced to 2.3 m, and the overall passability is improved. The spray system is the key system of a spray-rod sprayer, and its quality determines the application effect. The optimized design of the zonal double-spray system has the characteristics of "one machine double spray," which mainly includes a spray rod device, maize powder box, soybean medicine box, diaphragm pump, multi-way control valve, and pipeline, among other components. The spray rod device can be easily sprayed when it is expanded, easily shifted and carried when folded, and the spray rod can be adjusted to adapt to different maize and soybean spray heights (Fig. 11.1).

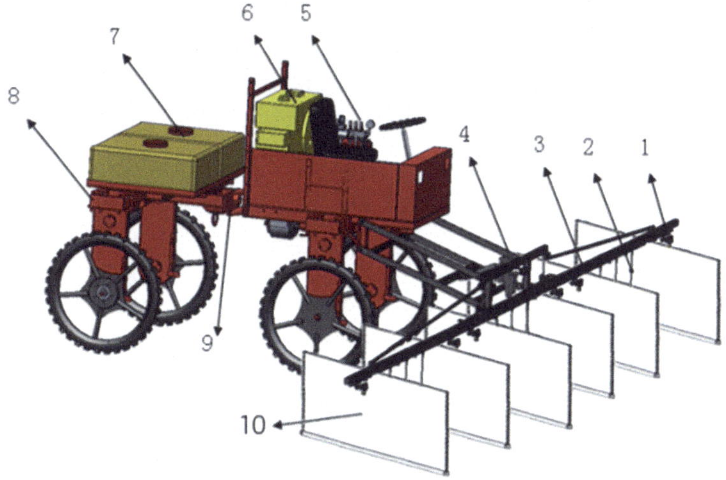

Fig. 11.1 Structural diagram of self-propelled strip ejector sprayer (Li et al. 2020c). 1. Plant protection sprinkler. 2. Hanging spray rod. 3. Three-section folding spray rod. 4. Parallel four-rod mechanism. 5. Variable control system. 6. Power output device. 7. Medicine Box. 8. Wheelbase adjustment device. 9. Folding steering mechanism. 10. Divider anti-interference partition

11.2.1.2 Main Design Parameters

The main performance parameters of the system are listed in Table 11.1.

11.2.1.3 Key Component Design

The sprayer is mainly composed of a driving device, chassis system, and spraying system. Using the 3WPZ-1000 strip spray rod sprayer as an example, the design of the key parts is introduced next.

1. Power configuration: The engine primarily provides power for vehicle driving and the application of two parts. The driving part is for the engine to connect the gearbox and bridge box through the drive shaft to move the vehicle forward and backward. The power required for driving is related to the working environment (wheel sinking depth), vehicle weight, and driving speed. The application part drives the diaphragm pump through the transfer box connection, providing jet stirring when the medicine liquid is mixed in two 500-L medicine boxes and the instantaneous flow of 12 nozzles. This machine adopts the 4L68 model four-cylinder water-cooled diesel engine (as shown in Fig. 11.2) produced by Changchai, China, which is lightweight and exhibits low fuel consumption and stable performance. Furthermore, its emission standard meets the requirements of the third stage of the GB 20891-2014 "Emission Limits and Measurement Methods of Diesel Exhaust Pollutants for Non-road mobile Machinery (China's third and fourth stages)." Its maximum static torque is 175 N·m/1680 rpm and its rated net power is 36.8 kW/2400 rpm.
   The mixing of liquid medicine generally requires a flow of 5–10% of the volume of the medicine box to be stirred evenly. The machine uses two symmetrically placed 500-L medicine boxes, and takes 7% of the flow for jet mixing, that is, 35 L. The flow rate of the largest sprinkler is 1.53 L·min^{-1} at the highest pressure, and the flow rate required for a total of 12 sprinkler heads is 30.6 L, so the flow rate cannot be less than 65.6 L during spraying.

Table 11.1 Technical parameters of the 3WPZ-1000 self-propelled strip ejector sprayer

Parameter	Unit	Specification
Soybean medicine box capacity	L	200
Maize box capacity	L	200
Overall dimensions (L × W × H)	mm	4820 × 2040 × 2970
Wheel span	mm	1600–1900 (adjustable)
Wheelbase	mm	1600
Empty weight	kg	2109
Drive type	/	Four-wheel drive
Spray rod spray width	m	9
Rated power	kW	16.8
Number of sprinkler heads	/	12
Ground clearance	mm	1100

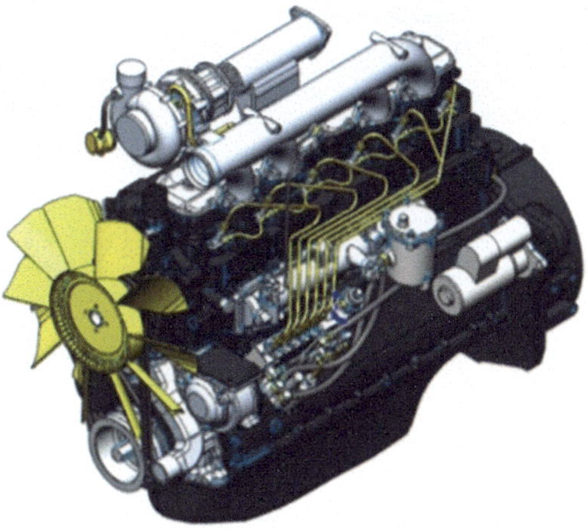

Fig. 11.2 Diesel engine

2. A high-gap gantry drive axle design works often over ridge climbing, which requires greater torque and four-wheel drive technology while reducing crop crushing. To improve the adaptability of the machine, the drive axle using bending steering technology can achieve the same return and distance and reduce the turning radius. The design method of gantry bridges can improve the clearance from the ground, increase the clearance between the mechanical chassis and the ground, and adapt to the prevention and control of diseases, insects, and grasses during different growth periods of maize and soybeans.
3. Transmission function analysis and design: The self-propelled spray rod sprayer requires low speed and large torque when working between paddy fields, and sets a low-speed gear. The working speed can be improved appropriately when working in dry fields. Low- or high-speed gears can be selected. With three forward gears and one backward gear, the secondary transmission can achieve high-/low-speed switching, for a total of six forward gears and two backward gears. The driver can arbitrarily choose between eight speed ranges according to the field management condition of the vehicle, the change of speed, and the slope of the road. Finally, the clutch is a single-piece friction clutch.
4. Tire design and selection: According to the maize and soybean growth height, the self-propelled spray rod sprayer was designed to have a ground clearance of 110 cm, the diameter of which was designed to be 95 cm according to the bridge box and transmission coefficient. The tire width was designed to be an 10-cm high-flower solid tire according to the row distance of the paddy field. This type of tire has a strong grip and small rutting in paddy fields. The phenomenon of winding grass seedlings during use was avoided, and a tire protection board was designed. Entangling the spokes of the wheel should be avoided, as shown in Fig. 11.3.
5. The spray rod of the spray system was designed according to the vehicle height, the position of the lifting mechanism, and the installation of the spray system. The vertical lifting and lifting of the spray system should be controlled within

Fig. 11.3 Wheel diagram (Cheng 2022)

50–180 cm, and the horizontal direction of the spray system should be closed quickly during the field operation field. Furthermore, the problem of transferring in the transportation state should be considered. Controlling the spraying system directly affects the pesticide spraying uniformity. To reduce the weight of the spray system, tubular steel was used to design the truss structure and reduce the influence of the swinging of the spray rod with the vehicle on droplet distribution uniformity.

6. The two sides of the double-medicine-box design (i.e., the maize and soybean medicine boxes) were symmetrically placed on both sides of the strip spray rod spray machine so that the center of gravity of the machine was centered and the space layout was more compact. The maize and soybean medicine box was designed with a narrow top and wide trapezoid bottom so that no local eddy current is generated when the medicine liquid is mixed. The bottom of the medicine box has a liquid tank. The application process does not breathe air into the pipeline, and it is easy to clean the medicine box. There are two jet agitators inside the medicine box, and the jet adopts a double-Venturi tube loop to ensure that the medicine liquid in the application process is fully mixed and uniform, thereby reducing water consumption and preventing foam formation.
7. Nozzle bodies are generally divided into two types: external and internal water flows. The number of installed nozzles can be divided into one-way, three-way, and five-way nozzle bodies. The nozzle body is the base of the installation nozzle, which affects the overlapping state of the droplets and has a significant impact on spray uniformity. The machine uses a three-way nozzle body (Fig. 11.4), which is used to install three different nozzle functions simultaneously to select the nozzle in a timely manner according to different diseases and pests.

Fig. 11.4 Three-body anti-drift sprinkler head (He et al. 2019)

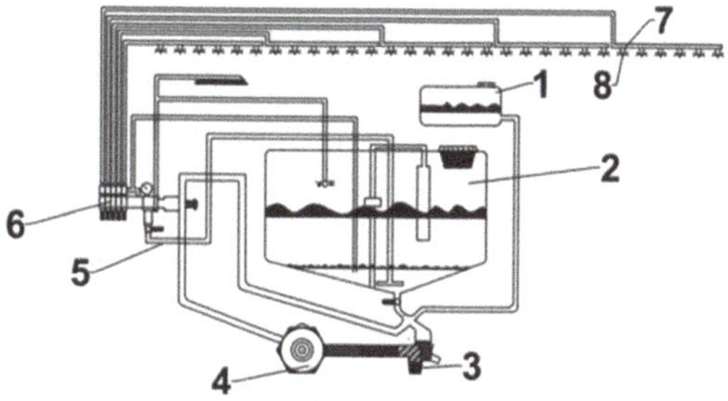

Fig. 11.5 Components of the spray system (He et al. 2019). 1. Clean-water tank. 2. Liquid medicine box. 3. Filter. 4. Liquid pump. 5. Reflux stirring. 6. Pipe control components. 7. Spray rod truss. 8. Nozzle

11.2.1.4 Main Structure of the Spray System

The double-spray system of the self-propelled strip sprayer is mainly composed of a liquid pump, liquid tank, nozzle, anti-drip device, agitator, spray rod, truss mechanism, and pipe strip control components. In the dual-spray system, the spray operation parameters can be controlled independently in different sections according to different operation requirements (Fig. 11.5).

1. Liquid pump spray rod spray machines primarily use diaphragm pumps. The working principle of a single-cylinder diaphragm pump is that when the engine drives the eccentric mechanism to rotate, the eccentric mechanism drives the diaphragm for reciprocating deformation movement, thereby changing the volume between the diaphragm and the pump. When the connecting rod moves down, the volume of the pump cylinder increases, the inlet valve is opened, the outlet valve is closed, and the liquid is sucked out under the action of a pressure

difference. In contrast, liquid medicine was eliminated. This reciprocates the completion of the suction and discharge process.

2. Liquid medicine box: The liquid medicine box contains liquid medicine, and its volume is 200–2000 L. A liquid inlet is opened above the liquid tank, a liquid-inlet filter screen is provided, a liquid outlet is placed below the box, and a stirrer is placed inside the box. A liquid medicine tank is typically made of pesticide-resistant glass-fiber-reinforced plastic or polyethylene.
3. The anti-drift nozzles take advantage of the fact that large-diameter droplets do not easily float away when the target is running, the atomized droplets of the nozzles are larger, and the droplet spectrum is narrow. In particular, the flight-releasing and anti-drift nozzles adopt the jet principle, and the gas–liquid two-phase flow is mixed in the nozzle bodies, and the "small balloons" containing liquid gas are eared through the nozzles. When each of these small balloons reaches the target, smaller droplets and a larger covering density are obtained via secondary atomization under the action of cilia punctures on the leaf surface and the momentum of the leaf surface.
4. Anti-drip device: To eliminate the drug damage caused by liquid dripping along the nozzle under the action of residual pressure when the spray is stopped, an anti-drip device is used. The working principle is that when the spray pressure exceeds the opening pressure of the anti-drip valve, the liquid can open the cutoff diaphragm and flow to the nozzle for spraying. When the liquid path is cut off, the nozzle continues to spray under the action of the residual pressure in the pipeline; however, the pressure in the pipeline drops sharply. When the pressure drops to the closing pressure of the anti-drip valve, the cut-off diaphragm quickly closes the liquid outlet under the action of a spring, effectively preventing the residual liquid in the pipeline from dripping along the nozzle.
5. Churn-dasher: During the sprayer operation, to make the agent in the liquid medicine tank fully mix with water, prevent the agent (such as wettable powder) from settling, and ensuring that the sprayed liquid medicine has a uniform and consistent concentration, the spray rod sprayer is equipped with a blender. Agitators are of mechanical, pneumatic, and hydraulic types. A spray-rod sprayer is commonly used in a hydraulic agitator, which introduces part of the liquid flow into the liquid tank through the mixing nozzle or the nozzle of the jet pump with water to stir.
6. Spray rod truss mechanism: The function of the spray rod truss is to install a nozzle and achieve wide and uniform spraying after deployment. Depending on the length of the spray rod, the spray rod truss can have three, five, or seven sections, except for the central spray rod, which can be folded back, up, or to the sides for easy transportation and parking.

   In addition, the spray rod sprayer often travels on uneven terrain when working in the field, which makes the tractor inevitably appear as irregular left and right swings, owing to the large width of the spray rod. Even if the tractor is slightly shaken, it will cause the spray rod end to swing greatly, thus affecting the spray quality. To overcome this disadvantage, some products are equipped with an isosceles trapezoid four-link hanging mechanism. Two points, A and D, on the

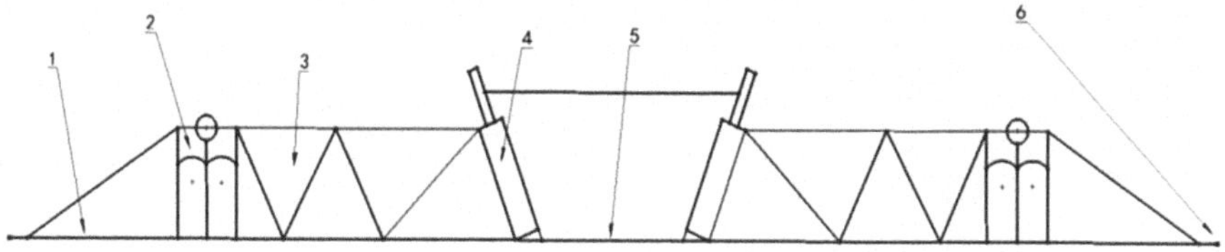

Fig. 11.6 Copying and avoidance device of the spray rod. 1. External spray rod. 2. Spring. 3. Middle spray rod. 4. CAM mechanism. 5. Central spray rod. 6. Copying ring

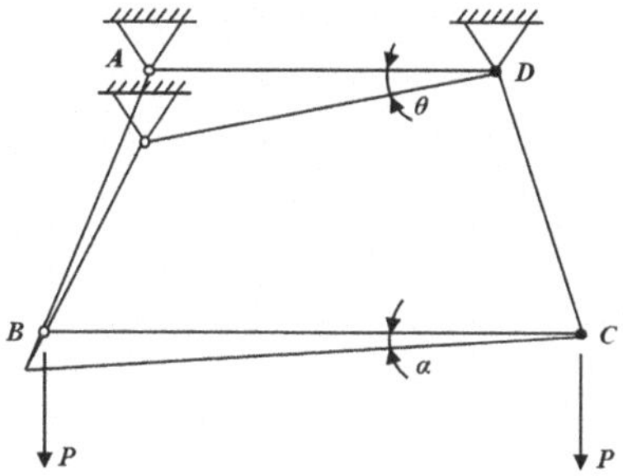

Fig. 11.7 Isosceles trapezoidal four-rod hanging mechanism

spray rod truss, are installed on the frame, and the spray rod is connected to two points, B and C, on the four-link mechanism. When the tractor tilts at an θ angle, the spray rod tilts at an α angle, which is much smaller than θ. Meanwhile, because of the inertia of the spray rod truss itself, the tractor's irregular shaking only slightly affects the spray rod.

7. Pipe control components: The pipe control components of a spray rod spray machine are generally composed of a pressure regulator, safety valve, cutoff valve, distribution valve, and pressure gauge. The pipe control components are installed in a position that the driver can easily access and is easy to operate. The main function of the dispensing valve is to evenly distribute the liquid from the pump to each section of the spray rod to ensure spray uniformity. There are maize multi-channel control valves and soybean control valves, and two sets of multi-channel control valves corresponding to the maize and soybean belt in the spray width, through the multi-channel valve opening and closing, can achieve all the nozzle spray and achieve maize belt or soybean belt spray. Independent application of maize or simultaneous application of maize and soybean can be achieved (Figs. 11.6 and 11.7).

11.2.1.5 Working Principle

The engine drives the gearbox to operate through the transmission shaft, and the output power of the gearbox is divided in two ways. To realize the walking function, the gearbox transmits power to the front- and rear-axle boxes through the transmission shaft and to the walking wheel through the internal gear of the bridge. The transmission transmits part of the power to the transfer box, which transmits the

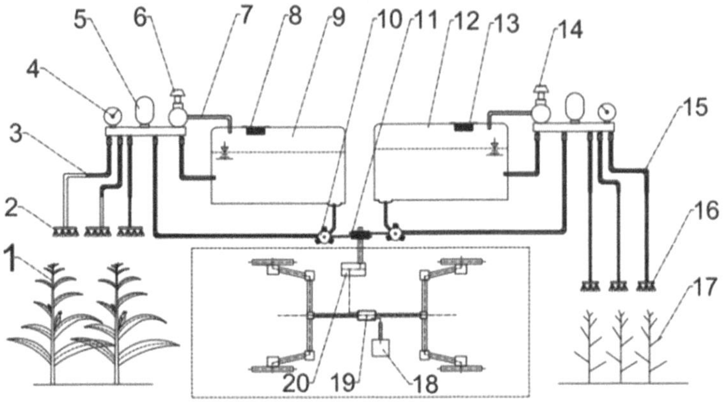

Fig. 11.8 Dual spray system (Cheng 2022). 1. Maize plant. 2. Maize nozzle. 3. Pipe. 4. Pressure gauge. 5. Gas chamber. 6. Pressure regulator. 7. Return pipe. 8. Medicine box filter. 9. Maize medicine box. 10. Maize liquid pump. 11. Double-belt wheel. 12. Soybean medicine box. 13. Filter. 14. Pressure regulator. 15. Liquid. 16. Soybean nozzle. 17. Soybean plant. 18. Engine. 19. Transmission. 20. Transfer box

power to the diaphragm pump, which in turn performs the application operation to automatically add water and stir. Variable control is carried out by the precision variable application system, and the spray is sprayed into the maize and soybean belt (Fig. 11.8).

11.2.2 Other Self-Propelled Belt-Spray Boom Sprayers

11.2.2.1 3WPZ-200 Electric-Driven Self-Propelled Belt-Spray Boom Sprayer

To meet the requirements of mechanized pest control operations in small-scale maize–soybean intercropping fields, a 3WPZ-200 electric-driven self-propelled belt-spray boom sprayer was designed. The machine mainly consists of spraying and self-propulsion systems. The spraying system includes a boom device, lifting mechanism, chemical tank, liquid pump, and liquid circuit system. The boom device and anti-interference belt divider are installed in front of the chassis, while the battery is located in the driver's compartment. The maize and soybean chemical tanks are arranged at the rear of each machine. The self-propulsion system is composed of a chassis, front- and rear-drive axles, steering wheel, and front and rear motors. The front and rear motors provide torque to the corresponding drive axles, enabling the four-wheel drive of the belt-spray boom sprayer. To reduce crop damage during turns, the front and rear Ackermann linkages are connected through a steering rod to achieve four-wheel steering. The machine operates with a 60-V battery-powered motor, featuring a lightweight structure and simple operation. The variable spraying system was designed to adjust the spray flow rate according to

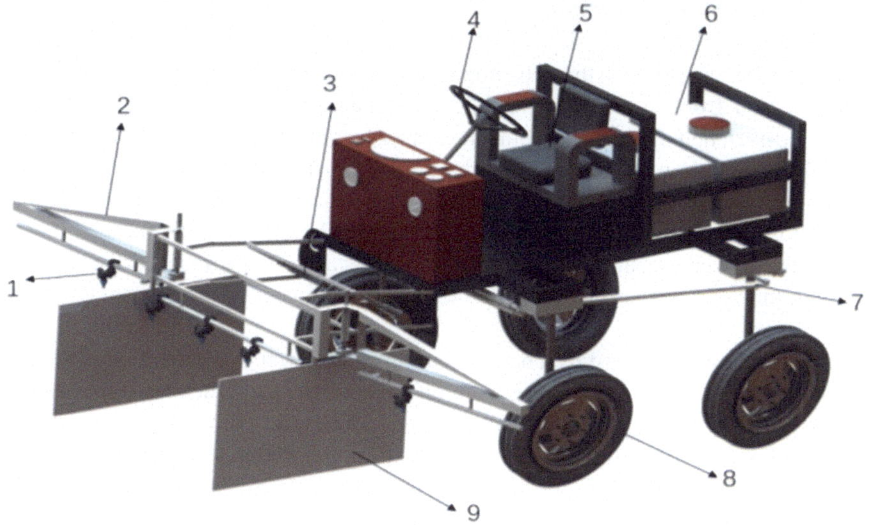

Fig. 11.9 3D schematic of the 3WPZ-200 electric-driven self-propelled belt-spray boom sprayer. 1. Spray nozzle. 2. Spray boom. 3. Lifting frame. 4. Steering wheel. 5. Driver's seat. 6. Chemical tank. 7. Steering rod. 8. Drive wheel. 9. Anti-interference belt-divider

Fig. 11.10 3WPZ Electric-driven belt-spray boom sprayer (Shang 2021)

vehicle speed, thereby ensuring the uniformity and efficiency of pesticide spraying. Therefore, this machine is suitable for pest-control operations in mid- and small-sized fields (Figs. 11.9 and 11.10).

11.2.2.2 Working Principle

The energy from the battery is primarily provided to the self-propulsion and spraying systems. The structure of the self-propulsion system mainly consists of motors, a gearbox, and front- and rear-drive axles. To enhance the maneuverability of the electric-driven belt-spray boom sprayer, dual motors are employed for front and rear propulsion. As shown in Fig. 11.11, the power transmission path begins with the battery providing power to a high-speed rotating motor. The gearbox amplifies the motor torque and transfers it to the countershaft. Through a set of meshed bevel gears, the power transmission direction changes, and finally reaches the wheels through the vertical shaft, completing the traveling process of the belt-spray boom sprayer.

The machine is powered by a battery that supplies power to the diaphragm pump. By activating the spray operation switch on the control panel, the diaphragm pump draws liquid from the chemical tank through a filter and converts it into a high-pressure liquid. Part of the liquid enters the spray nozzles through the connecting pipes, while the remaining portion returns to the chemical tank, completing the pesticide spraying process.

11.2.2.3 Main Design Parameters

The main performance parameters of the 3WPZ-200 electric-driven belt-spray boom sprayer are listed in Table 11.2.

11.2.3 3WPZ-100 Belt-Spray Boom Sprayer Design

11.2.3.1 Overall Machine Design

The 3WPZ-100 belt-spray boom sprayer consists of spraying and self-propulsion systems. The self-propulsion system includes a chassis with a two-wheel drive, gasoline engine, gearbox, and drive axle. The key components of the spraying system include a boom device, lifting mechanism, swinging mechanism, chemical tank, and liquid pump. The boom device unfolds during spraying and folds during movement. The height of the spray boom can be adjusted to adapt to the different maize and soybean growth stages for pesticide spraying. The boom device, swinging boom, and accessories are installed in front of the chassis, while the driver's seat is located in the rear. The chemical tank and plunger pump are arranged in sequence

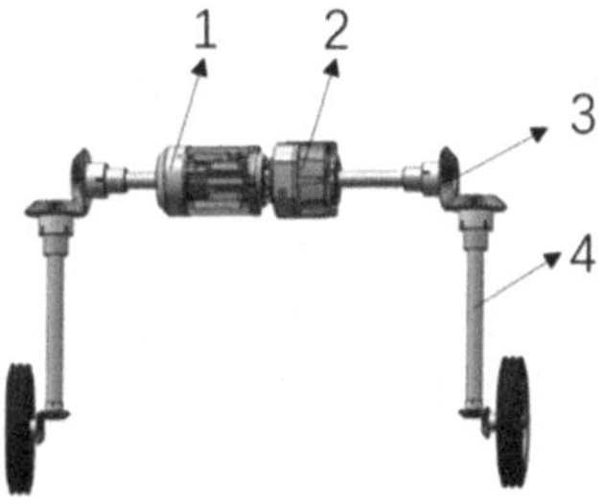

Fig. 11.11 Electric-driven self-propulsion system (Shang 2021). 1. Motor. 2. Gearbox. 3. Bevel gear. 4. Vertical shaft

Table 11.2 Technical parameters of the 3WPZ-200 electric-driven belt-spray boom sprayer

Parameter	Unit	Specification
Soybean tank capacity	L	100
Maize tank capacity	L	100
Overall dimensions (L × W × H)	mm	2000 × 1400 × 1500
Wheelbase	mm	1350
Axle distance	mm	1000
Empty weight	kg	150
Drive mode/four-wheel	/	Drive
Nozzle height adjustment range	mm	500~1200
Spray boom width	m	3
Working pressure	MPa	0.3~0.6
Rated power	kW	3
Number of spray heads	/	6

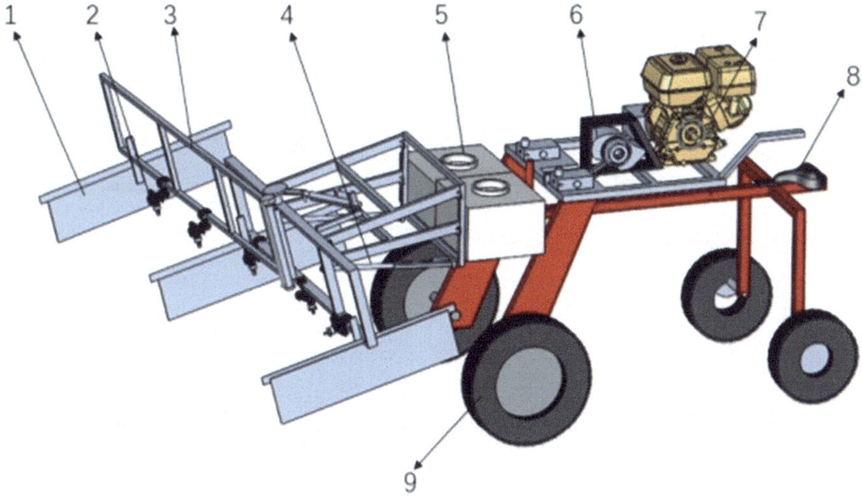

Fig. 11.12 3D diagram of the 3WPZ-100 belt-spray boom sprayer (Li et al. 2020d). 1. Anti-interference belt divider. 2. Spray nozzle. 3. Swinging boom. 4. Lifting rod. 5. Chemical tank. 6. Plunger pump. 7. Gasoline engine. 8. Driver's seat. 9. Drive wheel

between the boom device and the driver's seat. The machine adopts a new type of articulated steering structure that provides a small turning radius and strong maneuverability (Figs. 11.12 and 11.13).

The swinging boom is driven by an electric actuator that rotates the swinging sprocket, which, in turn, drives the chain. This swinging boom addresses the issue of spray leakage during maize spraying, improving the mechanization of pest control for small-scale maize–soybean intercropping (Fig. 11.14).

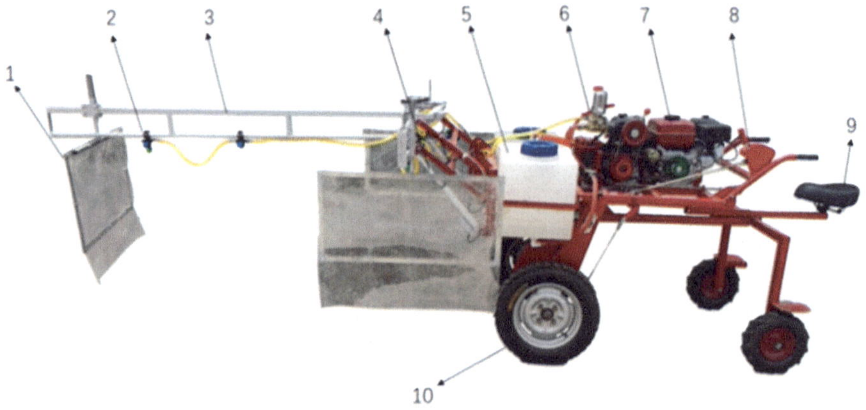

Fig. 11.13 3WPZ-100 belt-spray boom sprayer (Qin 2019). 1. Anti-interference belt divider. 2. Spray nozzle. 3. Swinging boom. 4. Lifting rod. 5. Chemical tank. 6. Plunger pump. 7. Gasoline engine. 8. Brake lever. 9. Driver's seat. 10. Drive wheel

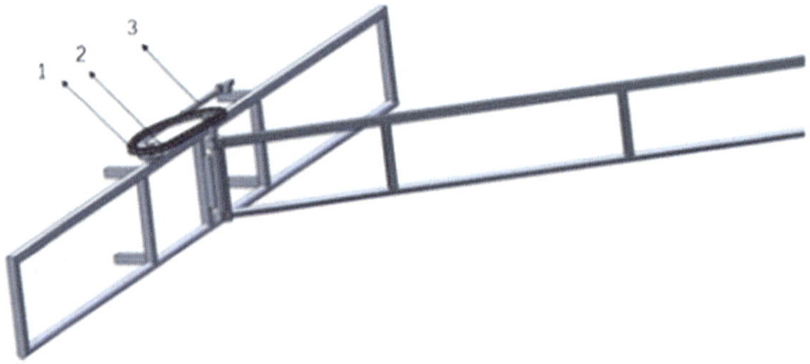

Fig. 11.14 Schematic of the swinging boom device (Li et al. 2020d). 1. Swinging chain. 2. Swinging sprocket. 3. Electric actuator

11.2.3.2 Working Principle

The gasoline engine operates and transfers power to the gearbox and the three-cylinder plunger pump via belts. The gearbox amplifies the torque and transmits the integrated power to the front axle, which drives the wheels through the chain and sprocket. Another portion of the engine power drives the three-cylinder plunger pump for spraying and agitation via belt transmission. The diaphragm pump draws liquid from the chemical tank through a filter and converts it into a high-pressure liquid. Some of the liquid enters the spray nozzles through the connecting pipes, while the remaining portion returns to the tank, completing the liquid-filling process. During field operations, the three-cylinder plunger pump draws liquid from the tank through a filter. In the soybean spraying system, some of the liquid enters the intermediate boom spraying tube through a ball valve and is sprayed out after atomization by the spray nozzles; the remaining portion flows back to the tank. In

Table 11.3 Technical parameters of the 3WPZ-100 belt-spray boom sprayer

Parameter	Unit	Specification
Soybean tank capacity	L	60
Maize tank capacity	L	40
Overall dimensions (L × W × H)	mm	1800 × 1200 × 1200
Wheelbase	mm	1150
Axle distance	mm	900
Drive mode	/	Two-wheel drive
Nozzle height adjustment range	mm	500–1200
Spray boom width	m	2
Operating pressure	MPa	0.3–0.6
Rated power	kW	3.05
Number of spray nozzles	pcs	5
Swinging boom swing range	degrees	0–180

the maize spraying system, some of the liquid enters the outer movable-boom spraying tube through a ball valve and is sprayed out after atomization by the spray nozzles; the remaining portion flows back to the tank.

11.2.3.3 Main Design Parameters

The main performance parameters of the 3WPZ-100 belt-spray boom sprayer are listed in Table 11.3.

11.3 Key Technologies

11.3.1 Dual-Spray System Technology

Considering the different types of pesticides used to spray maize and soybeans, a mutually independent dual-spray system was designed. This system includes a maize tank, soybean tank, dual diaphragm pumps, liquid pipelines, multiway control valves, and drift-reducing nozzles. During spraying, the soybean pesticide is pressurized using a diaphragm pump and enters the soybean pipeline system. Then, through the multiway control valve, the soybean pesticide is atomized by the target spray nozzles and deposited on the target leaves. The operation of the maize spray system is based on the same principle as the soybean spray system (Fig. 11.15).

The maize and soybean tanks are symmetrically distributed on both sides of the spray boom, resulting in a more compact spatial distribution and a significant improvement in maneuverability with a shortened wheelbase. Jet stream mixers inside the maize and soybean tanks ensure a thorough and uniform pesticide mixing. It is worth noting that the same pesticide can be used in both tanks, allowing the sprayer to spray the same pesticide and enhancing the versatility of the split boom sprayer.

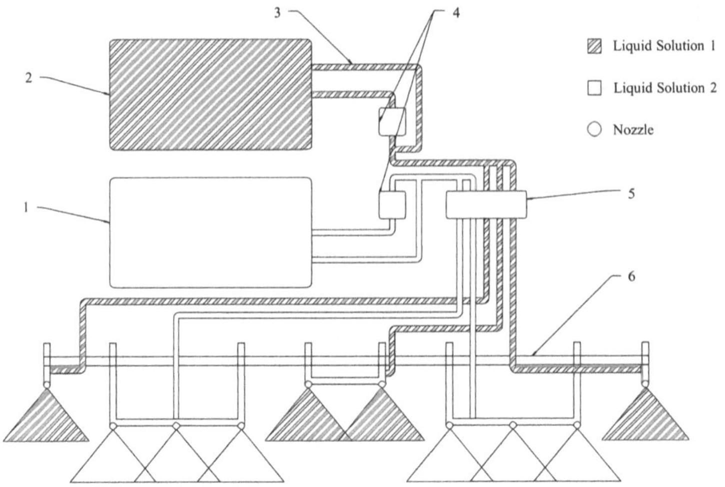

Fig. 11.15 Schematic of the dual-spray system pipelines (Cheng 2022). 1. Soybean tank. 2. Maize tank. 3. Pipelines. 4. Diaphragm pumps. 5. Multiway control valves. 6. Spray boom

11.3.2 Interference Prevention Partition

In the strip cultivation of maize and soybeans, pesticide drift across strips can cause crop damage. Therefore, effective partitioning measures are necessary to prevent interference from different pesticides. Through experimental testing of different partitioning methods, a physically resilient partitioning panel was selected that could effectively suppress droplet drift while avoiding crop damage. The recommended size for the panel is 1.3 × 0.6 m. The designed foldable interference-prevention partition can be folded for convenient transportation and field transition.

11.3.3 Adjustable Spray Boom Structure

There is a significant difference in height between the maize and soybean plants during the mid- to late-growth stages. When the soybeans reach a height of approximately 80 cm, the maize plants reach a height of 2 m. In such situations, traditional spray booms and unmanned aerial vehicles for crop protection cannot always ensure a satisfactory spray quality. To overcome this limitation, an adjustable spray-boom structure was designed to accommodate the varying maize and soybean plant heights, enabling mechanized crop-protection operations throughout the growth process (Fig. 11.16).

The adaptable spray boom has two operating configurations: I-shaped and inverted L-shaped configurations. The I-shaped configuration is suitable for spraying during the early stages of maize and soybean growth. The inverted L-shaped configuration is suitable for mid- to late-stage crop-protecting spraying. This allows

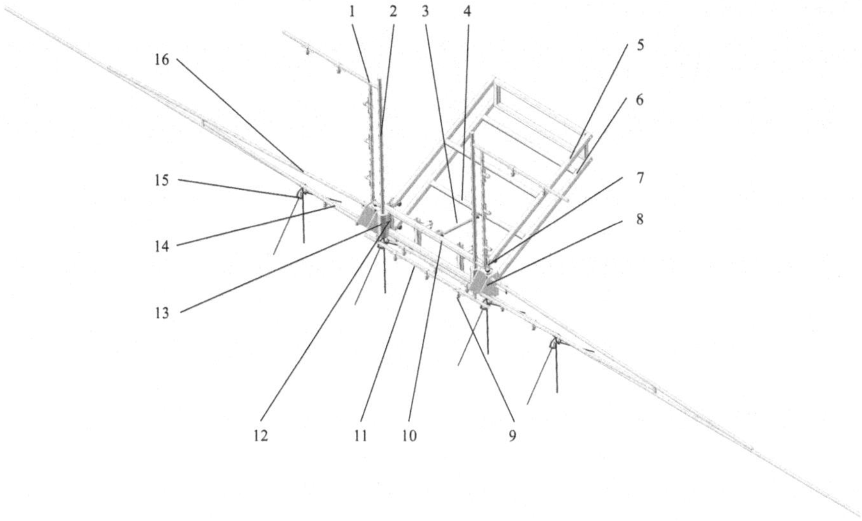

Fig. 11.16 Integrated spray boom truss (Li et al. 2021a). 1. L-shaped vertical spray boom. 2. Vertical connecting rod. 3. Hydraulic lift cylinder. 4. Upper rear pivot shaft. 5. Upper swing arm. 6. Lift frame. 7. Lower swing arm. 8. Reinforcement rib plate. 9. Triple nozzle. 10. Parallel quadrilateral lift adjustment mechanism. 11. Cantilever swing adjustment mechanism. 12. Vertical connecting rod locking bolt. 13. Vertical sliding sleeve. 14. Side spray boom. 15. Interference prevention partition. 16. Rib rod

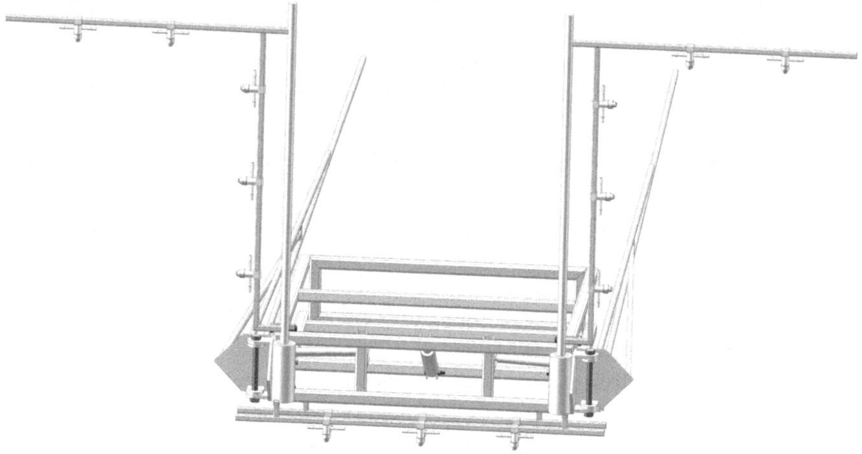

Fig. 11.17 Inverted L-shaped spray boom configuration (Li et al. 2021a)

for comprehensive spraying from the top and sides of maize plants, increasing spray penetration into the canopy, and achieving better spray coverage. During these transitions, the spray boom could be folded and retracted for convenient transportation (Fig. 11.17).

11.3.4 Chassis Adjustment

11.3.4.1 Waist Steering

The split-row spray-boom sprayer utilizes hydraulic power-assisted waist steering, which allows for a small turning radius and improves maneuverability during spraying. Traditional spray machines with conventional steering mechanisms have large turning radii, which can damage crops and result in economic losses when turning at the end of each crop row. The chosen waist-steering mechanism ensures both climbing and crossing abilities while minimizing seedling damage. The vehicle direction is changed by the articulated chassis units. Waist steering provides better coordination during turns and is superior to skid steering, as it maintains the maximum driving force provided by the propulsion components during turning.

11.3.4.2 Wheelbase Adjustment

To accommodate different maize and soybean inter-row spacings and plant heights, a self-propelled high-ground-clearance chassis with a hydraulically adjustable wheelbase was designed. The wheelbase adjustment range is 1.6–1.9 m, effectively improving the efficiency of split-row crop-protection operations in 5/6-row maize and soybean planting systems. The wheelbase-adjustment function is hydraulically driven. As the engine runs, the wheelbase of the front and rear wheels can be adjusted separately by manipulating the hydraulic control lever (Fig. 11.18).

Two horizontal hydraulic cylinders are installed on the beam frame of the vehicle chassis. The piston rods of the hydraulic cylinders are connected to the inner side of the drive-axle gearbox, and the upper end of the gearbox is connected to an adjustment sleeve. A distance sensor is installed on the inner side of the gearbox to provide real-time feedback for wheelbase adjustment. The lower end of the hydraulic cylinder is horizontally equipped with a drive-shaft assembly, which includes a drive shaft, follower shaft, output shaft, and a double sprocket. The drive shaft is connected to the differential output half-shaft through a double sprocket; similarly, the follower shaft is connected to the output shaft through a double sprocket. A sliding block-type adjustment mechanism with a simple structure and easy operation was adopted. The wheelbase can be adjusted synchronously or asynchronously according to operational needs. Continuous wheelbase adjustment is achieved through hydraulic cylinder actuation. The drive shaft is connected using double sprockets, which facilitate installation and disassembly for future maintenance. The wheelbase can be automatically adjusted based on operational needs through automatic control mechanisms, such as distance sensors and controllers, with real-time monitoring during the adjustment process.

11.3.5 Optimization and Selection of the Spray Head Position

In the maize and soybean strip intercropping mode, the maize-to-soybean row ratio is often 2:3–2:6, resulting in significant differences in pesticide application rates during crop protection operations.

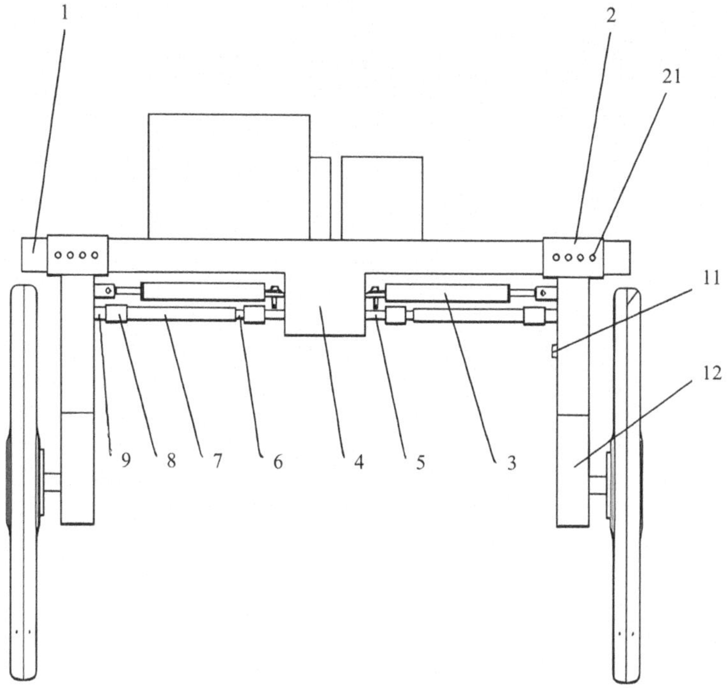

Fig. 11.18 Schematic of the adjustable-wheelbase chassis structure (Li et al. 2019c). 1. Beam frame. 2. Adjustment sleeve. 3. Horizontal hydraulic cylinder. 4. Differential gearbox. 5. Differential output shaft. 6. Drive shaft. 61. Prismatic sleeve. 7.Follower shaft. 8. Double sprocket. 9. Output shaft. 10. Transmission gearbox. 11. Distance sensor. 12. Controller

The main purpose of post-seeding band weeding is to control the spray mist surface of herbicides. In addition to the requirement for complete and uniform coverage, high drift prevention performance is necessary to reduce the interference between different chemicals and minimize potential crop damage. Moreover, to adapt to different intercropping patterns and ensure the effectiveness of herbicide application, the spray boom must freely adjust the position of the spray heads. Therefore, the new spray boom design adopts a flexible adjustable structure, allowing for the adjustment of the height, spray width, and number and spacing of the spray heads according to the actual operating conditions.

Similar to most spray booms, the normal working pressure range for spray heads is 2.5–3.5 MPa. Nozzles with anti-drift fan-shaped patterns, such as that containing 3–5 nozzles, were selected based on the desired application flow rate. Compared with ordinary fan nozzles, air-induced nozzles introduce air bubbles into the spray droplets, increasing the droplet volume and improving the drift-prevention performance while maintaining uniformity.

Spray heads are critical components of crop-protection machinery and can be classified as hydraulic, pneumatic, centrifugal, and powder heads. Slot-type

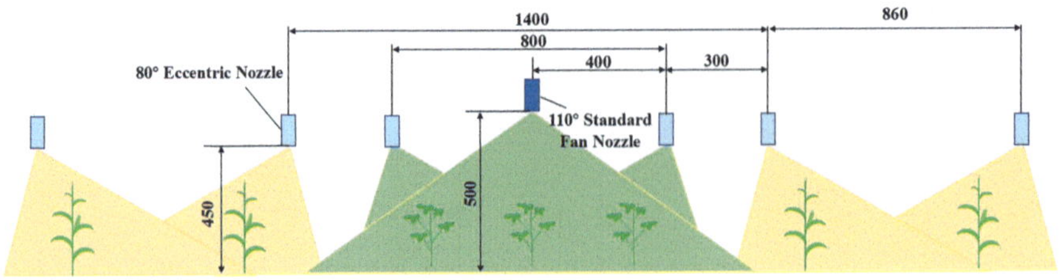

Fig. 11.19 Spray angle of the selected fan nozzle configuration (Cheng 2022)

hydraulic heads are commonly used to spray plant protectants and growth regulators. The atomization principle of slot-type heads involves the sprayed liquid forming a flat film under the compression of a groove-shaped wedge surface, which is then broken and atomized upon impact by stationary air. Because of their fan-shaped spray pattern, they are referred to as fan nozzles. When spray isolation devices are used to prevent interference, the sprayed liquid may accumulate, causing waste and excessive application in certain areas. To minimize the negative effects of isolation devices, it is necessary to optimize the spray head design to maximize sprayed-liquid utilization.

Spray heads were selected and adjusted accordingly to minimize the interference between different chemicals in the intercropped areas of maize and soybeans. The outermost spray heads in each group are equipped with 80° off-center nozzles to reduce the outward spray width and minimize interference between groups while ensuring uniformity within each group. Foldable spray-isolation devices were added between each spray head group to completely limit spray drift within the maize and soybean areas.

After multiple experiments and explorations, a combination of spray heads with excellent uniformity and minimal interference was obtained under the given conditions. This combination includes four 80° off-center nozzles and one 110° nozzle. The 110° nozzle is placed symmetrically on both sides, with the combination pattern shown in Fig. 11.19; and the configuration is repeated accordingly.

11.4 Operational Effect and Outlook

Large-scale mechanization of crop-protection machinery is essential for the development of modern agriculture. The spray-boom sprayer, which incorporates multiple nozzles mounted on a spray boom for spraying operations, significantly enhances crop protection efficiency. However, outdated machinery or spraying techniques often result in uneven spray distribution and excessive application. Spray distribution uniformity along the boom is a crucial criterion for evaluating the spray quality of a self-propelled boom sprayer. Through operational tests of the performance of the strip spray-boom sprayer, the optimal parameters applicable to this equipment can be determined.

11.4.1 Use of the Spray Rod Sprayer

11.4.1.1 Equipment Preparation

Before operation, the machine is repaired and maintained so that all parts of the machine are in a good technical state, the connected parts are smooth and not leaking, the switch is flexible, the atomization is good, and the necessary lubrication is carried out according to requirements.

According to the normal working spray pressure and the determined spray-hole spray volume, the spray volume of a single nozzle ($L\cdot min^{-1}$) is measured, and the total spray volume of the spray rod is equal to the sum of the spray volume of all nozzles.

For test spraying, it should be checked whether the pressure-indicating device and safety valve work normally, assess the spray pressure and atomization quality, and check whether the connection of each part is leaking.

11.4.1.2 Spray Operation

(1) Dosage adjustment: The dosage of effective agents applied per unit area is determined via agricultural technical requirements, and can be adjusted by adjusting the spray dosage, liquid medicine concentration, and operating speed during spray operations. If the concentration is high, the spray amount can be reduced, and vice versa. The spray amount should also consider the plot length, medicine box volume, and the influence of different concentrations on drug efficacy. The spray amount is related to the spray pressure, and a small amount of change can be made by fine-tuning the spray pressure. However, the spray pressure directly affects the atomization quality, which is generally determined before operation and must not be changed at will during operation.
Accordingly, the forward speed v of the unit can be determined:

$$v = \frac{40}{BQ} \times g$$

where v is the unit forward speed, $km\cdot h^{-1}$; g is the spray quantity of the sprayer, $L\cdot min^{-1}$; Q is the dosage per unit area, $L\cdot acre^{-1}$; and B is the spray width of the sprayer, m. v should not be too high; otherwise, the machine turbulence affects the spraying uniformity. After the driving speed is determined, it should not be arbitrarily changed during operation.

(2) Spray rod height adjustment: When spraying herbicides to the ground, the spray rod height plays an important role in determining the uniformity of the drug amount on the ground. A fan-spray nozzle should be used to spray the herbicide, and the spray rod height should make the adjacent spray surfaces cross and overlap so that the ground is evenly coated, as shown in Fig. 11.20. At a fan spray angle of 110°, the distance between the nozzle and spray rod is 50 cm, and the spray rod height is 50 cm (Fig. 11.21). As shown in Fig. 11.22, when the spray rod height is too low (as in (a)), the adjacent spray surface will leak owing to the failure to connect; when the overlap is insufficient (as in (b)), the uneven

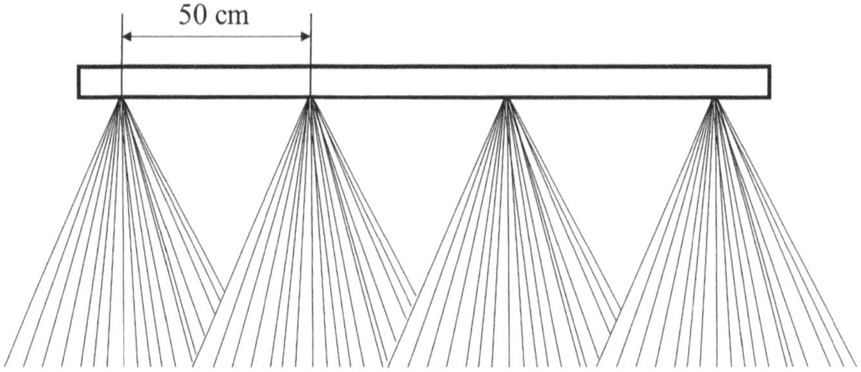

Fig. 11.20 Spacing between sprinkler heads

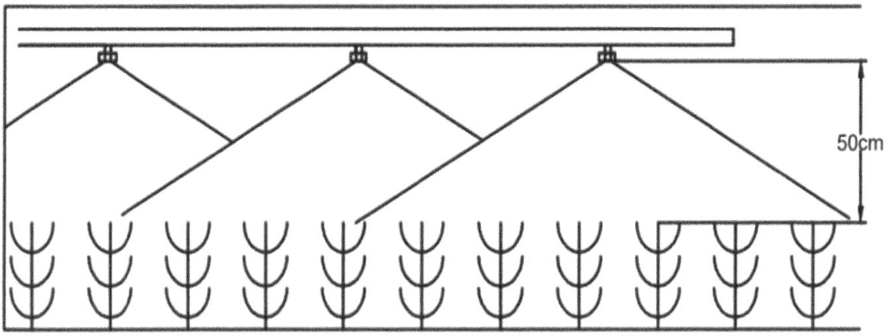

Fig. 11.21 Proper height configuration between sprinkler heads and crops

ground will cause spray leakage; when the height is appropriate (as in (c)), the ground will receive the most uniform drug distribution. Additionally, the spray rod should always be parallel to the ground.

The boom sprayer is mainly used to spray insecticides on cotton and other crops; therefore, a hollow conical fog nozzle is usually installed at the top of the cotton plant from top to bottom, and several similar nozzles are installed on the boom according to the actual situation of the cotton plant. The spray angle can be adjusted arbitrarily according to specific needs so that the front and back of the entire cotton plant can be sprayed with the fog drops. Thus, a three-dimensional spray is used to achieve the best pest control.

(3) Walking is generally fusiform. When carrying out comprehensive noncrop spraying, there should be clear signs indicating the route to prevent re-spraying or missing spraying. When spraying between rows of field crops, the tractor spraying path should be set aside during seeding to ensure that the adjacent spray working ranges are connected. The operator must keep the

Fig. 11.22 Influence of spray rod height on spray quality

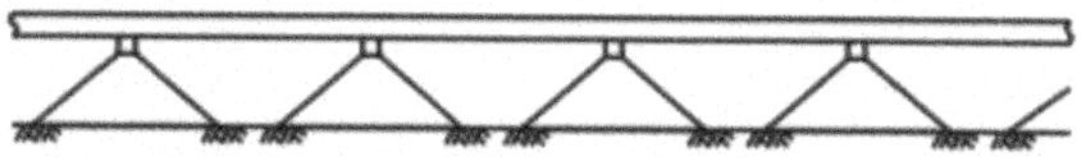

(a) Extremely low position

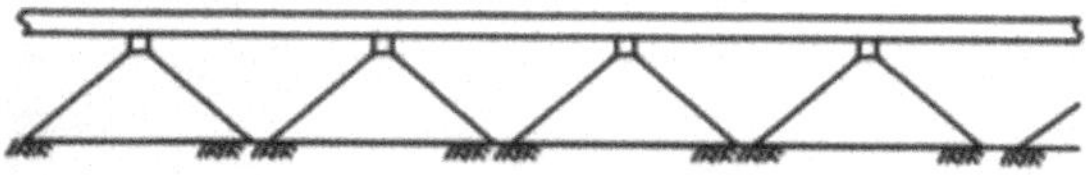

(b) Critical position

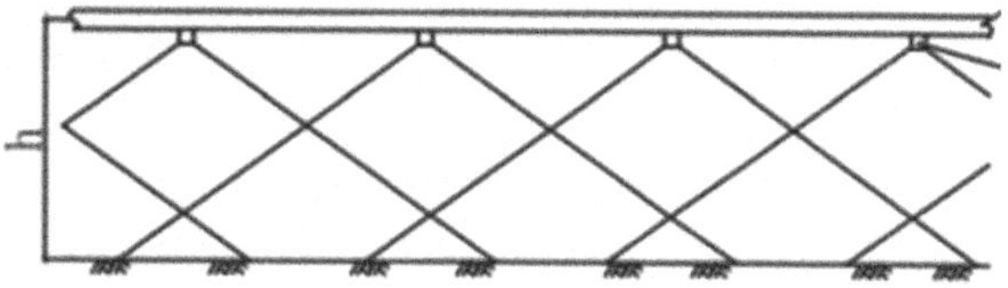

(c) Suitable position

forward speed and working pressure stable and also pay attention to prevent nozzle blockage and leakage, the liquid medicine tank from becoming empty, which causes the pump to dehydrate and run, and spray rod collision obstacles.

(4) Safety technology: During operation, the necessary safety protection equipment should be worn; do not eat or drink; hands should be thoroughly washed before eating; pay attention to the change in wind direction; and when troubleshooting in the field, the pressure should be relieved first and then disassembled; pay attention to prevent spattering when dosing; liquid medicine containers should be treated in a centralized manner; among other measures.

(5) The sprayer must be carefully cleaned after each pesticide spray, after the end of the spraying season, or before repairing the sprayer. After each dose, the pesticides that splashed on the outside surface of the sprayer must be removed immediately. The sprayer should be thoroughly cleaned with soapy water or a neutral detergent, and rinsed with water. The solid–liquid deposits can be removed using a stiff brush.

11.4.2 Spray Effect Test

11.4.2.1 Sprinkler Pressure and Flow Test

The spraying process is complex. The movement of droplets to a target is affected by the application machinery, pesticide dosage form, application technology, and environmental conditions. Many factors influence the plant-protection machinery

Fig. 11.23 Traffic test diagram (Shang 2021)

and application technology, such as whether the liquid in the medicine box is evenly mixed, application pressure, application flow rate, and spray uniformity during application.

The flow from each nozzle is a key factor affecting spray uniformity, and the nozzle is installed on the spray rod because the liquid flow process is affected by the friction of the entire pipeline. The liquid flow with its own resistance produces nozzle blockage, flow loss, pressure reduction, and other phenomena when arriving at the nozzle, resulting in inconsistent pressure at each nozzle and a large flow difference. First, it is necessary to ensure that the difference in each nozzle is controlled within the qualified range, regulate the flow rate of a single nozzle in a timely manner, install multiple nozzles neatly on the spray rod, and make consistent and accurate regulations along the spray rod direction. Therefore, studying the spray droplets from each nozzle is important for studying the uniformity of the spray rod. The flow rate and pressure of each nozzle were tested according to the test method of the Spray Rod Sprayer (GB/T 24677.2-2009) and the Technical Conditions of the Spray Rod Sprayer (GB/T24677.1-2009). The spray machine was adjusted to achieve the same flow rate, and the error was within 15%. The flow test of the sprinkler head used an AMS-Salvarani BVBV, a Belgian high-precision sprinkler flow tester, to measure the flow of all types of sprinkler heads installed on the sprayer. The accuracy of the AAMS nozzle flow tester was 1%. The measurement range was 0.4–5 $L \cdot min^{-1}$, and up to ten sets of data of 100 nozzles can be recorded to the monitor memory and transmitted to the computer.

First, according to the vehicle direction, the number of mechanical nozzles ranges from left to right. The flow tester was calibrated, and a cone adapter was placed under the nozzle to collect the liquid flowing out of the nozzles. After the test data are stable, the "confirm" button is clicked to test the nozzles successively. This procedure is repeated three times for each nozzle; then, the average of the three measurements is taken as the final data of each nozzle. Finally, the data are analyzed and the spray-rod sprayer is adjusted after each test until the nozzle flow satisfies the standard requirements (Figs. 11.23 and 11.24).

Fig. 11.24 Spray pressure test (Shang 2021)

Table 11.4 Test result comparison for sprinkler flow rate for each group

Sort	Grouping			
	1	2	3	4
Average spray volume of the sprinkler head/L·min^{-1}	1.256	1.242	1.302	1.306
Total spray volume/L·min^{-1}	25.110	24.840	26.040	26.120
Standard deviation	0.103	0.072	0.076	0.024
Variable coefficient/%	8.2	5.8	5.9	1.8
Average nozzle pressure/bar	3.240	3.120	3.240	3.555
Standard deviation	0.578	0.263	0.300	0.051
Variable coefficient/%	17.8	8.4	9.3	1.4

The nozzle pressure test used an imported AAMS nozzle pressure detector that met the EU EN13790 requirements for spray detection. The pressure value can be accurately read at the nozzle installation position or at the nozzle on the nozzle body. The nozzle pressure detector is equipped with a pressure gauge with a diameter of 100 mm and an accuracy of 1.0. First, the same model nozzle was installed under the pressure gauge and a spray cap connector was installed at the nozzle; then, the pressure value was read after the flow stabilized. The test results show that the variability of a single sprinkler in terms of flow rate change was within 15% (see Table 11.4).

11.4.2.2 Spray Uniformity Test Along the Spray Rod Direction

The deposition variation coefficient, that is, the distribution uniformity, of the liquid sprayed by the spray-rod sprayer under different pressures and spray heights determines the law of spray distribution uniformity with pressure change. The smaller the CV value, the better the spray uniformity. In general, a CV value of less than or equal to 10% is accepted worldwide; however, some exceptions exist. For instance, Germany and other countries require 7% or less, China requires no more than 15%. More than 15% do not meet the needs of prevention and control, drug damage, and other phenomena.

An AMS-Salvarani horizontal droplet distribution scanner imported from Belgium was used. The horizontal droplet distribution scanner is an electronic measurement device used to assess droplet distribution. The working principle of the horizontal droplet distribution scanner conforms to the international statutory spray test specifications for field sprays (e.g., EN13790 and EN12761). The measuring device consists of nine guides (each 3.2 m long) and a scanner device. With the help of the motor torque, the device can move the guide rail horizontally to automatically stop 800 mm below the spray rod to measure the overall distribution of fog droplets under the spray rod. The spray liquid was collected in a sink and then introduced into a measuring cup. The measurement time between the upper and lower contacts in the measuring cup enabled accurate flow measurements. The droplet distribution measurements were stored on the memory card of the horizontal droplet distribution scanner and could be synchronously transmitted to a computer via a wireless connection. The measured data can be easily converted into graphs, charts, and tables using computer software. The horizontal droplet distribution scanner automatically moved under the spray rod with a maximum operating width of 99.9 m.

During the measurement process, droplets were collected in the tank and introduced into the measurement system. Each cylinder is equipped with an electrode plastic tube to detect the water supply. When the water content in the cylinder increases, it affects the lower electrode to begin the time measurement. When the water level reached the upper electrode position in the relevant measuring cylinder, the measurement was stopped. After the final electrode detects that the cylinder is full, the catheter opens, water is discharged, and the AMS-Salvarani horizontal droplet distribution scanner moves to the next measurement position. The filling times measured at each measurement point were stored on a memory card. Beams with metal contacts were installed on the guide rail every 800 mm. A position sensor detects the metal contact and accurately positions the measurement device. In this process, because there were no gaps between the spray rods, the entire spray rod could be measured. The transmission system was mounted on one drive wheel and four support wheels to place the scanner in a straight line on the guide rail.

An ST110-03 test nozzle model was used. The CV values of the nozzle were tested at heights of 30, 50, and 60 cm from the scanner at a pressure of 1.5, 3, and 5 bar. Each treatment was repeated three times. Table 11.4 shows the spray uniformity variation coefficients of the spray-rod sprayer at different heights and pressures. When the nozzle was 30 cm away from the scanner and the pressure was 1.5 bar, its variation coefficient was large, with little difference being observed in the other position tests. Furthermore, the spray uniformity tended to be stable with an increase in height. The effect of spray height was greater than that of spray pressure, and the working height of the self-propelled precision variable sprayer in field operation was between 50 and 60 cm.

11.4.2.3 Determination of the Droplet Deposition Characteristics

The distribution uniformity of pesticide droplet deposition refers to the distribution uniformity of droplet deposition at different levels along the crop growth height. In

Fig. 11.25 Field droplet-drift characteristics test site (Cheng 2022)

Table 11.5 Performance parameters of fog drop separation

Index	Speed/km/h				
	4	5	6	7	8
Droplet deposition covering density/(unit/cm^2)	1.56	8.56	15.39	21.56	43.52
Droplet deposit coverage (%)	0.09	0.21	0.37	0.62	1.05

pesticide-spraying operations, most of the pesticide droplets are often sprayed and deposited on the top part of the crown of the crop, whereas the droplets distributed in the mid and lower layers of the crop and inside the crop are relatively few (Fig. 11.25). This cannot effectively control the diseases and pests existing in the mid and lower layers of the crop, thereby reducing the utilization rate of the pesticide spray.

Before the field experiment, a self-propelled strip sprayer was used to test the effects of maize on soybean crops. The strip sprayer sprayed distilled water at speeds of 4, 5, 6, 7, and 8 km/h to simulate the spray operation. The vertical spray-rod sprayer was evenly arranged in four rows of water-sensitive test paper and in five rows of soybeans at 30-cm intervals in the driving direction. Three sets of experiments were conducted at 20, 25, and 30 m in front of the strip sprayer. After spraying, the water-sensitive test papers were collected. Finally, the mist drop density and coverage rate were measured using DepositScan software.

The data measured after several field tests are listed in Table 11.5. Under a northeasterly wind with a speed of 1.9 m/s, the drift of droplets in the separation zone increased with an increase in traveling speed. When the advance speed of the zonal spray rod sprayer reached 8 km/h, the weed-control drugs used in soybean belt drifted significantly toward the maize crops, causing drug damage to the maize. The weeding speed of this planting mode should not exceed 7 km/h.

Harvesting Machinery in Maize–Soybean Strip Intercropping System

12

Lihua Zhang, Xiaorong Lv, Yang Zhou, Junshan Nie, Qiqiang Li, Linfeng Chen, Qi Chen, and Yuanfeng Xiao

Abstract

In this chapter, three harvesting methods are proposed based on the maturity sequence of corn and soybeans in the strip intercropping system of corn and soybeans. These methods include harvesting corn first, harvesting soybeans first, and harvesting corn and soybeans simultaneously. At the same time, in response to the high loss rate and low efficiency of machine harvesting of corn and soybeans under the strip composite planting mode, key technologies, such as narrow-range anti-blocking, low-damage ear picking, lightweight cutting table, low-loss high-pass threshing, and efficient dual air duct combination cleaning, were tackled. The key devices developed include a lightweight narrow-width corn header, a single-action low-position soybean header, a low-loss high-efficiency threshing and cleaning device, and an HST full hydraulic track drive chassis. These key devices are ultimately integrated into the 4YZP-2D narrow-width crawler-type corn harvester, GY4D-2 soybean harvester, and a small-sized crawler-type combined harvester with a universal chassis. According to field trials, the grain loss rate of corn is 2.24% (≤3%), the cob loss rate is 3.37% (≤4%), and the husk stripping rate is 92.93% (≥85%). The grain breakage rate of soybeans is 1.81% (≤5%), the impurity rate is 2.02% (≤3%), and the loss rate is 0.52% (≤3%). The operational parameters of the machinery mentioned above are all better than the national standards. They have successfully achieved efficient and low-loss mechanized harvesting for the strip intercropping system of corn and soybeans.

L. Zhang (✉) · X. Lv · Y. Zhou · J. Nie · Q. Li · L. Chen · Q. Chen · Y. Xiao
College of Mechanical and Electrical Engineering, Sichuan Agricultural University, Yaan, China
e-mail: 2021217003@stu.sicau.edu.cn; 2021317023@stu.sicau.edu.cn; 2021317021@stu.sicau.edu.cn; 2022317021@stu.sicau.edu.cn; 2022217004@stu.sicau.edu.cn; 2022317014@stu.sicau.edu.cn

W. Yang et al. (eds.), *Theories and Application in Maize-Soybean Strip Intercropping System*, https://doi.org/10.1007/978-981-97-8723-4_12

Keywords

Maize · Soybean · Intercropping · Harvesting machine

12.1 Mechanized Harvesting Mode

In the maize–soybean strip intercropping system, owing to the different ripening orders of maize and soybean, the corresponding harvesting mode of the machinery is also different and can be divided into three modes: maize-first, soybean-first, and maize–soybean simultaneous harvesting.

1. **Maize-first harvesting mode.** This harvesting mode is suitable for areas where maize ripens before soybeans, which are mainly distributed in the belt intercropping area of Southwest China. The narrow two-row maize combine or high-gap cross-row maize combine is used to harvest maize first; then, commonly used models are used to harvest soybeans after soybean ripening, as shown in Fig. 12.1.
2. **Soybean-first harvesting mode.** This harvesting mode is suitable for banded intercropping areas where soybeans mature before maize. The narrow soybean combine is used to harvest soybeans first; then, commonly used models are used to harvest maize after the maize matures, as shown in Fig. 12.2.
3. **Maize–soybean simultaneous harvesting.** This harvesting mode is suitable for zonal intercropping with the same maturity period for maize and soybeans, and there are two kinds of harvesting patterns simultaneously. First, maize and soybean models commonly used in local production are used to harvest maize and soybeans simultaneously, as shown in Fig. 12.3a. Second, ensilage maize and soybeans are simultaneously harvested and crushed by a silage harvester, as shown in Fig. 12.3b.

Fig. 12.1 Maize-first harvesting sequence model

Fig. 12.2 Soybean-first harvesting sequence model

Fig. 12.3 Maize–soybean simultaneous harvesting model

Because of the high-density planting characteristics of small maize plants, as well as the influence of bandwidth and spacing, the effective working space between maize belts is reduced to a minimum of 1.6 m. Additionally, the unit feeding amount of the cutting table is increased by 2–3 times, resulting in problems such as large harvester power consumption, easy blockage of the cutting table and threshing device, and high loss and crushing rates during harvesting. Therefore, commonly used maize and soybean harvesters cannot be well adapted to the existing maize–soybean strip planting harvest operations, especially when the maize-first or soybean-first harvesting modes are selected. Therefore, it is necessary to develop specific maize and soybean harvesters according to the agronomical requirements of maize and soybeans.

12.2 Maize Combine Harvester

12.2.1 Agronomic Requirements

According to the maize and soybean field configuration, the total width of the maize harvester is less than 1.6 m, the length of the entire machine is less than 6 m, the row spacing of the two cutting tables of the harvester is 0.4–0.6 m, the row spacing of the maize planting is 0.40 m, and the hole spacing is 0.08–0.12 m, while the strip composite planting density doubles the feeding amount of the harvest. The picking, conveying, and peeling efficiency of the harvester must be greater than that of the conventional maize grain harvester. In addition, according to the relevant standards of maize-harvesting machinery (GB/T 21962-2020)' under the designated continuous operation amount, the grain water content of the ear at harvest is 25–35%, the plant lodging rate is less than 5%, the ear sag rate is less than 15%, and the minimum ear height is greater than 35 cm. The operating indicators must comply with the provisions listed in Table 12.1.

12.2.2 Design Ideas

12.2.2.1 4YZP-2D Narrow-Crawler-Type Maize Harvester Overall Design

Currently, there are some problems with harvesters, such as difficult row-spacing matching, large unit feeding amount, easy plugging of the cutting table, low peeling efficiency of the harvester, and high ear damage rates. To meet the requirements of mechanized maize harvesting under the maize–soybean strip intercropping mode, a narrow-crawler maize-ear harvester was designed combined with a cutting table row spacing of 0.40–0.60 m. The cutting table width of the harvester is 1.6 m, the spacing between the left and right ears is 0.41 m, the harvester length is less than 6 m, and the weight of the entire machine is less than 4 t. The designed harvester is mainly composed of a maize-cutting table, ear elevator, peeling device, ear box, straw shredding, and return device (Shang et al. 2002), as shown in Fig. 12.4.

The harvester adopts a crawler-type walking chassis, which has a large driving force, can realize in situ steering, and has good fuselage flexibility, field passability, and stability, which ensures that the harvester does not roll over easily under narrow cutting width and high-speed working conditions. The harvester chassis design was

Table 12.1 Main performance indicators

Parameter	Maize–ear harvester	Maize–grain harvester
Productivity/(hm^2/h)	Not less than the nominal productivity	
Total loss rate/%	≤3.5	≤4
Grain-crushing rate/%	≤0.8	≤5
Ear impurity content/%	≤1	–
Grain impurity content/%	–	≤2.5
Bract stripping rate/%	≥85	–

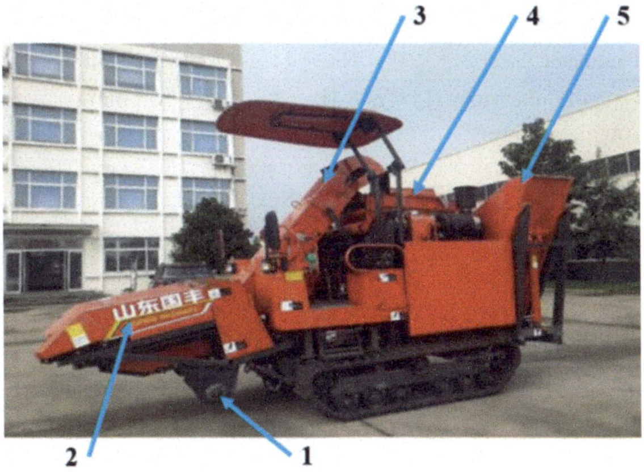

Fig. 12.4 4YZP-2D narrow-crawler maize harvester. 1. Straw-crushing device. 2. Header. 3. Ear elevator. 4. Peeling device. 5. Ear box

optimized to reduce the chassis weight by 13% compared with conventional chassis, further reducing the energy consumption of the entire machine.

In the design, a single-chain plucking device and a grain-feeding device were adopted, and the drive system was optimized to achieve a narrow-width cutting table design to meet the requirements of tightly spaced, densely planted maize. The weight of the cutting table was controlled within 300 kg. By increasing the rotational speed of the reel chain, the problem of block cutting during large-flux feeding was solved. Vibration-reducing flexible picking technology was introduced to reduce the ear damage rate. An all-rubber peeling roller and an adjustable-pressure feeding mechanism were used to improve the stripping rate of the ear bract, and a grain-recovery device was used to further reduce the post-natal maize loss.

During the harvesting operation, the cutting table cuts the maize stalk from the bottom and sends it backward; the maize ear is picked and sent to the peeling device through the conveying device. Finally, the peeled maize ears are sent directly to the ear box, and the maize stalks are chopped and returned to the field.

12.2.2.2 Main Design Parameters

The main design parameters of the 4YZP-2D narrow-crawler maize harvester are listed in Table 12.2.

12.2.2.3 Key Component Design

12.2.2.3.1 Maize Header

According to the field configuration characteristics of the maize–soybean strip intercropping system, a maize picking and cutting platform with a total width of 1.35 m and a row spacing of 0.30–0.60 m was designed via the lightweight optimization method. The structure of the maize picking and cutting table is shown in Fig. 12.5 and is mainly composed of a divider, picking conveyor, picking device, ear-conveying device, and straw-crushing device. During maize harvesting, the maize stalk enters the ear-picking device under the action of the divider and plucking

Table 12.2 Design parameters of the 4YZP-2D narrow-crawler maize harvester

Parameter	Unit	Specification
Overall dimensions (L × W × H)	mm	5660 × 1600 × 2770
Auxiliary power	kW	48
Overall quality	kg	2990
Number of working rows	/	2
Adaptive row spacing	mm	300~600
Hourly production	hm^2/h	0.16~0.35
Total loss rate	%	≤4
Grain crushing rate	%	≤1
Ear impurity content	%	≤1.5

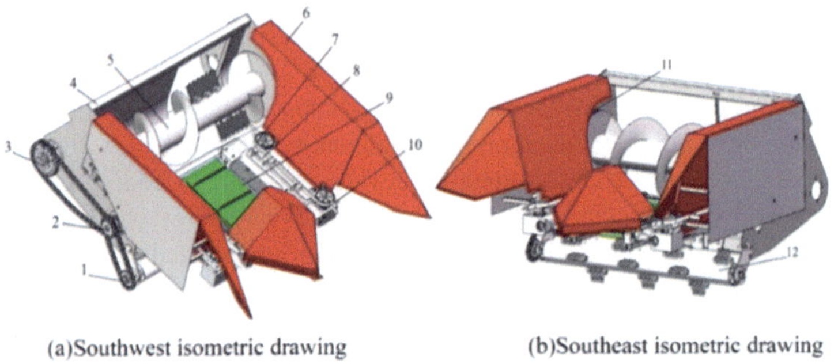

Fig. 12.5 Overall structure of the cutting table. 1. Stubble cutter transmission device. 2. Driven shaft. 3. Driving shaft. 4. Rack. 5. Auger. 6. Body covering. 7. Single-chain picking device. 8. Flexible knife-plate-picking device. 9. Ear conveyor belt. 10. Ear conveyor belt. 11. Agitator transmission device. 12. Stubble cleaner

device, and the ear-picking device picks the maize ear and transfers it back. Due to the narrow row spacing and high planting density of maize and soybeans, the unit feeding amount doubled. To meet the requirements of mechanized maize harvesting, a two-row maize picking and cutting table designed with a narrow width and an anti-blocking structure obtains better operational quality (Liu 2014).

Picking Device Selection

(i) Roller-type ear-picking device

A roller-type ear-picking device mainly relies on a pair of ear-picking rollers to squeeze and pull down the maize stalk while rotating and realizes ear-picking by squeezing the root of the maize ear with the opposite roller. The main feature of the roller picker is that the stalk compression degree is small, so the power consumption is also small. However, there is the problem of nibbling; therefore, the grain loss is large, which mainly occurs in simple maize harvesters (Jin 2011).

(ii) Ear-picking device combined with an ear-picking plate and stem-pulling roller

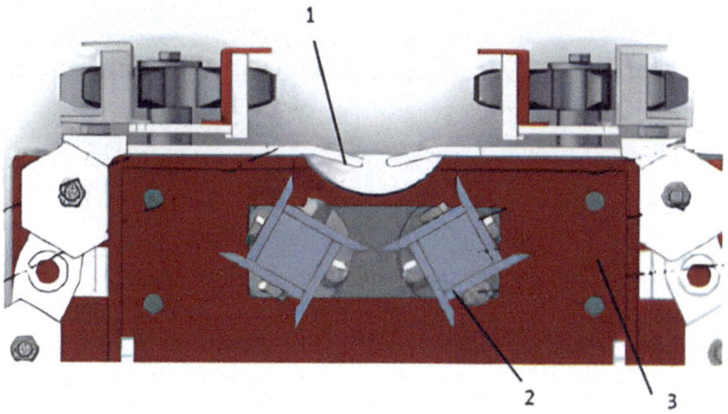

Fig. 12.6 Combined picking device with a stem-pulling knife roller picking plate. 1. Stripper plate. 2. Stem-pulling knife roller. 3. Rack

The picking device combining a picking board and pulling roller holds and pulls the stalk down using a pair of rotating pulling rollers during picking, and the maize ear is forced-picked after being blocked by a pair of picking plates. The main feature of the combined picking device is that there is no contact between the ear and the pulling roll, resulting in minimal damage to the ear. However, when harvesting green wet maize, the upper part of the stalk is prone to breakage, and the broken stalk after being pulled is difficult to deal with in the subsequent workflow, which causes blockage of subsequent work parts (Wang et al. 2014; Guan and Zhao 2009). A picking device with this structure is suitable for harvesting maize with a low grain water content.

(iii) Ear-picking device combined with a stem-pulling knife roller and ear-picking plate

The combination of a stem-pulling cutter roller and picking board is mainly composed of a stem-pulling cutter roller and picking board, as shown in Fig. 12.6. When the maize stalk enters the picking channel, the stem-pulling knife roller cuts and pulls the stalk downward. When the ear contacts the picking board, its diameter is larger than the gap between the picking boards, and the maize stalk is plucked (Chen 2014; Guan et al. 2014). The main features of this structure are as follows: ear damage is minimal, the maize stalk can be crushed, its structure is simple, and its installation is compact. However, this device can only cut the straw and return it to the field; it cannot recycle it.

Based on the analysis and comparison of the aforementioned picking devices, combined with the characteristics of narrow-width ear harvesters requiring a small overall size, the combined picking device with a stem cutter roller and picking board was selected as the picking device for this header (Li et al. 2011). This not only meets the requirements of small size and complete function but also meets the requirements of minimal ear damage and low power consumption.

Straw-Crushing Equipment Selection

The function of the maize straw-returning device is to cut the stalk and scatter it in the field. There are two types of straw-returning devices: vertical and horizontal.

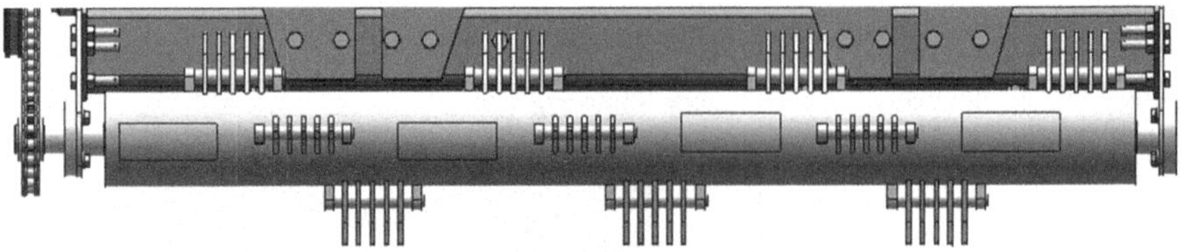

Fig. 12.7 Straw-crushing device

Fig. 12.8 Flat roller configuration

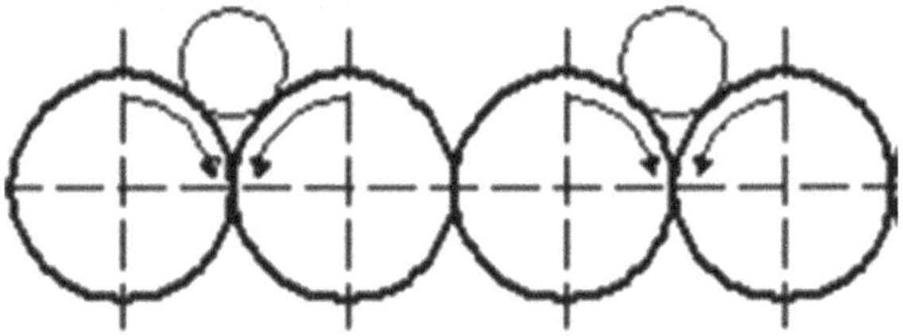

Horizontal straw-returning devices are divided into drum- and disc-cutter-type devices according to their structure.

Considering the small overall structure of the small maize harvester, a small-diameter high-speed crushing device was selected as the shredding device of the harvester, which meets the requirements of simple structure, compact installation, small mass, and low power consumption. The diameter of the crushing device is 130 mm and its structure is shown in Fig. 12.7.

Ear-Peeling Device

To make the ear come in close contact with the peeling roller and prevent beating, a peeling roller was installed above the press feeder to adjust the lower press angle. The underside of the peeling roller is fitted with a grass scraper to prevent stems and leaves from becoming entangled in the peeling roller and causing blockages. The installation requirement of the peeling roller device is high at the front and low at the back, which is conducive to ear flow. During operation, the press feeder presses the ear, while the broken stalks and leaves are sent toward the peeling roller and slide downward. A mixture of ear bracts, stems, and leaves is discharged to the ground through the gap in the peeling roller under the action of several pairs of rotating rollers in opposite directions, and the ear slides into the ear box.

Peeling Roller Configuration Selection

A peeling roller group with a flat roller configuration has a simple structure, low cost, and high working efficiency, as shown in Fig. 12.8. In field operation, the grain loss and crushing rates are low, and it is not easy to "nibble" the ear and grain, but the stripping rate of the ear with higher water content is not high. A V-shaped peeling device with a peeling roller has a simple structure, as shown in Fig. 12.9. The V-shaped arrangement has only one groove, so the distance between the two upper rollers is not large, requiring the ear to be sent strictly in a certain direction. When the

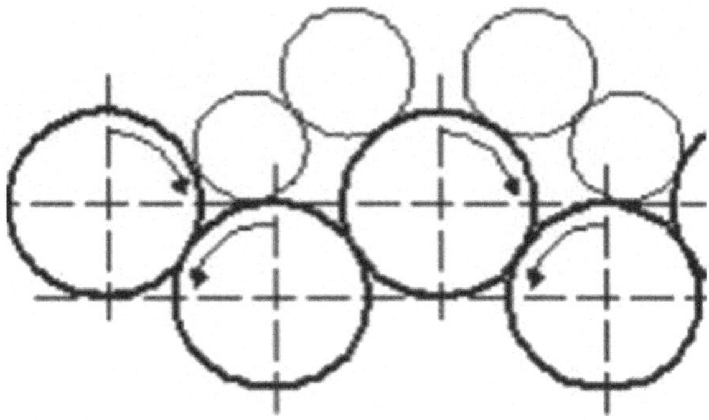

Fig. 12.9 V-shaped arrangement

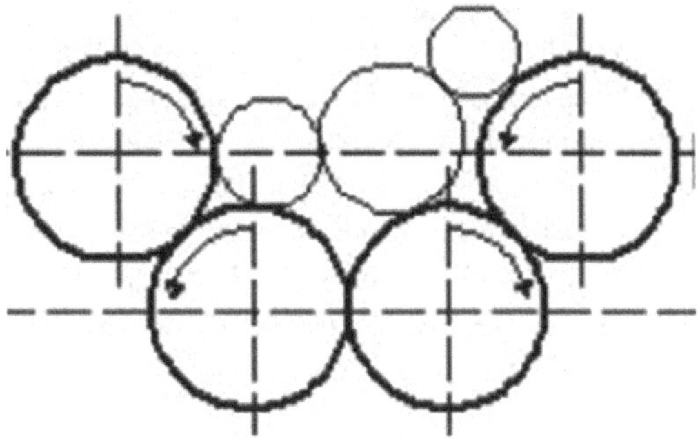

Fig. 12.10 Slot arrangement

ear is sent to the mechanism in no particular order, the ear will shift to one side of the peeling roller group, such that one side of the peeling roller overloads, while the load of the other side is not satisfied. This significantly reduces the efficiency of the peeling device, and the peeling effect is poor and the stripping rate is low. The stripping roller set has a groove configuration with two grooves, as shown in Fig. 12.10. The groove configuration eliminates the phenomenon in which one side of the V-shaped peeling roller set is overloaded and the other side is not satisfied with the load; thus, the ear rotates relatively on the roller and is placed in the groove of the peeling device, which is conducive to improving the peeling effect and efficiency. A stripper equipped with a stripper roller in the trough can achieve a higher stripping rate during the operation process (Zhang et al. 2007; Yan et al. 2008).

Based on the analysis and comparison of the aforementioned peeling roller configurations, the groove arrangement mode was selected as the peeling roller layout to improve the peeling effect and efficiency.

Material and Surface Structure of the Peeling Roller

Commonly used maize-husking roller materials are cast iron and rubber. Rubber rollers are generally used to make rings and are installed on the surface of steel rods to reduce damage to the fruit ears. There are four main maize-husking roller structures, as shown in Fig. 12.11.

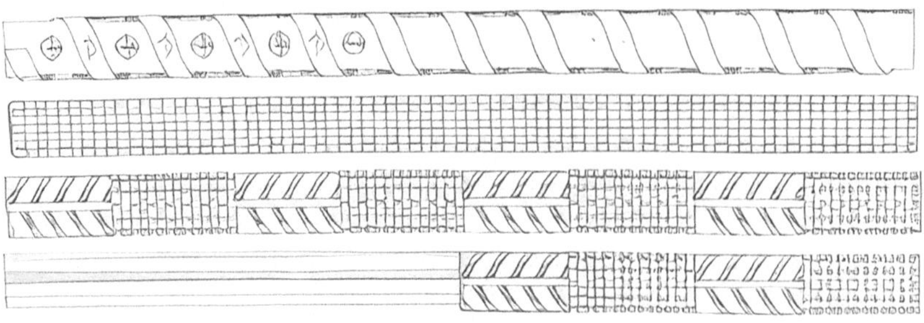

Fig. 12.11 Main structural forms of peeling rollers

(a) Spiral rollers. Spiral rollers are made of HT200, and the roll surface is spiral-shaped. These peeling rollers have the characteristics of low cost, convenient manufacturing, good wear resistance, and a long service life. They are primarily used in general peeling.
(b) Rubber rollers. A rubber roller is composed of several rubber sleeve rings with protrudes installed on a steel core shaft. The surface of the rubber ring is uneven, the groove is approximately 2–3 mm, and the friction between the peeling roller and the ear is increased by the convex structure, improving the peeling efficiency.
(c) Peeling rollers arranged alternately with spiral and rubber sections. The spiral section is made of cast steel, while the rubber section is made of wear-resistant rubber. Two types of bushings are alternately arranged on the steel core shaft. This type of peeling roller exhibits both the wear resistance of the spiral roller and the high friction force of the rubber roller.
(d) The peeling roller is a front flange, and the back-end spiral and rubber sections are alternately arranged. This type of peeling roller can effectively remove broken stems and residual leaves mixed in the fruit ear and prevent clogging of the peeling device.

The grain moisture content of the maize ear is approximately 30%; therefore, an all-rubber roller was selected as the main peeling roller structure, and seven pairs of star-wheel press feeds could achieve low ear damage and a high stripping rate.

12.2.3 Key Technologies

12.2.3.1 Key Technology of the Cutting Table with Narrow-Width, Anti-blocking, and Low-Damage Picking

12.2.3.1.1 Narrow-Width Anti-blocking Technology

High-Speed Single-Chain Plucking Device

To address the problems of the bloated structure of the existing double-harvesting chain, wide width of the entire machine, large ear damage, and high straw-breaking rate under high-speed harvesting operation, the harvesting structure was designed in a light and simple manner. The principle of straw breaking and ear damage caused by the traditional rigid harvesting under high-speed harvesting was analyzed, optimized, and improved to improve the harvesting efficiency. Through analysis and research, it was found that the instant impulse of the straw-touching finger is very large, which leads to straw breaking on both sides and causes cutting table blockage, especially when the harvest is not in line. When the working speed of the plucking chain is increased, the breaking probability of the straw is reduced, and the plucking efficiency is improved by changing the material and structure of the plucking finger. Lengthening thumb and single-chain structures were designed to effectively reduce the transverse size of the cutting table and realize a narrow cutting table design.

Traditional rigid fingers for grain harvesting are made of carbon steel, which has a high hardness; however, direct contact with the ear or straw can easily cause straw breakage and ear damage and is not suitable for picking operations. By testing different flexible materials, it was found that adding a nylon plate to the plucking finger can greatly reduce the ear damage and straw breakage rates. A traditional rigid rotary chain is shown in Fig. 12.12, while a flexible extended rotary chain is shown in Fig. 12.13.

In addition, a rubber-toothed threshing chain made of rubber material can be used to reduce impact damage to maize ears after they are picked. A schematic of a rubber-toothed threshing chain is shown in Fig. 12.14. To further minimize grain damage during transportation, a flexible conveyor belt was installed at the center of the cutterbar, as shown in Fig. 12.15.

Header blockages during high-volume feeding can be avoided by increasing the reel chain speed. The designed reel chain speed is more than 30% faster than that of a conventional reel chain, ranging between 250 and 350 r/min.

Adaptive Adjustment Mechanism for the Header Row Clearance

An adaptive adjustment mechanism was designed for the header row clearance, which adjusts the gap between the picking plates based on the diameter of the straw to solve the problems of grain ear nibbling and stalk blockage, as shown in Fig. 12.16. The adaptive adjustment mechanism for the header row clearance primarily consists of fixed picking plates, adjustable picking plates, a guiding structure, and torsion and buffer springs.

Fig. 12.12 Traditional rigid finger for grain harvesting

The process of adjusting the gap between the picking plates is as follows. During the harvesting process, the pressure from the maize stalks increases the gap between the adjustable picking plates, which improves the passage of the maize stalks. As the cross-sectional area of the maize stalks gradually decreases from the root, a torsion spring generates a restoring force during stalk movement, allowing the adjustable picking plates to bounce back in a timely manner. A limiting and guiding device was used to control the maximum gap between the adjustable picking plates, avoiding an excessively large gap that may cause stalk knives to damage the maize cobs and reduce the risk of header blockage.

Based on the parameters of maize plants, the diameter of maize stalks ranges from 18.58 to 28.97 mm with an average of 25.70 mm, while the diameter of stalks at the base ranges from 22 to 34.45 mm with an average of 27.6 mm. To ensure good passage of the maize stalks and prevent the large end of the maize cobs from being stuck and damaged by the picking plate, the preset adjustment range for the entrance gap of the picking plate was set to 24–46 mm, and the adjustment range for the gap between picking plates was 15–40 mm.

The force analysis of a maize stalk entering the picking plate is shown in Fig. 12.17. During the process of feeding maize stalks into the picking plate until the picking is complete, the picking plate is primarily subjected to compression and friction forces from the stalks. Additionally, a spring reaction force is generated on the picking plate when it is compressed by the stalks.

The resultant force satisfies the x- and y-axis directions as follows:

Fig. 12.13 Flexible extended finger for grain harvesting

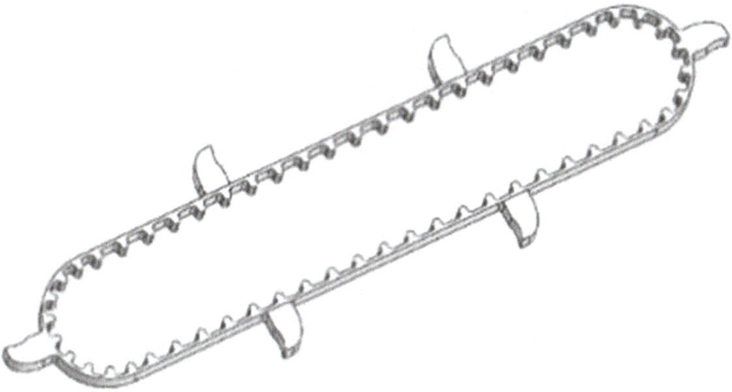

Fig. 12.14 Rubber-toothed threshing chain

$$\begin{cases} \sum x = F_1 \cos\theta_1 + f_1 \sin\theta_2 + F_2 \cos\theta_3 \\ \sum y = F_1 \sin\theta_1 + f_1 \cos\theta_2 + F_2 \sin\theta_3 \end{cases}, \tag{12.1}$$

where F_1 represents the stem pressure; F_{1x}, F_{1y} represent the components of F_1 in the x- and y-directions, respectively; f_1 represents the frictional force between the stem

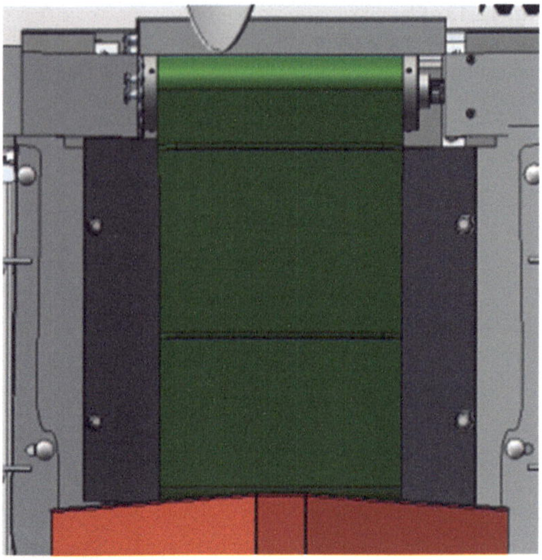

Fig. 12.15 Flexible conveyor belt

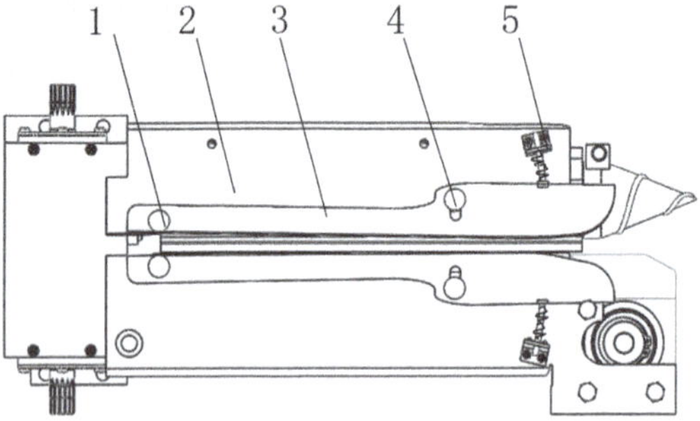

Fig. 12.16 Schematic of the gap-adjustment structure for the picking plate. 1. Restoring torsion spring. 2. Fixed picking plate. 3. Adjustable picking plate. 4. Limiting device. 5. Guiding device (including the buffer spring)

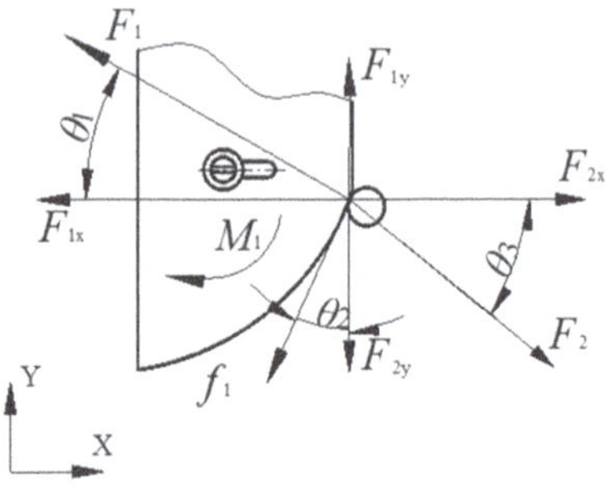

Fig. 12.17 Stress analysis of a maize stalk and ear-picking board

and header plate; F_2 represents the reaction force of the spring; F_{2x}, F_{2y} represent the components of F_2 in the x- and y-directions, respectively; θ_1 represents the angle between the stem pressure and the x-axis; θ_2 represents the angular displacement of the spring; and θ_3 represents the angle between the tangent line of the header plate and the y-axis.

The force on the header plate satisfies the following equation:

$$\begin{cases} F_1 = \dfrac{M}{l_1} \\ F_2 = kx_2 \\ M = Nl_1 \sin\theta \\ J = \dfrac{1}{3}ml_1{}^2 \end{cases} \tag{12.2}$$

where M represents the torque applied to the stem (N·m); l_1 is the distance between the collision point of the stem and the header plate and the base of the stem (mm); N is the normal pressure exerted by the snapping teeth on the maize stalk; θ is the angle between the stem and the ground; k is the spring stiffness (N/mm); x_2 is the displacement of the spring (mm); and J represents the rotational inertia of the stem (kg·m^2).

Low-Damage Harvesting

A cantilevered header plate structure was designed based on the structural characteristics of the adjustable device for the gap between the header plates. The header plate adopted a 3-mm thin steel plate cantilever beam design and was covered with a layer of 2-mm rubber on the surface. The impact caused by the collision of the crop with the header plate is absorbed through rubber cushioning and slight vibrations of the header plate, thereby reducing crop damage. The crop collision process with a vibrating header plate is shown in Fig. 12.18.

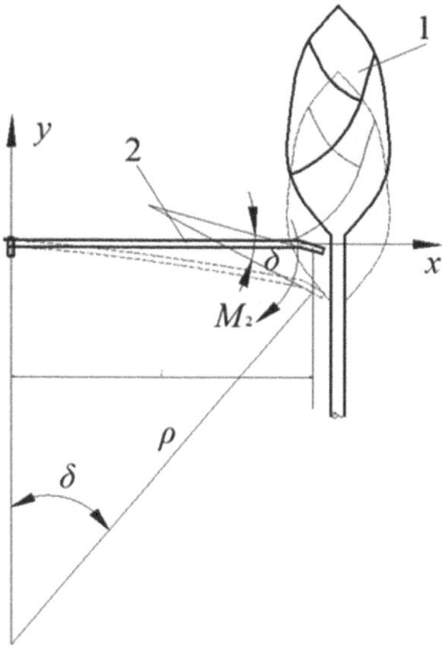

Fig. 12.18 Vibration harvesting principle. 1. Header plate. 2. Fruit spike. (Note: l is the harvesting plate length (mm); ρ is the flexural curve radius (mm); σ is the angular displacement at the top of the harvesting plate (degrees); M_2 is the torque generated by the effective force of the harvesting action (N·m); The y-axis direction represents the direction of the effective force of the harvesting action)

12.2.3.2 Key Technologies for Light and Simple Cutting Table Optimization

12.2.3.2.1 Transmission System Optimization

A conventional cutting table transmission device has a complex structure design. In such a device, a worm gear and pull stem cutter roller integrated design is used, and the transmission of the drawing device is shown in Fig. 12.19.

A structural diagram of the improved picking device is shown in Fig. 12.20. During operation, the output power of the gearbox drives the cutting roller and worm to rotate, while the worm wheel drives the driving wheel of the plucking device and the plucking chain to rotate around the slave wheel at a high speed.

After optimizing the cutting table transmission box design, there are only two pairs of bevel gear meshes in the transmission box, which greatly reduces the size of the gearbox, reduces the width of the cutting table by 30 cm, and reduces the mass of the conventional gearbox by 50%. The transmission box of the cutting table after optimization is shown in Fig. 12.21.

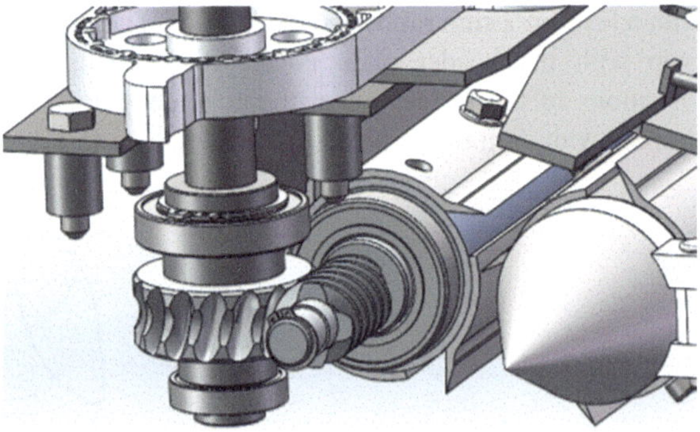

Fig. 12.19 Dipper drive

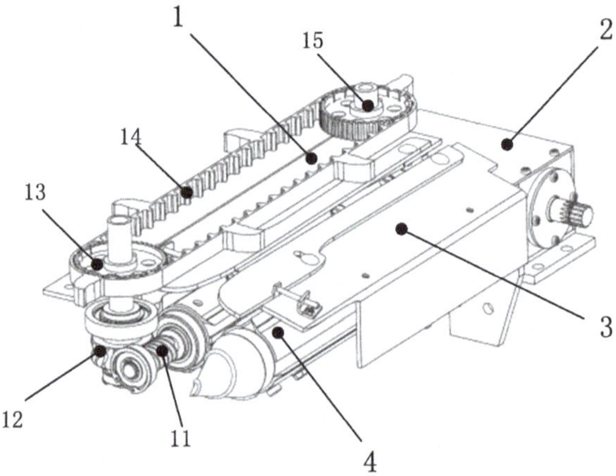

Fig. 12.20 Structural diagram of the picking device. 1. Single-sided plucking device (11. Worm; 12. Turbine; 13. Capstan; 14. Conveying chain; 15. Driven wheel); 2. Gearbox. 3. Ear-picking board. 4. Stem-pulling knife roller

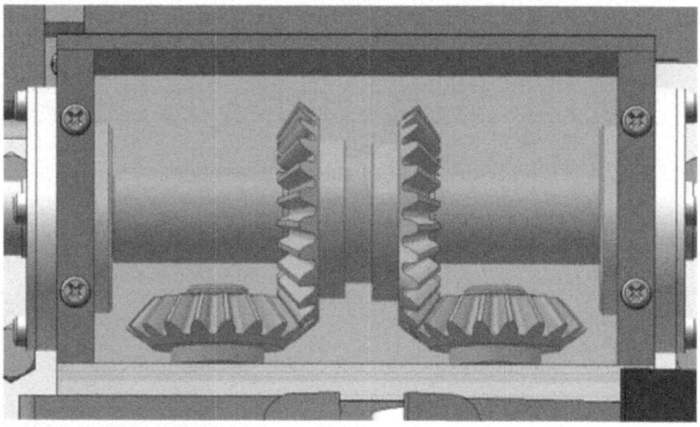

Fig. 12.21 Optimized gear box

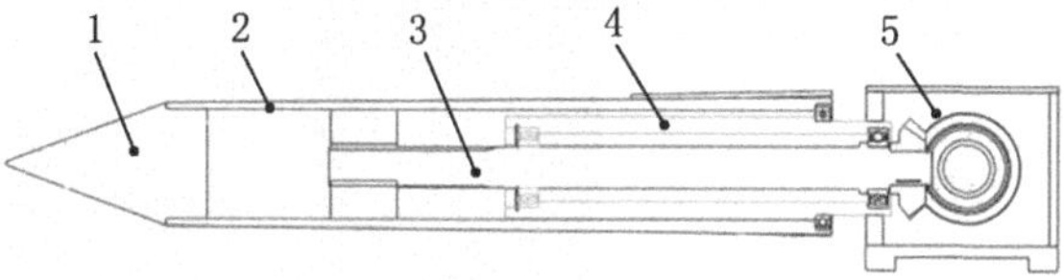

Fig. 12.22 Traditional stem-pulling roller structure. 1. Guide cone. 2. Stem-pulling roller body. 3. Roller shaft. 4. Roller shaft sleeve. 5. Gearbox

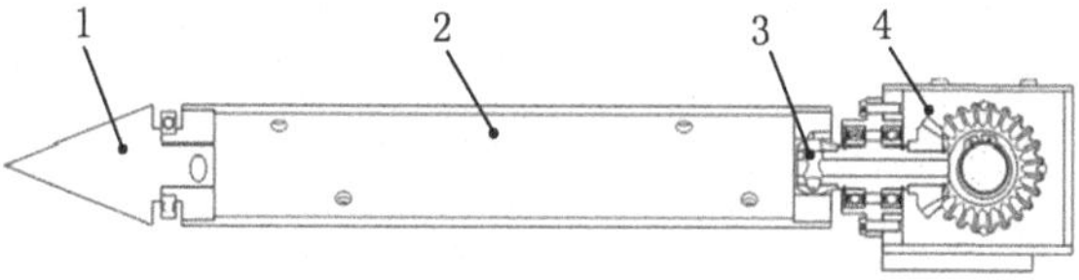

Fig. 12.23 Improved stem-pulling roller structure. 1. Guide cone. 2. Stem-pulling roller body. 3. Spherical connection mechanism. 4. Gearbox

12.2.3.2.2 Stem-Pulling Roller Structure Improvement

The structure of a conventional stem-pulling roller supported by a transmission shaft and a roller shaft sleeve is shown in Fig. 12.22. The material used for the transmission and support in a stem-pulling roller body accounts for approximately half or more of the entire internal space. The structure of the stem-pulling knife roller was redesigned, and the improved structure of the stem-pulling knife roller body has a hollow structure (as shown in Fig. 12.23). The improved structure is mainly composed of a guide cone, stem-pulling roller body, spherical connection mechanism, and gearbox. The weight of the improved ear-picking device was reduced by approximately 12 kg.

12.2.3.2.3 Peeling Device Key Technologies

Full-Rubber Roller Application

The application of a full-rubber roller solves the problems of extensive damage and low peeling rates in conventional peeling machines. Full-rubber rollers were adopted

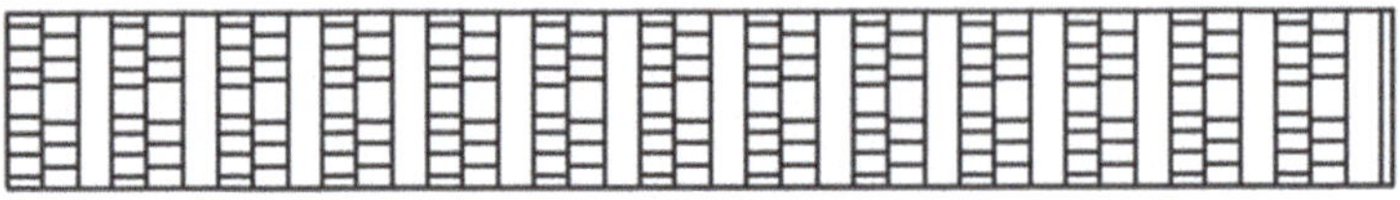

Fig. 12.24 Overall structure of the rubber rollers

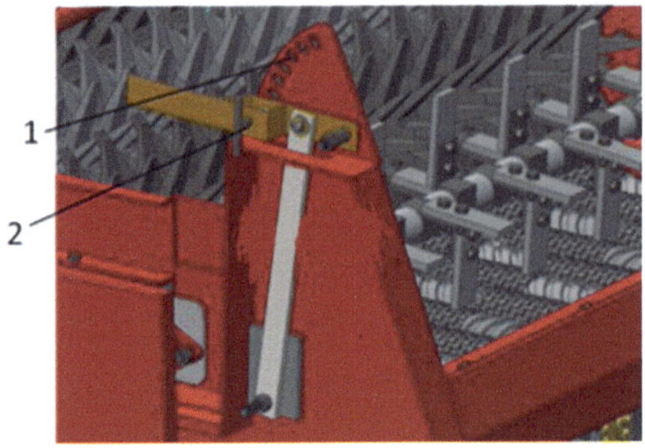

Fig. 12.25 Conical rubber roller. 1. Adjusting seat. 2. Adjusting bolts

to reduce ear damage and, based on conventional rubber rollers, improve the peeling rate by changing the surface shape of the peeling roller. The protrusions were formed using an alternating arrangement of three different protrusion heights. The structure of the full-rubber roller is shown in Fig. 12.24.

Gap Adjustment Mechanism Between the Press and Stripping Roller

The upper and lower clearance between the star-wheel press and the peeling roller can be adjusted according to the diameter of the fruit ear; the adjustment mechanism is shown in Fig. 12.25. The adjustment mechanism can achieve precision adjustment in the range of 0°–60°, and when the ear is large, a large gear can be used to improve the throughput of the ear. By contrast, using a small scale can increase the ear stripping rate of small-ear-diameter fruit.

Stripping Roller Spacing Adjustment Mechanism

The gap between the stripping rollers is one of the key factors affecting the stripping efficiency. The peeling rollers are arranged in a groove shape, and the gap between the rollers is automatically adjusted through a spring sliding rod mechanism, which meets the peeling requirements of fruit ears of different sizes. This principle is shown in Fig. 12.26.

12.2.4 Job Effects and Prospects

12.2.4.1 Operational Effect of the Maize-Ear Harvester

For the operational effect of the 4YZP-2D narrow-width tracked maize harvester, the machine has a total width of less than 1.6 m and a total length of less than 6 m, and it

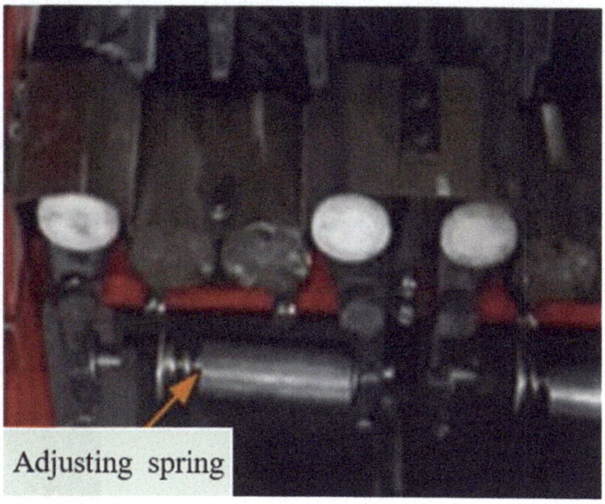

Fig. 12.26 Adjustment mechanism for the spacing between stripping rollers

can perform maize harvesting operations in intercropping modes with a row spacing of 0.30–0.60 m and a width of 1.5 m (GB/T 21961-2017). According to the National Jade Standard, the maize-harvesting machinery test method (GB/T 21961-2008) was used to test the operating performance of the 4YZP-2D narrow caterpillar maize harvester jointly developed by Shandong Guofeng Machinery Co., Ltd. and Sichuan Agricultural University, and the harvest effect was investigated at Duoying Farm, which is part of the Maize Research Institute of Sichuan Agricultural University. The maize variety used in this study was Yayu 988. The average grain loss rate of the 4YZP-2D narrow-track maize harvester obtained from the experiment was 2.24% (≤3%), the average ear loss rate was 3.37% (≤4%), and the average stripping rate of bracts did not exceed 92.93% (≥85%); all of these results comply with national standards (Fig. 12.27).

12.2.4.2 Prospects for Maize-Ear Harvesters

1. Development toward high flux, low damage, and low loss. With the continuous increase in the area and planting density of maize–soybean belt composite planting in China, to improve the harvest efficiency and adaptability of the maize harvester, research on the key technologies of high-throughput, low damage, low loss, and other maize-harvesting features should start with the comprehensive control of maize picking, reduce the fragmentation rate and impurity content, and strengthen intelligent research and development to achieve efficient and low-loss combined harvesting.
2. Development toward energy conservation and environmental protection strategies. Maize harvesters should develop toward high efficiency. China's research on energy utilization efficiency for maize harvesters began relatively late; presently, technical research is still limited to the functional field. However, with the deepening of the concepts of energy conservation and environmental protection, structural optimization and functional upgrading will significantly reduce the load on maize-harvesting machinery operations and reduce energy consumption and pollution by improving operational efficiency.

Fig. 12.27 Image of the 4YZP-2D narrow-width two-row tracked maize harvester

3. Automation and intelligence. Modern high-tech technologies, such as mechatronics, artificial intelligence, and unmanned driving of maize harvesters, significantly reduce the occurrence of faults through testing equipment, further improving the precision operation quality, performance, reliability, and service life of maize harvesters, achieving efficient and intelligent harvesting of maize harvesters, reducing energy consumption, and saving energy.
4. Automation and intelligence. Modern high-tech technologies, such as mechatronics, artificial intelligence, and unmanned driving of maize harvesters, significantly reduce the occurrence of faults through testing equipment, further improving the precision operation quality, performance, reliability, and service life of maize harvesters, achieving efficient and intelligent harvesting of maize harvesters and reducing energy consumption, and saving energy.
5. Light and simple harvesting equipment. A large maize harvester has a high working efficiency, but it has a large mass and high center of gravity; consequently, it is difficult for the machine to turn and has poor trafficability during harvesting. Therefore, large maize harvesters are not suitable for harvesting maize–soybean belt composite planting. Accordingly, the development of light and simple maize harvesters suitable for maize–soybean belt composite planting with a flexible harvesting operation will play an important role in promoting the mechanized development of maize–soybean belt composite planting.

12.3 Soybean Combine Harvester

12.3.1 Agronomic Requirements

In China, the row ratio of maize–soybean belt composite planting is 2:3~6 rows, the soybean row spacing is 30–40 cm, the soybean planting density is large, the plant spacing is basically maintained at 7–8 cm, and the minimum distance between maize belts is only 1.6 m. Therefore, it is required that the total width of the soybean combine harvester should not exceed 1.6 m, its cutting width should be less than 1.4 m, and the unit feeding amount should not be less than 2 kg/s.

The operating conditions should satisfy the technical requirements of mechanized harvesting. Mechanized harvesting should be performed in a timely manner during the soybean ripening period, and the operating time should be when the soybean is not lodged and the grain moisture content is 13–15%. Soybean harvesting should achieve a total loss rate of less than or equal to 3%, an impurity content rate of less than or equal to 5%, and a crushing rate of less than or equal to 5%.

12.3.2 Design Ideas

12.3.2.1 Overall Design Plan

To address the problems of small operating spaces and high crushing and loss rates in soybean harvesting, a small- and medium-sized tracked soybean combine harvester, the GY4D-2 (Fig. 12.28), was developed in collaboration with the industry (Zhang et al. 2015). This machine integrates a low-position single-action cutter,

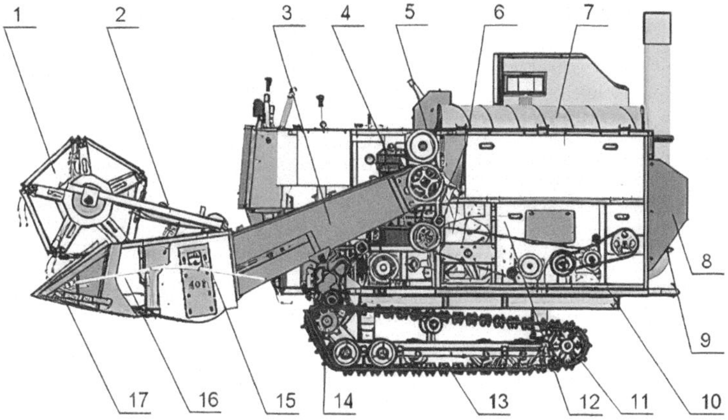

Fig. 12.28 GY4D-2 soybean combine harvester. 1. Reel. 2. Reel bracket. 3. Conveying bridge device. 4. Engine. 5. Threshing and cleaning assembly. 6. Concave plate. 7. Stir dragon. 8. Straw-collection assembly. 9. Straw-cutting device. 10. Rack. 11. Drive wheel. 12. Vortex fan. 13. Crawler-walking device. 14. Steering wheel. 15. Flexible header. 16. Header assembly. 17. Cutting knife. 18. Grain tank. 19. Nail tooth drum. 20. Air suction box

low-speed flexible threshing, ball column combination nail tooth threshing, and other technologies to form a narrow caterpillar, high-efficiency soybean combine harvester that can harvest soybeans in the maize belt. Field test results showed that the grain-crushing rate, impurity content, and loss rate of the machine were 1.81%, 2.02%, and 0.52%, respectively, all of which were superior to the relevant national standards.

The GY4D-2 soybean combine harvester mainly consists of a header assembly, conveying bridge device, threshing device, and cleaning device. A single-action low-level cutter header was used, and a hexagonal flexible reel with a stepless speed change was designed. The conveyor bridge device used a chain scraper conveyor and the threshing device adopts a ball-column combined-nail-tooth threshing structure equipped with a detachable steel wire rectangular screen. Additionally, the cleaning device adopts a dual-duct high-efficiency cleaning device. Meanwhile, the chassis adopts an HST fully hydraulic track drive, achieving a low center of gravity and good stability in the overall layout of the machine.

12.3.2.2 Main Design Parameters

The main design parameters of the GY4D-2 soybean combine harvester are listed in Table 12.3.

12.3.2.3 Key Component Design

12.3.2.3.1 Reel

The planting density of soybeans in strip compound planting is relatively high, and there are problems such as easy lodging of stems and broken pods. Additionally, soybean plants have lush branches and leaves, with an average plant height of 600–800 mm. As a direct power source for guiding soybean plants toward the cutting device, currently widely used pullers include pulleys and rollers. Owing to

Table 12.3 Main parameters of the GY4D-2 soybean combine harvester

Parameter	Unit	Specification
Structural style	/	Full feed track type
Header amplitude	mm	1200
Production efficiency	hm^2/h	0.25~0.4
Supporting power	kW	45
Overall dimensions (L × W × H)	mm	4230 × 1520 × 2300
Total mass of the host	kg	2100
Number of variable ranges	/	Hydrostatic continuously variable transmission
Fuel consumption per unit area	kg/hm^2	35
Operating mass of the entire machine	kg	2100
Working driving speed	km/h	Low speed: 0–2.7; medium speed: 0–3.6; high speed: 0–5.2

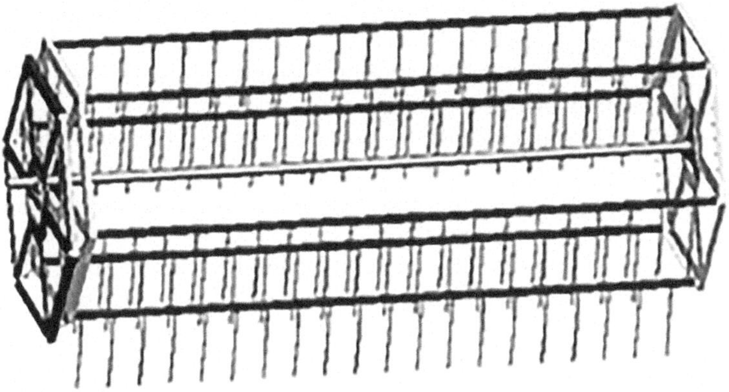

Fig. 12.29 Hexagonal eccentric spring-toothed reel

its simple structure, the reel is suitable for harvesting crops that are upright and not heavily lodged and is widely used in harvesters. Therefore, the GY4D-2 soybean combine harvester adopts an hexagonal eccentric spring-toothed reel based on the reel. The reel structure is illustrated in Fig. 12.29.

12.3.2.3.2 Cut-off Device

Soybean pods are prone to breakage, and their stems are highly flexible and difficult to cut. Cutting devices used in harvesting machinery worldwide are often reciprocating cutting devices. There are generally two types of reciprocating cutting knives: reciprocating single-action and double-action cutting knives. Double-action cutting knives use a moving blade to perform the reciprocating shear motion relative to the fixed blade on the blade guard to cut the crop. However, the reciprocating motion of the double-action blade generates a certain amount of inertial force and impact load, which have a certain impact on the life of various cutting tools and cutting table components, and an inertial force balance must be carried out (China Agricultural Mechanization Science Research Institute 2007). Reciprocating single-action cutting knives have a simple structure, a small blade spacing, and are not prone to knife blockage, making them suitable for cutting dense crops. Therefore, the GY4D-2 soybean harvester adopted a low-position single-action cutter device optimized by a reciprocating single-action cutter.

12.3.2.3.3 Conveying Device

Because soybeans are prone to pod explosions during transportation, the conveying bridge plays an important role in transporting soybeans as a transitional part that evenly and continuously transports crops from the header to the threshing device. Currently, intermediate conveying devices, such as chain rakes, belt rakes, and rotary wheels, are used in fully fed combine harvesters. The transmission device of the chain rake conveyor is a sprocket that has a good forced-feeding effect, strong adaptability to changes in grain thickness, and can effectively avoid blockages.

Therefore, the conveying device of the GY4D-2 soybean combine harvester is a chain-rake conveying device.

12.3.2.3.4 Threshing Device

Because of the high planting density of soybeans in strip compound planting, the unit threshing amount is 2–3 times that of the original net soybean, resulting in problems such as unclean threshing and clogged threshing equipment. As effective tools for improving the crop production efficiency, there are many different types of threshing devices worldwide. Based on the threshing tooth shape, threshing machines can be classified into three types: bar-, nail-tooth-, and bow-tooth-type devices (Wu et al. 2006). Among them, nail-tooth threshing devices are widely used in harvesters because of their high efficiency, good grasping ability during feeding, low damage to beans, and low crushing rate. Therefore, to improve the quality of the soybean harvest and reduce harvest losses, the threshing device of the GY4D-2 soybean combine harvester adopts a nail-toothed compound threshing device design.

12.3.2.3.5 Cleaning Device

At present, the increase in the unit feeding amount of soybeans in strip compound planting has led to a continuous increase in seed-cleaning power consumption and impurity content. Accordingly, a comparative analysis of the physical characteristics of soybean seeds and impurities after threshing revealed that the specific gravity and volume density of soybean seeds were relatively high, while the specific gravity and volume density of impurities, such as soybean husk, were relatively small, and the volume of seeds was much smaller than that of soybean husk. Therefore, based on the differences in these physical characteristics, the GY4D-2 soybean combine harvester uses a combination of air flow and sieve for separation to complete the cleaning of soybeans. Meanwhile, to improve the cleaning efficiency and reduce the impurity content, a high-efficiency dual-air-channel cleaning device was installed.

12.3.3 Key Technologies

12.3.3.1 Light and Simple Narrow Carriage Technology

The working space for soybeans in strip compound planting is limited, and the frame width of existing soybean harvesters exceeds 1.6 m, which cannot adapt to the limited effective soybean harvesting space. If the frame width is blindly reduced, the chassis becomes unstable, and the quality is too high to ensure the performance of operations in hilly and mountainous terrain. A finite element model was established based on the frame of a soybean harvester, and the simulation model was validated through sample modal tests. Additionally, the stress and weak areas of the chassis frame under different dynamic and static load conditions were analyzed. The thickness of the frame members was used as the design variable, and the vibration natural frequency and allowable stress were used as constraints. The goal was to minimize the mass of the frame, and lightweight optimization of the frame was performed.

The belt composite planting soybean harvester is different from ordinary road vehicles, requiring long-term operation in the field, with complex working conditions, severe road surface fluctuations, and overall significant negative loads (such as impact loads caused by sudden changes in driving speed and occasional natural environmental loads such as wind and rain). The soybean harvester was operated in the field. To simulate the real constraints of the soybean harvester as much as possible, full constraints in the *x*-, *y*-, and *z*-directions were applied to the front wheel of the frame. The rear wheel of the frame was constrained in the *x*- and *y*-directions, whereas the *z*-direction was free. The constraints and loading conditions are shown in Fig. 12.30.

Using Hypermes, a finite element preprocessing software, finite element mesh generation, and modal simulation analyses were performed on the frame. A modal test analysis was conducted on frame samples, and reasonable design dimensions were optimized, as shown in Table 12.4. The width of the machine tool was less than 1.6 m and its length was less than 4.3 m, making it convenient for wide maize operation.

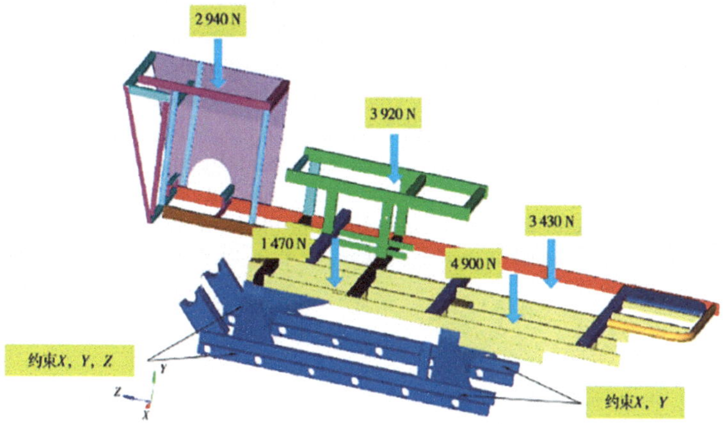

Fig. 12.30 Load distribution and constraint diagram of the harvester frame

Table 12.4 Dimensional comparison before and after optimization

Component	Before optimization/mm	After optimization/mm	Change
Chassis	4.0	3.0	1.0
Driver's cab longitudinal beam	3.0	3.2	−0.2
Engine room	4.0	3.0	1.0
Frame longitudinal beam	3.0	2.7	0.3
Frame width	>1600	<1600	

12.3.3.2 Narrow-Range Low-Loss Soybean Cutting Table

Existing harvester headers exhibit problems such as high power consumption, easy wear of the cutter, and easy damage to the header, and the harvest effect is not ideal. Moreover, because of the low pod setting position and high moisture content of the ribbon composite planting soybean, if the original soybean harvester is used to operate the header, the loss rate of the header is large and the straw cutting is not ideal. To improve the harvesting performance of low-pod-bearing plants, the cutting range of the cutting table was shortened to 1.2 m, the cutting device was optimized from a double blade to a single moving blade, and the height of the device was reduced from the ground to 0.05 m, reducing the loss of the cutting table. Meanwhile, to adapt to high-density soybean planting and prevent header blockage, the auger adopts a double-disc wire input auger, which increases the speed of the threshing, cutting blade, and auger (Fig. 12.31).

12.3.3.2.1 Single-Action Low-Position Cutting

Because of the long length and toughness of soybean stalks, which are generally curved, the cutting blade needs to be fast and stable to ensure a good cutting effect when cutting soybean stalks. Therefore, the GY4D-2 soybean harvester was designed with a reciprocating low-position single-blade cutter for processing on a type-II cutter. The sliding cutting angle diameter of the moving blade of the single-moving-blade cutter increases outward, achieving a height of 0.05 m above the ground for the cutting device. Additionally, the moving blade is arranged in a spiral shape on the blade shaft, and an elastic element is designed on the handle of the blade component to prevent straw blockage.

This design reduces the inertial force generated by the cutting device and improves the working effect of the header during the cutting process. By analyzing the mechanical characteristics of a single moving blade, it was found that when the reciprocating motion speed of the blade was 3.5 m/s, the cutter could complete the

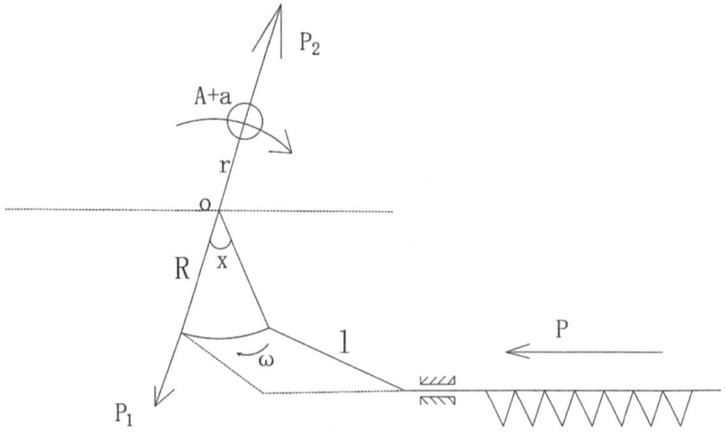

Fig. 12.31 Dynamics of single-moving-knife motion

cutting operation of the soybean stalks in one step. The experimental results showed that the shear angle had the greatest impact on the shear effect of soybean straw, while the blade angle had the smallest impact. Therefore, the moisture content of soybean straw, shear angle, and blade angle were analyzed as influencing factors; it was determined that the optimal design parameters were a shear angle of 60 ° and a blade angle of 44.99°.

12.3.3.2.2 Flexible Low-Loss Harvesting

During the soybean harvesting process, to ensure smooth completion of the cutting by the cutting table, it is necessary to guide the soybean to be cut toward the direction of the cutting blade by the harvester. Additionally, the straw should be straightened during the cutting process, and the cut soybeans should be pushed toward the conveying device to prevent it from accumulating on the cutting blade.

As shown in Fig. 12.32, the soybeans growing at point K are first directed toward the cutting blade using the first threshing board. At this time, if the speed of the threshing board is faster, when the axis of the wheel shaft tangential to the buckle is at point O_1, and the cutting blade is still at position C_1 without meeting the stem, the stem will experience “rebound” as threshing board 1 is lifted. To reach the second threshing board, the stem is re-supported and cut off at point C_2. If the cutting blade fails to cut off a bunch of crops integrated into the reel in a timely manner, this bunch of crops will “rebound” when the reel is lifted. When the reel speed ratio is too high, the crop will experience multiple “rebounds” before cutting, especially for crops with high spike quality, such as soybeans, which are prone to crop stem disorder and increased header loss. Therefore, the larger the tangent between the reel and the soybean, the more severe the rebound phenomenon; in contrast, the shorter the tangent, the more stable the plucking (Liang et al. 2015).

Fig. 12.32 Schematic of the plant rebound phenomenon when the cutting speed ratio is too large. (From Liang et al. 2015). Note: ω is the forward speed, m/s; v_m is the forward speed of the machine, m/s; H is the height from the center of the reel to the cutting blade, mm; L is the height of plant growth, mm; L_1 is the length of the cut plant stem, mm

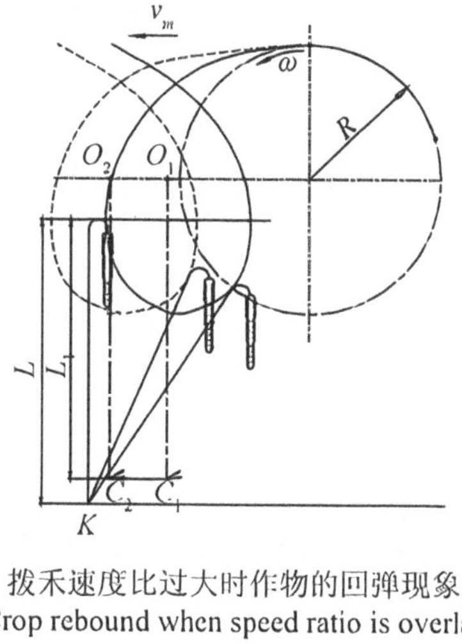

b. 拨禾速度比过大时作物的回弹现象
b. Crop rebound when speed ratio is overlarge

To meet the required conditions for work quality and prevent the occurrence of the "rebound" phenomenon, the GY4D-2 soybean combine harvester adopts a hexagonal eccentric spring-toothed reel. Meanwhile, to enhance the harvesting of upright and severely lodging soybean plants, its diameter is set at 760 mm and the rotational speed is 35.8 rpm.

12.3.3.2.3 Anti-blocking Conveying Technology

As the feeding amount of the reel increases, the auger on the header becomes prone to blockage and stem accumulation. Therefore, in the experiment, it was found that changing the auger diameter and adapting the telescopic finger could help prevent blockages and transportation to the header. The telescopic finger in the auger is usually fixed along with a fixed half shaft through a crank; therefore, there is an eccentricity between the center of the finger shaft and the center of the spiral tube, as shown in Fig. 12.33. When the finger is turned to the rear, it retracts into the body of the cylinder. However, to prevent the end of the finger from being worn, a 10-mm margin should be left outside the cylinder body. When the finger is turned to the front, it should extend out of the spiral blades of the cylinder body by 40–50 mm to achieve grasping with no blocking (Yang et al. 2010; Ji et al. 2012).

Because the gap between the fingers and the bottom plate is usually 10–15 mm, and it should not be less than 5–6 mm, to achieve harvesting efficiency and prevent crop blockage, the speed of the GY4D 2 combine harvester auger was set to 159 rpm, with a diameter of 900 mm, length of 163 mm, eccentricity of 21 mm, and gap of 15 mm between the auger and bottom plate.

12.3.3.3 Low-Loss High-Pass Threshing

The planting density of the strip intercropping system soybeans was high, and the unit feeding rate was 2–3 times that of net planting soybeans. If a net planting soybean threshing device is used, it can cause serious problems, such as unclean and blocked threshing. Moreover, most original soybean threshing devices are large

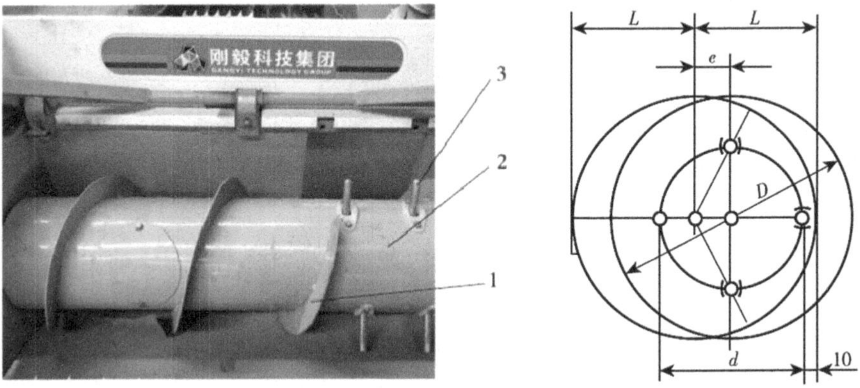

Fig. 12.33 Schematic of the Auger structure. 1. Auger blade. 2. Auger drum. 3. Finger picking

transverse axial-flow threshing devices, and the threshing effect is not ideal, resulting in a high harvest loss rate and low efficiency. The problem of blockage in the detachment device under high feed rates can be easily solved by increasing the straw introduction angle of the threshing device, installing guide plates on the inner side of the threshing cover, and increasing the number of straw discharge and threshing drum nail teeth. The structure of the threshing assembly is shown in Fig. 12.34.

A conical auger conveying end was designed at the front end of the threshing drum to increase the feeding rate of the threshing device and prevent soybean crops from accumulating and causing blockage. Generally, the number of spiral blades determines the conveying effect of a screw. If there are too few spiral blades, the conveying uniformity is poor; however, if there are too many spiral blades, delayed crop feeding can easily occur. Here, two spiral blades were used, and the length of the conveying spiral was 145 mm. The diameters of the large and small ends of the conical auger were 310 and 122 mm, respectively. This structure is shown in Fig. 12.35.

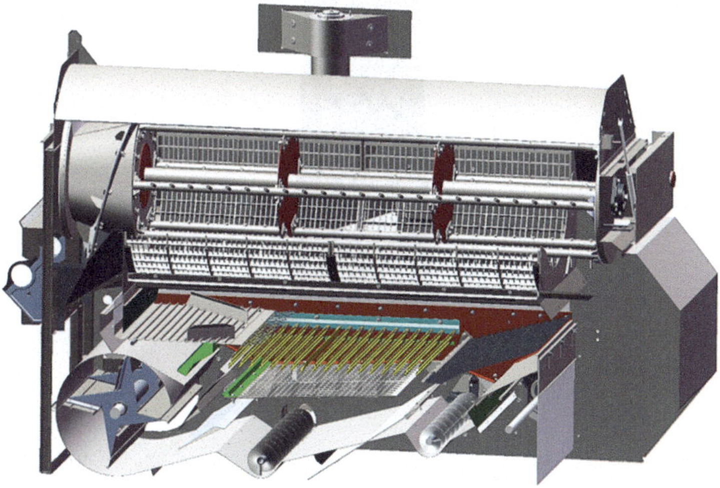

Fig. 12.34 Schematic of the threshing assembly structure

Fig. 12.35 Spiral conveyor structural diagram. 1. Conical head. 2. Spiral blade

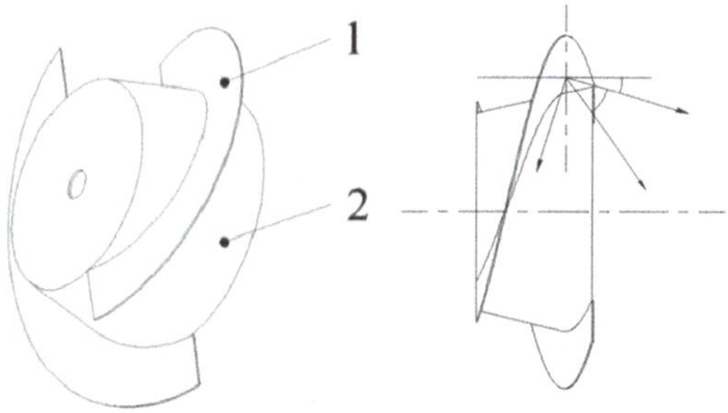

The strong impact force of the nail teeth can easily cause bean breakage or internal damage, thereby affecting the quality of soybeans. This can be addressed by reducing the spacing of the nail-tooth arrangement, changing the flat top to a spherical shape, increasing the number of threshing blows, reducing the impact pressure on soybeans, and solving the problem of grain fragmentation under high feed rates (Fig. 12.36).

Conventional threshing nail teeth are flat or trapezoidal, with a small impact cross-sectional area and high impact strength on seeds, while also producing a sliding effect, resulting in a high seed damage rate. Therefore, by increasing the impact cross-sectional area to reduce mechanical damage to the grains, the circular head nail teeth were in spherical contact with the grains, resulting in a low grain damage rate (Fan et al. 2019; Su et al. 2018) (Fig. 12.37).

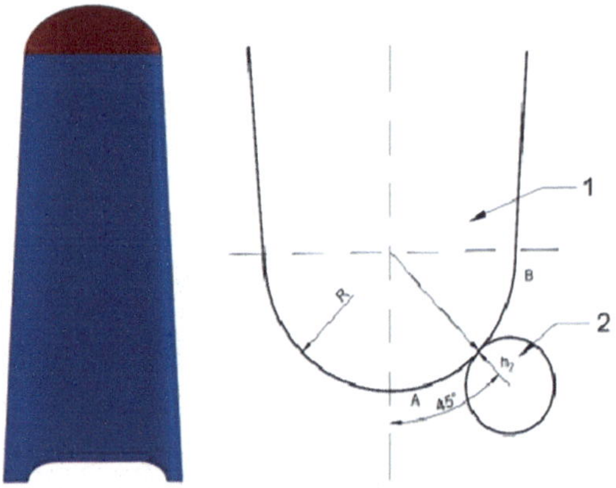

Fig. 12.36 Schematic of the spherical-cylindrical gear

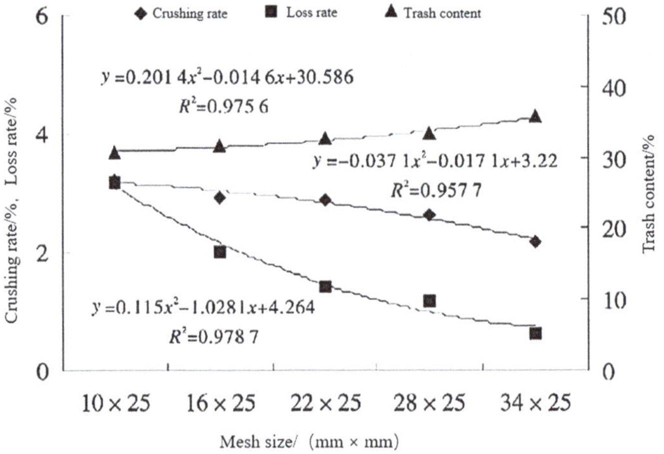

Fig. 12.37 Threshing effect of different mesh sizes. (From Yang et al. 2018)

Furthermore, it was found through experiments that with an increase in sieve size, the crushing and loss rates gradually decreased, while the impurity content exhibited an opposite trend (Yang et al. 2018). Therefore, a detachable steel wire rectangular screen was used, and after testing, it was determined that the drum speed was 460 rpm, the guide plate elevation angle was 11 °, and the screen hole size was 22 mm. At 25 mm, the threshing device exhibited excellent performance and anti-blocking effect, with corresponding performance indicators in terms of grain-crushing rate, impurity content, and loss rate of 1.81%, 25.02%, and 0.52%, respectively.

12.3.3.4 High-Efficiency Dual-Duct Combination Cleaning

At present, the most commonly used cleaning method in threshers and soybean harvesters is a grain-cleaning device combined with wind separation and screening. After threshing, the physical characteristics of soybean seeds and impurities were compared and analyzed. It was found that the specific gravity and bulk density of soybean seeds were larger, while the specific gravity and bulk density of impurities, such as the soybean skin, were smaller, and the volume of seeds was much smaller than that of soybean skin. Therefore, according to the differences in these physical characteristics, the transmission ratio of the fan in the GY4D-2 soybean combine under large-feeding-volume conditions was reduced, and dual-duct combined vibrating screen technology was adopted to solve the problem of high soybean impurity content under large feeding volumes (Fig. 12.38).

Additionally, the design and parameters of the cleaning device need to consider the amplitude, frequency, and vibration angle of the vibrating screen, and different jitter frequencies have an important influence on the cleaning degree. The general cleaning process is that the cut soybeans are separated from the stalk and enter the cleaning chamber along with the bean skin through the screen gap. Under the joint action of the fan and vibrating screen, the seeds are separated from the bean skin and transported to the grain box by a transverse agitator and elevator, and the bean skin is sucked away from the screen surface by the suction below to complete the cleaning

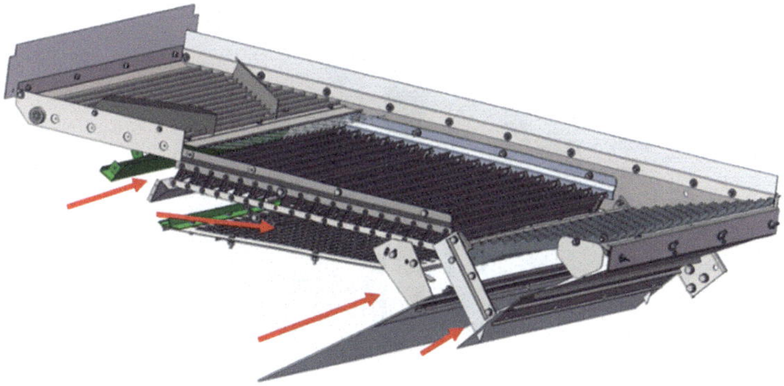

Fig. 12.38 Cleaning structure diagram

process (Cheng et al. 2019). It was found that to achieve a better cleaning effect, the following configuration is needed: a centrifugal fan with a diameter of 380 mm, a fan speed of 1837 rpm, a vibration frequency of 7 Hz, and a vibration angle of 60°.

12.3.4 Operational Effect and the Prospect of the Soybean Combine Harvester

12.3.4.1 Operational Effect

The GY4D-2 (Resolute) self-propelled soybean combine harvester was tested in the modern grain production demonstration base in Renshou County, Sichuan Province, China. The soybean variety used in the field test was Nandou 25, and maize–soybean belt composite planting was adopted. The soybean was planted on June 8. According to the soybean maturity, November 15 was selected as the test date, and the experiment was conducted according to a random block design, with a harvest length of 20 m. Field test results showed that the grain crushing, impurity, and loss rates were 1.81%, 2.02%, and 0.52%, respectively (Fig. 12.39 and 12.40).

12.3.4.2 Prospect

The future development of soybean harvesters should be toward high efficiency, lightweight, simplicity, energy-saving, and intelligence.

1. **High efficiency and low loss.** The soybean harvest times and terrain vary significantly in different regions in China, which leads to a decrease in the generality of harvesters. Soybean harvesting is characterized by the need to work during the best harvest period in the early morning and evening during

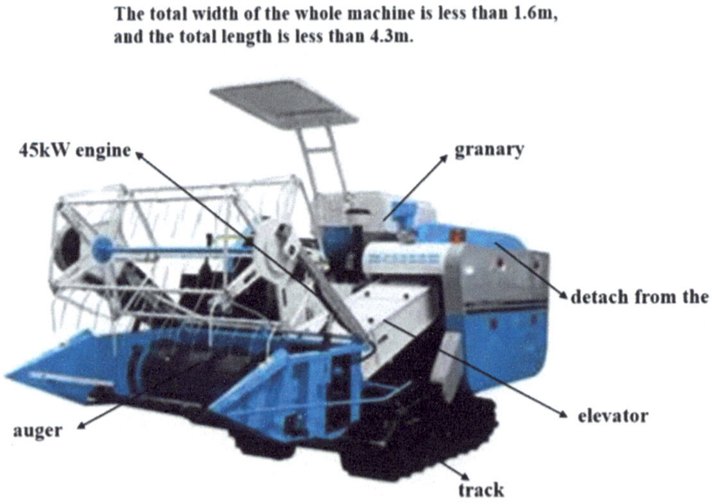

Fig. 12.39 Field operational diagram of the machinery used in the tests

Fig. 12.40 Schematic of the overall structure of the machine

the farming rush, and it is easy to fry pods and drop branches during the harvest process, resulting in soybean losses. Therefore, in the future, soybean harvesters should improve their applicability, efficiency, and loss prevention.

2. **Lightweight, simple, and energy-saving.** At present, there is a lack of research on the energy utilization of soybean harvesters in China, and the current research mainly focuses on their functionality (Lv et al. 2013). Through reasonable structure optimization and upgrading, harvesters can be slightly simplified, which can not only reduce the production cost of the harvester but also improve the adaptability and work efficiency of the machine. Moreover, this can also reduce the operational fuel consumption to achieve high efficiency, energy saving, and cost reduction.
3. **Intelligence.** Intelligence is the inevitable trend in the development history of harvesters. Information collection and automatic control technology will greatly improve the intelligence of harvesters. By installing a control system and detection components on soybean harvesters, real-time monitoring of the threshing drum speed, cutting table height, operating speed, and other parameters is possible (Lv et al. 2013, 2014). Then, by processing the collected information and adjusting the operating status of the drum speed accordingly, the stable operation of the soybean harvester could be ensured, as well as improving the working efficiency and quality of the harvester.

12.4 Universal Chassis for Tracked Combine Harvesters

12.4.1 Agronomic Requirements

Harvest is the final key to achieving a high yield of soybean and maize belt composite planting. The adoption of mechanized harvesting technology can not only improve labor productivity and achieve a timely harvest but also increase the soybean and maize yield, improve soybean quality, and increase economic benefits (Lv et al. 2014). However, the agronomic characteristics of strip composite planting require a certain soybean belt width and maize belt staggered planting. In this operational mode, the field steering space is small, and the harvester needs to make several reverse turns in the field to reach the working position in the next area; therefore, the harvester repeatedly rolls over crops while steering in the field, resulting in losses. For a successful harvester operation, achieving one-time turning in the field, and effectively reducing crushing losses, research on the general chassis of small-radius turning-in-place, which is suitable for soybean and maize belt composite planting harvesting, is needed.

12.4.2 Design Ideas

12.4.2.1 Universal Chassis Design Scheme of a Crawler-Type Combine Harvester

Based on the characteristics of soybean and maize belt composite planting, a small all-hydraulic general chassis was designed. The general structural design of the chassis is shown in Fig. 12.41, which mainly consists of a walking device, a control system, and a hydraulic system.

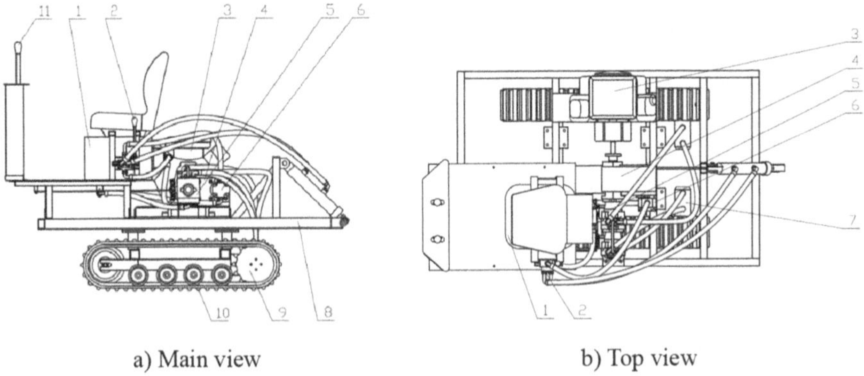

Fig. 12.41 General chassis structural diagram. 1. Bridge. 2. Hydraulic cylinder control hand valve. 3. Engine. 4. Transfer box. 5. Double pump. 6. Gear pump. 7. Motor. 8. Welded chassis bracket. 9. Drive system. 10. Track. 11. Control system

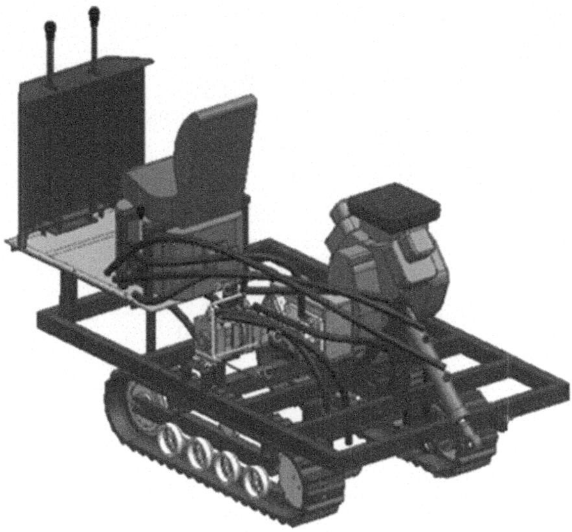

Fig. 12.42 Small multi-functional chassis virtual assembly

When the general chassis structure is working, the engine transmits power to the hydraulic transmission system through the transfer box, and the hydraulic transmission system transmits power to the walking system accordingly, so as to realize the walking movement of the machine (e.g., front, back, and steering). Through theoretical calculation and UG 3D prototyping, the complete machine structural design and key component parameters were determined. The overall virtual assembly of the chassis is shown in Fig. 12.42.

12.4.2.2 Main Design Parameters

The main technical parameters of the small multi-functional chassis are listed in Table 12.5.

12.4.2.3 Key Component Design

12.4.2.3.1 Running Gear

At present, most small agricultural machinery adopts a wheeled walking mechanism; however, wheeled machinery is not suitable for rolling terrain and sticky soil and is prone to slipping. Moreover, wheeled machinery has a large turning radius, which is not suitable for turning in small spaces. Nevertheless, designed as a tracked walking mechanism, it has a large grounding area, strong gripping ability, small subsidence, and good steering performance, and is suitable for strip composite field operations in different terrains (Lv et al. 2014).

The walking device of the small universal chassis consists of tracks, drive wheels, support wheels, guide wheels, and support frames. The left and right tracks are wrapped around the driving, supporting, and guide wheels, and are directly in contact with the ground. The structure of the walking device is shown in Fig. 12.43, and its working principle is that the motor outputs power to the driving

Table 12.5 Main technical parameters of the small multi-functional chassis

Serial No.		Parameter	Unit	Specification
1		Overall dimensions	mm	2000 × 1100 × 1520
2		Overall weight	kg	540
3		Gauge	mm	720
4		Track grounding length	mm	900
5		Track grounding length	mm	200
6		Connection to agricultural machinery	/	Suspension type
8		Operating speed	10^3 m/h	3 ~ 4
9	Engine	Rated power	kW	25
		Rated speed	r/min	3600
		Maximum torque	N.m/rpm	47/2600
10	Double-pump	Displacement	CC/r	12
		Maximum no-load revolutions	rpm	3600
		System operating pressure	MPa	16

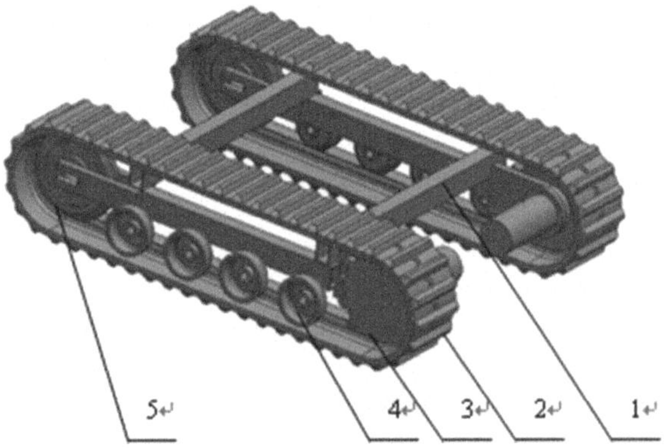

Fig. 12.43 Schematic of the walking device structure. 1. Support frame. 2. Track. 3. Drive wheel. 4. Support wheel. 5. Guide wheel

wheel while driving the track. The guide wheel guides the track to rotate correctly and tension the track to absorb vibrations and shock. The supporting wheel transfers the weight of the entire machine to the ground, rolls on the track surface, and prevents the lateral movement of the track. The weight of the entire vehicle passes through the frame, and the supporting wheel is transmitted to the lower track, causing it to press tightly against the ground.

12.4.2.3.2 Structural Design of Control Devices

Flexibility, accuracy, and convenience of operation are important indicators for evaluating the performance of agricultural machinery. Most existing self-propelled agricultural machinery adopts mechanical control, and gear-shifting devices are often used. Forces are typically transmitted through long steel rods, which have a

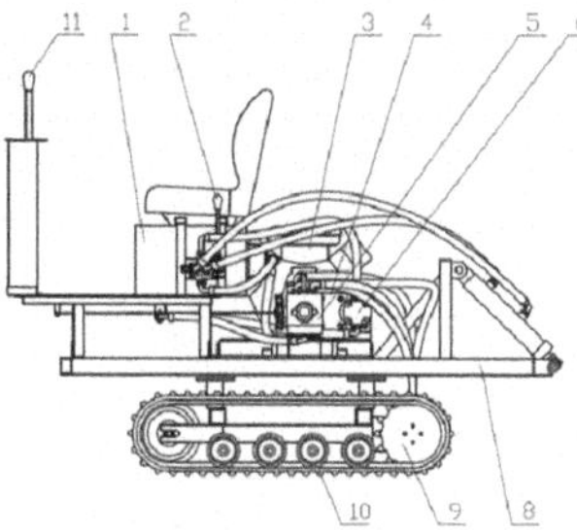

a) Main view

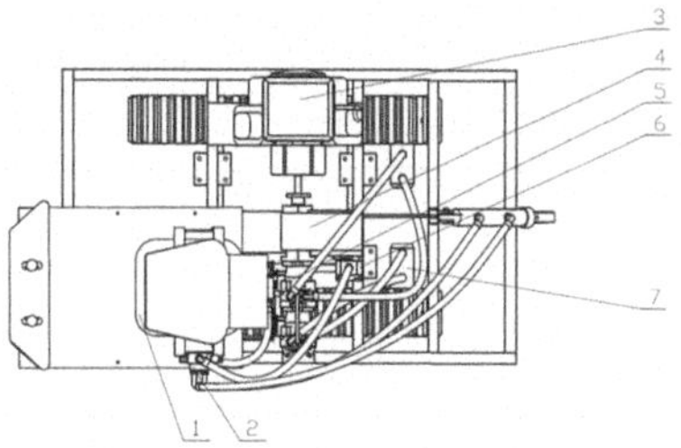

b) Top view

Fig. 12.44 Structural diagram of the control device. 1. Control handle. 2. Wire rope. 3. Wire rope guide mechanism. 4. Extension rod. 5. Dual pump

relatively complex structure and occupy a large space. Moreover, this mechanism has many steering joints, and the fit is mostly a clearance fit with a large cumulative error, making it difficult to ensure the operational accuracy of the machine. The control device of the universal chassis adopts steel wire rope control, which can effectively reduce the space occupation area and intermediate links of the mechanism, thereby ensuring control accuracy and satisfying the characteristics of a small and compact chassis structure space. The structure of the control device is shown in Fig. 12.44.

12.4.2.3.3 Full Hydraulic System Design

The following requirements are proposed for the hydraulic system design based on various investigations and fully combining the actual situation of belt composite planting: ①Initial mass of the entire vehicle: 600 kg; ② Engine power: 25 kW; ③ System working pressure: 16 MPa; ④ Adopting full hydraulic transmission and control, stepless speed regulation; ⑤ Operational speed: 3–4 km/h; ⑥ Rolling friction coefficient: 0.12; ⑦ Pavement condition: strip compound planting field.

The universal chassis adopts a fully hydraulic system with simple pipeline connections, reducing the installation space and effectively reducing the weight of the chassis. Hydraulic chassis transmission can easily achieve stepless speed regulation. Hydraulic components adopt standardized and serialized products, which are easy to design and manufacture, thus promoting their application. The full hydraulic chassis solves bottleneck problems such as small head space, difficult steering, and large loss of rolled crops in soybean–maize strip intercropping systems. The hydraulic components of a universal-chassis hydraulic system mainly include a dual pump, gear pump, walking motor, and hydraulic cylinder, among other components. The system transmission diagram is shown in Fig. 12.45. The hydraulic system of the designed and manufactured multi-functional chassis prototype is shown in Fig. 12.46.

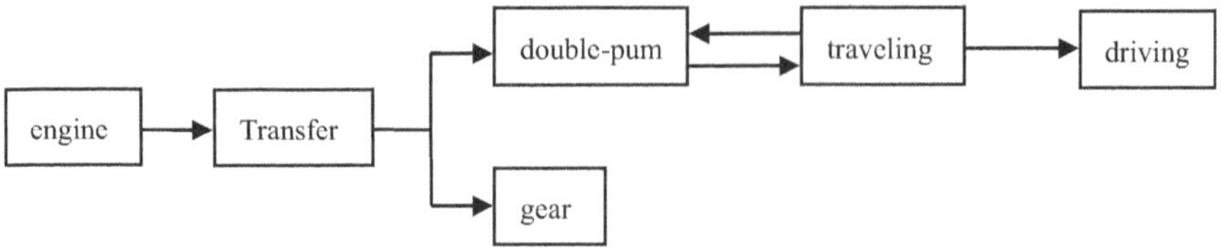

Fig. 12.45 Schematic of the hydraulic transmission system

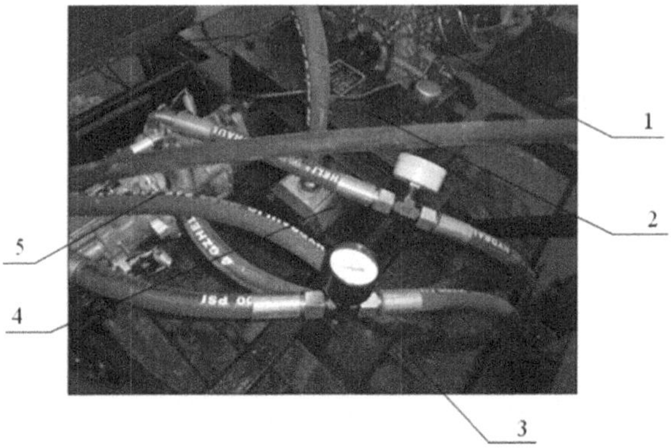

Fig. 12.46 Physical diagram of the hydraulic system. 1. Engine. 2. Transfer case. 3. Motor. 4. Gear pump. 5. Tandem pump

12.4.2.3.4 Brake System Design

The braking performance of a vehicle is one of the main indexes to evaluate its working performance. At present, the braking system of agricultural tracked vehicles mostly consists of a brake. However, the brake generally has one-way braking, that is, single-direction braking, and its two-way braking is very poor. Furthermore, the control end of the brake is generally mechanical, so the reaction is slow and the brake slips easily. To simplify the structure, some agricultural machinery with slow walking speed do not even install brakes; however, when the vehicle is parked on a sloping road surface, it may slide, raising serious safety risks. The brake device designed in this study is easy to implement with hydraulic braking. The hydraulic brake clamp is set to directly brake the driving wheel through the brake disc in the form of "clamp," which is simple in structure, reliable, safe, and can effectively prevent the slope sliding of the general chassis. The brake handle is fixed with the control handle on the chassis, which makes its operation very convenient to the operator and makes reasonable use of space, while being suitable for small track chassis with a slow walking speed. After the overall and part design, the parts of the brake device are processed and installed. The layout of the braking device after processing and installation is shown in Fig. 12.47.

a) Hydraulic brake mounting position

b) Brake operator mounting position

Fig. 12.47 Physical installation of the braking device

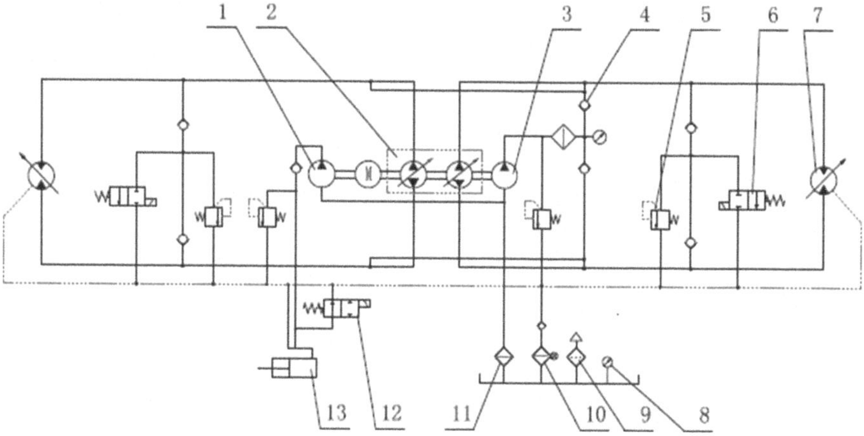

Fig. 12.48 Principle of the universal disc hydraulic system. 1. Gear pump. 2. Double pump. 3. Supplement oil pump. 4. Check valve. 5. Relief valve. 6. Normally closed unloading valve. 7. Motor. 8. Pressure gauge. 9. Air filter. 10. Return oil filter. 11. Oil suction filter. 12. Normally open unloading valve. 13. Hydraulic cylinder

12.4.3 Key Technologies

12.4.3.1 General Chassis Double-Sided Reverse Differential Steering

12.4.3.1.1 Two-Sided Reverse Differential Steering Principle

The bilateral reverse differential steering adopts full hydraulic technology and consists of two sets of independent walking systems, as shown in Fig. 12.48. Small-radius steering-in-place was realized through the speed difference between the positive and negative rotations of both sides of the track.

12.4.3.1.2 Simplified Model of the General Chassis Steering Process

The steering of the chassis is achieved via the speed difference between the two sides of the track, and the size of the speed difference determines the size of the steering radius to adapt to different curved roads. The steering process of the tracked chassis under simplified conditions does not consider that the inner and outer tracks do not produce slip on the ground. Because the tracked vehicle is subjected to greater resistance when steering, it needs to consume much more power than when driving straight; therefore, when the steering radius is different, the driving force required by the chassis inside and outside the track is also different. According to the chassis operational requirements, the force for chassis reverse differential steering, unilateral braking steering, and forward differential steering was analyzed.

Reverse Differential Steering

When the chassis reverses under differential steering, the tracks on both sides are in the driving state and the steering radius is $0 \leq R < B/2$. Here, the chassis stress situation of a special case of in situ steering ($R = 0$) was studied, as shown in Fig. 12.49.

Point O_1 in Fig. 12.49 is the steering center of the chassis, the outer track of the chassis is affected by the driving and friction forces (F_2 and F_{f2}, respectively), the inner track is affected by the braking and friction forces (F_1 and F_{f1}, respectively), the steering radius is $R = \frac{B}{2}$, the steering resistance moment of the inner and outer tracks is $M_1 = M_2 = \frac{GL\mu}{8}$ (where is the steering resistance coefficient), the outer track moves forward, the inner track moves backward to achieve reverse differential steering, and vice versa. After performing the force analysis of the chassis, the following relationship is obtained:

$$\begin{cases} F_2 + F_{f1} - F_1 - F_{f2} = 0 \\ (F_1 + F_2)\frac{B}{2} - \left(F_{f1} + F_{f2}\right)\frac{B}{2} - M_1 - M_2 = 0 \end{cases} \tag{12.3}$$

Based on the uniform steering condition, $F_{f1} = F_{f2} = 0.5F_f$ can be obtained, which is substituted into (12.1), and the driving force F_2 of the outer track is

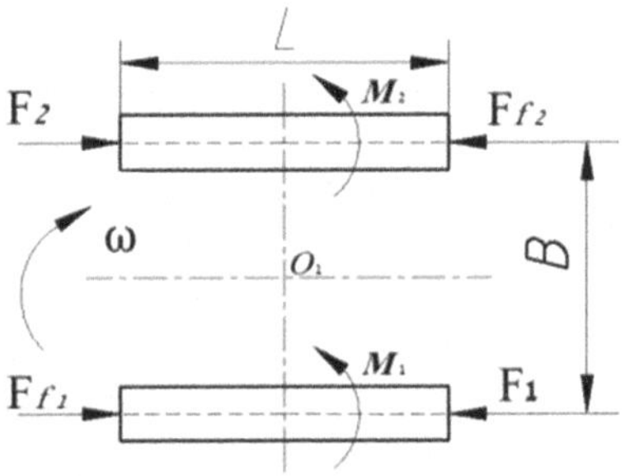

Fig. 12.49 Stress analysis diagram for in situ steering

$$F_2 = F_{f2} + \frac{2M_1}{B} = \frac{GF_f}{2} + \frac{GL\mu}{4B} \tag{12.4}$$

In this case, the driving force required by both tracks is equal. It can be seen from the previous analysis that when the steering radius is $R = 0$, the chassis can realize center steering in place, and the steering resistance moment of the track reaches the maximum at this time, requiring both the inner and outer motors to output a larger driving force. However, a smaller steering radius reduces the floor area of the chassis during steering and makes chassis steering more flexible.

Overall Design Plan

This steering of the chassis needs to be achieved by fully locking the inner shoe. During the steering process, the inner track drive motor has no output power, and only the outer track needs motor output power to apply the drive. The chassis is subjected to a force, as shown in Fig. 12.50.

Assuming that the crawler chassis rotates clockwise around the center point O_2 of the inner crawler and that the inner crawler has no power output, the rolling resistance is $F_{f1} = 0$, the driving force is $F_1 = 0$, and the steering resistance torque is $M_1 = M_2 = GL\mu/8$. From the force analysis, it can be concluded that

$$\begin{cases} F_2 + F_1 - F_{f_1} - F_{f_2} = 0 \\ (F_2 - F_{f_2})B - M_1 - M_2 = 0 \end{cases} \tag{12.5}$$

The outer track driving force is

$$F_2 = F_{f2} + \frac{M_1 + M_2}{B} = \frac{GF_f}{2} + \frac{GL\mu}{4B} \tag{12.6}$$

The resistance moment on the sprocket is

$$M_{R2} = M_{R1} = \frac{DF_2}{2\eta} = \frac{DG}{4\eta}\left(F_f + \frac{L\mu}{2B}\right) \tag{12.7}$$

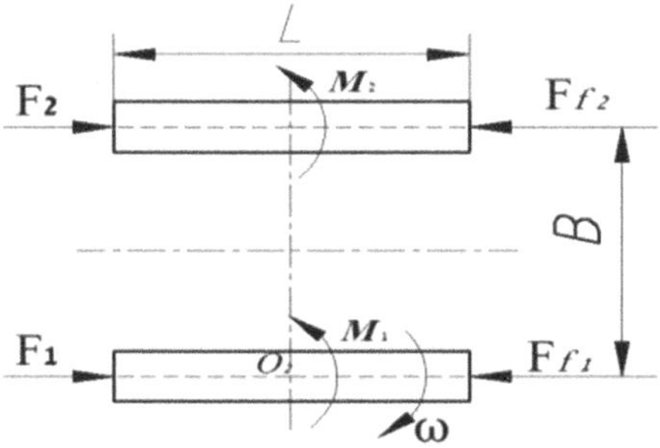

Fig. 12.50 Force analysis diagram of the B/2 steering

where η denotes the transmission efficiency. From this analysis, it can be seen that in this case, when the chassis is turning, the traction provided by the outer track should be greater than the steering resistance of the chassis $GF_f/2+ GL\mu/4B$.

Forward Differential Steering

To ensure that the chassis completes the effective steering process smoothly, the range of its steering radius is set to $B/2 \leq R \leq 2B$. The inner and outer tracks turn around the center point O_3 at different speeds, and the chassis is subjected to different forces, as shown in Fig. 12.51.

By conducting a force analysis on the chassis, it can be concluded that

$$F_1\left(R-\frac{B}{2}\right)+F_2\left(R+\frac{B}{2}\right)-F_{f_1}\left(R-\frac{B}{2}\right)-F_{f_2}\left(R+\frac{B}{2}\right)-M_1-M_2=0 \quad (12.8)$$

The driving force of the inner and outer tracks is calculated as follows:

$$F_2=F_1=F_{f2}+\frac{M_1+M_2}{B}=\frac{GF_f}{2}+\frac{GL\mu}{4B} \quad (12.9)$$

Therefore, the traction, braking, and steering resistance moments during steering could be determined using a simplified model.

Chassis ground mechanics relationship and road model establishment.

The force between the tracks and the ground varies according to ground type. In the simulation analysis model, the pressure between the track shoe and the hard ground is defined by the track contact, and the pressure between the track shoe and the soft road surface is realized by the contact force, that is, the shear stress generated by the interaction.

The contact collision force calculation in RecurDyn, a computer-aided engineering software package focused on multi-body dynamics, is as follows:

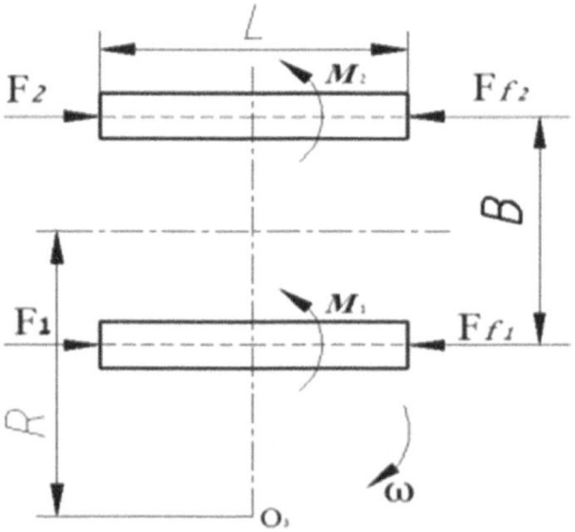

Fig. 12.51 Force analysis diagram for small-radius turning

$$F = -k(q - q_0)^n - cq \tag{12.10}$$

where F denotes the ground pressure, K is the stiffness coefficient, $q—q_0$ is the settlement amount, q is the deformation velocity, c is the damping coefficient, and n is the soil deformation index.

The soft ground model states that the soil has a “memory” function, and the positive pressure of the tracked chassis on the ground is based on the pressure settlement relationship proposed by American scholar M. G. Beck, with the following loading time relationship:

$$P_{di} = \left[\frac{k_c}{b} + k_\phi\right] \cdot z^n \tag{12.11}$$

When unloading, the following relationship is obtained:

$$P_{di} = P0_u \, max_{\, max \, max} \tag{12.12}$$

where P_{di} is the grounding pressure; k_c and k_φ are the soil cohesion and frictional deformation modulus, respectively; b is the width of the track shoe; Z is the deformation depth; n is the soil deformation index; Z_{max} is the maximum deformation depth; P_{max} is the maximum pressure; K_o and A_μ are characteristic soil parameters.

The horizontal force between the track and ground is calculated based on Baker’s theoretical shear stress–shear displacement relationship as follows:

$$\tau(j, z) = (c + P_{di} \tan \varphi)\left(1 - e^{-j/m}\right) \tag{12.13}$$

where τ is the shear stress; j is the shear displacement; φ is the shear resistance angle; and m is the shear deformation modulus.

Here, two common road conditions in small chassis operations, namely cement roads and loose ground, were considered for in situ steering simulation. The virtual steering simulation is shown in Fig. 12.52.

The simulation results for the two road conditions are shown in Figs. 12.53, 12.54, 12.55, 12.56 and 12.57. The results are presented in terms of changes in yaw rate, lateral acceleration response, torque output of the driving wheel, and tension on both sides of the tracked chassis during the steering process of the tracked chassis.

From the simulation results shown in Fig. 12.53, it can be observed that it is more difficult for the tracked chassis to turn on soft ground than on hard ground, exhibiting poor stability. In extreme cases, the tracked chassis can roll over.

As shown in Fig. 12.54, the lateral acceleration of the tracked chassis was relatively large when turning on sticky road surfaces, and the change rate trend became uneven. If the lateral acceleration changes significantly during the steering process, the steering process becomes unstable. In extreme cases, the tracked chassis can overturn.

Fig. 12.52 Virtual steering simulation

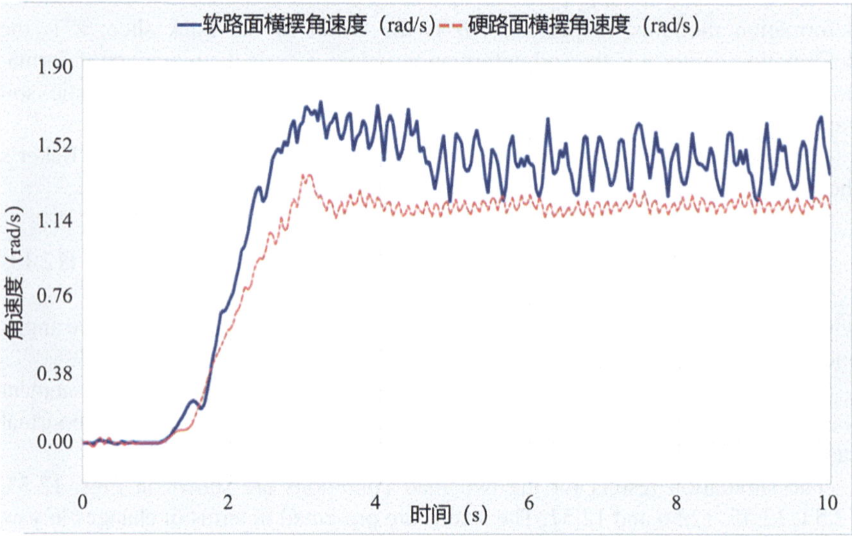

Fig. 12.53 Yaw rate response of the vehicle body center of mass

From Fig. 12.55, it can be observed that during the turning process of the tracked chassis in soft soil, the driving and braking torques of the driving wheels were greater than those on hard ground. To ensure stable driving during the turning process, it is necessary to apply greater driving and braking torques to the driving wheels.

From Figs. 12.56 and 12.57, it can be observed that the amplitude of the outer track changes significantly, and the response time is long when the tracked chassis turns on soft ground. Therefore, in the structural design of the tracked-chassis track-

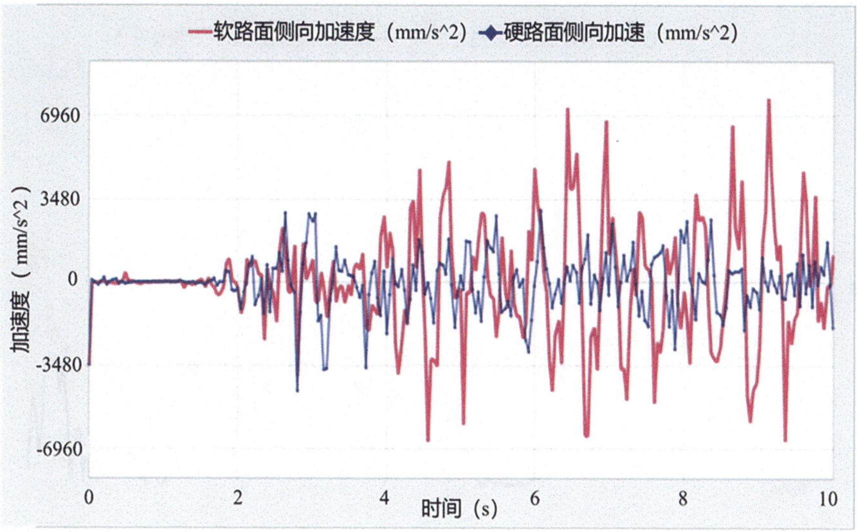

Fig. 12.54 Lateral acceleration response on two types of road for turning vehicle bodies

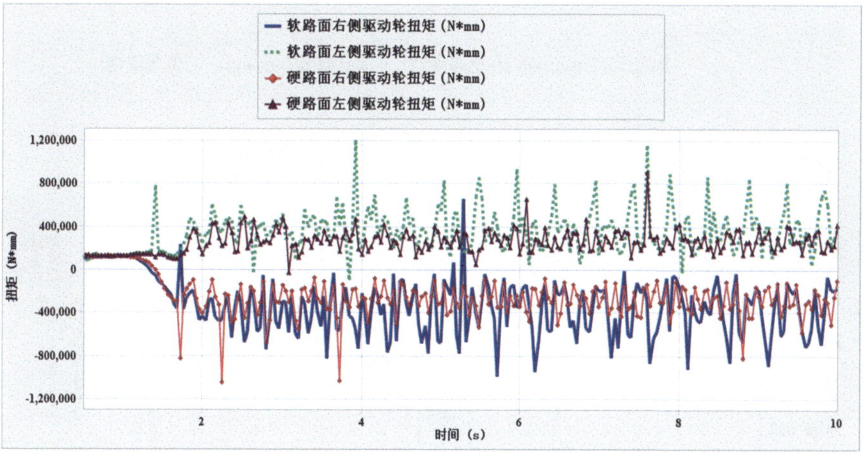

Fig. 12.55 Driving torque on two types of road for active steering wheels

walking device, it is necessary to appropriately set the initial tension to prevent track derailment.

After the calculation, analysis, and design, a universal chassis prototype was manufactured and tested with good results.

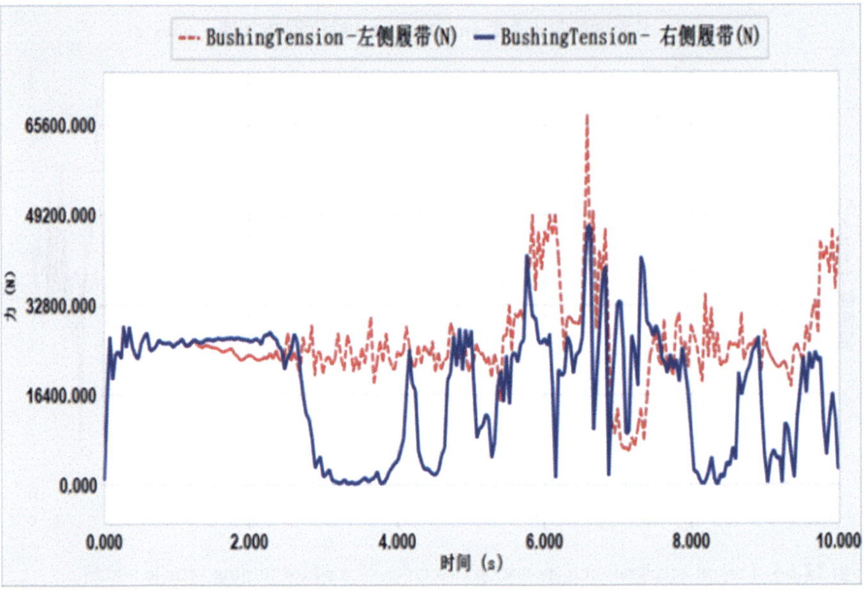

Fig. 12.56 Tensioning force of the steering track on a soft road surface

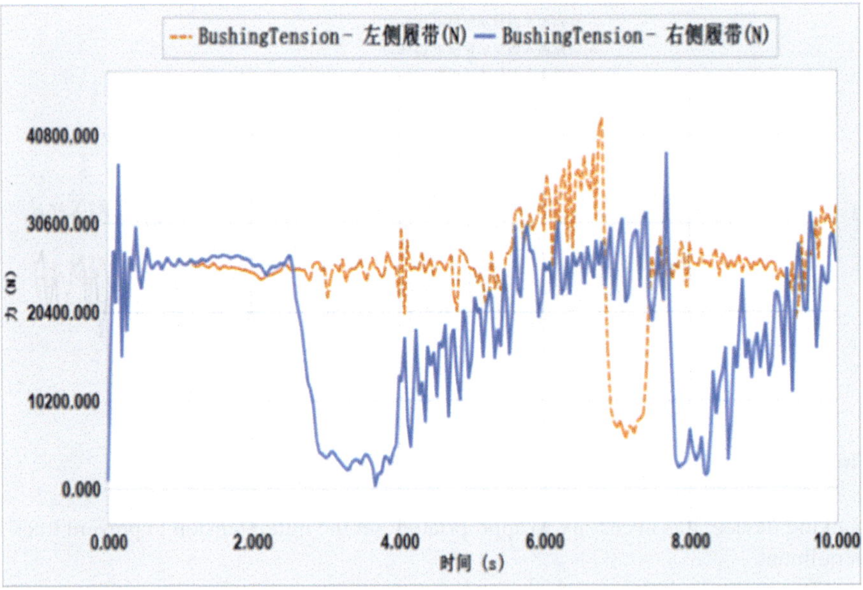

Fig. 12.57 Tensioning force of the steering track on a hard road surface

12.4.4 Operational Effect and Prospects for Universal Chassis Operation for Tracked Combine Harvesters

12.4.4.1 Operational Effect

In 2016, the working performance of a small general chassis was tested at the Sichuan Provincial Identification Station, and its scientific and technological achievements were identified and reported.

The small general chassis effectively realized small-radius steering in place, as shown in Figs. 12.58 and 12.59. Field test results show that the machine can be freely applied in strip compound planting fields, as shown in Figs. 12.60 and 12.61. Furthermore, the test results show that the general chassis runs smoothly and exhibits good working performance. It can effectively realize small-radius in-place turning in strip compound planting fields.

12.4.4.2 Prospects for Future Development

12.4.4.2.1 Intelligent Universal Chassis

With the integration of 5G mobile technology, the Internet, Big data, cloud computing, artificial intelligence, and other high-end technologies in modern manufacturing, the harvester general chassis is expected to become an intelligent chassis in the future. The goal of this development direction is to optimize the intelligence, reliability, safety, and universal chassis platform design of agricultural equipment chassis. For example, to improve the strip intercropping system for soybean and maize in hilly and mountainous areas, an intelligent leveling system

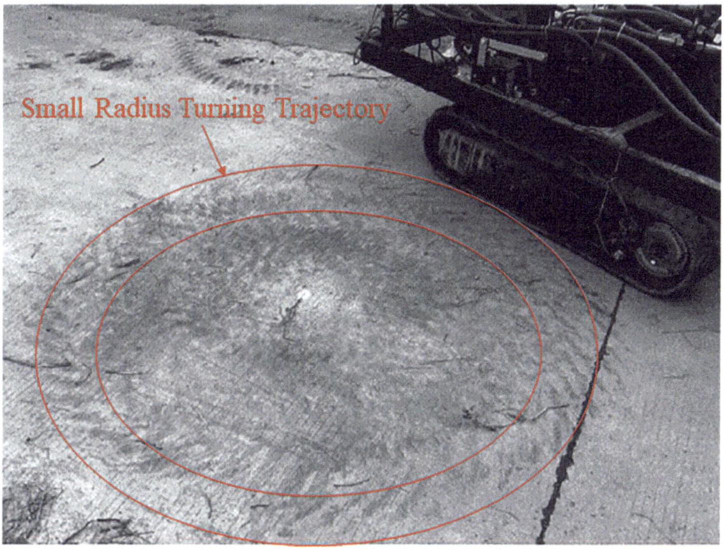

Fig. 12.58 Small-radius in-place turning on hard ground

Fig. 12.59 Small-radius in-place turning in the field

Fig. 12.60 Field intercropping conditions

for the universal chassis of the crawler combine harvester must be developed. This will effectively improve the quality, trafficability, and safety of mechanized strip intercropping systems and accelerate the development of mechanization in hilly areas.

Fig. 12.61 Field operation of multi-functional chassis intercropping

12.4.4.2.2 Autonomous Driving of the Universal Chassis

Unmanned agricultural machinery will be the focus of research in the agricultural field; accordingly, future agricultural machinery will develop toward unmanned driving. Beidou Navigation provides high-precision and highly reliable positioning, navigation, and timing services. Similarly, visual navigation is the use of cameras to collect image data, process the collected image data, and provide feedback on the processing results to control the machine. For example, the agronomic requirements of the strip intercropping system mode ensure straightness in crop planting, and the universal chassis of the crawler combine harvester carries out the remote control and unmanned driving systems in an orderly and phased manner. This will effectively improve the straightness of mechanized soybean–maize belt composite planting, ensure the accuracy of row spacing between crops, and improve the safety and trafficability of slope operations.

12.4.4.2.3 General Chassis Mechatronic and Hydraulic Integrated Control

By combining electronic and electrical technologies with hydrostatic transmission and intelligent control technologies, real-time detection, analysis, and processing of various parameters of the hydraulic system and chassis working components can be performed. Mechanization and automation of the production process for soybean and maize strip intercropping systems could be realized. Electromechanical–hydraulic integrated control technology can optimize the working parameters of various components of the hydraulic chassis through cooperation, thereby reducing costs, improving economic benefits and production efficiency, and reducing manual labor.

13 Regional Adaptation of the Best-Suited Technology Model According to Resources Availability and Specific Regional Ecological Conditions

Xiaochun Wang and Xuyang Zhao

Abstract

Regional crop varieties can be selected based on the specific resource endowment and production characteristics of different ecological regions using core and supporting technologies. Local adaptation of core technologies can be readily achieved through measures such as adjusting the number of soybean rows and row and plant spacing. Three specific sets of technological models have been shown suitable for different ecological regions in China: spring maize-summer soybean relay intercropping for the southwestern region, summer soybean-summer maize intercropping for the Huang-Huai-Hai region, and spring soybean-spring maize intercropping for the northwestern region. Furthermore, industry standards, including five provincial standards, have been formulated for intercropping. This technology has been continuously promoted in Sichuan Province, China for 13 years.

Keywords

Intercropping · Relay intercropping · Ecological regions · Spring maize/soybean · Summer maize/soybean

X. Wang (✉) · X. Zhao
College of Agronomy, Sichuan Agricultural University, Chengdu, China
e-mail: xchwang@sicau.edu.cn; zhao_xuyang@stu.sicau.edu.cn

W. Yang et al. (eds.), *Theories and Application in Maize-Soybean Strip Intercropping System*, https://doi.org/10.1007/978-981-97-8723-4_13

13.1 Spring Maize-Summer Soybean Relay Intercropping

13.1.1 Adaptation Regions and Current Situation of Maize and Soybean Production

Spring maize-summer soybeans relay intercropping is commonly practiced in the southwestern region of China. This region is bordered by the central and southern regions to the east and the northwest region to the north. It includes six provinces, municipalities, and autonomous regions: Sichuan, Chongqing, Yunnan, Guizhou, Tibet, and Guangxi. The topography of the region is complex and dominated by plateau and mountainous areas totaling 260.13 million ha, of which 13.75 million are cultivated land, making it one of the main agricultural production areas in China. The region is situated in the North Central-Asia Tropical Climate zone, with the Qinling and Dabashan mountain ranges keeping cold waves at bay. The average annual temperature is approximately 15 °C, and the accumulated temperature above 10 °C is 4000–5500 °C. The region typically receives approximately 1100 mm of rainfall annually, and the total solar annual radiation ranges between 376.74 and 1004.83 kJ/cm^2. Such widely ranging thermal conditions effectively allow the successful cultivation of two or even three crops per year.

In China, the Southwest hilly region is one of the three major maize-producing areas. Maize is the second-largest staple food crop in the region, after rice, and is a crucial feed source for livestock development and an important industrial raw material. In this region, maize occupies approximately 5 million ha of hilly dryland with poor soil quality and small land plots, comprising over 10% of the total maize cultivation area in the country. Maize yields range from 5000 to 5700 kg, with an overall total production of approximately 28 million tons, comprising roughly 10% of all maize produced nationally. Despite its significance, maize production in this region is constrained by a lack of mechanization, particularly for planting and harvesting, of which less than 10% is performed using machinery. Additionally, management costs are high, and seasonal drought is a common occurrence, in addition to summer drought, autumn flooding, and high temperatures during the middle and late crop growth stages. Altogether, these factors negatively affect summer maize production but are less detrimental to spring maize production, which is more reliable and productive.

In contrast, heat resources are abundant, and the frost-free period is longer in the southwestern region of China, making multi-cropping systems a prominent mode of dryland development. Thus, the dryland three-cropping system of "wheat/maize/sweet potato" has been widely applied in the Sichuan Basin, Chongqing, Yunnan-Guizhou plateau, and low-mountain areas since the 1980s. Although the system is labor-intensive and has limited economic benefits, it has remained stable for over 20 years, meeting the demand for food and feed despite continuous population growth. However, due to the improving living standards and dietary changes, sweet potato has shifted from being a staple crop to a miscellaneous grain crop. The high labor requirement at planting and harvesting, with a low efficiency rate and severe soil erosion, have resulted in a declining trend in the widespread adoption of the sweet potato crop. To address the most urgent needs of farmers for production technology, income, and higher efficiency in the production process, Professor Yang Wenyu of Sichuan Agricultural University has proposed a new dryland multi-cropping system of "wheat/maize/bean," which is ecological, high-efficiency, and water-saving. Hence, based on a summarized analysis of the existing dryland agricultural development models, the new model combines the natural characteristics and social needs of the southwestern hilly and mountainous areas. With continuous innovative research and promotion for over two decades, the demonstration and application of this system have continuously expanded in south-western areas such as Sichuan, Chongqing, and Yunnan, as well as in regions such as southern Shaanxi, western Hunan, western Hubei, northern Guizhou, Guangxi, and southern Yunnan, with the potential for spreading over a total area exceeding 4 million ha.

13.1.2 Key Technology

13.1.2.1 Variety Selection

For maize varieties, it is recommended to select compact or semi-compact, high-yielding, and stress-resistant varieties with a plant height of approximately 270 cm, such as Zhenghong 325, Nan N2090, or Miandan 53 (Table 13.1). With respect to soybean, it is deemed best to select high-yielding summer soybean varieties with high shading and population density tolerance, as well as lodging resistance, such as Gongqiudou No. 5, Gongxiadou 13, or Nanshadou 38. Alternatively, local varieties with strong shade tolerance might be selected (Table 13.2).

13.1.2.2 Proper Time for Sowing

Maize is sown from late March to early April, and soybeans from mid-to-late June. Planting should be performed using the 2BYFSF-2(3) type maize-soybean relay intercropping machine (a dual-purpose machine), or manually. The sowing depth for maize should be 4–5 cm, whereas it should be approximately 3 cm for soybean. Prior to planting soybeans, stubble, and wheat stubble should be treated through stubble-killer or rotary tillage. When sowing manually, attention should be paid to separating seed and fertilizer, as well as controlling the distance between hills.

13.1.2.3 Increase Light Through Intercropping

Plant maize in two rows 40 cm apart. Within the 180–200 cm interval between maize strips, sow three rows of soybeans 60 cm apart and with a strip interval of 60–70 cm (Table 13.3).

13.1.2.4 Reduce the Distance Between Plants to Ensure Density

For optimal yield, single-grain or double-grain maize should be sown per hill with 13–15 cm between plants (hill spacing of 26–30 cm) and a planting density of

Table 13.1 Yield and economic value of maize varieties in the southwestern region of China

Variety	Maize yield (kg/ha)	Soybean yield (kg/ha)	System output (yuan/ha)	Maize variety	Maize yield (kg/ha)	Soybean yield (kg/ha)	System output (yuan/ha)
Chengdan 306	7522	1122	27,795	Nan N2090	8805	1349	32,745
Chengdan 3211	7328	915	26,006	Nanyu 21	5734	1667	26,060
Chengdan 396	6811	1345	27,141	Rongyu	7744	890	27,024
Chengdan 716	7946	1236	29,664	Yidan 629	7866	1868	33,234
Chengdan 722	7063	1375	28,027	Zhenghong 325	9023	1723	35,599
Huashi 919	8030	2023	34,620	Zhenghong 507	7143	1604	29,624
Huashi 99	7958	1682	32,370	Zhongyu 3	7318	1310	28,350
Miandan 53	8132	1276	30,426				

Note: System output is calculated based on the unified prices released by the government at the time of harvest for the current season. Prices for maize and soybeans are 2800 and 6000 yuan/ton, respectively

Table 13.2 Crop yield and value of different soybean varieties in the southwestern region of China

Soybean variety	Soybean yield (kg/ha)	Maize yield (kg/ha)	System output (yuan/ha)	Soybean variety	Soybean yield (kg/ha)	Maize yield (kg/ha)	System output (yuan/ha)
Gongqiu 5	2483	7305	35,349	Nandou 12	1707	7425	31,032
Gongxia 13	2372	6627	32,784	Baimao 13	1254	9357	33,723
Nanxia 38	1979	7224	32,099	Nanxia 27	1278	8836	32,411
Cuandou 115	1803	7368	31,452	Nandou 24	990	9357	32,141

Note: System output is calculated based on the unified prices released by the government at the time of harvest for the current season. Prices for maize and soybeans are 2800 and 6000 yuan/ton, respectively. Lowercase letters indicate a significant difference at the $p < 0.05$ under all treatments

60,000 seeds/ha, which will result in an effective panicle count of over 52,500 panicles/ha. In comparison, for single-grain or double-grain soybean, a distance of 8–9 cm (16–18 cm between hills) and a planting density of 150,000 seeds/ha are recommended to achieve an effective count of approximately 105,000 plants/ha (Tables 13.4 and 13.5).

13.1.2.5 Reduce Fertilizer Application Rate and Integrate Fertilization

Fertilizer application should correspond with the requirements for monoculture maize. During sowing, apply 600–750 kg/ha of specialized controlled-release composite fertilizer for belt interplanting maize (such as $N\text{-}P_2O_5\text{-}K_2O = 28\text{-}8\text{-}6$). Subsequently, during tasseling and silking, apply 300 kg/ha of specialized compound fertilizer for legumes (such as $N\text{-}P_2O_5\text{-}K_2O = 14\text{-}15\text{-}14$) at a distance of 20–25 cm from the maize plants.

13.1.2.6 Chemical and Controlled Release of Growth Regulators

When maize plants have seven fully expanded leaves, spray a plant growth regulator, such as Paclobutrazol and Ethereal, to control plant height. For soybean, apply 5% Ethephon wettable powder at a rate of 300–750 g/ha during the third or fifth leaf period or at early flowering, and spray 450–750 kg/ha of water on the stem and leaves to control growth.

13.1.2.7 Green Prevention

Post-sowing and pre-emergence application of a 96% Propisochlor emulsion at a rate of 1.5 L/ha (15 bottles/ha) is recommended for weed control, particularly for S-metolachlor. Following emergence, selective herbicides such as 75% thifensulfuron-methyl for maize, 25% flumioxazin water-dispersible granules for soybean, or "Chu Ge Brothers" at a rate of 1 kit/ha, or the instructions recommended

Table 13.3 Yield and economic value of row ratio of spring maize-summer soybean relay-intercropping systems in the southwestern region of China

Row ratio	Maize yield and yield components				Soybean yield and yield components				Output (yuan/ha)
	Effective ear number (plant/ha)	Grain number per plant	1000-kernel weight(g)	Yield (kg/ha)	Effective plants number (plant/ha)	Seeds number per plant	100-seed weight (g)	Yield (kg/ha)	
3:2	56,820	509.8	323.0	9357	73,946	82.1	23.0	1094	32,760
4:2	40,473	644.1	285.8	7277	128,229	57.2	26.6	1725	30,725
6:2	57,975	480.8	232.0	6191	109,200	60.6	27.9	1775	27,978
4:3	57,570	485.9	239.3	6453	109,710	51.2	25.1	1349	26,160
6:3	57,150	477.6	226.6	6005	109,335	59.1	25.5	1578	26,285
4:4	56,970	527.5	238.2	6972	108,840	46.8	25.9	1275	27,171
6:4	56,835	513.4	218.5	6225	109,110	54.0	26.1	1478	26,297

Table 13.4 Crop yield and value of different maize yield levels in the spring maize-summer soybean relay-intercropping system in the southwestern region of China

Maize yield level (kg/ha)	Maize yield and yield components				Soybean yield and Yield components				Output (yuan/ha)
	Effective ear number (plant/ha)	Grain number per plant	Thousand kernel weight (g)	Yield (kg/ha)	Effective plant number (plant/ha)	Seeds number per plant	100-seed weight (g)	Yield (kg/ha)	
<7500	47,367	581.1	265.1	6944	120,926	56.1	26.4	1635.0	29,253
7500–9000	58,305	615.7	262.4	8966	109,335	33.0	24.9	862.5	30,282
9000–10,500	56,820	558.6	298.0	9357	73,946	82.1	23.0	1093.5	32,760

Table 13.5 Crop yield and value of different soybean yield levels in the spring maize-summer soybean relay-intercropping system in the southwestern region of China

Soybean yield level (kg/ha)	Soybean yield and yield components				Maize yield and yield components				Output (yuan/ha)
	Effective plant number (plant/ha)	Seed number per plant	100-seed weight (g)	Yield (kg/ha)	Effective ear number (plant/ha)	Grain number per plant	1000-kernel weight(g)	Yield (kg/ha)	
<1200	73,129	73.1	24	963	57,315	549.4	300.1	9227	31,613
1200–1500	109,853	52.7	25.1	1358	57,345	502.2	258.0	7181	28,249
1500–1800	109,268	59.8	26.7	1676	57,563	479.2	229.3	6098	27,131
>1800	121,025	66.8	26.0	1907	40,245	625.7	292.9	7377	32,094

by the manufacturers can be used for stem- and leaf-targeted weeding depending on plant type.

As for disease and insect pest prevention, various physical and chemical control techniques are recommended in combination with inducement and resistance strategies. Thus, to combat *Spodoptera frugiperda*, *Plutella xylostella*, *S. litura*, and Chrysomelidae pests, demonstration bases should consider installing smart LED integrated multi-spectral solar insecticidal-lamps and sex attractant lure and kill devices. Furthermore, *Ostrinia furnacalis*, *P. xylostella*, the northern corn leaf blight, and maize ear rot, all of which are severe threats to maize crops globally, can be controlled with 600–1200 g/ha of 70% methylthiophanate wettable powder or 25% triadimefon wettable powder. Additionally, 9–12 g/ha of 4% chlorfenapyr microemulsion or 900–1500 mL/ha of 1.8% avermectin emulsion, mixed with 750–1125 kg/ha of water for overall control are recommended for insect pest control. For soybean, 1500 mL/ha of 500 g/L methylthiophanate + 375 mL of 2.5% high-efficiency chlorofluazuron + 600 mL/ha of 12% methoxyfenozide should be applied in 450 kg/ha of water after flowering to control *S. frugiperda* and *A. gemmatalis*, among other pests.

13.1.2.8 Harvesting Pattern

In agricultural practice, maize is typically harvested during the last week of July or early in August with the utilization of narrow-row two-row maize harvesters, such as the 4YZ-2A self-propelled combine harvester or the 4YZP-2L tracked harvester. Subsequently, following manual maize harvesting, the straw is cut and immediately covered with soil. Conversely, soybean is harvested and threshed, and the straw is returned to the field toward the end of October using a GY4D-2 combine harvester or a local soybean combine harvester.

Attached: Field configuration diagram and management link diagram (Figs. 13.1, 13.2, 13.3, 13.4 and 13.5).

13.1.3 Technical Effects

Spring maize-summer soybean relay intercropping technology has been widely adopted in the southwestern region of China with significant impact. The annual intercropped area in Sichuan Province alone currently exceeds 330,000 ha. Furthermore, the Sichuan Provincial Department of Agriculture conducted on-site evaluations for two consecutive years in 2011–2012 in Zhujia Township, Renshou County, China. Maize yield during that time was 9316.5 kg/ha, while soybean yield was 1980 kg/ha, and the overall intercropping system output value stood at over 30,000 yuan/ha. Similarly, high-yielding fields for maize (soybean) were established in Renshou, Lezhi, Xichong, Rongxian, and Daying between 2013 and 2014, where maize yields above 9750 kg/ha and soybean yields of 2482.5 kg/ha were recorded, implying an increase in the economic benefit of over 7500 yuan/ha, compared with the traditional potato model. Recent experimental demonstrations indicate that soybean-maize relay intercropping produces maize yields comparable to

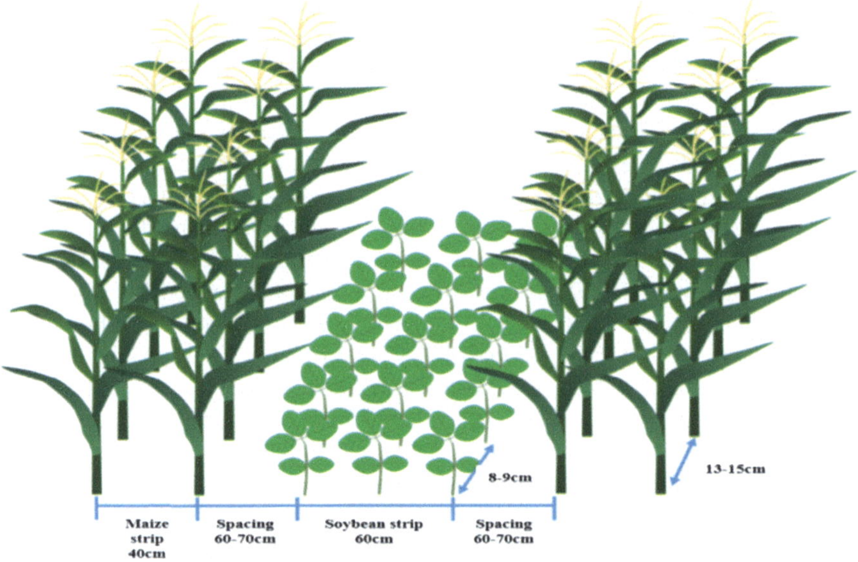

Fig. 13.1 Schematic representation of spring maize-summer soybean strip relay intercropping in Southwest China

Fig. 13.2 Machine-planting of relay intercropping maize

Fig. 13.3 Machine-planting of relay intercropping soybean

Fig. 13.4 Coexistence period of maize and soybeans in relay intercropping

Fig. 13.5 Machine-harvesting of relay intercropping soybean

Table 13.6 Yield of spring maize-summer soybean relay intercropping

Year	Location	Area of demonstration fields (ha)	Maize mean yield (kg/ha)	Soybean mean yield (kg/ha)
2011	Renshou	66.7	9573.0	1987.5
2012	Lezhi	33.3	9675.0	2400.0
2013	Renshou	6.7	10,566.0	1990.5
2014	Renshou	733.3	9315.0	2115.0
2015	Renshou	6.7	10,288.5	2803.5
2016	Xichong	33.3	9729.0	2209.5
2017	Renshou	333.3	9300.0	2058.0
2018	Renshou	66.7	7807.5	1992.0
2020	Renshou	333.3	7596.0	2089.5
2021	Renshou	333.3	8544.45	1834.5
2021	Rongxian	133.3	7983.0	1927.5
2021	Gongjing	666.7	7897.5	2029.5
2021	Anju	10.0	8301.0	2206.5
2022	Anju	80.0	9264.9	2703.0
2022	Renshou	73.3	8425.5	2407.5

monoculture, to which the benefit of a soybean crop is added with yields ranging from 1834.5 to 2703.0 kg/ha (Table 13.6). Thus, the land equivalent ratio reached 1.4 or higher, indicating that 1 ha of land produced 1.4 the economic output worth per ha. Therefore, the application of intercropping technology has resulted in the continuous expansion of soybean production in Sichuan, covering 61 counties (cities) and achieving near-universal coverage by 2018. The area cultivated with

Fig. 13.6 Demonstration spring maize-summer soybean intercropping fields in Renshou, Sichuan and Rongxian, Sichuan

soybeans reached 326,000 ha, accounting for 75% of the total soybean planting area in the province, with a total production of 9.105 million tons, representing 70% of the total national soybean output. The southwestern soybean-producing area has become a significant force in soybean production, whereas Sichuan has become the core area of the intercropping region of food-grade soybeans in Southwest China, significantly increasing the rate of self-sufficiency in soybean raw materials, which in turn has eased the domestic soybean supply-demand gap, and boosting the national self-sufficiency rate and international competitiveness in soybean production (Fig. 13.6). The promotion and application of soybean-maize relay intercropping technology has resulted in a significant change in the national soybean production pattern. Specifically, before 2000, the main soybean production areas were primarily concentrated in the Northeast high-oil soybean region. Then, approximately by 2009, the soybean planting area in Sichuan Province increased significantly, entering the list of major soybean-producing provinces for the first time. Some years later, in 2014, soybean gained comprehensive coverage in the hilly areas of Sichuan, and by the year 2021, soybean output from this province ranked third in the country. Thus, a new pattern comprising the Northeast high-oil area, the Huang-Huai-Hai high-protein area, and the maize-soybean intercropping area in the Southwest has emerged.

13.2 Summer Maize–Summer Soybean Intercropping

13.2.1 Adaptation Regions and Current Situation of Maize and Soybean Production

The summer soybean-summer maize relay intercropping pattern is predominantly utilized in the Huang-Huai-Hai summer maize planting region, which covers the provinces of Henan, Hebei, Shandong, Anhui, and Jiangsu. This region falls under the category of a warm, semi-humid climate with notable variations in topography and landforms, wide watersheds, an annual mean temperature of 10–14 °C, and a frost-free period ranging from 170 to 240 days from the north down toward the south. The annual accumulated temperature of ≥10 °C extends from 3600 to 4700 ° C, while its annual sunshine duration ranges from 2000 to 2800 h. The region receives an annual rainfall of 500–800 mm, primarily concentrated during the growth period of maize and soybean, with summer precipitation comprising over 70% of the total annual rainfall. The Huang-Huai-Hai region lies downstream of the Yellow River, the Huai River, and the Hai River system, with an irrigated area accounting for approximately 50% of the total cultivated area. The region is a prominent agricultural production base in China. Specifically, it ranks as the largest maize cultivation region in the country, being the only one with two summer maize crops per year. Further, the annual planting area in the region remains consistently over 15 million ha, thus contributing approximately 35% of the national maize production and more than 40% of the total grain production in China. The crop rotation of winter wheat and summer maize is the main form of cultivation, accounting for over 85% of total grain production. Winter wheat heavily constrains the growth of maize in this area, characterized by high evaporation, concentration of rainfall, and high temperatures, resulting in recurrent spring drought and summer flood events. Maize cultivation in this area is characterized by advanced mechanization levels, whereby more than 70% of the area undergoes no-till direct seeding, and more than 60% undergoes mechanical harvesting.

Additionally, the Huang-Huai-Hai region is the second-largest soybean-producing area in China, with a stable planting area of nearly 2.3 million ha in recent times, corresponding to approximately 30% of the total national soybean planting area. Within this region, soybeans are predominantly concentrated within Hebei, Shandong, Henan, and Anhui provinces, with Henan and Anhui being the most significant producing areas, each with an annual overall production of approximately 1 million tons. The growth period of summer soybeans is restricted by winter wheat, and its general growth period is 90–110 days, which is marginally longer in the southern region and shorter in the northern region. Soybeans are typically planted after wheat harvest, which spans from mid-September through early October, with abundant light and heat resources, abundant rainfall, high temperatures, and slight temperature differences during the growth period. Soybeans can adopt a two-year, three-harvest rotation or a one-year, two-harvest cycle, the latter being more common.

The Huang-Huai-Hai region boasts intense competition for land, being the primary maize and soybean-producing area in the country. The summer soybean-summer maize intercropping approach allows for two crops, maize and soybeans, to be integrated, planted, and harvested together, making it an effective and coordinated agro-management approach. Without affecting maize yield, this approach can generate significant economic benefits and highlight the potential for future development. A calculated soybean planting area of 7.5 million ha, based on half of the current maize planting area, can generate an extra 13.5 million tons of soybeans per year.

13.2.2 Key Technology

13.2.2.1 Variety Selection

Maize varieties for the region should be selected for compact plant types suitable for dense planting and mechanized harvesting with high productivity. Soybean varieties should be selected for their high yield, shade tolerance, and resistance to lodging. Recommended maize varieties include Mingtian 695, Denghai 511, Huangnuo 2, and Denghai 605 (Table 13.7), while recommended soybean varieties include Xudou 24, Cangdou 0734, Hedou 33, and Wando 37 (Table 13.8).

13.2.2.2 Selecting the Proper Time for Sowing

It is crucial to manage soil moisture when sowing after wheat is harvested from June 15 to 25. If the moisture content of the soil is between 60% and 65%, quick sowing is recommended. In contrast, for relatively dry or wet soils, sowing should be performed with the guidance of the weather forecast or with assistance by drip irrigation, but no later than June 25. Furrow irrigation should be employed if the soil is extremely dry before it is left to dry to an optimal moisture level for 3 to 5 days before sowing. Stubble that is over 15 cm high and straw over 10 cm long should be baled and removed before pre-sowing soil treatment. The 2BYCF-6 and 2BF-6 planting machines are suitable for sowing coupled with fertilization during planting, specifically for soybean-maize relay intercropping high-density planting.

13.2.2.3 Increase Light Through Intercropping

Maize should be planted with a spacing of 40 cm between rows and 210–230 cm between maize strips. Four rows of soybean may be planted in 90-cm wide strips. Intercropping with a spacing of 60–70 cm is applied between maize and soybean rows, and the best soybean-maize row ratio was 4:2 (Table 13.9).

13.2.2.4 Reduce Plant Spacing to Ensure High Plant-Population Density

Maize is sown at a density of a single seed hill per 10–11 cm, with approximately 67,500 seeds per hectare to ensure more than 60,000 effective spikes. Meanwhile, soybean is sown at a density of 10–11 cm between hills, with approximately 135,000 seeds per hectare to ensure more than 90,000 effective plants (Tables 13.10 and 13.11).

13.2.2.5 Fertilization and Control of Plant Height

For maize, specialized slow-release fertilizer should be applied at a rate of 750–1050 kg/ha ($N\text{-}P_2O_5\text{-}K_2O$ = 28-8-6). For soybean, we do not recommend using N fertilizer but specialized low-N compound fertilizer should be applied at a rate of 300 kg/ha ($N\text{-}P_2O_5\text{-}K_2O$ = 14-15-14). Furthermore, we suggest applying plant height regulators, such as paclobutrazol, uniconazole, and ethephon, at the 7-leaf stage of maize as appropriate. Alternatively, we suggest applying 5% mepiquat chloride wettable powder at the third- or fifth-leaf stage or early flowering at a rate of 300–750 g/ha (300 g for seedlings) in 450 kg of water to control soybean plant height.

Table 13.7 Crop yield and value of maize varieties in the Huang-Huai-Hai region of China

Maize variety	Maize yield (kg/ha)	Soybean yield (kg/ha)	System output (yuan/ha)	Maize variety	Maize yield (kg/ha)	Soybean yield (kg/ha)	System output (yuan/ha)
Mingtian 695	9198	1290	33,497	MY73	7902	1728	32,493
Denghai 511	9152	1698	35,813	Xiake 6	7887	1866	33,278
Xinshi 118	9012	2148	38,121	Zhongnongda 678	7800	2220	35,159
Denghai 605	8760	1400	32,924	Yufeng 620	7785	2325	35,748
Huangnuo 2	8708	2087	36,902	Quanke 789	7755	1830	32,694
Jiyuan 128	8511	1472	32,660	Luyu 9105	7734	2099	34,245
Liangyu 99	8537	1935	35,517	Jiyu 217	7649	1847	32,493
Jieyu 1606	8451	1868	34,868	Weike 908	7635	2240	34,812
Jietianyu 1606	8417	1277	31,223	Dedan 123	7613	1574	30,753
Baojing 186	8315	1571	32,708	Jiangyu 877	7430	2088	33,330
MC121	7953	1935	33,881	Denghai 553	7208	1782	30,875

Table 13.8 Crop yield and value of soybean varieties in the Huang-Huai-Hai region of China

Soybean variety	Soybean yield (kg/ha)	Maize yield (kg/ha)	System output (yuan/ha)	Soybean variety	Soybean yield (kg/ha)	Maize yield (kg/ha)	System output (yuan/ha)
Xudou 24	2586	7866	37,538	Cangdou 13	1706	5627	25,988
Cangdou 0734	2394	8342	37,718	Zhonghuang 310	1586	7430	30,311
Wanxia 33	2181	6524	31,347	Xiangfeng 4	1563	7790	31,193
Wandou 37	2160	7767	34,701	Huaidou 13	1559	6524	27,617
Jindou 99	2099	7836	34,530	Zheng 1307	1526	7649	30,567
Youchun 1204	2037	7557	33,375	Shidou 936	1472	8493	32,612
Hedou 12	2010	5858	28,460	Hedou 12	1185	7152	27,138
Zhonghuang 78	1877	8525	35,123	Xukedou 1	1160	6803	26,000
Andou 203	1782	7208	30,875	Handou 13	1086	8397	30,030
Qihuang 34	1740	8429	34,043				

Table 13.9 Yield and economic value of row ratio experiment of summer maize-summer soybean relay intercropping system in the Huang-Huai-Hai region of China

Row ratio	Maize yield and yield components				Soybean yield and yield components				Output (yuan/ha)
	Effective ear number (plant/ha)	Grain number per plant	Thousand kernel weight (g)	Yield (kg/ha)	Effective plants number (plant/ha)	Seeds number per plant	100-seed weight (g)	Yield (kg/ha)	
3:2	59,717	359.4	394.5	8354	84,654	53.8	23.9	1061	29,747
4:2	59,135	415.6	347.9	8114	94,467	76.4	22.9	1488	31,643
6:2	44,211	454.5	320.9	5945	66,686	132.3	21.3	1740	27,086
4:3	52,905	494.0	289.1	7556	53,970	91.2	25.7	1268	28,760
6:3	62,367	432.3	320.5	7814	102,246	69.6	26.5	1563	31,257
4:4	58,526	456.3	280.0	7236	82,512	114.0	19.4	1730	30,639
6:4	62,013	441.4	333.7	8057	89,778	91.6	23.6	1404	30,980

Table 13.10 Crop yield and value of maize yield levels in the summer maize-summer soybean intercropping system in the Huang-Huai-Hai region of China

Maize yield level (kg/ha)	Maize yield and yield components				Soybean yield and yield components				Output (yuan/ha)
	Effective ear number (plant/ha)	Grain number per plant	1000-kernel weight(g)	Yield (kg/ha)	Effective plant number (plant/ha)	Seeds number per plant	100-seed weight (g)	Yield (kg/ha)	
<7500	53,925	386.4	331.0	6303	88,982	89.3	20.8	1412	26,114
7500–9000	60,861	420.9	336.8	8174	93,260	87.3	23.2	1665	32,873
9000–10,500	65,081	440.6	349.4	9450	91,367	77.9	22.5	1347	34,548

Table 13.11 Crop yield and economic output value of varying soybean yield levels in the summer maize-summer soybean intercropping system in the Huang-Huai-Hai region of China

	Soybean yield and yield components				Maize yield and yield components				
Soybean yield level(kg/ha)	Effective plant number (plant/ha)	Seeds number per plant	100-seed weight (g)	Yield (kg/ha)	Effective ear number (plant/ha)	Grain number per plant	1000-kernel weight (g)	Yield (kg/ha)	Output (yuan/ha)
<1200	86,940	51.7	22.2	935	64,361	341.4	375.8	7956	27,887
1200–1500	95,861	67.8	22.4	1317	59,537	412.4	365.4	8339	31,247
1500–1800	96,492	81.6	23.7	1659	60,923	415.2	331.7	7943	32,189
>1800	92,637	111.7	22.7	2090	60,582	455.8	312.0	7929	34,743

13.2.2.6 Weed Control

To ensure effective weed control for maize and soybean, we recommend using a 96% quizalofop-p-ethyl emulsion (Goldner) prior to germination at a dose of 1200–1500 mL/ha. When administering post-emergent herbicides, it is essential to use suitable herbicides for the respective crop types, when using a physical barrier during spraying or using a dual-tank spray system. For maize, it is highly convenient to use 5% nicosulfuron and 20% prosulfuron in 225–450 kg water at a rate of 1.88–2.25 L/ha. Meanwhile, for soybean, using 25% fluorosulfonamide ether water agent at a rate of 1.2–1.5 kg/ha or 15% quizalofop-p-ethyl and 25% fluorosulfonamide ether (300 mL + 270 g type 1 set/ha) can be highly effective for weed control.

13.2.2.7 Disease and Pest Control

Insect resistance induction and chemical control techniques are commonly employed in pest management. To demonstrate the efficacy of this approach, an intelligent LED integrated-broadband solar insecticidal lamp + sex lure core device was installed in the demonstration base. The system could effectively kill insect pests such as *S. frugiperda*, *Euzophera perticella*, and Tortricidae. Drone-based pest control measures were performed three times throughout the maize growth cycle, including after 3–4 leaves and before tasseling, soybean pod-set, and the grain-filling stages. A five-in-one application of fungicides, insecticides, adjuvants, regulators, and micronutrients was used for optimal pest and disease control. For soybeans, during the pod-set and grain-filling stages, 1500 mL of 500 g/L methylthiophanate, 375 mL of 2.5% high-efficiency chloro(fluoro)acetic acid ester, and 600 mL of 12% methoxyfenozide per hectare were sprayed with 600–750 kg of water. Notably, the integrated pest-management approach should consider rust disease, root rot, *S. frugiperda*, *Chilo suppressalis*, *Lissorhoptrus oryzophilus*, *Bactrocera dorsalis*, and *Helicoverpa armigera* while planning management measures for corn pest and disease prevention.

13.2.2.8 Harvest

Soybeans and maize typically mature early in October. Depending on the available equipment and the specific requirements, either maize or soybean can be harvested first or simultaneously. If maize is harvested first, a narrow two-row maize harvester, such as the 4YZ-2A self-propelled combined harvester and the 4YZP-2L crawler harvester can be used between soybean rows. Local soybean harvesters can then be used to collect soybeans. To harvest soybeans first, use the GY4D-2 combined harvester to collect soybeans between maize rows, followed by local maize harvesters for collecting maize. Alternatively, soybean and maize harvesters can work at the same time. The appropriate width chilly harvester is necessary for making mixed green fodder out of soybean and maize. Finally, the 4QZ-2100 and 4QZ-3000 models are capable of harvesting both tall and short crops.

Attached: Field configuration diagram and management link diagram (Figs. 13.7, 13.8, 13.9, 13.10 and 13.11).

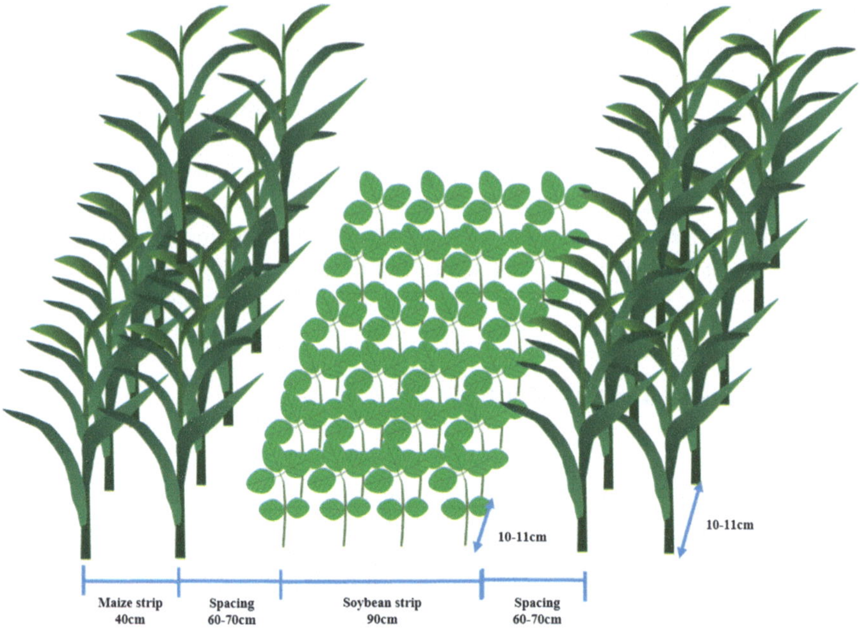

Fig. 13.7 Schematic representation of summer maize-summer soybean relay intercropping in the Huang-Huai-Hai region of China

Fig. 13.8 Maize and soybeans being simultaneously sowed without tillage after wheat harvest

Fig. 13.9 Post-emergence herbicide spraying targeted at stem and leaf

Fig. 13.10 Soybean harvest first

Fig. 13.11 Simultaneous harvest of soybeans and maize

13.2.3 Technical Effects

For the soybean-maize strip intercropping demonstration established in the Huang-Huai-Hai area in 2011, it is evident that the technical model has been increasingly standardized, and supporting machinery has been continually enhanced, resulting in significant improvements in both production and income. Table 13.12 exhibits the yields obtained at some demonstration sites, whereas fixed 1000-mu high standard-technology demonstration plots have been established in Gaocheng, Hebei Province, and Yucheng, Shandong Province (Fig. 13.12). In addition to illustrating the use of soybean-maize strip intercropping technology for grain production purposes, a 100-mu display of soybean-maize relay intercropping technology for silage was performed in recent years in Yucheng, Shandong Province, accompanied by experiments on mixed silage, cattle, and sheep feed. The yield of mixed maize and soybean silage was approximately 60 t/ha with a higher market price of 20 yuan/ton than maize alone, generating an added value per hectare of nearly 1200 yuan. When compared to maize silage alone, crude protein of maize and soybean mixed silage increased by 21.19%, while neutral and acid detergent fiber decreased by 14.2% and 18.7%, respectively. Further, after a 75-day feeding trial using cattle and sheep, the mixed maize and soybean silage lowered the use of maize concentrate by 37.8%, as compared to maize silage alone, and improved the daily weight gain of sheep and cattle by 13.6% and 29.3%, respectively.

Table 13.12 Yield of summer maize-summer soybean intercropping

Year	Location	Area of demonstration fields (ha)	Maize mean yield (kg/ha)	Soybean mean yield (kg/ha)
2015	Henan Yongcheng	13.3	9150.0	1200.0
2016	Henan Dangcheng	2.0	9817.5	1488.0
2016	Anhui Fuyang	6.7	9280.5	1684.5
2017	Shandong Yucheng	66.7	9148.5	1977.0
2017	Hebei Gaocheng	20.0	8340.0	2025.0
2018	Shandong Dezhou	66.7	8080.5	1941.0
2018	Hebei Gaocheng	66.7	7773.0	2935.5
2018	Anhui Linquan	6. 7	8767.5	1824.0
2019	Shandong Dezhou	100.0	8662.5	1813.5
2019	Shandong Dezhou	0.7	11,361.0	2412.0
2022	Shandong Yucheng	14.7	9506.7	2476.5
2015	Henan Yongcheng	13.3	9150.0	1200.0
2016	Henan Dangcheng	2.0	9817.5	1488.0
2016	Anhui Fuyang	6.7	9280.5	1684.5
2017	Shandong Yucheng	66.7	9148.5	1977.0
2019	Hebei Gaocheng	333.3	7762.5	1534.5
2019	Hebei Gaocheng	0.7	10,638.0	1968.0
2019	Henan Yongcheng	33.3	9066.0	1672.5
2020	Shandong Yucheng	66.7	8520.0	1825.5
2021	Shandong Weifang	3.3	8352.0	1896.0
2021	Shandong Yucheng	13.3	8242.5	1539.0
2021	Shandong Feicheng	69.3	8131.2	1715.4

(continued)

Table 13.12 (continued)

Year	Location	Area of demonstration fields (ha)	Maize mean yield (kg/ha)	Soybean mean yield (kg/ha)
2021	Hebei Gaocheng	533.3	7815.0	1555.5
2021	Henan Yongcheng	6.7	6590.6	1635.5
2021	Henan Linying	20.0	7758.0	1459.8

Fig. 13.12 Demonstration fields in Gaocheng, Hebei and Yucheng, Shandong

13.3 Spring Maize–Spring Soybean Intercropping

13.3.1 Adaptation Regions and Current Situation of Maize and Soybean Production

The spring soybean-spring maize intercropping system is predominantly suitable for planting areas in the northwest and northeast regions of China. The northwest region presents abundant solar energy resources, high temperatures, and drought resulting from low rainfall, which causes a high evaporative demand and high plant transpiration rates. Here, annual sunshine duration ranges from 1500 to 3400 h, while annual solar radiation in the Qing-Gan-Xin eastern Xinjiang and western Gansu border area falls between 6200 and 7000 MJ/m^2. Gansu Province shows an active accumulated temperature ≥10 °C that ranges from 3000 to 4900 °C, and Shaanxi Province shows an active accumulated temperature ≥5 °C, which is well over 3000 °C with an extended frost-free period of approximately 220 d. Further, the Hexi Corridor region also shows an active accumulated temperature of 2100–3300 °C. Similarly, the oasis agricultural area of southern Xinjiang boasts of an accumulated temperature exceeding 2500 °C and approximately 200 frost-free days, while Inner Mongolia, located in the middle temperate zone with a monsoon climate, has an accumulated temperature ≥10 °C that ranges from 1400 to 3400 °C with a frost-free period ranging from

100 to 140 days. The northeastern region of China predominantly consists of a temperate monsoon climate that shows abundant sunshine and high accumulated temperature, with a total annual solar radiation of 4400–5000 MJ/m^2 and an annual sunshine duration of 2100–3100 h. The mean frost-free period in this region ranges between 90 and 160 d, annual precipitation usually falls between 400 and 800 mm, and the active accumulated temperature ≥10 °C ranges from 180 to 2800 °C. Consequently, the one-crop-per-year system predominates in this area, with spring maize as the main crop.

The northwest region of China is a significant area for maize production due to its ample supply of solar and thermal resources, large day-night temperature range, and high temperatures. In 2018, maize was cultivated across 3.555 million ha of the northwest region, accounting for nearly 10% of the total national maize planting area. This region also produces approximately 9% of the national maize output, with more than 90% of the maize planting area in Shaanxi, Gansu, and Xinjiang provinces. In recent years, the issue of a supply-demand imbalance in the maize market has become increasingly apparent, prompting regions to adjust their maize industry structure. The hybrid maize seed production area in the Hexi Corridor has stabilized at approximately 70,000 ha, with moderation in the scale of maize grain production and an expansion in the planting area of silage maize. In 2018, the planting area for maize silage in Xinjiang reached approximately 40,000 ha. In 2019, it increased to 50,000 ha in Ningxia and 210,000 ha in Gansu. Additionally, due to fertile land and abundant land resources, the northeastern region has become a maize production hub, planting maize on approximately 200 million mu per year, representing approximately 31% of the national total.

Soybean production in the northwest region of China is primarily found in Shaanxi, Gansu, Ningxia, and Xinjiang, among which Shaanxi, Gansu, and Xinjiang collectively account for 98% of the total soybean planting area in the region. However, since 2010, there has been a decrease in the soybean planting area, falling to the lowest level in 2015, but has started to recover. In 2018, the soybean planting area in the aforementioned areas was 2.389 million ha, representing 3% of the total soybean planting area in the country. Additionally, soybean yield accounted for 2.5% of the overall national production. Compared with 2010, there was a loss of 1.46 million ha of soybean planting area and 51,800 tons of production. In contrast, the northeast region is the largest soybean production area in the country, with an annual sowing area exceeding 54 million mu, which accounts for 40% of the national soybean sowing area, and a high yield per unit area, totaling over 50% of the national soybean production.

In both northwest and northeast China, maize and soybean production are prevalent, and the government has proposed that during the period comprised of the "13th Five-Year Plan," crop rotation can contribute to the capacity for producing soybean together with maize. However, reconciling land-use conflicts between the two crops is difficult due to planting inefficiencies. One emerging option is adopting the spring maize-spring soybean relay intercropping model, which involves planting and managing maize and soybean together. This approach not only ensures maize area and yield but also increases farmer income as a result of the one-season soybean

Table 13.13 Crop yield and value of maize varieties in the northwestern region of China

Maize variety	Maize yield (kg/ha)	Soybean yield (kg/ha)	System output (yuan/ha)	Maize variety	Maize yield (kg/ha)	Soybean yield (kg/ha)	System output (yuan/ha)
Diyou 919	11,295	1097	38,199	Jini 336	10,148	2094	40,977
Kenyu 101	10,919	1193	37,727	Xianyu 698	9791	1859	38,565
Jinpingguo 619	10,779	1182	37,274	Xianyu 1483	9720	1265	34,803
Tie 391	10,556	1319	37,466	Xianyu 1225	9645	1112	33,672
Ruipu 686	10,496	1767	39,990	Jindan 73	9296	1104	32,652
Zhongxing 719	10,442	2238	42,665	Bosheng 818	9116	1323	33,458
Huamei 1	10,190	1259	36,077	Jinrun 919	8969	1434	33,719

crop. The spring maize-spring soybean intercropping model shows great potential for widespread adoption in northern and southwestern regions such as Yunnan, Guizhou, Hunan, and Jiangxi, where spring maize is grown.

13.3.2 Key Technology

13.3.2.1 Variety Selection

Selected maize varieties should have a growth period of approximately 100–150 days, compact plant type, and clear marginal advantages for high-yielding machine-harvestable varieties according to the thermal conditions of each region, such as Diyou 919, Kenyu 101, Zhongken 719, and Jenny 336 (Table 13.13). The selected soybean varieties should be high-yielding, machine-harvestable varieties that are tolerant to shade, lodging-resistant, and with a pod height greater than 12 cm and less than 20 cm, such as Suinong 2, Jinong 28, Qindou 2018, and Heilong 52 (Table 13.14).

13.3.2.2 Timely Sowing

The 2BYFSF-5 (4–8), a strip intercropping and precision seeding and fertilization machine, can be used for sowing and fertilizing both maize and soybean simultaneously in Northwest China, where additional equipment provided with a plastic film-laying apparatus can be used.

13.3.2.3 Increase Light Through Intercropping

The maize strip consists of two rows of maize 40 cm apart. The inter-row spacing between maize strips is 210–230 cm, and the inter-row spacing between maize and soybean strips is 0.60–0.70 m. The distance between soybean rows is 0.30 m, and the optimal soybean-to-maize ratio is 4:2 (Table 13.15).

Table 13.14 Crop yield and value of soybean varieties in the northwestern region of China

Soybean variety	Soybean yield (kg/ha)	Maize yield (kg/ha)	System output (yuan/ha)	Soybean variety	Soybean yield (kg/ha)	Maize yield (kg/ha)	System output (yuan/ha)
Suinong 2	2238	10,442	42,665	Shandou 125	1049	6462	24,386
Jinong 28	1859	9791	38,565	Jidou 19	1028	9029	31,445
Qindou 2018	1859	9702	38,316	Yudou 8	1019	7176	26,204
Heilong 52	1767	10,496	39,990	Kendou 38	987	12,335	40,458
Lindou 10	1578	5066	23,652	Liaodou 15	968	7790	27,618
Qihuang 34	1247	7415	28,244	Tiefeng 31	948	10,910	36,233
Fengdou 6	1236	10,871	37,854	Xidou 5	942	9480	32,192
Zhonghuang 30	1227	9288	33,368	Longhuang 2	885	7259	25,629
Kendou 95	1193	10,919	37,727	Kendou 62	867	12,309	39,668
Longhuang 3	1185	10,539	36,617	Dongdou 339	863	9720	32,393
Chengdou 3	1125	8495	30,533	Ningdou 7	785	9776	32,073
Ningdou 6	1119	10,391	35,810				

Table 13.15 Yield and economic value according to row ratio of spring soybean-spring maize intercropping in the northwestern region

Row ratio	Maize yield and yield components				Soybean yield and yield components				Output (yuan/ha)
	Effective ear number (plant/ha)	Grain number per plant	1000-kernel weight (g)	Yield (kg/ha)	Effective plants number (plant/ha)	Seeds number per plant	100-seed weight (g)	Yield (kg/ha)	
3:2	59,306	501.2	344.4	8355	93,812	66.9	22.4	1034	29,592
4:2	60,620	526.8	347.9	10,893	146,232	44.0	20.5	1277	38,162
6:2	56,570	418.7	360.7	8544	174,912	41.8	24.0	1757	34,458
4:3	47,592	641.3	347.5	8930	84,335	51.0	22.5	855	30,134
6:3	62,982	526.1	349.0	9942	70,178	89.5	21.4	1127	34,595
4:4	60,111	517.8	331.6	9521	77,432	73.0	22.9	993	32,613
6:4	43,011	540.2	347.1	7109	84,920	61.9	21.1	930	25,484

13.3.2.4 Reduce Plant Spacing to Ensure a High Plant-Population Density

Single seed drilling or double seed drilling machines can be used to plant maize with a distance of 9–10 cm (single seed) or 18–20 cm (double seed) between plants, using 75,000 seeds hm^{-2}. Approximately 67,500 effective ear numbers hm^{-2} can be ensured (Table 13.16). For soybean planting, distances of 8–9 cm (single seed) or 16–18 cm (double seed) between plants are required, using 165,000 seeds ha^{-1} and ensuring approximately 120,000 effective plant ha^{-1} (Table 13.17).

13.3.2.5 Integrated Fertilization

Maize and soybean in the northwest region of China have long growth periods and high yield levels. A total of 900–1200 kg of slow-releasing high-N fertilizer ($N\text{-}P_2O_5\text{-}K_2O = 28\text{-}8\text{-}6$) per ha is required for maize, while 300–375 kg of slow-releasing low-N fertilizer ($N\text{-}P_2O_5\text{-}K_2O = 14\text{-}15\text{-}14$) per ha is recommended for soybean. Furthermore, in the drip-irrigated land in the region, maize and soybean independent water-fertilizer integration systems are used for these two crops, separately.

13.3.2.6 Control of Plant Height and Prevention of Lodging

To optimize maize growth, growth regulators, such as paclobutrazol and Triacontanol (aminofat and Ethrel), should be applied when the seventh leaf emerges, considering plant height. In contrast, to enhance soybean stem and leaf development, 5% Ethephon (300–750 g) wettable powder must be sprayed per ha during the seedling, third and fifth-leaf stages, and at the initial flowering period. The quantity should be lowered to 300 g at the seedling phase and must be combined with 450 kg of water. Alternatively, it can be mixed with 16–20 mg of 5% Ethephon wettable powder per kg of seed.

13.3.2.7 Weed Control

In maize, timely application of 1200–1500 mL/hm^2 of 96% quizalofop-p-ethyl emulsion (Jinduer) is recommended before seedling emergence, whereas maize and soybean targeted herbicides are used after seedling emergence, taking care to use physical barriers to separate the two crops during application or using dual-tank spray equipment. For maize, 1875–2250 mL of 5% nicosulfuron+20% mesosulfuron-methyl per ha is used and added to water at a rate of 225–450 kg. Soybeans are treated with 1200–1500 g/hm^2 of 25% flufenacet ether water-dispersible granules or 15% quizalofop-p-ethyl with 25% flufenacet ether (300 mL + 270 g Type 1 set/hm^2). In the northwestern region, weed control is achieved by mulching and inter-row tillage.

13.3.2.8 Disease, Pest, and Weed Control

Combining physical, biological, and chemical control methods, the intelligent LED integrated-spectrum insecticidal lamp and sex pheromones are used to kill insect pests. With the assistance of unmanned aerial vehicles, pest and disease control is conducted three times during the maize monoculture stage, maize tassel period, and soybean pod filling period. The "fungicides, insecticides, synergists, regulators, and

Table 13.16 Crop yield and value according to maize yield level in the spring maize-spring soybean intercropping system in the northwest region of China

Maize yield level (kg/ha)	Maize yield and yield components				Soybean yield and yield components				Output (yuan/ha)
	Effective ear number (plant/ha)	Grain number per plant	1000-kernel weight (g)	Yield (kg/ha)	Effective plant number (plant/ha)	Seeds number per plant	100-seed weight (g)	Yield (kg/ha)	
<500	41,699	534.2	342.9	5946	78,315	70.4	21.2	953	22,368
500–600	54,093	514.9	360.6	8409	69,191	71.6	21.5	846	28,623
600–700	59,904	522.6	352.0	9636	98,589	72.2	21.8	1194	34,148
>700	62,934	583.6	362.7	12,110	76,163	71.6	20.5	956	39,636

Table 13.17 Crop yield and value according to soybean yield level in the spring maize-summer soybean intercropping system in the northwest region of China

Soybean yield level (kg/ha)	Soybean yield and yield components				Maize yield and yield components				Output (yuan/ha)
	Effective plant number (plant/ha)	Seeds number per plant	100-seed weight (g)	Yield (kg/ha)	Effective ear number (plant/ha)	Grain number per plant	1000-kernel weight (g)	Yield (kg/ha)	
<80	66,263	72.8	21.1	804	52,989	565	358.4	9246	30,717
80–100	95,024	77.6	20.9	1340	61,394	536.7	350.1	10,578	37,656
100–200	118,758	76.4	21.8	1676	61,349	455.9	343.1	7920	32,234
>200	125,522	81.3	20.8	2013	53,951	496.6	396.7	10,020	40,131

micronutrients" five-in-one package is applied to each ha using 1500 mL of 500 g/L methyl thiophanate, 375 mL of 2.5% efficient chlorfenapyr, and 600 mL of 12% Abamectin miticide during the soybean pod filling period, and mixed with 600–750 kg of water for spraying. This is performed in relay intercropping systems.

13.3.2.9 Harvest

Soybeans and maize typically reach maturity in mid to late September. Farmers can select their harvesting method based on suitable machinery available and their specific needs. Three options are available: harvesting soybeans first, harvesting maize first, or harvesting both crops simultaneously. If maize is harvested first, narrow-row maize harvesters with a body width of less than 1.8 m, such as the 4YZ-2A self-propelled combine harvester and the 4YZP-2L crawler combine harvester, can be used to harvest maize in the soybean rows. Then, local soybean harvesting machines can be used to harvest soybeans. If soybeans are harvested first, the GY4D-2 combine harvester can be used to harvest soybeans in between the maize rows. Next, local maize harvesting machines can be used to harvest maize. Both local soybean and maize harvesting machines can also be used to harvest both crops simultaneously. Additionally, farmers can opt for a mixed green silage of soybean and maize. In this case, a green silage harvesting machine with a suitable swath width, such as the 4QZ-2100 or 4QZ-3000 green silage forage harvester models capable of harvesting both tall and short stalks, can be selected.

Attached: Field configuration diagram and management link diagram (Figs. 13.13, 13.14, 13.15, 13.16 and 13.17).

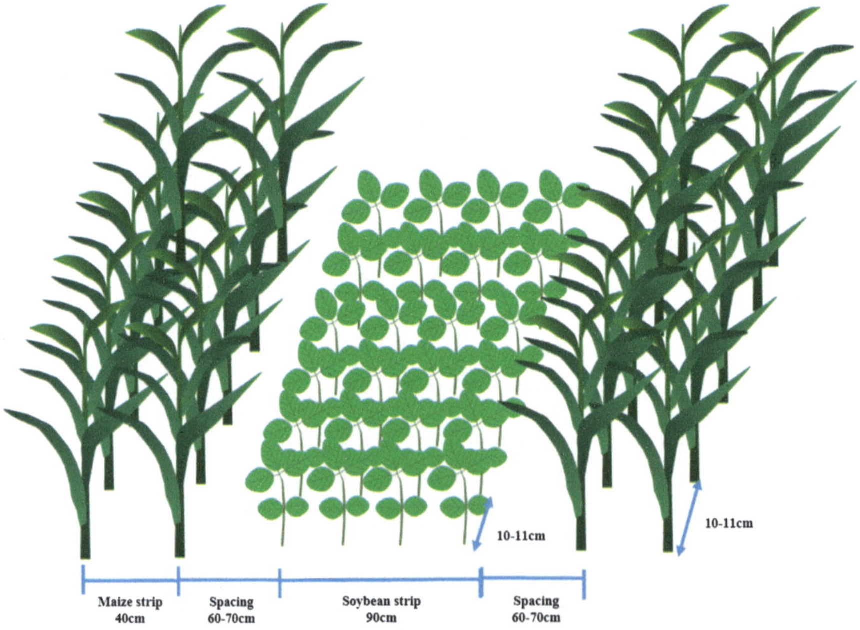

Fig. 13.13 Schematic representation of spring maize-summer soybean strip relay intercropping in the northwest and northeast regions of China

Fig. 13.14 Application of fertilizers and plastic mulching in spring soybean-spring maize intercropping in the northwest region of China

13.3.3 Technical Effects

Over the years, the cumulative planting area for soybean-maize relay intercropping in the northwest region of China has expanded to approximately 260 ha. The average maize yield is 12,109 kg/ha, which is comparable to that of monoculture maize. In 2017, the intercropping of soybeans and maize took place in several areas, including Zhangye, Gansu, Yinchuan, Ningxia, and Yan'an, Shaanxi. Among them, Yan'an registered a record high yield in a demonstration field of maize intercropped with soybeans, where soybean yield reached 1737 kg/ha, 18.2% and 24.3% higher than the yields in the demonstration areas established in Zhangye and Yinchuan, respectively (Fig. 13.18). The following year, Baotou, Inner Mongolia, was also included as a demonstration site and set a high record of 1987.5 kg/ha for soybean yield in intercropping, which was 11.2% higher than that recorded at Yan'an. Intercropping in Inner Mongolia continued to gain support, with 113.3 ha planted in 2019, while in

Fig. 13.15 Simultaneous sowing of spring soybean and maize in the northeast region of China

Fig. 13.16 Independent water and fertilizer integration management technique for soybean and maize

Fig. 13.17 Technique for harvesting mixed silages of soybean and maize

Fig. 13.18 Demonstration fields in Yinchuan, Ningxia and Baotou, Inner Mongolia

Table 13.18 Yield of spring maize-spring soybean intercropping

Year	Location	Area of demonstration fields (ha)	Maize mean yield (kg/ha)	Soybean mean yield (kg/ha)
2012	Liaoning Tiexi	13.3	8670.0	1005.0
2013	Jili Gongzhuling	6.7	10,410.0	1080.0
2013	Heilongjiang Fujin	20.0	10,902.0	2650.5
2016	Gansu Wuwei	40.0	12,510.0	1387.5
2017	Gansu Zhangye	66.7	12,750.0	1420.5
2017	Ningxia Yinchuan	10.0	14,055.0	1320.0
2017	Shanxi Yanan	13.3	7905.0	1740.0
2018	Gansu Zhangye	10.0	12,975.0	1404.0
2018	Ningxia Yinchuan	13.3	13,800.0	1365.0
2018	Shanxi Yanan	15.33	11,115.0	1770.0
2108	Inner Mongolia Baotu	16.0	11,835.0	1987.5
2019	Inner Mongolia Baotu	113.3	12,045.0	1852.5
2020	Inner Mongolia Baotu	1600.0	11,532.0	1237.5
2020	Gansu Wuwei	66.7	12,819.0	1494.0
2021	Inner Mongolia Baotu	1533.3	13,617.0	1159.5
2022	Inner Mongolia Baotu	13.3	10,605.0	1380.0

2020, 1600 ha were planted (Fig. 13.18). Additionally, during the same period, maize yields peaked at 12,405 kg ha^{-1} and 14,190 kg ha^{-1} in 2019 and 2021, respectively, and soybean yields fluctuated between 1380 and 1580 kg ha^{-1}. Finally, in 2022, the maize yield was 10,605 kg ha^{-1}, soybean yield was 1380 kg ha^{-1} (Table 13.18).

Mechanisms for Promoting the Achievements from Research and Technological Development for Strip Intercropping Systems

14

Taiwen Yong, Wenyu Yang, Xiaochun Wang, and Lihua Zhang

Abstract

This chapter presents a pioneering strategy to overcome farmer skepticism toward mechanized intercropping. The "Triple-Integration Transformation System," "Four-Level Promotion Network," and "Five-fold Combination Training Model" aim to transform research outcomes effectively. Collaborative efforts across disciplines and governmental levels have successfully promoted intercropping models, challenging long-standing beliefs. The strategy demonstrates a four-tier promotion network and a comprehensive training model, contributing to increased crop yields and farmer income.

Keywords

Intercropping · Mechanization · Collaborative efforts · Sustainable agriculture · Technology adoption

14.1 New Mechanism for Promoting Research Achievements: "Triple Integration, Four Levels, Five-fold Combination"

In response to the difficulty in promoting intercropping models due to the belief among farmers that such systems cannot be mechanized, the project team developed a new strategy for promoting achievements, which is based on the "Triple-Integration Transformation System," the "Four-Level Promotion Network," and the "Five-

T. Yong (✉) · W. Yang · X. Wang · L. Zhang
College of Agronomy, Sichuan Agricultural University, Chengdu, People's Republic of China

College of Mechanical and Electrical Engineering, Sichuan Agricultural University, Ya'an, People's Republic of China
e-mail: mssiyangwy@sicau.edu.cn; xchwang@sicau.edu.cn

W. Yang et al. (eds.), *Theories and Application in Maize-Soybean Strip Intercropping System*, https://doi.org/10.1007/978-981-97-8723-4_14

fold Combination Training Model," with demonstration templates as the starting point and a sense of responsibility and dedication as the bond. Altogether, these factors will ensure the effective transformation of the project outcome into a package that farmers are willing to adopt.

14.1.1 "Triple-Integration Transformation System"

With the leadership of science and technology at the center of the strategy, industrial demand as the orientation, and the increase of farmer income as the goal, with support from research funds of the Ministry of Science and Technology, the Ministry of Agriculture and Rural Affairs, and Sichuan Province, team members and collaborating units shoulder the responsibility to attain national food security, sustainable agricultural development, and to increase farmers income. Such commitment demands that team members rely on close collaboration among universities, research institutes, and enterprises, combining multiple disciplines such as crop science, agricultural engineering, and plant protection to comprehensively tackle and develop the project. Central and national departments, as well as local party committees and governments, attach great importance and share responsibilities in promoting the project on a large scale in 17 provinces and cities. Thus, through the collaboration of agricultural technology promotion departments at the national, provincial, and municipal (county) levels, system achievements are ultimately comprehensively transformed and transferred as a technological package to be implemented in various operating entities, including companies, large-scale plantations, and small-scale farmers in rural areas. This forms a "Triple-Integration Transformation System" of party committees and governments, research and promotion, and operating entities working together to accelerate the transformation process and achieve the desired comprehensive results.

This transformation system is crucial for effectively transferring technological achievements from research centers in Sichuan to the entire country. In addition to the Renshou core base, emerging models such as the Gaocheng, Feicheng, Yucheng, and Baotou Model have rapidly emerged. Hebei, Shandong, and Nei Monggol have quickly become major provinces for nationwide promotion, with areas committed to such promotion exceeding 1 million ha in 2022.

14.1.2 "Four-Level Promotion Network"

During the construction of the project, a promotion strategy was established, centered on "emphasizing dedication, setting examples, changing mindsets, and promoting applications." Considering high-yield demonstration templates as the starting point, a "field-template-plot-promotion area" network was formed, with outstanding high-yielding research fields, 1000-mu demonstration plots, 10,000-mu demonstration areas, and 100,000-mu promotion regions as the main components. This "Four-Level Promotion Network" is closely integrated with the program of the Ministry of Agriculture and Rural Affairs for fostering high crop yield and abundant grain production with the special demonstration and promotion plan of the Ministry of Science and Technology.

14.1.3 "Five-fold Combination Training Model"

Relying on the national soybean industry technology system, a research and development team was established, which integrates the "two systems," namely the maize and soybean industry systems, "six positions," namely genetics, breeding, agricultural machinery, cultivation, plant protection, and economic scientist positions, and "11 experimental stations," namely Nanchong, Yan'an, Yinchuan, Lanzhou, Guiyang, Fenyang, Nanning, Kunming, Dezhou, Jinan, and Xuzhou. Together, these elements conform to a comprehensive training team with "cross-system collaboration, station-to-station connection, station-to-station interaction, and regional cooperation." The team utilizes the official website of the Crop Relay-intercropping Planting Engineering Technology Research Center in Sichuan Province, the public account for Strip Intercropping Research and Application, the WeChat platform for soybean-maize strip intercropping planting technology exchange, and the specialized publication for soybean-maize relay-intercropping planting-promotion for real-time dissemination of technical guidelines and emergency techniques. Experts engage in online interactions to answer questions, establish QQ and WeChat groups with technical extension personnel and technology users, stay updated on field conditions, and conduct technical exchanges. The team of experts provides on-site guidance and face-to-face services to farmers through multiple "online and offline" channels and technical communication. A "Five-fold Combination" training model has been established, combining high-yield theories and practical techniques, project training and self-training, technical lectures and field demonstrations, party and government officials with scientific and technological personnel, and agricultural technical officers with farmers. Thus, the model maximizes the efficiency of the transformation process of the results of this project.

14.2 Characteristics of the New Mechanism of Promotion of Research Achievements

Due to the long-held notion by farmers that intercropping cannot be mechanized, there have been difficulties in promoting the widespread use of the technology. This project established a promotion strategy of "being a model, changing mindsets, and promoting application," forming a team strategy that integrates demonstration application to drive technological (machinery) research and basic research, mutually promoting and interdependent development. The goal of the strategy is to ensure that everyone understands, demonstrates, and promotes the technology. With the assistance of national, provincial, and municipal-level extension departments, a four-tier promotion network was established nationwide, with the "high-yield demonstration field, 1000-acre demonstration plot, 10,000-acre promotion area, and 100,000-acre promotion regions" as the main framework. Additionally, through the "Five-fold Combination Training Model" and various channels of technical exchange, the project strengthened technical training, significantly improving the adoption rate of the technology and achieving significant increases in crop yield and farmer income. Through various methods, including observation meetings, yield assessment meetings and on-site meetings, the technology has been promoted and received recognition from the government, experts, and farmers. Furthermore, it has gradually changed the long-lived mindset among farmers for intercropping. The project results have been rapidly promoted and applied in 20 provinces and cities nationwide. Concomitantly, this technology has also been applied on a large scale in countries along the Belt and Road Initiative, such as South Asia, Europe, and Africa, through international cooperation projects and student training, thereby invalidating the belief that intercropping should be discontinued. Moreover, it has come to fill in the gaps between intercropping theoretical research, technological development, machinery manufacturing, and the promotion of the application of intercropping systems.

14.2.1 Establishment of a Demonstration Base for the Application of the Technology

With the support of national, provincial, and municipal projects, after 15 years of development, a modern grain-industry (Renshou) demonstration park was established in Renshou County, Sichuan Province, one of the major grain-producing counties in the country. The park conducts technological research and development, demonstrations, and training, serving as a showcase for external displays, allowing the results to readily go beyond the campus and spread nationwide. According to incomplete statistics, 17 provincial-level and above on-site observation and training sessions have been held at the Renshou base, including six national conferences, with more than 3000 experts, scholars, and technical personnel visiting the base.

14.2.2 Establishment of an Application-Assurance System

Talent assurance: Relying on the national soybean and maize industry technology system, as well as provincial-level innovation teams for soybean and maize, and cooperation between agricultural research institutions and extension departments at the provincial and municipal levels, a technical service team consisting of "universities + research institutions + municipal (county) agricultural technicians" was formed to conduct technology training, guidance, and comprehensive monitoring and investigations. The 26 team members from the first completed unit of the project and over 300 graduate students were involved in on-site demonstrations and promotions, ensuring the technical needs of applying the results and improving the technology adoption rate.

Machinery assurance: Extensive cooperation with agricultural machinery companies has been maintained to develop and produce machinery, such as soybean-maize relay-intercropping planting seeders, dual-system zone spraying machines, narrow-track maize harvesters, and narrow-track soybean harvesters (Zhang et al. 2022). These machines have undergone years of multi-point trials and demonstrations in Southwest and Northwest China, and the Huang-Huai-Hai region. Companies provide full technical services, ensuring the demand for demonstration machinery and operational quality.

14.2.3 Construction of a New Application Pattern

Breaking away from the traditional "top-down control" or project-based promotion model, a new promotion mechanism of "Triple Integration, Four Circles, and Five-fold Combination" was created. This mechanism has formed a new application pattern with the Renshou base as the application showcase platform and the Sichuan central region, key regions in Southwest China, Northwest China, and the Huang-Huai-Hai region as the promotion areas. The promotion area has increased annually. Thus, the soybean cultivation area in Sichuan has risen from the 12th position in the country before 2000 to the fifth position, making the Southwest region one of the top three major soybean-producing areas in China and one of the three implementation regions of the National Soybean Revitalization Plan. This has promoted the large-scale promotion of the project results in Southwest China, Northwest China, and the Huang-Huai-Hai region.

14.3 Effectiveness of Application of the New Mechanism of Promotion

14.3.1 Effectiveness of Application

Since 2008, the integrated cropping system has been continuously selected as the key promoted technology by the Ministry of Agriculture and Rural Affairs and the Agriculture Department of Sichuan Province. In 2019, it was selected as a key technology promoted for the National Soybean Revitalization Plan, and in 2020, 2022, and 2023, it was included in the central government's important policy documents. In 2022, it was designated as the first recommended technology for the National Soybean and Oilseed Production Capacity-Enhancement Project.

From 2008 to 2022, the integrated cropping system has been promoted in more than 20 provinces (municipalities and autonomous regions), in regions such as the Southwest, the Northwest, and Huang-Huai-Hai region, with a total cumulative promoted area of over 106 million mu (approximately 7.07 million ha). This has resulted in an increase of 12.839 million tons in soybean production and a total economic benefit of 33.05 billion yuan. In 2022 alone, the system was promoted in 18.72 million mu (approximately 1.25 million ha) across 16 provinces (municipalities), producing 1.67 million tons of soybean and increasing the self-sufficiency rate by 1.5%. In Sichuan Province, the system has been promoted on a cumulative area of 55.03 million mu (approximately 3.67 million ha), achieving full-scale promotion in all 21 cities and prefectures and contributing to Sichuan Province becoming the fourth-largest soybean-producing province in the country (Yong and Yang 2022). The Southwest high-protein edible soybean production area, where the integrated cropping system has been implemented, has become one of the three major advantageous soybean production areas in China and one of the three major implementation regions for the National Soybean Revitalization Plan.

According to the characteristics of the regional production, the integrated cropping system has been implemented differently in different regions. Thus, in the spring maize regions of Southwest China, it is applied as a spring maize-summer soybean strip intercropping system. Meanwhile, in the spring maize regions of Northwest, Northeast, and the Yangtze River Basin, it is implemented as a spring maize spring-spring soybean strip intercropping system. In the summer maize regions of the Huang-Huai-Hai area, the system is practiced as a summer maize-summer soybean strip intercropping system. In all these regions, the integrated cropping system has shown high adaptability to the region's conditions, and typical high-yielding cases continue to emerge. In particular, in 2022, the national government proposed to promote the expansion of the soybean-maize strip intercropping system on 15.1 million mu (approximately 1.01 million ha) of land in 16 provinces. Research and development institutions fully leveraged the promotion mechanism of this technology and collaborated closely with the National Agricultural Technology Extension Center and local agricultural technology stations. Therefore, they jointly developed technical plans, conducted training programs, and built high-yield demonstration models. Technical guidance documents such as the "2022 National Soybean-Maize Strip Intercropping Technology Plan" and "2022 Technical Guidelines for Soybean-Maize Strip Intercropping in Southwest China" were successively released. A total of 18 provincial-level training sessions were conducted. The achievement of the "Integrated Innovation and Demonstration Promotion of Soybean-Maize Strip Intercropping Technology Model" won the "Agricultural Technology Promotion Cooperation Award" from the Ministry of Agriculture and Rural Affairs. Through mature promotion channels, the technical implementation rate and the completion of the implementation tasks in 1045 counties covering a total area of 16.45 million mu (approximately 1.10 million ha) were ensured, with Sichuan Province alone implementing 3.754 million mu (approximately 0.25 million ha), exceeding the established target by more than 20%.

In addition to ensuring the promoted area, the project team actively performed high-yield demonstrations through on-site services and provincial-level assignments. They established several replicable and promotable high-yield demonstration models in Sichuan, Shandong, Henan, Hebei, and Inner Mongolia. In the Anju District, Sichuan Province, Fengguangrong Planting Family Farm achieved a soybean yield of 180.2 kg per mu and a maize yield of 617.66 kg per mu. Similarly, in Xindian Town, Yucheng City, Shandong Province, Dazhou Village achieved a soybean yield of 165.1 kg per mu and a maize yield of 633.78 kg per mu. In turn, in

Jiangfeng Village, Guiren Town, Sihong County, Jiangsu Province, Song Yuli Grain Planting Family Farm achieved a soybean yield of 127.79 kg per mu and a maize yield of 521.87 kg per mu. Consistently, in Chenwei Village, Weixian Town, Lingbi County, Anhui Province, a soybean yield of 155.42 kg per mu and a maize yield of 518.3 kg per mu was achieved. In Zhangzhuang Village of Shangsi Township, Hanbei Town, Dayou Village, and Yuanyao Village of Dayou Township in Wuxiang County, Changzhi City, Shanxi Province, soybean yields ranged from 161.1 to 680.5 kg per mu, and maize yields ranged from 680.5 to 680.5 kg per mu. In 2022, the Ministry of Agriculture and Rural Affairs organized a national soybean high-yield competition, and 10 "Golden Soybean Kings (Strip Intercropping)" were selected. In conjunction with high-yield demonstrations, various regions have also innovated a series of promotion mechanisms, such as the "Anju Model" with the slogan "Technology Ensures Soybean Production and Promotes Abundance" and the "Five Strengthenings" model for achieving high yields of soybean and maize in Yucheng.

14.3.2 Typical Models

14.3.2.1 "Technology Ensures Soybean Production and Promotes Abundance" Anju Model

Anju District is located in the middle reaches of the Fu River in the Sichuan Basin and is a typical hilly area with low- and medium-height hills. The cultivated land covers an area of 843,200 mu (approximately 56,213 ha), with an annual grain planting area of over 1.16 million mu (approximately 77,333 ha). It is designated as a national safe agricultural machinery demonstration area and was selected as a national soybean technology self-strengthening demonstration county in 2022. In 2021, with the support of funds from Sichuan Province for the transfer of scientific and technological achievements and the coordinated promotion project of the Ministry of Agriculture and Rural Affairs, Anju District of Suining City, Sichuan Province, was included as a key demonstration area. It was jointly implemented by Sichuan Agricultural University, the General Station of Agricultural Technology Promotion of Sichuan Province, and local agricultural technology stations. Sichuan Agricultural University provided technical guidance and targeted technical formulas throughout the process, ensured the implementation of technical parameters at each step, and achieved standardized technical implementation through complete mechanization.

Emphasis was placed on the "**three-fold selection**." The region was selected first. Firstly, 500 mu (approximately 33.33 ha) of pilot fields and 3000 mu (approximately 200 ha) of demonstration plots were established in three towns with plentiful water resources, contiguous arable land, and high planting-density levels, namely Shidong, Baima, and Zhongxin Towns. The second selection was the main entities. New agricultural management entities such as Fengguangrong Planting Family Farm, Shenghe Junli Agricultural Company, Qiaonong Agricultural Machinery Professional Cooperative, and Jinqi Planting Family Farm constructed the pilot fields, demonstration plots, and demonstration areas of thousands of mu. The third selection step was the model. According to the recommendation of Professor Yong Taiwen, a scientist in the national soybean industry technology system, the selection was based on the mature belt intercropping of wheat (oil), maize, and soybean (or the mature belt intercropping of two mature crops and three crops). The spacing was 2.4 m (or 2.8 m), with 40 cm between maize rows and three soybean rows with a distance of 60 cm between them (or four rows of soybean 100 cm apart). Lastly, the distance between the maize and soybean strips was 70 cm.

Secondly, the "**three determinations**" were implemented. The first determination was the selection of varieties. Compact, dense, drought-resistant, lodging-resistant, and machine-harvestable maize varieties such as Zhongyu 3, Yunhai 365, and Zhenghong 6 were selected. Shade-tolerant, dense, and lodging-resistant summer soybean varieties such as Nanxia 38 and Gongqiu 5 were selected. The second determination was the selection of machinery. The 2BYFSF-3 (6) (Hebei Nonghaha) type of maize and soybean fertilization seeder was selected for sowing and fertilization. The 4YZ P-2685 or 4YZ-2A self-propelled two-row maize harvester was used for harvesting maize ears, while the GY4D-2 or 4LZ-3.0Z combined harvester was used for soybean harvesting, threshing, and incorporation of straw into the topsoil layer. Complete mechanization was achieved. The third determination was the selection of pesticides. Based on the trial parameters of the team in Suining City for many years, 96% metolachlor emulsifiable concentrate was used for pre-emergence weed control, targeted herbicides for maize and soybean were used after seedling emergence and effective control of weeds was achieved. "Integrated spraying of insecticides, fungicides, enhancers, foliar fertilizers, and growth regulators" was conducted at the 3–4 leaf stage of maize seedlings, tasseling stage, and pod formation stage of soybean to effectively prevent major insect pests and diseases incident on maize and soybean crops.

Thirdly, strict control was applied in the "**three stages**." First was the sowing stage. High-quality sowing is a prerequisite for high yields. Based on the selection of good varieties, Anju District strengthened technical training for agricultural management entities, agricultural technicians, and machine operators and provided technical guidelines. Additionally, relying on experienced farms (such as Fengguangrong Family Farm), high-yield demonstration fields were established for observation and demonstration of machinery selection, seeding depth, and sowing period to ensure uniform and vigorous seedlings. The second stage was pest and disease control. Early investigations were conducted in response to major pests and diseases, and preventive measures were formulated based on the incidence history. Early detection and prevention were implemented, and preventive measures were timely spread through platforms such as the Sichuan Rural Daily. The third stage was chemical control. Technical guidance was promptly provided to address issues such as excessive growth, lodging, and pod shedding of soybeans caused by seasonal drought and flooding in Sichuan. On-site visits by management entities were organized to effectively cope with the impact of adverse weather changes on soybean yields.

Under this guidance, the operating entity represented by the Fengguangrong Family Farm achieved significant results in the first year of system implementation. The average maize yield was 553.4 kg per mu, an increase of 171.4 kg per mu compared with large-scale production, with a growth rate of 44.9%. The average soybean yield was 150 kg per mu, an increase of over 35 kg per mu compared with large-scale production, with a growth rate of 30.4%. The combined increase in income per mu from the two crops exceeded 686.8 yuan compared with large-scale production.

In 2022, the Fengguangrong Family Farm expanded its planting scale to 1600 mu and continued to follow the guidance of the project team. Despite the record-high temperature and drought conditions in Sichuan in 60 years, the strip intercropping of soybean and maize achieved relatively high yields. Through yield measurements organized by the National Agricultural Technology Center, the average yield of strip intercropped spring maize was 617.66 kg per mu, and that of strip intercropped summer soybean was 180.2 kg per mu, setting a national record for the highest soybean yield in strip intercropping. The excellent demonstration effect was reported by CCTV's Financial Channel "Economic Half Hour," CCTV's "Secure China's Rice Bowl" series, People's Daily, Sichuan Daily, Farmers' Daily, and other media outlets on more than 20 occasions.

The Anju District was originally a small agricultural county in Sichuan Province. How did it develop maize-soybean strip intercropping into a model of high productivity in Sichuan over such a short period? Based on the summary, Anju District adopted the following typical practices:

First, Anju focused on expanding production and increasing yield while strengthening responsibility. Based on its positioning as a major grain-producing county in the country, the leading group for the demonstration and promotion of strip intercropping of soybean and maize was established, with the district head as the group leader and deputy district heads as deputy group leaders. The heads of agricultural, financial, development and reform departments, and other related departments, served as members. The group organized six special coordination meetings. Secretary Li Jiang of the Municipal Party Committee conducted two field research visits to the Anju District to investigate the maize-soybean strip intercropping performance. Three technical training sessions on strip intercropping of soybean and maize were held at the city level. Documents such as the "Notice on Stabilizing Grain Production in 2022" were issued, and 21 agricultural policies were formulated. The subsidy for purchasing maize and soybean seeding and single-seed planting machinery was increased to 40%, and that for purchasing specialized strip intercropping machinery and straw returning machinery for maize and soybean was increased to 50%. A total of 4 million yuan of local fiscal funds was allocated to subsidize strip intercropping of soybean and maize, with subsidies accumulating to 330 yuan per mu.

Second, Anju focused on improving quality and efficiency while strengthening technological innovation. Relying on new technologies and models for efficient production, agricultural scientific and technological achievements were promoted to achieve practical productivity. With Sichuan Agricultural University for technical support, innovative pilot projects such as "planting soybean in maize fields" were successfully performed in Shuangcitang Village in Shidong Town, and Tanjiaba Village and Baoquangou Village in Baima Town. The government purchased and distributed 40,000 kg of soybean seeds, including Nanxiadou 25 and other varieties. There are 25 agricultural professional cooperatives and 54,000 sets of agricultural machinery in the district. Personnel involved in this project included 85 technicians with expertise in mechanization technology for maize-soybean strip intercropping and 145 supporting staff for agricultural technology promotion. The technical promotion system is sound, and the service capacity is leading in the province. Explorations were performed on the integration of agricultural machinery and agronomy, green control of pests and diseases, and social services. A technical guidance service team for strip intercropping of soybean and maize was established, led by Professor Yong Taiwen, a scientist from the National Modern Agricultural Industry Technology System, and composed of municipal, district, and town-level agricultural technicians responsible for technical training and on-site guidance. Meanwhile, the district government attached great importance to the issue and conducted special seminars to discuss the problems and solutions for the ability of new types of operating entities and social service organizations to use new agricultural technologies, land scale circulation, agricultural mechanization, and improvement of production facilities. This leveraged the enthusiasm of new types of operating entities and social service organizations and fully utilized the role of fiscal funds.

Third, Anju focused on stabilizing production and increasing income while strengthening basic services. Over 50 million yuan of agricultural funds were integrated into the district for soybean industry development and infrastructure construction. Based on the pilot work of crop rotation and fallow in 2021, high-standard farmland construction, and suitable mechanization projects, the bases of the demonstration work were modified, and owners were introduced to conduct full-scale mechanized planting. The Suining Soybean Industry Development Consortium was established to deepen the industrial chain. With the Feng guangrong Family Farm as the model for demonstration planting, more than 50 large-scale demonstration planting households were introduced and cultivated to promote demonstration and extension among farmers. According to local conditions, different soybean expansion models were selected. The main focus was on promoting maize-soybean strip intercropping, with 10,000 mu of soybean intercropped with other crops, including 4000 mu of soybean intercropped with young fruit trees, 3000 mu of soybean intercropped with sorghum, and 3000 mu of soybean intercropped with sweet potatoes.

In 2017, Yucheng City in Shandong Province established the Yucheng Chuannong Rural Development Research Institute through a tripartite cooperation with Sichuan Agricultural University and Dezhou Agricultural Science Institute. For five consecutive years, they have been promoting the application of strip intercropping technology for soybean and maize, with a cumulative promotion area of over 10,000 acres. By employing techniques such as selecting suitable varieties, increasing spacing and illumination, regulating nutrition, and controlling pests and diseases, they have achieved the goal of "no yield reduction in maize, and getting the harvest from soybean." Particularly in 2022, they exceeded the target area of 100,000 acres and achieved a planting area of 134,600 acres, accounting for more than one-tenth of Shandong Province and ranking fourth among all counties and cities nationwide, as well as the first in Shandong Province. They also hosted field observation meetings on strip intercropping of soybean and maize in the Huang-huaihai region, herbicide screening for strip intercropping of soybean and maize, and comprehensive technical experiments and demonstrations for plant protection management, thereby promoting the experience and practices of strip intercropping and plant protection management in Yucheng.

The municipal party committee and government of Yucheng City have regarded the promotion of strip intercropping as a political task to ensure national food security. With the determination that "as long as our ideas do not falter, there will always be more solutions than difficulties," they have implemented the "five strengthenings" to ensure the implementation of strip intercropping.

The first target is to strengthen organizational leadership. It is often said at the grassroots level that "if the leader attaches importance, it won't be difficult." Yucheng City implements the principle of "the secretary is in charge of grain production" and assigns responsibilities at different levels. Village-level leaders focus on model fields, town-level leaders focus on demonstration plots, and city-level leaders focus on core areas. Municipal leaders are assigned to towns and streets, and the work of strip intercropping is included in the annual assessment of each township. The municipal party committee and government have issued the "Opinions on Implementing the" 1-2-3 "Project for Soybean Industry Revitalization," established a working group with the secretary and mayor as co-leaders, and the secretary and mayor conduct the inspection work on the progress in the fields at least once a month. They have also established a Soybean Industry Development Center, which is responsible for coordinating and dispatching the development of the soybean industry, as well as inspection and assessment. They have organized three city-wide observation events to assess the progress of strip intercropping in various townships, and the evaluation committee publicly announced the scores and rankings given on-site, summarizing experiences and identifying shortcomings in a timely manner.

The second target is to strengthen financial support. Funds are the most direct form of support; otherwise, "having countless ideas will not help if there is no money." Yucheng City established a 20 million yuan fund for the soybean industry development, specifically for subsidizing large-scale soybean cultivation, supporting leading soybean processing enterprises, and promoting the development of the entire soybean industry chain. In addition to the subsidies provided by the central, provincial, and municipal governments, they have raised funds to increase the subsidy standard for strip intercropping to 300 yuan per acre. For townships with a planting area of over 5000 acres, an additional 100,000 yuan in working funds is awarded for every additional 5000 acres. This effectively mobilizes the enthusiasm of townships to engage in the work.

The third target is to strengthen technical support. Strip intercropping requires new agricultural techniques, and "to find the right path, one must be guided by a master." Yucheng City fully implements the targets regarding selecting regions, subjects, and models. They focus on determining the varieties, agricultural machinery, and pesticides. They have established an expert guidance group and formed 13 scientific service teams. They have compiled technical manuals and hanging charts, issued the "Technical Recommendations for Mid- to Late-stage Production of Strip Intercropping of Soybean and Maize in Yucheng City," and provided comprehensive technical guidance and services throughout the process. Furthermore, they have organized over 130 training sessions on strip intercropping techniques, including planting, plant protection, machinery, and equipment, with 9988 participants trained. They invited six teams, including the Chinese Academy of Sciences, Sichuan Agricultural University, Shandong Agricultural University, Dezhou Agricultural Science Institute, and the China National Chemical Corporation, to conduct high-yield experiments. Among the participants, three teams from Sichuan Agricultural University, Shandong Agricultural University, and Dezhou Agricultural Science Institute, respectively, conducted full-scale mechanized experiments in a 50,000-acre demonstration field in Fangsi Town; additionally, the China National Chemical Corporation conducted full-scale mechanized experiments in a 500-acre test field in Shiliwang Town.

The fourth target is to strengthen social services. Ordinary farmers may have difficulty grasping new agricultural techniques, so "let professionals do professional work." They actively mobilize new types of operating entities to demonstrate and promote strip intercropping. They require that the planting area of grain growers, family farms, or farmer cooperatives participating in strip intercropping should not be less than 50 acres, and the land should be concentrated and contiguous. Ultimately, 681 new types of operating entities were identified to undertake the strip intercropping of 134,600 acres. Social services are provided at the township level, implementing the "five-fold unification:" unified stubble cleaning, unified ridging, unified soil preparation, unified seeding, and unified spraying. This improves operational efficiency while ensuring quality.

The fifth target is to strengthen income security. Apart from subsidies for seeds and agricultural machinery, the cost of strip intercropping is approximately 450 yuan per acre. Maize monoculture generates an income of approximately 1200 yuan per acre, whereas soybean monoculture generates approximately 800 yuan per acre. In contrast, intercropping generates an additional income of nearly 300 yuan per acre compared with maize monoculture. Traditional farmers are not prone to taking high risks; they need some assurance before making concerted efforts. As the first county-level city in Shandong Province to implement the "two guarantees and three fixings" policy, Yucheng City has innovated a "futures + insurance" mechanism. Soybean farmers pay a certain amount of money to ensure coverage for disasters, yield per acre, and prices. Representatives of the township planters have signed contracts with the Yuwang Group, the largest soybean protein processing company in China, stipulating a unified purchase price of 0.1 yuan/jin higher than the market price, relieving farmer concerns. Additionally, they have experimented with soybean straw-feeding techniques, where every 100 jin of soybean straw can replace six jin of high-quality feed, resulting in an additional income of 100 yuan per acre.

After 6 years of demonstration and promotion, Yucheng City achieved mechanization and a streamlined and efficient cultivation system throughout the process of seedling and plant cultivation management and harvest (Li et al. 2023a, b). They emphasized cost reduction and increased efficiency. Compared to maize monoculture, in maize-soybean strip intercropping systems, maize yield is almost unaffected (with yields of 500–650 kg per acre), while soybean yield has increased by 80–120 kg per acre, resulting in an additional income of nearly 200 yuan per acre. The efficient intercropping model of soybean and maize strips has become a new choice for the structural adjustment of the agricultural supply-side and a new driving force for rural revitalization.

Effects of the Application and Future Prospects for Maize–Soybean Strip Intercropping

15

Jiang Liu, Wenyu Yang, Yuanyuan He, Xiaochun Wang, Feng Yang, and Taiwen Yong

Abstract

This chapter provides an overview of the application and future prospects of maize-soybean strip intercropping. Over the past two decades, this technology has effectively addressed challenges in soybean cultivation, ensuring national food security and promoting the development of animal husbandry for high-quality meat, eggs, and milk production. Additionally, this method supports ecological health preservation through environmentally friendly fertilization, leading to improved fertilizer utilization rates and increased labor productivity. These developments also contribute to the revitalization of rural life and promote scientific and technological progress in the industry. Furthermore, this chapter examines the potential applications of maize-soybean strip intercropping, including advantages such as high industry competitiveness, low risk, great market potential, and policy guarantees. These strengths make it clear that this technology will be widely adopted in the future.

Keywords

Food security · Ecological health · Application prospects

J. Liu (✉) · W. Yang · Y. He · X. Wang · F. Yang · T. Yong
College of Agronomy, Sichuan Agricultural University, Chengdu, China
e-mail: jiangliu@sicau.edu.cn; mssiyangwy@sicau.edu.cn; 2020201031@stu.sicau.edu.cn; xchwang@sicau.edu.cn; f.yang@sicau.edu.cn

W. Yang et al. (eds.), *Theories and Application in Maize-Soybean Strip Intercropping System*, https://doi.org/10.1007/978-981-97-8723-4_15

More than 23 years of research, testing, and application have resulted in a series of significant technological innovations and achievements that have allowed the maize-soybean strip intercropping system to maintain high levels of maize productivity while concomitantly improving that of soybeans, hence increasing self-sufficiency rates, and fostering significant economic, social, and ecological benefits which, are reactivating rural life.

15.1 Effects of the Application of Maize-Soybean Strip Intercropping

15.1.1 Breaking Through the Soybean Cultivation Dilemma to Ensure National Food Security

Large-scale demonstration and promotion of the achievements realized during the process of development of the theory underlying current practice regarding the maize-soybean strip intercropping system have been performed in 20 provinces (municipalities and autonomous regions), including Southwest and Northwest China, as well as the Huang-Huai-Hai Plain, making Sichuan and the Southwest, the fifth and third largest soybean producing regions in China, respectively.

In 2019, the maize-soybean strip intercropping system was incorporated into the plan for soybean revitalization. Further, in 2020, the No.1 Central Document required to "increase the popularization and application of new agronomic techniques of maize and soybean intercropping." Subsequently, in 2021, the Central Rural Work Conference and the No.1 Central Document in 2022 pointed out the need to vigorously promote the maize-soybean strip intercropping system as a means to enhance maize and soybean-compatible development. Additionally, in 2022, the Ministry of Agriculture and Rural Affairs organized the promotion and application of the maize-soybean strip intercropping system in up to 15.5 million mu across 16 provinces (Municipalities and Autonomous regions) nationwide. More recently, in 2023, the No.1 Central Document calls for "fully and faithfully applying the maize-soybean strip intercropping system." Consistent with these official policies, the system has been included in the central financial subsidies for agricultural technology extension and service and agricultural machinery-purchase subsidies. Thus, according to the 14th Five-Year National Planting Development Plan, issued by the Ministry of Agriculture and Rural Affairs, 50 million mu of soybean-maize strip intercropping system will have been promoted by the year 2025. Under the premise of ensuring the absolute safety of rations, the decision to stabilize maize production capacity and substantially increase soybean production capacity is the most important strategic decision concerning national food security.

Driven by the above policies, this system has been rapidly demonstrated and popularized in nearly 50 counties and cities in 20 provinces and autonomous regions (namely Sichuan, Chongqing, Yunnan, Guizhou, Guangxi, Shaanxi, Gansu, Ningxia, Shanxi, Inner Mongolia, Hebei, Henan, Shandong, Jiangsu, Zhejiang, Jiangxi, Hunan, Hubei, Liaoning, and Jilin) in the Southwest, the Northwest, the

Huang-Huai-Hai Plain, the middle and lower reaches of the Yangtze River, and the Northeast. In 2020, the promotion area reached 9.58 million mu all over China, 4.9 million mu in Sichuan province alone, accounting for 75% of the total soybean planting area of the province. During 2001–2018, the soybean sowing area increased more than double in Sichuan, ranking No. 1 in the national list of provinces based on their growth. In 2022, the actual implementation area reached 16.45 million mu, increasing by 1 million mu over the target area established in the original plan.

The national soybean production patterns have changed with the promotion and application of the maize-soybean strip intercropping system. Before 2000, the main soybean production areas were mainly concentrated in Northeast China among high-oil soybean regions. Then, in 2009, with a significant increase in the soybean planting area, Sichuan province entered the ranks of the main soybean production provinces for the first time. Then, after 2014, soybean planting achieved full coverage in the hilly areas of Sichuan, and until 2018, the province ranked fourth in soybean production in the country, forming a new pattern of high-oil areas in the Northeast, high-protein areas in the Huang-Huai-Hai Plain, and maize-soybean intercropping areas in the Southwest.

15.1.2 Ensuring a Healthy Development of Animal Husbandry and the Production Capacity of High-Quality Meat, Eggs, and Milk

Maize and soybean are the main components of quality feed for domestic animals and fowl. The production capacity of maize and soybean is closely related to the healthy development of animal husbandry, whereby these two crops are paramount to guarantee the supply of "oil bottles," "meat plates," and "milk cans" for the people. In addition to their importance as human staple foods, maize, and soybean are the main source of feed for pigs, cattle, and sheep, but with the increasing trade friction between China and the United States and the impact of the COVID-19 pandemic, it has become increasingly challenging to fill the gap between supply and demand with imports, directly affecting food security and livestock industry development.

Due to the extremely limited cultivation-land resources, it is impossible to meet food demand through monoculture. Alternatively, the maize-soybean strip intercropping system developed from traditional maize and soybean intercropping represents a potential solution. With high output, mechanization, and sustainability, the new agronomy of maize and soybean intercropping has provided a new solution to realize the high-quality development of sustainable agriculture. A case in point, taking 80% of the existing maize planting area, the potential area in Sichuan province is up to 22.4 million mu/year, which should allow the production of 11.2 million tons of maize while increasing 2.69 million tons of soybeans. Altogether, this can help meet the basic demands for existing feed and bean processing. In addition, maize and soybean mixed silage can produce approximately 60 million tons of feed, thus playing a determinant and fundamental role in promoting the development of the grain, oil, and animal husbandry industries.

15.1.3 Ecological Health Preservation Through Environmentally Friendly Fertilization to Enhance Fertilizer Utilization Rates

Using no-tillage, straw mulching, nozzle-fixing, strip rotation, and other technologies, the maize-soybean strip intercropping system has effectively reduced energy consumption, greenhouse gas emissions, continuous cropping obstacles, and improved soil fertility. With the research and development of integrated fertilization with reduced application, chemical control to avoid loading, and green prevention and control supporting technology, nitrogen (N) and phosphorus (P) utilization rates have improved by 85% and 34%, respectively, and concomitantly, soybean yields have increased by 20%, and the use of chemicals has been reduced by more than 25%.

According to years of fixed-point observation, compared with the traditional maize-sweet potato intercropping model, soil losses and surface runoff have decreased by 10.6% and 85.1%, respectively, soil organic matter content has increased by 5.56%, and soil total N of maize and soybean strips have increased by 4.11% and 7.29%, respectively. Relay-intercropping soybean has reduced urea application by 4.4 kg/mu because of N fixation by root nodules and increased N fertilizer utilization rate by 39.21%, which has had a significant positive effect on protecting farmland and effectively promoting sustainable agricultural development.

15.1.4 Improving Labor Productivity While Promoting the Revitalization of Rural Life

The intercropping system has several important advantages: (1) double harvest in one field, (2) stable grain and increased soybean yields, (3) one planting with multi-effects, (4) crop combination using and conserving farmland, (5) multi-use of one machine, and (6) low-carbon input and high carbon efficiency. In 2021, experts conducted yield trials which showed that, in the strip intercropping system, 569.63 kg of maize per mu and 122.3 kg of soybean per mu were harvested in a 100-mu demonstrative field in Renshou, Sichuan; 542.08 kg of maize per mu and 114.36 kg of soybean per mu in Feicheng, Shandong; 521 kg of maize per mu and 103.7 kg of soybean per mu were harvested in a 1000-mu demonstrative field in Gaocheng district, Shijiazhuang, Hebei. The Department of Plantation Management of the Ministry of Agriculture and Rural Affairs organized the 2022 national soybean high-yield competition. Through a series of yield trials and review work, the top 10 soybean varieties with the highest yields were awarded the title "Golden Soybean King." The highest yield was 633.8 kg of maize per mu and 165.1 kg of soybean per mu (Fig. 15.1). The results obtained over more than 20 years of research suggest that the maize-soybean strip intercropping system has achieved an equivalent yield of

农业农村部种植业管理司

农农（油料）〔2022〕11 号

关于 2022 年全国大豆高产竞赛情况的通报

各省、自治区、直辖市及计划单列市农业农村（农牧）厅（局、委），新疆生产建设兵团农业农村局，北大荒农垦集团有限公司，有关科研院校、协会及企业：

今年是启动实施国家大豆和油料产能提升工程的第一年，为挖掘我国大豆增产潜力，探索提高单产路径，年初以来，农业农村部组织开展全国大豆高产竞赛，设"金豆王"、"奋豆者"、"豆明星"三大奖项，经各省推荐，全国大豆高产竞赛组委会复核，全国共评选出"金豆王"30 名，分清种春播、清种夏播和复合种植三种类型各 10 名，"奋豆者"30 名、"豆明星"50 名。

希望受表扬的单位、个人珍惜荣誉，再接再厉，开拓创新，充分发挥模范带头作用，力争取得更大成绩，为提高大豆产能和自给率作出新的更大贡献。同时，希望各地高度重视大豆生产，常抓不懈，加大力度，强化措施，努力提高大豆生产水平。

特此通报。

2022 年全国大豆高产竞赛"金豆王"（带状复合种植）名单

单位：亩，公斤/亩

排名	地 点	示范面积	实收面积	大豆	玉米	主体
1	山东禹城市	220	3.8	165.1	633.8	种植大户
2	安徽太和县	200	3.5	157.0	543.1	合作社
3	江苏睢宁县	160	7.2	155.0	675.0	合作社
4	山西翼城县	115	4.0	154.5	604.6	农业企业
5	湖南汨罗市	2080	3.8	140.9	486.9	种植大户
6	河北藁城区	180	3.0	139.5	586.1	家庭农场
7	安徽蒙城县	500	3.1	139.4	518.6	种植大户
8	山西武乡县	257	3.0	137.3	511.1	种植大户
9	河北无极县	120	3.1	135.7	697.1	家庭农场
10	内蒙古九原区	240	5.5	128.8	802.4	合作社

Fig. 15.1 The Information bulletin at the 2022 national soybean high-yield competition

maize and an extra yield of 100–150 kg of soybean per mu, while N application has been reduced by approximately 4–5 kg N per mu, whereas an increase in the economic performance of approximately 350 yuan per mu compared with that of maize monoculture, has been recorded, making it be loved by farmers, greatly improving labor productivity and becoming a powerful lever for promoting rural revitalization.

15.1.5 Promotion of Scientific and Technological Progress in the Industry

15.1.5.1 Promoting the Study and Application of Intercropping

Research and development and the application of the strip intercropping system have drawn global attention to intercropping research. Thus, 25 prestigious experts from the United States, the Netherlands, Australia, Japan, Canada, and other countries, and 15 domestic researchers and 336 colleagues serving as authorities have made field visits and provided full affirmation and high evaluation, promoting the technical innovation of traditional intercropping at home and abroad. The maize-soybean strip intercropping system has not only been vigorously promoted in China but has begun to spread widely abroad in recent years. Through technology exports, experimental demonstrations, network forums, and joint training of postgraduates, the R&D team has brought the technology to South Asia, Europe, and Africa and established demonstration sites in Pakistan, Sweden, and Ghana, setting off a wave of research on maize-soybean intercropping and application of strip intercropping in countries all over the world.

More recently, we achieved a "zero breakthrough" in soybean production in Pakistan through cooperation with major scientific research institutions, universities, and agricultural and animal husbandry companies in Pakistan. The effects of demonstration work conducted in 2020 highlighted the apparent technical advantages of strip-intercropping, which has since been widely reported on more than 20 occasions by the Pakistani national TV, Pakistan Independent News, Daily News, Islamabad Post, China Economic News, and CCTV, among other mass media agencies.

In addition, Stanford University and the University of Illinois, the Crop Society of Japan and Kyoto University, the Netherlands Organization for Scientific Research (NWO) and the University of Waringen, the Swedish Government Formas Sustainable Development Research Council, and Brazilian agricultural colleges and governmental foundations and educational and academic institutions have written to us seeking our reports of scientific research or technical cooperation. At the University of Wageningen in the Netherlands, the subject course of intercropping has been specially offered during the winter semester, and Prof. Wenyu Yang has been invited to teach a Ph.D. course on the maize-soybean intercropping system on their online education platform. The Formas and the NOW have specially invited Prof. Yang to serve as a professional reviewer for related scientific research projects toward international intercropping, effectively promoting the expansion of demonstration efforts of the maize-soybean intercropping system in Nordic countries, thereby promoting research and application of intercropping at home and abroad.

15.1.5.2 Enhancing the Support of National Project Funding

This project has been strongly supported by the National Natural Science Foundation of China, key R & D programs, and other sponsors of scientific research, with a total amount of 12.35 million yuan for 28 projects. During the "13th Five Year Plan," four national key R&D projects were obtained with a funding of 22 million yuan; additionally, in 2019 and 2020, the project received two regional innovation science and technology assistance projects in Sichuan Province, funded with 2 million yuan, and received more than 10 million yuan from the "14th Five Year Plan" national project. Thus, the research team at the Sichuan Provincial Science and Technology Department has become a key supported team for scientific and technological innovation and regional cooperation, which has opened the door to the opportunity for the application of this technology in western China.

15.1.5.3 Establishing National- and Province-Level Scientific Research Platforms

During the research period, two provincial key laboratories and two provincial engineering technology research centers/engineering laboratories have been applied for and built. As a result, The Crop Environmental Response and Physiology Function Laboratory has already become an important part of the national key laboratory network.

15.1.5.4 Cultivating a Large Number of Industry Specialized Talents

We formed a group of 223 Master's and Doctoral students majoring in maize-soybean strip intercropping systems, ranking first in the world. The project leader is the representative of the 20th Party Congress and the head of the expert guidance group for the soybean and oilseed capacity enhancement project of the soybean-maize strip intercropping system. Additionally, two members of the research team have become members of the expert group, three more have been selected as post-experts of the National Modern Agricultural Industrial Technology System, three more have been selected as academic and technical leaders of Sichuan Province, and 10 have been promoted to professor.

15.1.5.5 Promoting Theoretical and Technological Progress of Intercropping Systems Both Domestically and Internationally

The theoretical system of "two synergies and one regulation" formed by the research fills the international gap and lays the theoretical foundation for the improvement of the production capacity of intercropping between tall and short crops; the core technology of "selecting varieties with suitable plant types, optimizing the configuration of strip width and row spacing, and reducing seed distance to keep high population density," and its parameter mode, provide technical support for the field configuration of tall and short position crops.

The theory and technology of maize-soybean strip intercropping have been compiled into five national textbooks, including Crop Cultivation (General Theory) and Tillage, the Chinese Encyclopedia of Agriculture (2019 edition), and two popular science books, enriching the disciplinary connotation of crop cultivation and tillage.

During the research period, the number of SCI papers, Chinese journal papers, and approved (applied for) patents on intercropping research at home and abroad increased from 32.7 to 333.0 pieces and from 1.6 items per year from 1997 to 2007 to 404.8, 809.0, and 60.5 items per year from 2008 to 2020, and the total number of papers and approved patents published by this project team in the field of intercropping have been right up in the front row in the world.

15.2 Application Prospects

15.2.1 High Industry Competitiveness

Under the condition of maintaining maize yield, the maize-soybean strip intercropping system causes the soybean yield to exceed the average level of the main soybean monoculture-producing areas in the north, and the comparative benefits are evident, which are well received by most farmers. In recent years, due to the advantages of high protein and non-GMO, Sichuan soybeans have been in the market for both supply and sale, and the price has risen, in sharp contrast to soybeans produced in the north. Simultaneously, planting soybeans can reduce the amount of N fertilizer (urea) application by 4–6 kg per mu, and the effect of land cultivation is evident. In addition, the special maize and soybean varieties selected for breeding, and the special sowing and harvesting machinery for compound planting have brought huge economic benefits to seed and agricultural machinery enterprises. Due to the promotion and application of this technology, the disadvantage of soybean raw material sources from Sichuan, mainly relying on the supply of other provinces, has ended. Furthermore, the gap between soybean supply and demand has been alleviated, the cost of raw material transportation has been greatly reduced, and the sustainable development of the Sichuan soybean product-processing industry, animal husbandry, and many other industries has been effectively promoted, with attractive economic benefits having been created.

15.2.2 Low Risk

The maize-soybean strip intercropping system is the crystallization of more than 20 years of hard work of scientific and technological personnel dealing with advanced and practical technology with wide adaptability, strong operability, and support aiming at compliance with the standards promoted by the Ministry of Agriculture industry and Sichuan Province local standards as guidance, high degree of technical standardization, and easy acceptance by business entities and farmers. According to the characteristics of regional production, the proposed three technical systems, namely, "spring maize-summer soybean strip intercropping technology, spring maize-spring soybean intercropping technology, and summer maize-summer soybean intercropping technology" are the major targets. Large-scale demonstrations showed good adaptability and high-yield models continue to emerge, which meet the needs of corn and soybean production and development in Southwest and Northwest China and the Huang-Huai-Hai Plain.

Biodiversity is the most effective means of responding to natural disasters. Consistent with that principle, the maize-soybean strip intercropping technology developed in this project, which integrates the two major crops among legumes and grasses, has outstanding ecological niche complementary functions in natural risks, especially in drought tolerance, high-temperature resistance, pest resistance, wind resistance, high yield, sustainability, mechanization, and low risk. Through the zoning application and regional layout of the three major technology models, the technical requirements of different ecological and production conditions in the southwest, northwest, and Huang-Huai-Hai Plain can be met and have been promoted in more than 20 provinces (municipalities directly under the central government and autonomous regions), with low natural risks.

15.2.3 Great Market Potential

According to 80% of the existing corn planting area (625 million mu) in the country, strip compound planting is conducted with an average annual potential area of 500 million mu, a total maize output of 250 million tons, and a total output of 60 million tons of soybeans. Furthermore, according to the 20% target, the average annual potential area can reach 125 million mu, the total output of maize can reach 62.5 million tons, and the total output of soybean can reach 15 million tons. The "China Agricultural Outlook Report (2019–2028)" highlighted that in the next decade, the total consumption of corn and soybeans will reach 328 and 118.82 million tons, respectively. With a continuous increase in demand for corn and soybeans, the pattern of insufficient production and demand will continue to remain at a high level, and the domestic demand gap for corn and soybeans rapidly increase to unprecedented levels. Taking Sichuan Province as an example, the annual demand for soybeans in this province is more than 2.3 million tons, whereas the annual output of soybeans is 880,000 tons. The market demand gap is 1.5 million tons, feed enterprises need 1.95 million tons of soybean meal, of which 1.25 million tons need to be transferred into soybean meal, and the huge soybean and soybean meal demand gap is the driving force for the development of Sichuan soybean production. At the same time, corn and soybeans have a variety of uses, such as grain, fresh food, and green storage, with large market demand and broad prospects; the grain and straw of corn and soybean are excellent feed sources for cattle and sheep breeding, and the development of maize-soybean strip intercropping combines land-consuming crops (maize) with land crops (soybeans), roughage and concentrate feed, and high starch and high protein crops, and the corn soybean grains and straw per mu of land can meet the feed needs of 5–6 sheep in captivity. In addition, the maize-soybean belt compound planting technology considers both maize and soybean crops, which can effectively avoid the risk of price reduction for a single crop, and the market potential is enormous.

15.2.4 Policy Guarantees

The maize-soybean strip intercropping technology inherits the essence of traditional intercropping and has many advantages, including high yield, ecological sustainability (fertilizer fertility), high labor productivity (mechanization), multi-purpose, low risk, and conducive to planting and breeding cycle. In the past 10 years, after the promotion and application of four typical ecological regions and more than 10 demonstration sites in southwestern and northwestern China, the Huang-Huai-Hai Plain and the middle and lower reaches of the Yangtze River, maize-soybean strip intercropping technology has become increasingly mature. This technology was listed in the 2019 "Soybean Revitalization Plan Implementation Plan," and has been listed as the primary agricultural technology of the whole country/Sichuan Province for 13 consecutive years. Furthermore, in 2020, 2022, and 2023, the "maize-soybean strip intercropping technology" was incorporated into Central Document No. 1, which provides a policy guarantee for the application of maize-soybean strip intercropping technology. In addition, maize-soybean strip intercropping has been included in the central financial agricultural technology extension and service subsidies and agricultural machinery purchase subsidies. According to the "14th Five-Year Plan" National Planting Industry Development Plan issued by the Ministry of Agriculture and Rural Affairs, the maize-soybean strip intercropping area will have been promoted to 50 million mu by the year 2025. Under the premise of ensuring the absolute safety of food rations, stabilizing corn production capacity, and greatly increasing soybean production capacity will be a strategic decision for national food security.

In summary, the application of the maize-soybean strip intercropping technology has a strong policy guarantee and broad market potential, which provides a new way for the development of the national soybean industry, a new choice for agricultural restructuring, while it improves the international competitiveness of the national soybean industry.

References

Bao G, Yao W (2019) The current situation and main risks of grain production in China. J South China Agric Univ (Social Science Edition) 18(6):111–120. (in Chinese with English abstract)

Bu WZ (2015) Study on the morphological response and dry matter accumulation differences of different soybean varieties to intercropping shade. Sichuan Agricultural University Press, Chengdu. (in Chinese)

Cai L, Liu WG, Li Q et al (2016) Effects of maize-soybean strip intercropping on soybean protein characteristics. Chin J Oil Crops 38(03):328–335. (in Chinese with English abstract)

Chen ZQ (2011a) Effects of molybdenum on nitrogen fixation, character of photosynthesis, yield and quality of soybean in sole-cropping and relay strip inter-cropping system. Sichuan Agricultural University, Sichuan. (in Chinese)

Chen D (2011b) Study on flexible truss boom system design and dynamic simulation. Doctoral dissertation, Chinese Academy of Agricultural Mechanization Sciences. (in Chinese)

Chen Z (2014) Key technology and equipment for full value harvesting of corn. Science Press, Beijing. (in Chinese)

Chen P, Liu XM, Song C et al (2017) Study on nitrogen utilization characteristics in a maize-soybean relay intercropping system by labeled 15N. J Sichuan Agric Univ 3:300–308. (in Chinese)

Chen J, Feng L, Gao J et al (2019) Effects of light intensity on stomatal characteristics and photosynthetic characteristics of soybean leaves in seedling stage. Sci Agric Sin 52(21): 3773–3781. (in Chinese with English abstract)

Cheng SS (2022) Design and experimental study of striping boom sprayer for soybean-maize strip intercropping. Master dissertation, Henan Agricultural University. (in Chinese)

Cheng C, Fu J, Chen Z et al (2019) Parameter optimization test of corn grain harvester cleaning device. J Agric Mach 50(07):151–158. (in Chinese)

China Agricultural Mechanization Science Research Institute (2007) Agricultural machinery design manual volume II. China Agricultural Science and Technology Publishing House, Beijing, pp 912–913. (in Chinese)

Cui L, Su BY, Yang F et al (2014) Effects of photosynthetically active radiation intensity distribution characteristics on soybean photosynthetic characteristics and yield under different maize-soybean strip relay-intercropping combination conditions. Sci Agric Sin 47(08):1489–1501. (in Chinese with English abstract)

Ding WM (2010) Agricultural machinery. China Agriculture Press, Beijing. (In Chinese with English abstract)

Ding GH (2017) The design and parameter study on the maize and soybean intercropping seeder. Sichuan Agricultural University, Yaan. (In Chinese with English abstract)

Ding L, Yang L, Liu SR et al (2018) Design of air suction high speed precision maize seed metering device with assistant seed filling plate. Trans Chin Soc Agric Eng 34(22):1–11. (In Chinese with English abstract)

W. Yang et al. (eds.), *Theories and Application in Maize-Soybean Strip Intercropping System*, https://doi.org/10.1007/978-981-97-8723-4

Du Q, Wang QM, Chen P et al (2017) Screening of herbicides in maize-soybean relay strip intercropping system. Soybean Science 36(01):98–103. (In Chinese with English abstract)
Du JB, Han TF, Gai JY et al (2018) Maize-soybean strip intercropping: achieved a balance between high productivity and sustainability. J Integr Agric 17:60345–60347
Fan Y, Yang F, Liu Q et al (2017) Effects of intercropping shading on leaf structure and photosynthetic fluorescence characteristics of soybean seedlings. Acta Agron Sin 43(02):277–285. (in Chinese with English abstract)
Fan CL, Cui T, Zhang DX et al (2019) Design and test of low damage combined corn threshing and separating device. J Agric Mach 50(04):113–123. (in Chinese)
Fang P, Liu WG, Liu XD et al (2016) Effect of maize-soybean intercropping on quality of vegetable soybean. J Zhejiang Univ (Agric Life Sci) 42(05):556–564. (in Chinese with English abstract)
Fu ZD, Zhou L, Chen P et al (2017) Effects of nitrogen application rate on soil microbial quantity and soil enzymes activities in maize/soybean intercropping systems. Chin J Eco-Agric 25(10): 1463–1474
Gao R, Yang F, Liao D et al (2015) Effects of row spacing on light environment and morphological characteristics and yield of intercropped soybean canopy. Soybean Sci 34(4):611–615. (in Chinese with English abstract)
Gao YY, Zhai CY, Yang S et al (2020) Development of CAN-based downforce and sowing depth monitoring and evaluation system for precision planter. Trans Chin Soc Agric Mach 51(06): 15–28. (In Chinese with English abstract)
Guan GJ, Zhao H (2009) Vigorously promote the mechanization technology of corn harvesting to increase farmers' production and income. Use and maintenance of agricultural machinery, 4–20. (in Chinese)
Guan CK, Wang YY, Wang DW et al (2014) Partial design of 4YZ-2 horizontal corn cutting table. Mechanization Res 36(11):66–68. (in Chinese)
Han DD, He B, Zhou Y et al (2023) Design and experiment of air suction type dense planting precision seed metering device. J Huazhong Agric Univ 42(01):237–247. (In Chinese with English abstract)
He XK, Zeng AJ, Liu YJ et al (2019) Pesticide application machinery and technology. Chemical Industry Press. (in Chinese)
Hu Y (2021) Effects of density and fertilizer on yield under high-density conditions of monoculture and strip relay-intercropping maize. Sichuan Agricultural University Press, Chengdu. (in Chinese)
Ji B, Yin J, Liu WB et al (2012) Modeling and simulation of harvester reel based on ADAMS and SolidWorks. Modern Mach 2012(04):52–53+89. (in Chinese)
Jin CQ (2011) Corn production mechanization technology. China Agricultural Publishing House, Beijing. (in Chinese)
Knrzer H, Graeff-hnninger S, Guo B et al (2009) The rediscovery of intercropping in China: a traditional cropping system for future Chinese agriculture–a review. In: Lichtfouse E (ed) Climate change, intercropping. Pest control and beneficial microorganisms, vol 2. Springer, Dordrecht, pp 13–44. https://doi.org/10.1007/978-90-481-2716-0_3
Li BF (2003) Agricultural machinery. China Agriculture Press, Beijing. (In Chinese with English abstract)
Li L (2013) Principles and applications of nitrogen fixation ecology of legumes in intercropping systems. China Agricultural University Press, Beijing. (in Chinese with English abstract)
Li GL, Qin YY, Zhang XR et al (2011) Optimization experiment of clamping and feeding device of corn combine harvester. Jiangsu Agric Sci 5:511–512. (in Chinese)
Li S, Tan T, Fan Y et al (2019a) Effects of maize shading on photosynthetic characteristics and vein and stomatal traits of soybean. Sci Agric Sin 52(21):3782–3793. (in Chinese with English abstract)
Li YH, Yang L, Zhang DX et al (2019b) Design and experiment of pneumatic precision seed-metering device with single seed-metering plate for double-row. Trans Chin Soc Agric Mach 50(7):61–73. (In Chinese with English abstract)

Li H, Zhang K, Zhang Z et al (2019c) The utility model relates to a wheel distance adjustment device for plant protection in a high gap field. Chinese Patent, CN209096388U. (in Chinese)

Li HJ, Li DX, Chen SY et al (2020a) Mutual infestation of *Fusarium oxysporum*, the causal agent of root rot of soybean and stem rot of maize in set crops. J Sichuan Agric Univ 38(5):551–557

Li L, Chen GP, Pu T et al (2020b) Material accumulation and distribution characteristics of corn varieties suitable for relay-intercropping. J Nucl Agric Sci 34(3):592–600. (in Chinese with English abstract)

Li H, Zhao YQ, Qin CB et al (2020c) Design and test of control system of articulated steering unmanned plant protection vehicle. Trans Chin Soc Agric Mach 51(S2):544–553. (in Chinese)

Li H, Zhang Y, Cheng SS et al (2020d) The utility model relates to a striping boom sprayer. Chinese Patent, CN112088855A. (in Chinese)

Li H, Cheng SS, Zhang Y et al (2021a) The invention relates to a whole plant protection spray system for corn and soybean belt compound planting. Chinese Patent, CN112514872A. (in Chinese)

Li YL, Peng XH, Chen P et al (2022) Effect of reducing nitrogen application on leaf stay-green, photosynthetic characteristic and system yield in maize-soybean relay strip intercropping. Chin Agric Sci 55(09):1749–1762. (in Chinese)

Li C, Stomph TJ, Makowski D et al (2023a) The productive performance of intercropping. PNAS 120(2):e2201886120

Li X, Zhang X, Chen GP et al (2023b) Cost and comprehensive benefit analysis of corn soybean strip intercropping in Southwest China. Chin J Oil Crops 45(02):427–436. (in Chinese)

Liang SN, Jin CQ, Zhang FF et al (2015) Design and test of 4LZG-3.0 type millet combine harvester. J Agric Eng 31(263):31–38. (in Chinese)

Liao DP (2015) Systematic evaluation of different row spacing configurations under maize-soybean strip relay-intercropping. Sichuan Agricultural University Press, Chengdu. (in Chinese)

Lin J, Gong YJ, Li GJ et al (2004) Discussion on the development advantages and prospects of precision seeder. Farm Mach 04:43–44. (In Chinese with English abstract)

Liu Z (2014) Practical technology for high yield of corn. Anhui University Press, Hefei. (in Chinese)

Liu X (2016) Light distribution, interception and utilization of soybean-maize strip intercropping system. Doctoral dissertation, Sichuan Agricultural University. (in Chinese)

Liu DL (2019) QTL analysis for the protein content related traits in soybean seeds under monoculture and relay intercropping. Sichuan Agricultural University, Sichuan. (in Chinese)

Liu FL, Zhang XH, Ma WW et al (2010) Current development status of foreign big-scale protection machinery and application technology. J Agric Mech Res 32(3):246–248+252. (in Chinese)

Liu J, Jin CQ, Liang SN et al (2017a) Research status of soybean mechanical harvesting loss. Mech Res 39(7):1–9, 15. (in Chinese)

Liu X, Rahman T, Yang F et al (2017b) PAR interception and utilization in different maize and soybean intercropping patterns. PLoS One 12:e0169218

Liu X, Rahman T, Song C et al (2017c) Changes in light environment, morphology, growth and yield of soybean in maize-soybean intercropping systems. Field Crop Res 200:38–46

Liu X, Rahman T, Song C et al (2018) Relationships among light distribution, radiation use efficiency and land equivalent ratio in maize-soybean strip intercropping. Field Crop Res 224: 91–101

Liu R, Li YJ, Liu ZJ et al (2021) Analysis and calibration of discrete element parameters of coated maize seed. Trans Chin Soc Agric Mach 52(S1):1–8+18 (In Chinese with English abstract)

Lou Y (2014) Effects of strip relay-intercropping row spacing configuration on root system characteristics of soybeans and maize. Sichuan Agricultural University Press, Chengdu. (in Chinese)

Lu FZ (2014) Effects of different bandwidths on crop nutrient accumulation competition and yield in maize-soybean strip relay-intercropping system. Sichuan Agricultural University Press, Chengdu. (in Chinese)

Lv XR, Lv XL, Zhang LH (2013) Design on small multi-function chassis manipulator device. Adv Mater Res 397-400:172–175

Lv XR, Ding WM, Lv XL (2014) Design of hydraulic system for small multi-function chassis in hilly area. J Huazhong Agric Univ 33(02):128–132. (in Chinese)

Ma YW (2018) Optimization and experimental research on opener parameters of seeder. Henan University of Science and Technology, Luoyang. (In Chinese with English abstract)

Ma YW, Pu T, Chen GP et al (2019a) Study on the physiological characteristics of high-yield maize leaves in maize-soybean strip relay-intercropping. J Sichuan Agric Univ 37(05):611–616. (in Chinese with English abstract)

Ma YW, Pu T, Li L et al (2019b) Morphological characteristics of high-yield maize varieties in maize-soybean strip relay-intercropping. Mazie Sci 27(4):93–99. (in Chinese with English abstract)

Munz S, Graeff-Hönninger S, Lizaso JI et al (2014) Modeling light availability for a subordinate crop within a strip–intercropping system. Field Crop Res 155:77–89

Peng XH, Chen P, Du Q et al (2022) Effects of reduced nitrogen application on soil aeration and root nodule growth of relay strip intercropping soybean. J Crop 48(5):1199–1209. (in Chinese)

Qin CB (2019) Study on structure design of autonomous walking plant-protection vehicle and its control system. Master dissertation, Henan Agricultural University. (in Chinese)

Qu Z, Yu Y, Li H (2014) Research on 2BJYM-4 type precision seeder for interplanting corn with soybean. Soybean Sci 33(1):127–131. (In Chinese with English abstract)

Ren J (2019) The effect of nitrogen and phosphorus combined application on material accumulation and nitrogen and phosphorus absorption and utilization in intercropping soybean [D]. Sichuan Agricultural University. (in Chinese)

Ren YF, Chen GP, Pu T et al (2019a) Response of photosynthetic characteristics of maize leaves in narrow rows with high light efficiency in maize-soybean strip relay-intercropping to weak light stress. Acta Agron Sin 45(5):728–739. (in Chinese with English abstract)

Ren L, Zhang LH, Ding GH et al (2019b) Design and experiment of a type of 2BF-5 maize-soybean strip intercropping precision seeder. J Henan Agric Univ 53(2):207–212+226. (In Chinese with English abstract)

Ren JB, Yang XL, Chen P et al (2022) Effects of interspecific distance on soil physicochemical properties and root spatial distribution of maize-soybean relay strip intercropping system. Chin Agric Sci 55(10):1903–1916

Shang ZQ (2021) Design and experimental study on small electric self-propelled striping boom sprayer. Master dissertation, Henan Agricultural University. (in Chinese)

Shang SQ, Dong YF, Jiang XD et al (2002) Use and maintenance of corn combine harvesters. China Agricultural Publishing House, Beijing. (in Chinese)

Sheng K (1995) Study on the configuration of profiling wheel of profiling mechanism of seeder. J Changchun Univ Technol 04:21–27. (In Chinese with English abstract)

Shi S, Zhou JL, Liu H et al (2019) Design and experiment of pneumatic precision seed-metering device with guided assistant seed-filling. Trans Chin Soc Agric Mach 50(5):61–70. (In Chinese with English abstract)

Song YX (2009) Effects of shading and restoration light intensity on photosynthetic characteristics, nitrogen metabolism, yield and quality in various varieties relay cropping soybean. Sichuan Agricultural University, Sichuan. (in Chinese)

Su Y, Liu H, Xu Y et al (2018) Optimization and test of nail tooth element of axial flow corn threshing device. J Agric Mach 49(S1):258–265. (in Chinese)

Tao J (2013) The effect of relay-strip-cropping on the grain quality of soybean (Glycine max) and differential proteomic analysis of it. Sichuan Agricultural University, Sichuan. (in Chinese)

Tsubo M, Walker S (2002) A model of radiation interception and use by a maize–bean intercrop canopy. Agric For Meteorol 110:203–215

Wang L (2020) Structural design and finite element analysis of trencher. Farm Mach 12:79–87. (In Chinese with English abstract)

Wang ZS, Tian JH, Wang HQ (2014) General textbook for cultivating new professional farmers modern corn production practical technology. China Agricultural Science and Technology Publishing House, Beijing. (in Chinese)

Wang Z, Zhao X, Wu P et al (2015) Radiation interception and utilization by wheat/maize strip intercropping systems. Agric For Meteorol 204:58–66

Wang Y, Zhang X, Yang WY et al (2016) Effect of shading on soybean leaf photosynthesis and chlorophyll fluorescence characteristics at different growth stages. Sci Agric Sin 49(11): 2072–2081

Wen X, Wang X, Deng X, et al (2015) The effect of nitrogen fertilizer management on maize yield, dry matter accumulation and transport under maize soybean intercropping mode [J]. Acta Agronomica Sinica 41(03):448–457. (in Chinese)

Willey R (1990) Resource use in intercropping systems. Agric Water Manag 17(3):215–231

Wu DF, Xu F, Yuan CS (2006) Performance comparison of plate tooth type and nail tooth type corn threshing machine. Mech Res 10:78–80. (in Chinese)

Xiang DB (2012) Studies on effect and mechanism of potassium on lodging-resistance and yield improvement in relay strip intercropped soybean. Sichuan Agricultural University, Sichuan. (in Chinese)

Xiao XL (2020) Research on the regulation pattern of spatio-temporal shade on the chemical quality of soybean seeds. Sichuan Agricultural University, Sichuan. (in Chinese)

Xu T, Yong TW, Yang WY et al (2014) Effects of sowing date and density on soybean plants, dry matter accumulation and yield under maize-soybean relay-intercropping. Chin J Oil Crop Sci 36(5):593–601. (in Chinese with English abstract)

Yan HY, Chen XG, Wu WF (2008) Design of vertical roller corn harvester test bench. Mech Res 12:76–78. (in Chinese)

Yan L, He HY, Chen HB et al (2018) Trapping effects of different LED single-wavelength insecticidal lamps on major pests within the corn-soybean strip-crop pattern. Chin J Appl Entomol 55(5):904–911

Yang WY (2017) Inheritance and innovative development of traditional intercropping of corn and soybean. In: Baoding: 2017, Annual conference of The Crop Science Society of China. (In Chinese with English abstract)

Yang WY, Yang F (2019a) Developing strip intercropping of corn and soybean to ensure national food security. Sci Agric Sin 52(21):3748–3750. (in Chinese with English abstract)

Yang WY, Yang F (2019b) Developing maize-soybean strip intercropping for demand security of national food. Sci Agric Sin 52(21):3748–3750. (In Chinese with English abstract)

Yang SC, Yang SM, She YW (2010) Computer simulation of the motion trajectory of the reel. Mech Res 32(12):141–145. (in Chinese)

Yang F, Cui L, Huang S et al (2015a) Dynamic of growth environment and population yield of different plant type maize intercropped with soybean. Soybean Sci 34(03):402–407. (in Chinese with English abstract)

Yang F, Lou Y, Liao DP et al (2015b) Effects of maize-soybean strip relay-intercropping row spacing configuration on crop biomass, root morphology and yield. Acta Agron Sin 14(4): 642–650. (in Chinese with English abstract)

Yang L, Yan BX, Zhang DX et al (2016) Research progress on precision planting technology of maize. Trans Chin Soc Agric Mach 47(11):38–48. (In Chinese with English abstract)

Yang H, Du YL, Chen P et al (2018) Parameter optimization of vertical axial flow threshing device of small feed soybean harvester. J Gansu Agric Univ 53(04):184–189+196. (in Chinese)

Yang WY, Yong TW, Wang XC et al (2021) Maize-soybean strip intercropping technology. Science Press, Beijing. (in Chinese with English abstract)

Ye MY (2010) Isoflavones accumulation pattern in growth period and dynamic change during air dry phase of soybean under sole cropping and relay-cropping conditions. Sichuan Agricultural University, Sichuan. (in Chinese)

Ye L, Yang F, Su B et al (2015) Effects of different field configurations on soil moisture distribution and water use efficiency of soybean-maize strip intercropping system. Agric Res Arid Areas 33(04):41–48. (in Chinese with English abstract)

Yong TW, Yang WY (2022) Advantages, effects, and development suggestions of corn and soybean strip composite planting technology. China Farm Cooperative 03:20–22. (in Chinese)

Yu XB (2009) Comparative study on photosynthesis, nitrogen fixation characteristics and yield quality of different genotypes of soybean under monoculture and relay intercropping. Sichuan Agricultural University Press, Sichuan. (in Chinese)

Zeng J (2018a) Photosynthetic characteristics of high light efficiency maize in maize-soybean strip intercropping system. Sichuan Agricultural University. (in Chinese)

Zeng JX (2018b) Research on the photosynthetic characteristics of high-light-efficiency in maize-soybean strip relay-intercropping. Sichuan Agricultural University Press, Chengdu. (in Chinese)

Zeng TR, Li XL, Guan H et al (2020) Dynamic microbial diversity and fermentation quality of the mixed silage of corn and soybean grown in strip intercropping system. Bioresour Technol 313: 123655

Zhang ZY (2008) Effects of different densities and field configurations on the yield and quality of strip relay-intercropping soybeans. Sichuan Agricultural University Press, Chengdu. (in Chinese)

Zhang LJ (2014) Design and performance analysis of self-propelled paddy field boom sprayer. Master dissertation, Jiangsu University. (in Chinese)

Zhang XR, Dong YF, Zhang DL et al (2007) Research on clamping and conveying device of corn harvester. Agric Equip Vehicle Eng 45(10):9–10, 17. (in Chinese)

Zhang LH, Yang WY, Dai JW et al (2015) Small soybean combine harvester: China, 201510505965.0.2015-11-11. (in Chinese)

Zhang LH, Nie JS, Han DD et al (2022) Specialized seeders: key to improving quality and efficiency of strip composite planting. Agric Mach Market 04:25–26. (in Chinese)

Zhou T, Wang L, Sun X, Wang X, Pu T, Yang H, Rengel Z, Liu W, Yang W (2021) Improved postsilking light interception increases yield and P-use efficiency of maize in maize/soybean relay strip intercropping. Field Crops Research 262:108054

Zou JL, Liu WG, Yuan J et al (2015) Relationship between stem lignin synthesis and lodging resistance in relay-intercropping soybean seedling stage. Acta Agron Sin 23(7):1098–1104. (in Chinese with English abstract)

The manufacturer's authorised representative in the EU is Springer Nature Customer Service Centre GmbH, Europaplatz 3, 69115 Heidelberg, Germany. If you have any concerns regarding our products, please contact ProductSafety@springernature.com

Printed and bound by CPI Group (UK) Ltd, Croydon, CR0 4YY

01/10/2026

02225921-0019